# ABSTRACT ALGEBRA:
# A Geometric Approach

**Theodore Shifrin**
*University of Georgia*

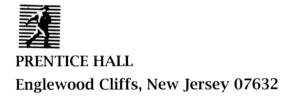

PRENTICE HALL
Englewood Cliffs, New Jersey 07632

*Library of Congress Cataloging-in-Publication Data*
Shifrin, Theodore
  Abstract algebra: a geometric approach / Theodore Shifrin
    p.     cm.
  Includes bibliographical references and index.
  ISBN 0-13-319831-6
  1. Algebra, Abstract.     I. Title.
QA162.S485   1996                          95-20079
412'.02—dc20                               CIP

Acquisitions Editor: **George Lobell**
Production Editor: **Barbara Mack**
Managing Editor: **Jeanne Hoeting**
Director of production and manufacturing: **David W. Riccardi**
Cover design: **Bruce Kenselaar**
Manufacturing buyer: **Alan Fischer**
Cover photo: Kilim, 19th century. West-Persia, Shahsavan-nomads, Khaladj-tribe $161 \times 120$ cm from: Sadighi, H. and K. Hawkes. *Kelims der Nomaden und Bauern Persiens*. Galerie Neiriz: Berlin, 1990. Plate 24.

Produced from the author's $\mathcal{A}_{\mathcal{M}}S$-LATEX file, using the Lucida fonts designed by Bigelow & Holmes, Inc.
Printed in the United States of America

10  9  8  7  6  5  4  3  2  1

ISBN 0-13-319831-6

PRENTICE-HALL INTERNATIONAL (UK) LIMITED, LONDON
PRENTICE-HALL OF AUSTRALIA PTY. LIMITED, SYDNEY
PRENTICE-HALL CANADA INC., TORONTO
PRENTICE-HALL HISPANOAMERICANA, S.A., MEXICO
PRENTICE-HALL OF INDIA PRIVATE LIMITED, NEW DELHI
PRENTICE-HALL OF JAPAN, INC., TOKYO
SIMON & SCHUSTER PTE. LTD., SINGAPORE
EDITORA PRENTICE-HALL DO BRASIL, LTDA., RIO DE JANEIRO

# CONTENTS

## APPENDICES

# PREFACE

When I began writing this book, I was teaching high school mathematics teachers hands-on "abstract algebra," emphasizing first the more intuitive concepts of high school algebra (the various number systems, number theory, and polynomials, before groups) and then the interplay between algebra and geometry—symmetries of the plane, compass and straightedge constructions, group actions, the regular polyhedra, and then some projective geometry. As time passed, I realized this course is really ideally suited to most mathematics majors as well. All mathematics students should be exposed to the basic ideas of modern algebra, its problem-solving skills and basic proof techniques, and certainly to some of its elegant applications. The link with geometry, as Felix Klein espoused in his *Erlanger Programm*, comes from understanding a geometry by means of its group of motions. And we should not always forget the historical development of a subject: groups arose originally in Galois' study of the "symmetries" of polynomials.

I have adopted a somewhat more concrete and example-oriented approach to the subject of modern algebra than is usual. The student must be active, and should learn to experiment with mathematics. We tell students that to learn mathematics effectively, they must read with pencil in hand, working through the examples in the text for themselves, and then attack lots of exercises—both concrete and theoretical. To this end, there are over 225 substantial examples, and more than 750 exercises, many having multiple parts. These include computations, proofs varying from the routine to the challenging, and open-ended problems ("Prove or give a counterexample ... "). But there is, I believe, another ingredient here. Students need to learn to relate ideas—whether in the text or in exercises—to what has come before. To understand a result, it is important to try examples: what's going on for specific values of $n$, for specific fields or groups, and so on? I hope this text will help them learn to do this.

This book is appropriate for students who have some experience with linear algebra (with exposure to matrix algebra and naïve vector space theory) and the language of proofs. For reference, I have included a brief summary of "foundational material" (sets, functions, equivalence relations, and basic logic) in Appendix A, as well as some linear algebra (a discussion of the matrix of a linear transformation, determinants, and eigenvalues and eigenvectors) in Appendix B.

I have *not* tried to avoid all references to calculus. In several places, proofs are given using calculus—e.g., discussion of the discriminant of a cubic polynomial, the fundamental theorem of algebra, Descartes' Rule of Signs, and Liouville's Theorem. In addition, a bit of calculus appears in the final discussions of non-Euclidean geometry. Some other unusual attributes of this book are the following:

• Elementary number theory and the Chinese Remainder Theorem are emphasized early in the text. Modular arithmetic is used to motivate the basic concepts in ring theory.
• There is a careful review of the basic number systems (including the algebra and geometry of the complex numbers) in Chapter 2.
• Geometry enters in Chapter 2, with a detailed discussion of the isometries of the complex plane. Symmetries of the regular polyhedra are a major theme in the later work on group theory.
• Adjunction of elements to a field is introduced, and the splitting field of a polynomial defined, in Chapter 3.
• The ring of Gaussian integers is introduced in Chapter 4, both to provide some interesting and natural examples of quotient rings, and to prove one of the classic theorems in elementary number theory.
• Group actions are emphasized throughout Chapter 7. Applications include: Burnside's theorem to count orbits, symmetries

and the classification of the regular polyhedra, finite subgroups of $SO(3)$, and the more standard topics—Sylow theorems and Galois theory.

• A constructive treatment of projective geometry, based on group theory and elementary linear algebra, is included in Chapter 8, including the standard classical theorems. The spectral theorem for symmetric matrices is proved and applied to give the classification of quadratic forms. The book concludes with a concrete treatment of elliptic and hyperbolic geometry.

The material of the first three chapters is the cornerstone of modern algebra at both the secondary and college levels—as well as the key to many interesting applications. Chapter 1 provides a detailed study of the arithmetic of the integers, beginning with mathematical induction and the Euclidean algorithm. Using the latter, we develop the ideas of greatest common divisor, solutions of diophantine equations, and modular arithmetic (with an exercise giving a basic application to public-key cryptography). This leads naturally to the ring $\mathbb{Z}_m$ and a discussion of rings, integral domains, and fields.

In Chapter 2, we move from the integers to the fields of rational, real, and complex numbers. We emphasize the interplay between the geometry and algebra of complex numbers, culminating in the classification of the isometries of the real line and complex plane.

The Euclidean algorithm returns in Chapter 3 as the linchpin in the study of polynomials. We all know how to rationalize the denominator of $1/(a + b\sqrt{2})$, but it is not nearly so obvious what to do with $1/(a + b\sqrt[3]{2} + c\sqrt[3]{4})$. As a consequence of the Euclidean algorithm, we obtain a field upon adjoining a root of any irreducible polynomial. The partial fraction decomposition of rational functions is also given as a significant application. In Section 2, we introduce splitting fields, as well as the Fundamental Theorem of Algebra and Descartes' Rule of Signs. The standard methods for determining irreducibility of a polynomial with integer coefficients (rational root test, undetermined coefficients, reduction mod $p$, and Eisenstein's criterion) appear in Section 3.

It is in Chapter 4 that the mathematician's "usual" course in modern algebra makes its appearance. We define the usual structures of modern algebra: ring homomorphisms, ideals, quotient

rings, isomorphisms, and the first of the fundamental homomorphism theorems. Here we treat the adjunction of a root more formally, prove the existence of splitting fields, and formalize the difference between algebraic and transcendental numbers. The last
section of the chapter presents a study of the Gaussian integers,
starting once again with the Euclidean algorithm. We give several
nontrivial examples of quotient rings in this setting, and conclude
with a proof that a prime number $p$ is the sum of two squares if
and only if it is of the form $p = 4k + 1$ for some natural number $k$.

Chapter 5 includes a review of the standard material on vector spaces, basis, and dimension. The multiplicativity of degree
of field extensions provides a nice tool for dealing with splitting
fields. In Section 2, we present the classical material on compass
and straightedge constructions; and, in Section 3, a very brief (and
optional) introduction to finite fields.

The treatment of group theory in Chapter 6 is reasonably standard: the symmetry group of the triangle and square, as well as the
permutation groups, appear as the basic examples. We include the
standard material on homomorphisms, cosets, normal subgroups,
and quotient groups. In Section 4, we treat the symmetric group
$S_n$ rather thoroughly, proving that $A_n$ is simple for $n \geq 5$, and discussing the solution of the fifteen puzzle along the way. Inciden-
tally, I highly recommend that both the instructor and the students
have plenty of geometric models with which to play and to discover the relations between group theory and geometry: a triangle,
square, a tetrahedron, and a cube are essential, and models of all
the Platonic solids are wonderful fun!

Chapters 7 and 8 are the culmination of the course: the true
integration of geometry and algebra. In Section 1 of Chapter 7, we
treat group actions and the associated counting formulas (involving orbits and stabilizers). We then apply this material to study
the symmetry groups of the regular polyhedra in Section 2 and
Burnside-Pólya's Theorem in combinatorics in Section 3. The three
remaining sections of Chapter 7 are independent of each other and
may be covered according to the taste of the instructor. In Section
4, we classify the isometries of $\mathbb{R}^3$ and give the group-actions classification of the regular polyhedra (by classifying the finite subgroups
of the rotation group, as H. Weyl does in his lovely book *Symmetry*).
In Section 5, we return to more algebraic topics, discussing direct

products of groups, the standard results on $p$-groups, and the Sylow theorems (with some of the easier applications). In Section 6, we give a snappy treatment of Galois theory, proving the Fundamental Theorem of Galois theory, and treating in some detail the *casus irreducibilis* and the insolvability of the quintic. This is an elegant and accessible application of group actions and the basic group theory that has appeared up to this point.

In Chapter 8, we give a heavily linear-algebraic treatment of non-Euclidean geometries. We begin with a brief discussion of affine geometry, a venerable subject that has all but disappeared from modern-day mathematics. However, it is the right way to understand certain results from Euclidean geometry; it also sets the scene for using the group of motions to make problems more tractable. In Section 2, we develop projective geometry in one and two dimensions, including the classical Desargues' and Pappus' Theorems, as well as an introduction to the theory of conics. In Section 3, we study the standard topic of quadratic forms and the Spectral Theorem for self-adjoint (real) linear transformations. This is applied in Section 4, where we discuss three-dimensional projective geometry, quadric surfaces, and an all-time favorite problem: Given four lines in general position in three-space, how many lines intersect them all? (Experience shows that this problem can capture the imagination of some good high school mathematics students, and is certainly worth a careful treatment here.) In Section 5, we give a brief treatment of elliptic and hyperbolic geometry from the Kleinian viewpoint, i.e., making use of their isometry groups. A smattering of spherical and hyperbolic trigonometry is included, and angle excess (or defect) is related to area. We then close with some brief remarks on differential geometry, explaining how curvature accounts for the differences we've observed in our three classical models of geometry—Euclidean, elliptic, and hyperbolic.

For a reasonable one-semester course, I would suggest the following: Chapter 1 (treating §1 lightly and skipping the Chinese Remainder Theorem in §3), Chapter 3, Chapter 4 (§§1,2), Chapter 6 (§§1,2,3), and Chapter 7 (§§1,2,3). For a full-year course covering the standard algebraic material, most instructors will wish to cover Chapters 1–7 (emphasizing §5 and §6 of Chapter 7) and treating Chapter 8 lightly. On the other hand, a more geometric course would treat Chapters 1–7 (deleting Chapter 5, §3, most of Chapter 6, §4, and Chapter 7, §§5-6) and give a thorough treatment of Chapter 8. A brief Instructor's Manual, including more detailed suggestions for instructors and specific comments on exercises, is available from the publisher.

I would like to thank my many students, colleagues, and friends who have tirelessly suggested ways to improve this text. I would like to give special thanks to M. Adams, E. Azoff, T. Banchoff, B. Boe, J. Hollingsworth, D. Lorenzini, C. Paul, D. Penney, M. Saade, D. Slutzky, and R. Varley. I would like to thank all those who have reviewed the manuscript, among them:

| | |
|---|---|
| J. P. Anderson, | University of Houston, ClearLake |
| M. Bergvelt, | University of Illinois, Champaign-Urbana |
| R. Biggs, | University of Western Ontario |
| S. Chapman, | Trinity University |
| J. Havlicek, | Michigan State University |
| T.-Y. Lam, | University of California, Berkeley |
| E. Previato, | Boston University |
| Z. Reichstein, | Oregon State University |
| E. Stitzinger, | North Carolina State University |
| R. Willard, | University of Waterloo. |

I would also like to express my appreciation to Michael Artin, who taught me most of the algebra I know; Arthur Mattuck, who taught me early what teaching should be; and, above all, S.-S. Chern, who showed me that I could be a mathematician. I am very grateful to George Lobell, Mathematics Editor at Prentice Hall, for his enthusiasm and support.

On the next page is a chart of the interdependencies of the various chapters and sections of the text. Those appearing in "bold" boxes are considered more or less essential to a year-long course with a moderately geometric bias.

I welcome any comments and suggestions. Please address any e-mail correspondence to `shifrin@math.uga.edu`.

# CHARTING THE COURSE

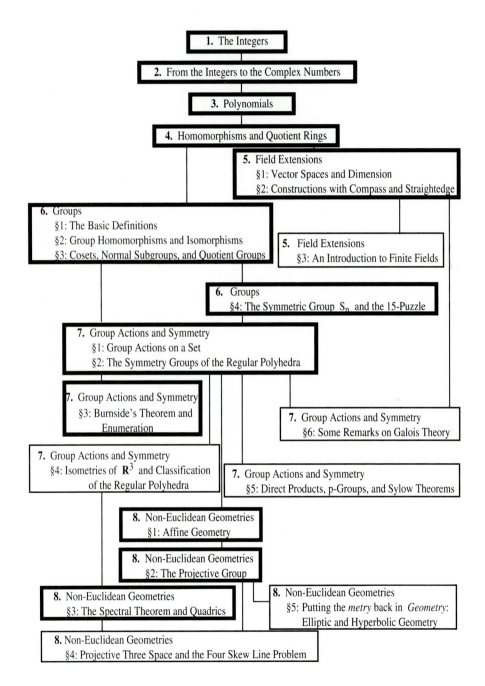

**1.** The Integers

**2.** From the Integers to the Complex Numbers

**3.** Polynomials

**4.** Homomorphisms and Quotient Rings

**5.** Field Extensions
§1: Vector Spaces and Dimension
§2: Constructions with Compass and Straightedge

**6.** Groups
§1: The Basic Definitions
§2: Group Homomorphisms and Isomorphisms
§3: Cosets, Normal Subgroups, and Quotient Groups

**5.** Field Extensions
§3: An Introduction to Finite Fields

**6.** Groups
§4: The Symmetric Group $S_n$ and the 15-Puzzle

**7.** Group Actions and Symmetry
§1: Group Actions on a Set
§2: The Symmetry Groups of the Regular Polyhedra

**7.** Group Actions and Symmetry
§3: Burnside's Theorem and
Enumeration

**7.** Group Actions and Symmetry
§6: Some Remarks on Galois Theory

**7.** Group Actions and Symmetry
§4: Isometries of $\mathbf{R}^3$ and Classification
of the Regular Polyhedra

**7.** Group Actions and Symmetry
§5: Direct Products, p-Groups, and Sylow Theorems

**8.** Non-Euclidean Geometries
§1: Affine Geometry

**8.** Non-Euclidean Geometries
§2: The Projective Group

**8.** Non-Euclidean Geometries
§3: The Spectral Theorem and Quadrics

**8.** Non-Euclidean Geometries
§5: Putting the *metry* back in *Geometry*:
Elliptic and Hyperbolic Geometry

**8.** Non-Euclidean Geometries
§4: Projective Three Space and the Four Skew Line Problem

# CHAPTER 1

# The Integers

Algebra grew out of both arithmetic and geometry. The Pythagoreans dealt with whole numbers in geometric configurations and considered ratios of lengths of sides of triangles; they understood that, from their perspective, $\sqrt{2}$ could not be a number. Somewhat later, Euclid also developed more algebra and geometry simultaneously; indeed, his famous treatise *Elements* consists of thirteen volumes and covers Euclidean geometry, algebra, and the elementary number theory that we shall soon discuss. We begin this text by developing the structure of the classic number systems, focusing on the integers in Chapter 1. We recapitulate Euclid's algorithm for finding the greatest common divisor of two integers, as well as his treatment of primes and the factorization of integers as products of prime numbers. The remainder of the first chapter is devoted to the study of the arithmetic of congruences (a sophisticated version of "clock arithmetic"), developed in the eighteenth century by Euler and Lagrange, and refined by Gauss in the nineteenth. This leads naturally to the discussion of the structure of a ring (the "obvious" generalization of the algebraic properties of the integers). The ideas introduced in this chapter will provide both a foundation and motivation for much of our remaining study in this text.

## 1. Integers, Mathematical Induction, and the Binomial Theorem

We are all acquainted, of course, with the integers: $\dots, -3, -2, -1, 0, 1, 2, 3, \dots$. Traditionally, we start with the natural numbers (or *positive* integers) $1, 2, 3, \dots$, and then append 0 and the negative integers. We begin by recalling some properties of integer arithmetic. Let $\mathbb{Z}$ denote the set of integers; integers can be added and multiplied, subject to the following eight algebraic laws. For any $a$, $b$, $c \in \mathbb{Z}$,

(1) $a + b = b + a$   (commutative law of addition)
(2) $(a + b) + c = a + (b + c)$   (associative law of addition)
(3) $0 + a = a$   (0 is the additive identity)
(4) $a + (-a) = 0$   ($-a$ is the additive inverse of $a$)
(5) $a \times b = b \times a$   (commutative law of multiplication)
(6) $(a \times b) \times c = a \times (b \times c)$   (associative law of multiplication)
(7) $a \times 1 = a$   (1 is the multiplicative identity)
(8) $a \times (b + c) = (a \times b) + (a \times c)$   (distributive law)

Of course, we take these properties—and many others—for granted. For example, in doing arithmetic and algebra, we casually write "$a + b + c$" with no parentheses, knowing that the associative law guarantees us the same result when we compute $(a + b) + c$ and $a + (b + c)$; the same is true of products.

We next show that a few familiar algebraic properties of the integers are consequences of these eight basic laws.

**Lemma 1.1.** $0 \times a = 0$ *for any integer* $a$.

**Proof.** Since 0 is the additive identity, we know that $0 + 0 = 0$. Multiplying both sides of the equation by $a$, we obtain $(0 + 0) \times a = 0 \times a$. Expanding the left-hand side by the distributive law, we find $(0 \times a) + (0 \times a) = 0 \times a$; now we add the additive inverse of $0 \times a$ to both sides, obtaining $0 \times a = 0$, as required.   □

Next, we deduce the commonly used fact that "negative times negative is positive":

**Lemma 1.2.** *Additive inverses are unique. Moreover, for any integer* $a$, $-1 \times a = -a$ *and* $-(-a) = a$; *i.e.,* $a$ *is the additive inverse of* $-a$.

**Proof.** Suppose $b$ and $c$ are additive inverses of $a$: i.e., $a + b = 0$ and $a + c = 0$. Adding $b$ to both sides of the latter equation, we obtain $b + (a + c) = b$; and so, using the associative and commutative laws of addition, $b + (a + c) = (b + a) + c = 0 + c = c$, from which we conclude that $b = c$. This shows that $a$ has a unique additive inverse.

Multiply the equation $1 + (-1) = 0$ by $a$, obtaining $a + ((-1) \times a) = 0$, by Lemma 1.1; this says that $-1 \times a$ is the additive inverse of $a$.

Since additive inverses are unique, we must have $-1 \times a = -a$, as required. Lastly, the fact that $a$ is the additive inverse of $-a$ follows simply from the uniqueness of the additive inverse and the equation $a + (-a) = 0$. □

Certainly, $\mathbb{Z}$ is not the only collection of "numbers" that has properties (1) through (8). For example, the set $\mathbb{Q}$ of rational numbers does as well, and at the end of this chapter we shall meet several others. But $\mathbb{Z}$ has two additional characteristics: it has an ordering relation $<$ (which we shall emphasize in Chapter 2); and it also has the property that whenever $k$ is an element of $\mathbb{Z}$, so is $k+1$. Now, $\mathbb{Z}$ is in some sense characterized as the "smallest" number system to have both these properties. The latter property should remind you of the Principle of Mathematical Induction: Let $\mathbb{N}$ denote the set of natural numbers, i.e., the set of positive integers.

**The Principle of Mathematical Induction.** If $S \subset \mathbb{N}$ satisfies the two properties:

      (1) $1 \in S$,
      (2) if $k \in S$, then $k + 1 \in S$,

then $S = \mathbb{N}$.

In other words, if 1 belongs to $S$, and if, given any positive integer belonging to $S$, we are guaranteed that the subsequent integer also belongs to $S$, then $S$ consists of all the positive integers.

It will be convenient to rephrase the Principle of Mathematical Induction as follows.

**Proposition 1.3** (The Well-Ordering Principle). *Every nonempty set $T$ of positive integers has a least element.*

**Proof.** If $1 \in T$, then clearly 1 is the least element, and we are done. If not, let $S = \{n \in \mathbb{N} : \text{none of the numbers } 1, 2, \ldots, n$ belongs to $T\}$. Suppose $k \in S$; then if $k + 1 \notin S$, this means that $k + 1$ is the least element of $T$ and again we are done. If, however, for all $k \in S$ it is true that $k + 1 \in S$ as well, then by the Principle of Mathematical Induction, $S = \mathbb{N}$, and so $T$ is empty. □

Here is yet another way of stating the Principle of Mathematical Induction; it is required for certain of the proofs we'll do later.

**The Principle of Complete Mathematical Induction.** If $\Sigma \subset \mathbb{N}$ satisfies the two properties:

$(1')$ $1 \in \Sigma$,
$(2')$ if $1, 2, \ldots, k \in \Sigma$, then $k + 1 \in \Sigma$,

then $\Sigma = \mathbb{N}$.

The Well-Ordering Principle and the two versions of the Principle of Mathematical Induction are all equivalent; see Exercise 9.

**Remark.** In practice, mathematicians do not usually bother with the set $S$ appearing in the statement of the Principle of Mathematical Induction. Rather, we proceed a bit more informally: If we are given a statement $P(n)$, depending on a positive integer $n$, whose validity we wish to establish for all $n \in \mathbb{N}$, we argue that

(1) $P(1)$ holds, and that

(2) if $P(k)$ holds for an arbitrary $k \in \mathbb{N}$, then $P(k+1)$ holds.

To make it official, we could define
$$S = \{k \in \mathbb{N} : \text{the statement } P(k) \text{ is valid}\}.$$

Here now are a few "down to earth" examples of proofs by mathematical induction.

**Example 1.** We say an integer $n$ is **even** if there is an integer $m$ such that $n = 2m$; we say $n$ is **odd** if there is an integer $m$ such that $n = 2m + 1$. We now prove that every natural number is either even or odd.

First, 1 is odd, since $1 = 2 \times 0 + 1$. Now we suppose we know that $k \in \mathbb{N}$ must be either even or odd, and we attempt to prove that the same is true of $k + 1$. If $k$ is even, then $k = 2m$, whence $k + 1 = (2m) + 1$ is odd; if $k$ is odd, then $k = 2m + 1$, whence $k + 1 = (2m + 1) + 1 = 2(m + 1)$ is even. In either event, then, we have completed our task.

**Example 2.** For any real number $r \neq 1$ and $n \in \mathbb{N}$, we derive the formula (used in the discussion of geometric series)
$$1 + r + r^2 + \cdots + r^n = \frac{r^{n+1} - 1}{r - 1}.$$

It is important not to get confused here: $r$ is fixed, and the induction is on the natural number $n$. When $n = 1$ the formula reads
$$1 + r = \frac{r^2 - 1}{r - 1},$$

which is of course valid. Now suppose we know the formula holds when $n = k$; we must prove that it holds when $n = k + 1$. That is, given the formula

$$(*) \qquad 1 + r + r^2 + \cdots + r^k = \frac{r^{k+1} - 1}{r - 1},$$

we must prove

$$(**) \qquad 1 + r + r^2 + \cdots + r^{k+1} = \frac{r^{k+2} - 1}{r - 1}.$$

We proceed as follows: we add $r^{k+1}$ to both sides of equality $(*)$, obtaining

$$1 + r + r^2 + \cdots + r^k + r^{k+1} = \frac{r^{k+1} - 1}{r - 1} + r^{k+1}$$
$$= \frac{r^{k+1} - 1 + r^{k+1}(r - 1)}{r - 1} = \frac{r^{k+2} - 1}{r - 1},$$

which is equation $(**)$, as required. By the Principle of Mathematical Induction, the desired formula holds for all $n \in \mathbb{N}$.

**Remark.** We can equally well apply the Principle of Mathematical Induction with a slight modification. If $S' \subset \{n \in \mathbb{N} : n \geq n_0\}$ has the properties that

(i) $n_o \in S'$, and
(ii) if $k \in S'$, then $k + 1 \in S'$,

then $S' = \{n \in \mathbb{N} : n \geq n_o\}$. If we define

$$S = \{m \in \mathbb{N} : m + (n_o - 1) \in S'\},$$

we see that $1 \in S$ and $k \in S \implies k + 1 \in S$, and so $S = \mathbb{N}$. Thus,

$$S' = \{n \in \mathbb{N} : n = n_o + (m - 1) \text{ for some } m \in \mathbb{N}\} = \{n \in \mathbb{N} : n \geq n_o\},$$

as desired.

**Example 3.** We prove that for all integers $n \geq 3$, $2^{n+1} > 5n$. First, since $2^4 = 16 > 15 = 5 \cdot 3$, the statement is correct when $n = 3$. Next, suppose $2^{k+1} > 5k$. Multiplying the inequality by 2, we obtain $2^{k+2} > 2(5k) \geq 5k + 5$, since $5k \geq 5$ when $k \geq 1$. Rewriting this, we have $2^{(k+1)+1} > 5(k+1)$, as desired. By the slightly modified Principle of Mathematical Induction, the inequality holds for all integers $n \geq 3$.

Our last example (for now) of a use of mathematical induction is a cornerstone of high school and college mathematics, the binomial theorem, which gives a convenient formula for $(x + y)^n$ for any $n \in \mathbb{N}$. We begin with the following definition.

**Definition.** For any $n \in \mathbb{N}$ and any integer $k$ satisfying $0 \leq k \leq n$, we define $\binom{n}{k}$ (often read "$n$ choose $k$") to be the number of $k$-element subsets of the set $\{1, 2, \ldots, n\}$.

The first order of business is to derive a formula for $\binom{n}{k}$. We need to determine the number of ways that we can choose a $k$-element subset of $\{1, 2, \ldots, n\}$. Let's pick the $k$ elements one at a time: there are $n$ choices for the first, $n - 1$ choices for the second, $\ldots$, and $n - k + 1$ choices for the $k^{\text{th}}$; so, there are $n(n - 1) \cdot \ldots \cdot (n - k + 1)$ possible ways of selecting these $k$ numbers. However, we obtain the same $k$-element subset regardless of the order in which we choose the $k$ numbers; since there are $k!$ ways of choosing the same $k$-element subset, there are therefore

$$\binom{n}{k} = \frac{n(n - 1) \cdot \ldots \cdot (n - k + 1)}{k!} = \frac{n!}{(n - k)!k!}$$

possible different $k$-element subsets. (Recall that $n!$ (read "$n$ factorial") is equal to the product $n(n - 1)(n - 2) \cdot \ldots \cdot 2 \cdot 1$, and that, by convention, $0! = 1$.)

Note next that $\binom{n}{k} = \binom{n-1}{k} + \binom{n-1}{k-1}$. We start by calculating the right-hand side:

$$\binom{n - 1}{k} + \binom{n - 1}{k - 1} = \frac{(n - 1)!}{k!(n - k - 1)!} + \frac{(n - 1)!}{(k - 1)!(n - k)!}$$

$$= (n - 1)! \left( \frac{1}{k!(n - k - 1)!} + \frac{1}{(k - 1)!(n - k)!} \right)$$

$$= (n - 1)! \left( \frac{(n - k) + k}{k!(n - k)!} \right)$$

$$= (n - 1)! \left( \frac{n}{k!(n - k)!} \right) = \frac{n!}{k!(n - k)!} = \binom{n}{k},$$

as required. By the way, this is the origin of Pascal's triangle:

$$
\begin{array}{ccccccccccc}
 & & & & 1 & & 1 & & & & \\
 & & & 1 & & 2 & & 1 & & & \\
 & & 1 & & 3 & & 3 & & 1 & & \\
 & 1 & & 4 & & 6 & & 4 & & 1 & \\
1 & & 5 & & 10 & & 10 & & 5 & & 1 \\
\end{array}
$$

$$\vdots$$

We now come to the binomial theorem; a less technical and more intuitive proof is suggested in Exercise 10.

**Theorem 1.4** (Binomial Theorem). *For any $n \in \mathbb{N}$,*

(†)
$$(x + y)^n = \sum_{k=0}^{n} \binom{n}{k} x^{n-k} y^k.$$

**Proof.** To prove the theorem, we observe first that $(x + y)^1 = \binom{1}{0} x^1 y^0 + \binom{1}{1} x^0 y^1$, so (†) holds for $n = 1$. Now suppose (†) holds when $n = j$: i.e.,

$$(x + y)^j = \sum_{k=0}^{j} \binom{j}{k} x^{j-k} y^k.$$

We must prove that

$$(x + y)^{j+1} = \sum_{k=0}^{j+1} \binom{j+1}{k} x^{j+1-k} y^k.$$

We begin by multiplying both sides of the given equation by $x + y$:

$$(x + y)^{j+1} = (x + y)^j (x + y) = \left( \sum_{k=0}^{j} \binom{j}{k} x^{j-k} y^k \right) (x + y)$$

$$= \sum_{k=0}^{j} \binom{j}{k} x^{j-k+1} y^k + \sum_{k=0}^{j} \binom{j}{k} x^{j-k} y^{k+1}$$

(letting $k = \ell - 1$ in the second sum)

$$= \sum_{k=0}^{j} \binom{j}{k} x^{j-k+1} y^k + \sum_{\ell=1}^{j+1} \binom{j}{\ell-1} x^{j-\ell+1} y^{\ell}$$

(now replacing $\ell$ by $k$)

$$= \sum_{k=0}^{j} \binom{j}{k} x^{j-k+1} y^k + \sum_{k=1}^{j+1} \binom{j}{k-1} x^{j-k+1} y^k$$

$$= x^{j+1} + y^{j+1} + \sum_{k=1}^{j} \left( \binom{j}{k} + \binom{j}{k-1} \right) x^{j-k+1} y^k$$

$$= \sum_{k=0}^{j+1} \binom{j+1}{k} x^{j+1-k} y^k,$$

concluding the proof. □

We close this first section with some optional remarks on **recursive definition**, an application of the Principle of Mathematical Induction. Suppose we wanted to explain exponentiation to a novice. We might proceed as follows: $a^1$ is obviously just $a$, $a^2 = a \cdot a$ ($a$ multiplied by itself 2 times), $a^3 = a \cdot a \cdot a$ ($a$ multiplied by itself 3 times), etc., assuming that our listener could extrapolate and figure out for himself what $a^{97}$ means. To give a complete definition, however, it seems that an inductive argument is needed. We might try this: $a^1 = a$; and supposing that we've defined $a^n$ for some $n \in \mathbb{N}$, we define $a^{n+1}$ to be $a^n \cdot a$. (This is the recursion: we've defined the function value at $n + 1$ in terms of the function value at $n$.) The point is this: by the Principle of Mathematical Induction, we've determined a unique value of $a^n$ for each natural number $n$. The proof of this goes as follows. Consider all $n \in \mathbb{N}$ such that $a^k$ is uniquely defined for all $k \le n$. Clearly, $n = 1$ is fine, since we know that $a^1 = a$. If we suppose that $a^k$ is uniquely defined for all $k \le n$, it follows that it is uniquely defined for all $k \le n + 1$, since, by our recursive formula, $a^{n+1} = a^n \cdot a$. So, the Principle of Mathematical Induction tells us that $a^n$ is now uniquely defined for all $n \in \mathbb{N}$.

## EXERCISES 1.1

1. Use your calculator to compute the exact value of $814,235 \times 24,486$. (Here, I am assuming your calculator screen will, like mine, show at most ten digits. But the point is to find the exact value.)

2. A druggist has the five weights of $1, 3, 9, 27$, and $81$ ounces, and a two-pan balance. Show that he can weigh any integral amount up to and including 121 ounces. How can you generalize this result?

3. Prove that the square of an even number is even and the square of an odd number is odd.

4. Prove the following by induction:
   a. The sum of the first $n$ positive integers is $\frac{n(n+1)}{2}$.
   b. The sum of the first $n$ odd integers is $n^2$.
   c. The sum of the squares of the first $n$ positive integers is $\frac{n(n+1)(2n+1)}{6}$.

    d.   For $n \geq 1$, $n^3 - n$ is divisible by 3.

    e.   For $n \geq 3$, $n + 4 < 2^n$.

    f.   For all $n \in \mathbb{N}$, $1 + 5 + 9 + \cdots + (4n + 1) = (2n + 1)(n + 1)$.

    g.   For any positive integer $n$, one of $n, n + 2, n + 4$ must be divisible by 3.

    h.   $3^n \geq 2n + 1$ for all $n \in \mathbb{N}$.

    i.   For all $n \in \mathbb{N}$, $\frac{1}{n+1} + \frac{1}{n+2} + \cdots + \frac{1}{2n} = 1 - \frac{1}{2} + \frac{1}{3} - \cdots + \frac{1}{2n-1} - \frac{1}{2n}$.

    j.   Prove that for any $x > -1$ and any $n \in \mathbb{N}$, $(1 + x)^n \geq 1 + nx$.

5.   Prove that for every $n \in \mathbb{N}$, $\sum_{j=0}^{n} \binom{n}{j} = 2^n$. (What is the number of subsets of an $n$-element set?)

6.   a.   Using the original definition of $\binom{n}{k}$, prove that

$$\binom{n}{k} = \binom{n-1}{k} + \binom{n-1}{k-1}.$$

      (Hint: Consider two cases—either the number $n$ belongs to your particular $k$-element subset or it does not.)

    b.   Prove by induction on $n$ that $\binom{n}{k} = \frac{n!}{k!(n-k)!}$.

7.   The Fibonacci sequence 1, 1, 2, 3, 5, 8, 13, 21, 34, 55, 89, ... is obtained by the following recursive formula: $a_1 = 1$, $a_2 = 1$, $a_{n+1} = a_n + a_{n-1}$. Prove by induction that

$$a_n = \frac{1}{\sqrt{5}} \left( \left( \frac{1 + \sqrt{5}}{2} \right)^n - \left( \frac{1 - \sqrt{5}}{2} \right)^n \right).$$

(Hint: Either use complete induction, or consider those $n \geq 2$ so that the desired formula holds for both $a_n$ and $a_{n-1}$.)

    **Remark.** To see how to find such a formula, we claim that there are numbers $C$ and $D$ so that $a_n = C \left( \frac{1+\sqrt{5}}{2} \right)^n + D \left( \frac{1-\sqrt{5}}{2} \right)^n$. Given the values of $a_1$ and $a_2$, solve for $C$ and $D$. The claim comes from considering the problem as a linear algebra problem as follows: let $A = \begin{bmatrix} 0 & 1 \\ 1 & 1 \end{bmatrix}$. Let $\mathbf{x}_n = \begin{bmatrix} a_n \\ a_{n+1} \end{bmatrix}$. The formula for the sequence tells us that $\mathbf{x}_{n+1} = A\mathbf{x}_n$, whence $\mathbf{x}_n = A^{n-1}\mathbf{x}_1$. Thus one must compute the powers of the matrix $A$; this is most easily done by diagonalizing $A$. But it is easy to check that the eigenvalues of $A$ are $\frac{1 \pm \sqrt{5}}{2}$, and then the claim follows.

8.  a.   Prove that $\binom{2n}{n} = \sum_{j=0}^{n} \binom{n}{j}^2$. (Hint: Use Theorem 1.4 to compute $(1 + x)^{2n}$ two ways.)

    b.   Prove, more generally, that when $0 \le m \le n$,

    $$\binom{n+m}{m} = \sum_{j=0}^{m} \binom{m}{j}\binom{n}{j}.$$

9.  Prove that the Principle of Mathematical Induction, the Well-Ordering Principle, and the Principle of Complete Mathematical Induction are logically equivalent, as follows.

    a.   Prove that the Well-Ordering Principle implies the Principle of Complete Mathematical Induction. (Hint: Given $S \subset \mathbb{N}$ with $1 \in S$ and $1,\ldots,k \in S \implies k+1 \in S$, let $T = \{k \in \mathbb{N} : k \notin S\}$.)

    b.   Prove that the Principle of Complete Mathematical Induction implies the Principle of Mathematical Induction. (Hint: Let $\Sigma = \{k \in \mathbb{N} : 1, 2, \ldots, k \in S\}$.)

    c.   Using Proposition 1.3, finish the proof of equivalence.

10. Give an alternative proof of the binomial theorem, Theorem 1.4, by counting the number of ways the monomial $x^{n-k}y^k$ can appear in the product

    $$(x + y)^n = \underbrace{(x + y)(x + y)\ldots(x + y)}_{n \text{ times}}.$$

11. Prove by induction that every positive integer $n$ has a unique expression of the form

    $$n = b_k 2^k + b_{k-1} 2^{k-1} + \cdots + b_1 2 + b_0,$$

    where the integers $b_0,\ldots,b_{k-1}$ are either 0 or 1, and $b_k = 1$. (More generally, if $m$ is any integer greater than 1, every positive integer $n$ has a unique expression of the form

    $$n = b_k m^k + b_{k-1} m^{k-1} + \cdots + b_1 m + b_0,$$

    where the integers $b_0,\ldots,b_k$ satisfy $0 \le b_j < m$, $b_k \ne 0$.)

12. Rephrase the definition of $n!$ carefully, using recursive definition.

13. Recalling that exponentiation is defined recursively, prove:

    a.   for any numbers $a$ and $b$, $(ab)^n = a^n b^n$;

    b.   for any number $a$ and any $m, n \in \mathbb{Z}$, $a^m a^n = a^{m+n}$;

    c.   for any number $a$ and any $m, n \in \mathbb{Z}$, $(a^m)^n = a^{mn}$. (Hint: fix $m$ and proceed by induction on $n$.)

14.   a.   Show that the formulas $f(1) = 5$, $f(n+1) = \sqrt{f(n) + 1}$ define a unique function $f : \mathbb{N} \to \mathbb{R}$.

      b.   Show that the formulas $f(1) = 5$, $f(n+1) = \sqrt{f(n) - 1}$ do not define a function $f : \mathbb{N} \to \mathbb{R}$.

      c.   Explain why recursive definition cannot be applied successfully to b.

15.   Let $a_m$ be (real) numbers for all $m \in \mathbb{N}$. Use recursive definition to make sense of $\sum\limits_{m=1}^{n} a_m$.

16.   Let $n \in \mathbb{N}$, and suppose you choose $n+1$ distinct numbers from the list $1, 2, 3, \ldots, 2n - 1, 2n$. Will it always be the case that among the numbers you've chosen, you can find a pair with the property that one divides the other? Provide a proof or counterexample.

17.   If 76 women play a round-robin tennis tournament (each woman playing each other woman once), is it always possible to arrange their names in a vertical list so that each player on the list has beaten the person just beneath her on the list? (Hint: Start off with a more manageable number of participants.)

18.   Prove by induction that for any $k \in \mathbb{N}$ the product of $k$ consecutive integers is divisible by $k!$. (Hint: Prove that for all $k \in \mathbb{N}$, for all $n \in \mathbb{N}$, the number $n(n+1) \cdots (n+k-1)$ is divisible by $k!$.)

19.   Let $a_n$ be the fraction of the entries in the first $n$ rows of Pascal's triangle that are odd. Calculate $\lim\limits_{n \to \infty} a_n$. (It may be suggestive to draw a large Pascal's triangle, coloring the odd entries black.)

## 2. The Euclidean Algorithm, Prime Numbers, and Factorization

We turn next to the divisibility properties of integers. We say $d$ is a **divisor** of $a$ (written $d \mid a$) if $a = md$ for some integer $m$. An immediate consequence of this definition is the following lemma.

**Lemma 2.1.** *Let a, b, and d be integers. If $d|a$ and $d|b$, then $d|(ma + nb)$ for any integers $m, n$.*

**Proof.** By hypothesis, there are integers $x$ and $y$ so that $a = xd$ and $b = yd$. Therefore,

$$ma + nb = m(xd) + n(yd) = (mx + ny)d,$$

and since $mx + ny$ is again an integer, $d|(ma + nb)$, as required. $\square$

On the other hand, if $b$ is not a divisor of $a$, we ought to be able to perform division with remainder. We formalize this process in the following result.

**Proposition 2.2** (Division Algorithm). *Given $a, b \in \mathbb{N}$, there are integers q (for "quotient") and r (for "remainder") so that*

$$a = qb + r, \quad with \quad 0 \le r < b.$$

We will leave it to the reader (see Exercise 21) to check that there is an obvious generalization to all integers. Another question is also worth pondering: are the integers $q$ and $r$ specified by Proposition 2.2 unique? (See Exercise 5.) The division algorithm is normally taken as being self-evident, but we'll give a proof by induction on $a$.

**Proof.** If $a = 1$, when $b = 1$ take $q = 1$ and $r = 0$; and when $b > 1$, take $q = 0$ and $r = a = 1$. Suppose we know the division algorithm holds for $a = k$, so that for any $b \in \mathbb{N}$, there are integers $q$ and $r$ so that $k = qb + r$, $0 \le r < b$. Consider now the integer $a = k + 1$. If $r < b - 1$, then we may write

$$k + 1 = qb + (r + 1), \quad where \quad r + 1 < b;$$

whereas if $r = b - 1$, then we write

$$k + 1 = (q + 1)b$$

(with the remainder term zero). $\square$

We next come to one of the central ideas of this course. Suppose we start with two integers $a$ and $b$, say $a = 20$ and $b = 36$, and consider all possible sums of their multiples. After some experimentation, we find that 4 can be written as $4 = 2 \times \underline{20} + (-1) \times \underline{36}$, and, therefore, every multiple of 4 can be written as a sum of multiples of 20 and 36 (why?). What about 6? Suppose we had $6 = 20m + 36n$ for some integers $m$ and $n$; since 4 divides both 20 and 36, by Lemma 2.1, 4 would have to divide 6. (Similarly, no positive integer smaller than 4 can be written in the form $20m + 36n$.) Such reasoning suggests that the set of all sums of multiples of 20 and 36 coincides with the set of multiples of 4. In general, we obtain the following theorem.

**Theorem 2.3.** *Given $a, b \in \mathbb{Z}$, not both zero, the set $S = \{ma + nb : m, n \in \mathbb{Z}\}$ consists of all multiples of some positive integer $d$. This integer $d$ divides both $a$ and $b$; it also has the property that if an integer $e$ divides both $a$ and $b$, then $e$ divides $d$.*

**Proof.** First, we may assume $a, b \geq 0$. Consider the set $S^+ = S \cap \mathbb{N}$. By the Well-Ordering Principle, Proposition 1.3, $S^+$ has a smallest element $d = m_0 a + n_0 b$. (It is crucial here that we are considering only those combinations $ma + nb$ which are positive integers!) We now claim that $S$ consists of all multiples of $d$. Note first that every multiple of $d$ is indeed an element of $S$ (why?).

Next we check that $d$ divides $a$ and $b$. Applying the division algorithm, Proposition 2.2, divide $a$ by $d$; if the division results in a nonzero remainder $r$, then $r = a - qd = a - q(m_0 a + n_0 b) = (1 - qm_0)a + (-qn_0)b$ belongs to $S^+$ and is smaller than $d$. This result contradicts the fact that $d$ is the smallest element of $S^+$. Thus $d|a$. By an identical argument, $d|b$. It now follows from Lemma 2.1 that $d$ divides every element of $S$, completing the proof that $S$ consists of all multiples of $d$.

Lastly, we must check that if $e|a$ and $e|b$, then $e|d$. This follows again from Lemma 2.1, since $d = m_0 a + n_0 b$. This concludes the proof. □

This theorem leads us to make the following definition. Given two integers $a, b$, not both 0, we define their **greatest common divisor** $\gcd(a, b)$ to be the unique number $d \in \mathbb{N}$ having the following properties:

    (i) $d|a$ and $d|b$, and
    (ii) if $e \in \mathbb{Z}$, $e|a$ and $e|b \implies e|d$.

In other words, $d$ is a divisor of both $a$ and $b$, and it is divisible by any other such common divisor. Thus, it is the greatest (largest) common divisor (see also Exercise 15). We often abbreviate the greatest common divisor as the "g.c.d."

**Examples 1.** $\gcd(35, 154) = 7$, $\gcd(57843, 296304) = 3$, and $\gcd(56, 697) = 1$.

One way to find the g.c.d. is to factor both numbers into smaller, easily recognizable numbers (for example, primes); but, as the numbers get larger, this is not a very appetizing thought. The trick is to use the division algorithm to produce the g.c.d. Consider the first example above: we wish to find the g.c.d. of 35 and 154. We begin by dividing 154 by 35; if the division were to yield zero remainder, then 35 would divide 154 evenly, and so the g.c.d. would clearly be 35. Well, this is not the case; but we repeat the process, replacing 35 and 154, respectively, by the remainder, 14, and 35. We continue until the remainder is zero. At this point, the last divisor is the candidate for our greatest common divisor.

$$
\begin{array}{ccc}
\;\;\;\underline{\;\;4\;\;} & \;\;\;\underline{\;\;2\;\;} & \;\;\;\underline{\;\;2\;\;} \\
35\,)\,154 & 14\,)\,35 & 7\,)\,14 \\
\;\underline{140} & \;\underline{28} & \;\underline{14} \\
\;\;14 & \;\;7 & \;\;0
\end{array}
$$

Thus, 7 is the g.c.d. of 35 and 154. We display these calculations in a more convenient form:

(A) $\qquad\qquad\qquad 154 = 35 \times 4 + 14$

(B) $\qquad\qquad\qquad 35 = 2 \times 14 + 7$

(C) $\qquad\qquad\qquad 14 = 2 \times 7 + 0$

Here is the reason that $7 = \gcd(35, 154)$: first, we check that 7 divides both 35 and 154 by working our way backwards through these equations. Since 7 divides 14 (by (C)), and therefore divides 35 (by (B)), it now follows that 7 divides 154 as well (by (A)). On the other hand, if $e$ is any divisor of 35 and 154, by (A) $e$ is a divisor of 14, therefore, by (B), a divisor of 7. This shows that, indeed, $7 = \gcd(35, 154)$.

What's more, these calculations show us how to compute the expression $d = m_o a + n_o b$ guaranteed by the theorem: starting with (B) and proceeding upwards to (A), we obtain

$$
\begin{aligned}
7 &= 35 - 2 \times 14 \\
&= 35 - 2 \times (154 - 4 \times 35) \\
&= (9 \times \underline{35}) + (-2 \times \underline{154}),
\end{aligned}
$$

as required.

**Example 2.** $\gcd(56, 697) = 1$; we leave it to the reader to check that $1 = 9 \times \underline{697} + (-112) \times \underline{56}$:

$$
\begin{array}{r}
12 \\
\hline
56\,)\overline{697} \\
56 \\
\hline
137 \\
112 \\
\hline
25
\end{array}
\qquad
\begin{array}{r}
2 \\
\hline
25\,)\overline{56} \\
50 \\
\hline
6
\end{array}
\qquad
\begin{array}{r}
4 \\
\hline
6\,)\overline{25} \\
24 \\
\hline
1
\end{array}
$$

This procedure for computing the g.c.d. of two numbers—and the resulting expression $d = m_0 a + n_0 b$—is due to Euclid and is usually called the **Euclidean algorithm**. (See Exercise 22 for a proof.)

**Remark.** It is interesting to examine the set of *positive* combinations of $a$ and $b$, namely, $\{ma + nb : m, n \in \mathbb{N}\}$. In particular, this leads to the classic "postage stamp problem"; see Exercises 2 and 20.

We say $p \in \mathbb{N}$ is a **prime number** if $p > 1$ and its only positive divisors are 1 and $p$. The first prime numbers are 2, 3, 5, 7, 11, 13, 17, and 19. We say two integers $a$ and $b$ are **relatively prime** if $\gcd(a,b) = 1$. For example, 56 and 697 are relatively prime. We say that $n \in \mathbb{N}$ is **composite** if $n > 1$ and $n$ is not prime.

We first point out the following very important consequence of Theorem 2.3, apparently first discovered by Euclid.

**Corollary 2.4.** *a and b are relatively prime if and only if there are integers m and n so that* $1 = ma + nb$.

**Proof.** Both implications follow from Theorem 2.3. Certainly, if $\gcd(a,b) = 1$, then 1 has such an expression. Conversely, if $1 = ma + nb$, then 1 is clearly the smallest positive number so expressible, and it is therefore the g.c.d. of $a$ and $b$.  □

We observe that 6 can divide $4 \times 3$ without dividing either factor; but when the prime number 2 divides $4 \times 3$, it *must* divide one of the factors. This observation is the key to much of our work to come.

**Proposition 2.5.** *Suppose p is prime and $p|ab$. Then p must divide a or b.*

**Proof.** Suppose $p|ab$ and $p \nmid b$ (we use the symbol "$\nmid$" to denote "does not divide"). We must show then that $p|a$. Since $p$ is prime and is not a factor of $b$, it follows that $\gcd(p,b) = 1$. Therefore, by Corollary 2.4, we may write $1 = mp + nb$ for appropriate $m, n \in \mathbb{Z}$.

Multiplying this equation by $a$, we obtain $a = (mp+nb)a = (ma)p + n(ab)$. Since $p$ divides $p$ and $p$ divides $ab$, it divides the right-hand side of this equality; therefore, $p$ divides $a$, as required. $\square$

One important application of this idea, which we shall use frequently, is the following corollary.

**Corollary 2.6.** *Suppose $p$ is a prime number. If $0 < k < p$, then* $p \mid \binom{p}{k}$.

**Proof.** Intuitively, since $\binom{p}{k} = \frac{p!}{k!(p-k)!}$, the factor of $p$ in the numerator cannot be "canceled out." To prove this, let $a = (p-1)!$ and $b = k!(p-k)!$, so $\binom{p}{k} = p\frac{a}{b}$. This means $pa = b\binom{p}{k}$, whence $p \mid b\binom{p}{k}$. By Proposition 2.5, $p$ must divide $b$ or it must divide $\binom{p}{k}$. But $b$ is a product of factors each one of which is a natural number less than $p$ (and hence not divisible by $p$); it follows again from Proposition 2.5 that $p$ cannot divide $b$. Thus, we conclude that $p \mid \binom{p}{k}$, as required. $\square$

It is essential that $p$ be prime for the result (and the proof!) we've just given to be valid. (For example, $\binom{4}{2} = 6$ and $4 \nmid 6$.) We now use Proposition 2.5 to prove the classic result that every positive integer can be factored (uniquely) as a product of prime numbers.

**Theorem 2.7** (Fundamental Theorem of Arithmetic). *Let $n > 1$ be an integer. Then there are prime numbers $p_1, \dots, p_m$ and positive integers $v_1, \dots, v_m$ so that*

$$n = p_1^{v_1} p_2^{v_2} \cdots p_m^{v_m}.$$

*Moreover, this expression is unique (except for the possibility of rearranging the factors).*

**Proof.** We use proof by complete induction. For starters, 2 can be factored uniquely as a product of primes. Suppose now that all integers less than or equal to $n$ can be factored uniquely as a product of primes. We now prove that $N = n+1$ can be factored uniquely as a product of primes. If $N$ is prime, we're done. If not, there are integers $a$ and $b$, $2 \le a, b \le n$, so that $N = ab$. By assumption, each of $a$ and $b$ can be written as a product of primes; concatenating the two products, $N$ is in turn expressed as a product of primes. (More concretely, let $p_1, \dots, p_m$ be a list of all the primes appearing in the factorizations of $a$ and $b$, and put

$$a = p_1^{\mu_1} \cdots p_m^{\mu_m}, \qquad b = p_1^{v_1} \cdots p_m^{v_m}, \qquad \mu_i, v_i \ge 0.$$

Then $N = p_1^{\mu_1 + \nu_1} \cdots p_m^{\mu_m + \nu_m}$, as required.)

We now turn to the question of uniqueness. Suppose (now using the Well-Ordering Principle) that $N$ is the smallest integer having two distinct expressions as products of primes. In particular, suppose

$$N = p_1^{\nu_1} \cdots p_m^{\nu_m} = q_1^{\sigma_1} \cdots q_\ell^{\sigma_\ell}, \qquad p_i, q_j \text{ primes}, \qquad \nu_i, \sigma_j \in \mathbb{N}.$$

By Exercise 9, the prime $p_1$ divides $N = q_1^{\sigma_1} \cdots q_\ell^{\sigma_\ell}$, and therefore must divide one of $q_1^{\sigma_1}, \ldots, q_\ell^{\sigma_\ell}$; by the same exercise, it follows that $p_1$ must divide one of the primes $q_1, \ldots, q_\ell$. The only way this can happen is for one of the primes $q_j$ to equal $p_1$. Dividing $N$ by $p_1 = q_j$, we now obtain a number smaller than $N$ with distinct expressions as products of primes. This contradiction completes the proof. $\square$

We would be remiss if we did not give Euclid's famous proof of the following theorem.

**Theorem 2.8.** *There are infinitely many prime numbers.*

**Proof.** Suppose there were only finitely many; let $p_1, p_2, \ldots, p_m$ be a complete list. Now consider the number $N = p_1 p_2 \cdots p_m + 1$. Since $N$ is not on our complete list of primes, by Theorem 2.7 it can be factored as a product of primes. But none of the primes $p_1, \ldots, p_m$ divides $N$; this contradiction completes the proof. $\square$

## EXERCISES 1.2

1. For each of the following pairs of numbers $a$ and $b$, find $d = \gcd(a, b)$ and express $d$ in the form $ma + nb$ for suitable integers $m$ and $n$.
   a. 14, 35
   b. 56, 77
   c. 618, 336
   d. 2873, 6643
   e. 512, 360
   f. 4432, 1080

2. You have at your disposal arbitrarily many 4-cent stamps and 7-cent stamps. What are the postages you can pay? Show in particular that you can pay all postages greater than 17 cents.

3.  Prove that whenever $m \neq 0$, $\gcd(0, m) = m$.

4.  a.  Prove that if $a|x$ and $b|y$, then $ab|xy$.
    b.  Prove that if $d = \gcd(a, b)$, then $\gcd(\frac{a}{d}, \frac{b}{d}) = 1$.

5.  Prove or give a counterexample: the integers $q$ and $r$ guaranteed by the division algorithm, Theorem 2.2, are unique.

6.  Prove or give a counterexample. Let $a, b \in \mathbb{Z}$. If there are integers $m$ and $n$ so that $d = am + bn$, then $d = \gcd(a, b)$.

7.  Generalize Proposition 2.5: if $\gcd(m, c) = 1$ and $m|cz$, then prove $m|z$.

8.  Suppose $a, b, n \in \mathbb{N}$, $\gcd(a, n) = 1$, and $\gcd(b, n) = 1$. Prove or give a counterexample: $\gcd(ab, n) = 1$.

9.  Prove that if $p$ is prime and $p|(a_1 a_2 \ldots a_n)$, then $p|a_j$ for some $j$, $1 \leq j \leq n$. (Hint: Use Proposition 2.5 and induction.)

10. Given a positive integer $n$, find $n$ consecutive composite numbers.

11. Prove that there are no integers $m, n$ so that $\left(\frac{m}{n}\right)^2 = 2$. (Hint: You may start by assuming $m$ and $n$ are relatively prime. Why? Then use Exercise 1.1.3.)

12. Find all rectangles whose sides have integral lengths and whose area and perimeter are equal.

13. Given two nonzero integers $a, b$, in analogy with the definition of $\gcd(a, b)$, we define the **least common multiple** $\text{lcm}(a, b)$ to be the positive number $\mu$ with the properties:
    (i)   $a|\mu$ and $b|\mu$, and
    (ii)  if $s \in \mathbb{Z}$, $a|s$ and $b|s \implies \mu|s$.
    Prove that
    a.  if $\gcd(a, b) = 1$, then $\mu = ab$. (Hint: If $\gcd(a, b) = 1$, then there are integers $m$ and $n$ so that $1 = ma + nb$; therefore, $s = mas + nbs$.)
    b.  more generally, if $\gcd(a, b) = d$, then $\mu = ab/d$.

14. See Exercise 13 for the definition of $\text{lcm}(a, b)$. Given prime factorizations $a = p_1^{\mu_1} \cdots p_m^{\mu_m}$ and $b = p_1^{\nu_1} \cdots p_m^{\nu_m}$, with $\mu_i, \nu_i \geq 0$, express $\gcd(a, b)$ and $\text{lcm}(a, b)$ in terms of $p_1, \ldots, p_m$. Prove that your answers are correct.

15.    Note that in our definition of g.c.d. we never literally said $d$ was the largest possible common divisor. Show that if we define $\delta > 0$ by the properties:

   (1′)    $\delta | a$ and $\delta | b$, and

   (2′)    $e | a$ and $e | b \implies |e| \le \delta$,

then $\delta = \gcd(a, b)$. (The reason for our original choice is that it is by far more convenient to work with. On the other hand, it is clearer that $\delta$ exists: Make a list of the common divisors, and take the largest one.)

16.    The point of this exercise is to show that we should not take the unique factorization property of $\mathbb{N}$ for granted.

   a.    Let $T = \{\text{primes}\} \cup \{2^\nu : \nu \text{ prime}\}$. Then we clearly can factor all natural numbers greater than or equal to 2 as products of elements of $T$; can we do so uniquely?

   b.    Consider the set of numbers $T = \{1, 4, 7, 10, 13, \ldots, 3n + 1, \ldots\}$. Show that $T$ is closed under multiplication. Call $p \in T$ a mock-prime if the only factors of $p$ in $T$ are 1 and $p$, $p > 1$. Show that every number in $T$ can be factored as a product of mock-primes, but not necessarily uniquely.

17.    Prove that whenever $n > 1$ is odd, $2^{mn} + 1$ is a composite number. (Hint: Use Example 2 of Section 1 to prove that $2^m + 1$ is a factor.)

   **Remark.** It follows that a number of the form $2^s + 1$ can be prime only when $s$ is a power of 2. Such prime numbers are called *Fermat primes*. Be warned! $2^{2^5} + 1 = 4,294,967,297 = 641 \times 6,700,417$ is *not* prime. Indeed, Mathematica indicates that the numbers $2^{2^n} + 1$ are composite for $n = 5, 6, \ldots, 12$, and it is currently conjectured that all numbers $2^{2^n} + 1$ are composite for $n \ge 5$.

18.    Write an algorithm (or preferably a computer program) that generates $\gcd(a, b)$ using the division algorithm and expresses $d$ in the form $d = ma + nb$ for suitable integers $m$ and $n$. (To do the latter most effectively, you should ideally find a way of doing the algebra "from the top down," rather than "from the bottom up.")

19.    Show that for any positive integers $a$ and $b$, the set $\{ma + nb : m, n \ge 0\}$ includes all multiples of $\gcd(a, b)$ larger than $ab$.

20. Generalize the result of Exercise 2 to stamps of value $a$ and $b$. The case where $\gcd(a, b) = 1$ is of particular interest.

21. State and prove a generalization of the division algorithm (Theorem 2.2) when $a, b \in \mathbb{Z}$ ($b \neq 0$).

22. Give an official proof of the Euclidean algorithm. That is, suppose $a, b \in \mathbb{N}$, $a > b$, and

$$a = q_0 b + r_1$$
$$b = q_1 r_1 + r_2$$
$$r_1 = q_2 r_2 + r_3$$
$$\vdots$$
$$r_{n-2} = q_{n-1} r_{n-1} + r_n$$
$$r_{n-1} = q_n r_n.$$

Prove that $d = r_n$ is $\gcd(a, b)$. (Hint: Show that $d | r_n$, $d | r_{n-1}$, $d | r_{n-2}$, ..., $d | r_1$, $d | b$, $d | a$; and that if $e | a$ and $e | b$, then $e | r_1$, $e | r_2$, ..., $e | r_n$.)

## 3. Modular Arithmetic and Solving Congruences

Let $m \in \mathbb{N}$. We next introduce "modular arithmetic," or integer arithmetic **modulo** $m$. Given two integers $a$ and $b$, we say that $a \equiv b \pmod{m}$ (read "$a$ is congruent to $b$ mod $m$") if $m | (a - b)$. For example, $76 \equiv 22 \pmod{18}$ and $2 \equiv -13 \pmod{3}$. Congruence mod $m$ is an equivalence relation on $\mathbb{Z}$ (see Appendix A):

(1) $a \equiv a \pmod{m}$    (reflexive property)
(2) $a \equiv b \pmod{m} \implies b \equiv a \pmod{m}$    (symmetric property)
(3) $a \equiv b \pmod{m}, b \equiv c \pmod{m} \implies a \equiv c \pmod{m}$
(transitive property)

Note, in particular, that $a \equiv 0 \pmod{m}$ if and only if $m | a$. We can also say that $a \equiv b \pmod{m}$ if $a$ and $b$ yield the same remainder when divided by $m$. For example, if we divide 76 by 18, we get a remainder of 4, as we do when we divide 22 by 18. More formally, if $a = qm + r$ and $b = q'm + r$, where $0 \leq r < m$, then $a - b = (q - q')m$; so $a \equiv b \pmod{m}$. Conversely, if $a \equiv b \pmod{m}$, then $a - b = sm$ for some $s \in \mathbb{Z}$; if $a = qm + r$, where $0 \leq r < m$, then

$b = a - sm = (q - s)m + r$, as required. Thus, every integer $a$ is equivalent (mod $m$) to exactly one of the numbers $0, 1, \ldots, m - 1$; that is, $a \equiv r$ (mod $m$), where $r$ is the remainder when we divide $a$ by $m$).

As we now show, equivalence mod $m$ respects the algebraic operations on $\mathbb{Z}$.

**Proposition 3.1.** *If $a_1 \equiv b_1$ (mod $m$) and $a_2 \equiv b_2$ (mod $m$), then*

(1) $a_1 + a_2 \equiv b_1 + b_2$ (mod $m$);
(2) $ca_1 \equiv cb_1$ (mod $m$); *and, more generally,*
(3) $a_1 a_2 \equiv b_1 b_2$ (mod $m$).

**Proof.** Suppose $a_1 - b_1 = xm$ and $a_2 - b_2 = ym$ for some $x, y \in \mathbb{Z}$. Then $(a_1 + a_2) - (b_1 + b_2) = (a_1 - b_1) + (a_2 - b_2) = xm + ym = (x + y)m$. This proves (1). $ca_1 - cb_1 = c(a_1 - b_1) = c(xm) = (cx)m$, proving (2). Lastly, $a_1 a_2 - b_1 b_2 = (b_1 + xm)(b_2 + ym) - b_1 b_2 = m(xb_2 + yb_1 + xym)$, proving (3). $\square$

For our applications, it will be necessary to observe that the analogous results hold for sums and products of any number of integers. The fastidious reader may provide a proof by induction.

We apply Proposition 3.1 to derive the rather well-known tests for divisibility by the integers 2, 3, 4, 5, 7, 9, and 11.

**Proposition 3.2** (Divisibility Criteria). *Let $n = \sum\limits_{i=0}^{k} a_i 10^i$, $0 \le a_i \le$*
*9, be a positive integer. (The $a_i$'s are the digits of the base 10 representation of $n$.) Then*

(a) *$2 \mid n$ if and only if $2 \mid a_0$;*

(b) *$3 \mid n$ if and only if $3 \mid \sum\limits_{i=0}^{k} a_i$;*

(c) *$4 \mid n$ if and only if $4 \mid (10a_1 + a_0)$;*

(d) *$5 \mid n$ if and only if $5 \mid a_0$;*

(e) *$7 \mid n$ if and only if $7 \mid (\sum\limits_{i=1}^{k} a_i 10^{i-1} - 2a_0)$;*

(f) *$9 \mid n$ if and only if $9 \mid \sum\limits_{i=0}^{k} a_i$;*

(g) *$11 \mid n$ if and only if $11 \mid \sum\limits_{i=0}^{k} (-1)^i a_i$.*

***Proof.*** We will indicate proofs of (d), (e), and (f). For (d), note that for $i \geq 1$, $10^i \equiv 0 \pmod 5$, and so $n \equiv a_0 \pmod 5$. For (f), note that for $i \geq 1$, $10^i \equiv 1^i = 1 \pmod 9$; so $n \equiv \sum_{i=0}^{k} a_i \pmod 9$, as required.

Before proving the rule for 7, we give an illustration:

$$7 \mid 31934 \iff 7 \mid (3193 - 8) = 3185 \iff 7 \mid (318 - 10) = 308$$
$$\iff 7 \mid (30 - 16) = 14.$$

In a more graphical form,

$$
\begin{array}{r|l}
3\ 1\ 9\ 3 & 4 \\
8 & \\
\hline
3\ 1\ 8 & 5 \\
1\ 0 & \\
\hline
3\ 0 & 8 \\
1\ 6 & \\
\hline
1\ 4 &
\end{array}
$$

Now, we must show that

$$n \equiv 0 \pmod 7 \iff n' = \sum_{i=1}^{k} a_i 10^{i-1} - 2a_0 \equiv 0 \pmod 7.$$

This all becomes clearer if we observe that

$$n - 21a_0 = \sum_{i=0}^{k} a_i 10^i - 21a_0 = \sum_{i=1}^{k} a_i 10^i - 20a_0 = 10n'.$$

Thus, $n \equiv 10n' \pmod 7$, and so $n' \equiv 0 \pmod 7 \implies n \equiv 0 \pmod 7$. Conversely, if $n \equiv 0 \pmod 7$, then $5n \equiv 0 \pmod 7$ as well; but $5n \equiv 50n' \equiv n' \pmod 7$, so $n' \equiv 0 \pmod 7$ as well. Thus, $7 \mid n \iff 7 \mid n'$, and this completes the proof of (e). The remaining parts are left as exercises for the reader (see Exercise 2). □

We give some further elementary applications of modular arithmetic.

**Example 1.** Could the number $m = 3,241,594,226$ be a perfect square? Let's see how we can use arithmetic mod 4 to detect perfect squares. We start by making a table of the possible values of $n$ and

$n^2$ mod 4:

| $n$ (mod 4) | $n^2$ (mod 4) |
|:---:|:---:|
| 0 | 0 |
| 1 | 1 |
| 2 | 0 |
| 3 | 1 |

(Note that we could save a little work by noting that $3 \equiv -1$ (mod 4), so that $3^2 \equiv (-1)^2 \equiv 1^2$ (mod 4).) The upshot is that any perfect square must be congruent to either 0 or 1 (mod 4). Our integer $m$ is congruent to 2 (mod 4). (The reader should figure out an easy way to see this without actually dividing $m$ by 4.) In conclusion, $m$ cannot be a perfect square.

Notice that working mod 10 (as we are wont to do, since we are used to base 10 arithmetic) will be inconclusive, because the square of a number whose units digit is 4 or 6 will have a units digit of 6.

**Example 2.** How can the units digit of the sum of two squares be equal to 5 if neither is divisible by 5? (One obvious solution is suggested by the famous Pythagorean equality $3^2 + 4^2 = 5^2$.) We proceed as in the preceding example, first making a table of the values of squares mod 10:

| $n$ (mod 10) | $n^2$ (mod 10) |
|:---:|:---:|
| 0 | 0 |
| $\pm 1$ | 1 |
| $\pm 2$ | 4 |
| $\pm 3$ | 9 |
| $\pm 4$ | 6 |
| 5 | 5 |

If neither number is divisible by 5, we discard the values 0 and 5, and consider the sums of the remaining possible values of squares (working mod 10):

| + | 1 | 4 | 6 | 9 |
|:---:|:---:|:---:|:---:|:---:|
| 1 | 2 | 5 | 7 | 0 |
| 4 |   | 8 | 0 | 3 |
| 6 |   |   | 2 | 5 |
| 9 |   |   |   | 8 |

(Note that since addition is commutative, we need only fill in the upper portion of the table.) We see that a units digit of 5 can arise in two ways: either by adding the square of a number ending in 1 or 9 to the square of a number ending in 2 or 8, or by adding the square of a number ending in 4 or 6 to the square of a number ending in 3 or 7. <u>Query</u>: Why did we not try to solve this problem by working mod 5?

We next prove Fermat's little theorem, a useful tool in the theory of codes. It will also arise in our work many times.

**Proposition 3.3** (Fermat). *If $p$ is any prime number $p$ and $n$ is any integer, then $n^p \equiv n$ (mod $p$).*

***Proof.*** Here is the first of many proofs of this result that we'll encounter. We proceed by induction on $n$ (thus establishing the result only for positive $n$; see Exercise 6). The result is obvious when $n = 1$. Now suppose we know it to be true when $n = k$; we must then deduce its validity when $n = k + 1$. That is, given that $k^p \equiv k$ (mod $p$), we are to prove that $(k + 1)^p \equiv k + 1$ (mod $p$). We expand $(k + 1)^p$ using the binomial theorem (Theorem 1.4):

$$(k+1)^p = k^p + \binom{p}{1}k^{p-1} + \binom{p}{2}k^{p-2} + \cdots + \binom{p}{p-1}k + 1 \equiv k + 1 \pmod{p}.$$

Here, we are using Corollary 2.6: when $p$ is prime and $1 \le j \le p - 1$, the binomial coefficient $\binom{p}{j}$ is divisible by $p$. □

Next, we will refine Euclid's observation that there are infinitely many primes, Theorem 2.8. Note that 2 is the only even prime; the remaining primes, being odd, must be congruent to either 1 or 3 (mod 4). It is a natural question to ask whether there are infinitely many of both types. We can now settle the question for 3.

**Proposition 3.4.** *There are infinitely many primes of the form $4k + 3$, $k \in \mathbb{N}$.*

***Proof.*** Suppose, as in Euclid's argument, that there were finitely many, and list them all: $p_1, p_2, \ldots, p_m$. Consider the number $N = 4p_1 p_2 \cdots p_m - 1$. Note that $N \equiv -1 \equiv 3$ (mod 4). (We might have chosen $N = 4p_1 p_2 \cdots p_m + 3$; but then the prime 3 would divide $N$, introducing complications, as we'll soon see.) Since $N$ is an odd number, its prime factorization consists only of odd primes. Consider such a factorization: $N = \pi_1^{\alpha_1} \pi_2^{\alpha_2} \cdots \pi_q^{\alpha_q}$. If all the primes $\pi_j \equiv 1$ (mod 4), then by Proposition 3.1, $N \equiv 1$ (mod 4). Since $N \equiv -1$ (mod 4), it follows that at least one prime $\pi_j$ must be congruent to $-1$ (mod 4); i.e., it is one of the primes $p_1, \ldots, p_m$. Since no $p_i$ can divide $N$, we have reached a contradiction. Thus, there must be infinitely many such primes.  □

One of the first orders of business when we all learned elementary algebra was to set up and solve equations. We now use the algebra modulo $m$ to solve these congruences:

$$cx \equiv b \pmod{m} \qquad \text{and} \qquad \begin{cases} x \equiv b_1 \pmod{m_1} \\ x \equiv b_2 \pmod{m_2} \end{cases}.$$

We will see in Section 4 how to interpret them as *equations* in an appropriate setting. We begin with the following proposition.

**Proposition 3.5.** *Suppose* $\gcd(c, m) = 1$. *If* $cx \equiv cy \pmod{m}$, *then* $x \equiv y \pmod{m}$; *moreover, for any* $b$, *the congruence* $cx \equiv b \pmod{m}$ *has a solution, and any two solutions are congruent* (mod $m$).

**Proof.** Suppose $cx \equiv cy \pmod{m}$. This means that $m|(cx - cy)$, and so $m|c(x - y)$. Now we use the result of Exercise 1.2.7: if $m|cz$ and $\gcd(m, c) = 1$, then $m|z$. To prove this, we write $1 = sm + tc$, for some $s, t \in \mathbb{Z}$, and so $z = smz + tcz$. Since $m$ divides $cz$, $m$ divides the right-hand side of this equation, and so $m|z$, as required. Applying this in our case, we conclude that $m|(x - y)$, and so $x \equiv y \pmod{m}$. (Note that the conclusion fails without the proviso that $c$ and $m$ are to be relatively prime: $4 \times 3 \equiv 4 \times 0 \pmod{6}$, but it is not true that $3 \equiv 0 \pmod{6}$.)

To solve the congruence $cx \equiv b \pmod{m}$, given that $\gcd(c, m) = 1$, we proceed similarly. Using $1 = sm + tc$, we have $b = smb + tcb$, and so $b \equiv tcb \pmod{m}$. Thus, taking $x = tb$ gives a solution. Why is this solution unique mod $m$? This is the point of the first statement of the proposition: if $x$ and $y$ are two integral solutions of the congruence, then $cx \equiv cy \pmod{m}$, since both are congruent to $b$, and so $x \equiv y \pmod{m}$.  □

**Corollary 3.6.** *If* $p$ *is prime and* $p$ *does not divide* $c$, *then the congruence* $cx \equiv b \pmod{p}$ *always has a solution (which is unique mod* $p$).  □

**Example 3.** Solve $8x \equiv 11 \pmod{35}$. We proceed as in the proof, writing $1 = (-13) \times \underline{8} + 3 \times \underline{35}$. Therefore, $1 \equiv -13 \times 8 \pmod{35}$, and so $x \equiv (-13 \times 8)x = (-13) \times (8x) \equiv -13 \times 11 \pmod{35}$; thus, $x \equiv -143 \equiv -3 \pmod{35}$.

More generally, we can solve the congruence $cx \equiv b \pmod{m}$ whenever $\gcd(c, m)$ divides $b$. Divide through by the g.c.d. $d$: let $y = c/d$, $\beta = b/d$, and $\mu = m/d$; then note that $yx \equiv \beta \pmod{\mu} \iff cx \equiv b \pmod{m}$. (See Exercise 25.)

**Example 4.** To illustrate this technique, let's solve the congruence $6x \equiv 9 \pmod{75}$. Note that $\gcd(6, 75) = 3$. Dividing by 3, we consider the modified problem: solve $2x \equiv 3 \pmod{25}$. Well, $\gcd(2, 25) = 1$, and $1 = (-12) \times \underline{2} + 1 \times \underline{25}$. Therefore, $1 \equiv -12 \times 2 \pmod{25}$, and so we take $x \equiv (-12 \times 2)x \equiv -12 \times 3 = -36 \equiv 14 \pmod{25}$. To check, $2x \equiv 28 \equiv 3 \pmod{25}$, as required, and, indeed, $6x \equiv 9 \pmod{75}$. The solutions are $x \equiv 14, 39, 64 \pmod{75}$.

We next turn to the solution of simultaneous congruences. From thirteenth-century Chinese literature comes the following:

> A military unit won a battle. At 5 a.m., immediately following the victory, the commander sent three express messengers to the capitol. Oddly, they arrived at different times to announce the news. Messenger A arrived several days ago, at 5 p.m.; messenger B, a few days later at 2 p.m.; and messenger C arrived this morning at 9 a.m. If messenger A ran at a pace of 300 li/day, B 240 li/day, and C 180 li/day, find the distance (in li) from the battle site to the capitol and figure out how many days ago the battle was won.

You are asked in Exercise 24 to solve this problem, but we might stop here to suggest why the solution involves congruences at all. We are interested in knowing the distance $x$ (in li) from the battle site to the capitol. Messenger A, we are told, ran some number of days *plus* one half a day, and so he covered a distance of some multiple of 300 li *plus* 150 li. Thus, $x \equiv 150 \pmod{300}$, and so on.

Because of the ancestry of this problem, the next result is usually called the Chinese Remainder Theorem. Suppose for starters that $m_1$ and $m_2$ are relatively prime; we wish to find an integer $x$ with the property that

$$\text{(A)} \qquad x \equiv b_1 \pmod{m_1} \qquad \text{and} \qquad x \equiv b_2 \pmod{m_2}.$$

As usual, we begin by writing

$$1 = a_1 m_1 + a_2 m_2, \quad \text{so}$$

(B) $$b_1 = a_1 b_1 m_1 + a_2 b_1 m_2 \quad \text{and}$$

(C) $$b_2 = a_1 b_2 m_1 + a_2 b_2 m_2.$$

Here comes the clever idea: we set

(D) $$x = a_2 b_1 m_2 + a_1 b_2 m_1.$$

Then $x \equiv a_2 b_1 m_2 \pmod{m_1}$ and $x \equiv a_1 b_2 m_1 \pmod{m_2}$; now we see from (B) that $x \equiv b_1 \pmod{m_1}$, and from (C) that $x \equiv b_2 \pmod{m_2}$.

**Remark.** The origin of the inspired guess (D) is the principle of superposition familiar to all students of physics. We write $x = x_1 + x_2$, where $x_1$ and $x_2$ are solutions of the corresponding problems

$$\begin{cases} x_1 \equiv b_1 \pmod{m_1} \\ x_1 \equiv 0 \pmod{m_2} \end{cases} \quad \text{and} \quad \begin{cases} x_2 \equiv 0 \pmod{m_1} \\ x_2 \equiv b_2 \pmod{m_2} \end{cases}.$$

We see from (B) and (C), respectively, that we may take $x_1 = a_2 b_1 m_2$ and $x_2 = a_1 b_2 m_1$, as above.

**Example 5.** Solve:

(E) $$x \equiv 16 \pmod{35} \quad \text{and} \quad x \equiv 27 \pmod{64}.$$

We start by using the Euclidean algorithm to find that $1 = 11 \times \underline{35} - 6 \times \underline{64}$. Thus, $1 \equiv -6 \times 64 \pmod{35}$, and so $16 \equiv (16 \times -6) \times 64 \pmod{35}$; note that this is a multiple of 64. Next, $1 \equiv 11 \times 35 \pmod{64}$, whence $27 \equiv (27 \times 11) \times 35 \pmod{64}$; note that this is a multiple of 35. So, we take $x = (16 \times -6) \times 64 + (27 \times 11) \times 35 = 4251$, as prescribed by (D); we leave it to you to check that $x$ satisfies the requisite congruences.

The rather large answer we just obtained brings up the following question: is this the smallest possible positive integer satisfying congruence (E)? We can find (by clever experimentation, or by applying the reasoning soon to come) that the number 2011 is also a solution. But is there any smaller one? And how many solutions are there altogether?

To attack this problem, we begin with the homogeneous case, i.e., the analogous problem when $b_1 = b_2 = 0$. Given $m_1$ and $m_2$ as above, what are all the solutions to

(F) $$x \equiv 0 \pmod{m_1} \quad \text{and} \quad x \equiv 0 \pmod{m_2}?$$

The answer is clearly the set of all integer multiples of $m = m_1 m_2$, the least common multiple of $m_1$ and $m_2$ (see Exercise 2.13). But, returning to the original question, given one solution to (A), we can add to it any multiple of $m$ and obtain another solution. Conversely, given two solutions $x$ and $x'$ of (A), their difference $y = x - x'$ is a solution of (F); and so $x$ and $x'$ differ by a multiple of $m$. We have now proved the following theorem.

**Theorem 3.7** (Chinese Remainder Theorem). *Let $m_1$ and $m_2$ be relatively prime. The solutions of the congruence problem*

$$x \equiv b_1 \pmod{m_1} \quad \text{and} \quad x \equiv b_2 \pmod{m_2}$$

*are of the form*

$$x \equiv a_2 b_1 m_2 + a_1 b_2 m_1 \pmod{m_1 m_2},$$

*where $1 = a_1 m_1 + a_2 m_2$.*  □

**Example 6.** Continuing the previous example, we now see that to 4251 we may add any multiple of 2240, the least common multiple of 35 and 64. Thus the smallest positive solution is $4251 - 2240 = 2011$, as promised.

**Example 7.** A dentist has a collection of candies to hand out to his patients. He typically has between fifteen and twenty patients a day. He figures out that if he gives three candies to each patient, he will have two candies left over, whereas if he gives out five candies at a time, he will run out of candy earlier and have four left over. With how many candies did he begin?

Solution. Let $x$ denote the number of candies. Then we know that $x \equiv 2 \pmod 3$ and $x \equiv 4 \pmod 5$. It is an easy application of the Chinese Remainder Theorem to find that $x \equiv 14 \pmod{15}$, and so the possible values of $x$ are as follows: 14, 29, 44, 59, 74, 89, 104, .... On the other hand, he must have started with between $15 \times 3 + 2 = 47$ and $20 \times 3 + 2 = 62$ candies, so he originally had 59 candies.

Now we introduce a new wrinkle, solving three simultaneous congruences.

**Example 8.** Solve:

$$x \equiv 7 \pmod 9$$
$$x \equiv 5 \pmod{11}$$
$$x \equiv 18 \pmod{26}.$$

Solution. The trick is to solve a pair first. Applying the Chinese Remainder Theorem, if $x \equiv 7 \pmod 9$ and $x \equiv 5 \pmod{11}$, then $x \equiv 16 \pmod{99}$. Next solve the new pair of congruences:

$$x \equiv 16 \pmod{99}$$
$$x \equiv 18 \pmod{26}.$$

Since $1 = 5 \times \underline{99} - 19 \times \underline{26}$, $x \equiv 1006 \pmod{2574}$.

We began this discussion with the stipulation that $m_1$ and $m_2$ be relatively prime. Suppose we now lift this requirement; it is natural then to ask for which $b_1$ and $b_2$ the problem (A) has a solution. If we examine the proof of the Chinese Remainder Theorem, we see that we began by using the Euclidean algorithm and the fact that $m_1$ and $m_2$ were relatively prime to write $1 = a_1 m_1 + a_2 m_2$. Now the best we'll be able to do is to write $d$, the g.c.d. of $m_1$ and $m_2$, in such a form; and so we suspect that $b_1$ and $b_2$ must now be divisible by $d$. Fortunately, we can do better than this.

**Theorem 3.8** (Improved Chinese Remainder Theorem). *Given integers $m_1$ and $m_2$ with $\gcd(m_1, m_2) = d$, and integers $b_1$ and $b_2$ so that $b_1 \equiv b_2 \pmod d$, the solutions of the congruence problem*

$$x \equiv b_1 \pmod{m_1} \qquad and \qquad x \equiv b_2 \pmod{m_2}$$

*are of the form*

$$x \equiv c a_2 m_2 + b_2 \pmod{\frac{m_1 m_2}{d}},$$

*where $d = a_1 m_1 + a_2 m_2$ and $b_1 - b_2 = cd$.*

**Remark.** The condition $b_1 \equiv b_2 \pmod d$ is also necessary for the congruence problem to have a solution, as the reader should check. That is, if (A) has a solution, then $b_1 \equiv b_2 \pmod d$. Note also that we obtain congruence mod $\frac{m_1 m_2}{d}$; this number is the least common multiple of $m_1$ and $m_2$ (see Exercise 1.2.13).

***Proof.*** Let $y = x - b_2$. Then we wish to solve the congruence problem

$$y \equiv b_1 - b_2 \pmod{m_1} \qquad and \qquad y \equiv 0 \pmod{m_2}.$$

Note that $b_1 - b_2$, $m_1$, 0, and $m_2$ are all divisible by $d$; so let's divide through by $d$ and consider the resulting congruence problem:

(G)          $z \equiv c \pmod{\frac{m_1}{d}}$     and     $z \equiv 0 \pmod{\frac{m_2}{d}}$.

If we let $\frac{m_1}{d} = \mu_1$ and $\frac{m_2}{d} = \mu_2$, then $\mu_1$ and $\mu_2$ are relatively prime and $1 = a_1\mu_1 + a_2\mu_2$; so a general solution of (G) is given by Theorem 3.7:

$$z \equiv a_2(c)\mu_2 + a_1(0)\mu_1 \pmod{\mu_1\mu_2}.$$

Since we now recover $y$ by multiplying by $d$ (cf. Exercise 25), and since $\mu_1\mu_2 d = \frac{m_1 m_2}{d}$,

$$y \equiv a_2 c m_2 \pmod{\tfrac{m_1 m_2}{d}}, \quad \text{whence} \quad x \equiv a_2 c m_2 + b_2 \pmod{\tfrac{m_1 m_2}{d}},$$

as desired. □

**Example 9.** Solve

$$x \equiv 45 \pmod{70} \quad \text{and} \quad x \equiv 3 \pmod{28}.$$

Solution. Here, $\gcd(28, 70) = 14$ and $14 \mid (45 - 3)$. Following the proof of Theorem 3.8, we first set $y = x - 3$, and then divide through by 14. So we must solve

$$z \equiv 3 \pmod{5} \quad \text{and} \quad z \equiv 0 \pmod{2}.$$

Using Theorem 3.7, $1 = 1 \times 5 - 2 \times 2$, so $z \equiv -4 \times 3 + 1 \times 0 = -12 \pmod{10}$, i.e., $z \equiv 8 \pmod{10}$. Therefore, $y \equiv 112 \pmod{140}$, or $x \equiv 115 \pmod{140}$.

## EXERCISES 1.3

1.  Check that congruence mod $m$ is indeed an equivalence relation.

2.  Prove parts (a), (b), (c), and (g) of Proposition 3.2.

3.  Find a divisibility test for $2^j$, $j \geq 3$.

4.  Here is a "real-world application" of modular arithmetic appropriate for the absent-minded author of this text. In tennis, the players switch sides after the odd games. One hot Sunday the author found himself on the side where he had begun the match a while earlier. He knew the game score was either 6-2, 4-3 or 6-2, 5-4. Which was it?

5.  Prove that for any integer $n$, $n^2 \equiv 0$ or $1 \pmod{3}$, and $n^2 \equiv 0, 1,$ or $4 \pmod{5}$.

6.  We proved Proposition 3.3 for positive $n$ by induction. Finish the proof by establishing its validity for $n \le 0$ as well.

7.  Use Proposition 3.3 to show that 65 is not a prime. (Hint: Let $n = 2$ and $p = 65$, and compute $2^{65}$ (mod 65) efficiently.)

8.  Determine the last digit of $3^{400}$; then the last two digits. Determine the last digit of $7^{99}$.

9.  Show that if an integer is a sum of two fourth powers, then its units digit must be 0, 1, 2, 5, 6, or 7.

10. Let $n = \sum_{i=0}^{k} a_i 10^i$. Prove that $13|n \iff 13|(\sum_{i=1}^{k} a_i 10^{i-1} + 4a_0)$.
    (Hint: Cf. the divisibility test for 7 in Proposition 3.2.)

11. a.  Suppose $p_1, \ldots, p_k$ are distinct primes. Prove that if $a \equiv b$ (mod $p_j$) for all $j = 1, \ldots, k$, then $a \equiv b$ (mod $p_1 p_2 \cdots p_k$).
    b.  Prove that $2^{30} \equiv 4$ (mod 15) and $24^{12} \equiv 1$ (mod 35).

12. Suppose $p$ is prime. Prove that if $a^2 \equiv b^2$ (mod $p$), then $a \equiv b$ (mod $p$) or $a \equiv -b$ (mod $p$).

13. Prove the principle of "casting out nines," which is useful for checking addition: Given integers $n_1, \ldots, n_k$ with $n = \sum_{j=1}^{k} n_j$, the sum of the digits of $n$ is congruent (mod 9) to the sum of the digits of all the numbers $n_1, \ldots, n_k$. Illustrate this principle with an example.

14. Prove that if $x^2 \equiv n$ (mod 65) has a solution, then so does $x^2 \equiv -n$ (mod 65).

15. Show that if $n \equiv 7$ (mod 8), then $n$ is not the sum of three squares.

16. Prove that if $p \ge 5$ is prime, then $p^2 + 2$ is composite.

17. Suppose $m$ and $n$ are positive integers. Show that $3^m + 3^n + 1$ cannot be a perfect square. (Hint: It may help to work mod 8.)

18. a.  Suppose $2 \nmid n$. Prove that $8 \mid (n^2 - 1)$. (Hint: What are the possible values of $n$ (mod 8)?)
    b.  Suppose $2 \nmid n$ and $3 \nmid n$. Prove that $24|(n^2 - 1)$.

19. Prove the following corollary of Proposition 3.3: If $p$ is prime and $p \nmid n$, then $n^{p-1} \equiv 1 \pmod{p}$.

20. Solve the following congruences:
    a.  $3x \equiv 2 \pmod{5}$
    b.  $6x + 3 \equiv 1 \pmod{10}$
    c.  $243x + 17 \equiv 101 \pmod{725}$
    d.  $20x \equiv 4 \pmod{30}$
    e.  $20x \equiv 30 \pmod{4}$
    f.  $49x \equiv 4000 \pmod{999}$
    g.  $15x \equiv 25 \pmod{35}$
    h.  $15x \equiv 24 \pmod{35}$

21. Use the Chinese Remainder Theorem, Theorem 3.7 (or its improvement, Theorem 3.8), to solve the following simultaneous congruences:
    a.  $x \equiv 1 \pmod{3}$, $\equiv 1 \pmod{5}$
    b.  $x \equiv 1 \pmod{4}$, $\equiv 7 \pmod{13}$
    c.  $x \equiv 3 \pmod{4}$, $\equiv 4 \pmod{5}$, $\equiv 3 \pmod{7}$
    d.  $19x \equiv 1 \pmod{140}$ (Hint: factor 140 and see Exercise 11.)
    e.  $x \equiv 3 \pmod{9}$, $\equiv 18 \pmod{24}$
    f.  $x \equiv 4 \pmod{105}$, $\equiv 29 \pmod{80}$
    g.  $x \equiv 11 \pmod{142}$, $\equiv 25 \pmod{86}$

22. Find all integers that give the remainders 1, 2, and 3 when divided by 3, 4, and 5, respectively.

23. If $a$ is selected at random from 1, 2, ..., 11, and $b$ is selected at random from 1, 2, ..., 12, what is the probability that $ax \equiv b \pmod{12}$ has at least one solution? (Answer: approx. 57.6%.)

24. Solve the military question posed in the text: How far was it from the battle site to the capitol, and for how long did each messenger run?

25. Prove that one can solve the congruence $cx \equiv b \pmod{m}$ if and only if $\gcd(c, m) | b$. Show, moreover, that the answer is unique mod $m / \gcd(c, m)$.

26. Show that another way of solving (A) is to solve the simultaneous congruence $y \equiv 0 \pmod{m_2}$ and $y \equiv (b_1 - b_2) \pmod{m_1}$, and then set $x = y + b_2$. Indeed, show that this gives precisely the same formula as in Theorem 3.7.

27. Check that if $x \equiv b_1 \pmod{m_1}$, $x \equiv b_2 \pmod{m_2}$ has a solution, then $b_1 \equiv b_2 \pmod{\gcd(m_1, m_2)}$.

28. Mimic the proof of Proposition 3.4 to show there are infinitely many prime numbers $p \equiv 5 \pmod 6$. (Be sure to consider all possible values of $\pi \pmod 6$ for the factors $\pi$ of the $N$ you concoct.)

29. ("The characteristic $p$ binomial theorem")
    a. Prove that $(a + b)^p \equiv a^p + b^p \pmod p$ for any integers $a, b$, provided $p$ is prime.
    b. More generally, prove that if $p$ is prime and $q = p^n, n \in \mathbb{N}$, then $(a + b)^q \equiv a^q + b^q \pmod p$ for any integers $a, b$. (Hint: Induction.)

30. For what positive integers $n$ is the fraction $\frac{n^2 + 12}{n + 4}$ not in lowest terms?

31. Suppose $S$ is a set of 34 integers with the property that no two differ by a multiple of 65. Prove that there is a pair of elements of $S$ whose sum is a multiple of 65.

32. Let $p$ be prime. Prove that $(p+1)^{p^{n-1}} \equiv 1 \pmod{p^n}$ for all $n \in \mathbb{N}$. (Hint: Use Theorem 1.4.)

33. One of your friends has found a three-digit number that, when divided by the product of its digits, yields as quotient the hundreds digit. Find her number (in particular, can there be only one?).

34. a. Suppose $\gcd(a, b) = 1$ and $ab$ is a perfect square. Prove $a$ and $b$ are each perfect squares.
    b. Suppose $x$, $y$, and $z \in \mathbb{N}$ have no common factor except 1, and that $x^2 + y^2 = z^2$. Prove first that $x$ and $y$ cannot both be odd or even.
    c. Continuing b., suppose that $x$ is even and $y$ is odd. Prove that there are integers $m, n$ such that $y = m^2 - n^2$, $z = m^2 + n^2$, and $x = 2mn$. (Hint: First show that $\gcd(z+y, z-y) = 2$, and then apply a. to $\frac{z+y}{2}$ and $\frac{z-y}{2}$.)
    d. Show that one of $x$ and $y$ is divisible by 3; and show that one of $x$, $y$, and $z$ is divisible by 5.

**Remark.** The numbers $x$, $y$, and $z$ comprise a Pythagorean triple, i.e., a triple of integral lengths of the sides of a right triangle. For a geometric interpretation of this problem, cf. Exercise 8.2.7.

35.  On a desert island, five men and a monkey gather coconuts all day; then the men go to sleep, leaving the monkey to guard their stash. The first man awakens and decides to take his share. He divides the coconuts into five equal shares, finding that there is one left over; this he throws to the monkey (as hush coconut). He then hides his share of the coconuts and goes back to sleep. The second man awakens a little bit later and similarly decides to take his share; he repeats the scenario (likewise finding one extra coconut, which is provided the monkey for his silence). Each of the three remaining men does the same in turn. When they all awaken in the morning, the pile contains a multiple of five coconuts, and the monkey is too stuffed to divulge the night's activities. What is the minimum number of coconuts originally present?

36.  Suppose $p$ and $q$ are distinct prime numbers. Prove that for any $n \in \mathbb{N}$ with $\gcd(p,n) = \gcd(q,n) = 1$, we have $n^{(p-1)(q-1)} \equiv 1 \pmod{pq}$. (Hint: Use the Chinese Remainder Theorem and apply Exercise 19.)

37.  Let $p$ and $q$ be distinct primes and let $N = pq$. Let $k = (p-1)(q-1)$. For each $e$ with $\gcd(e,k) = 1$, by Proposition 3.5 there is $d$ satisfying $de \equiv 1 \pmod{k}$. Prove that $n^{de} \equiv n \pmod{N}$ for all $n \in \mathbb{N}$.

38.  Here is an application of Exercise 37 to public-key codes.* The values of $N$ and $e$ are made known to the public, but the values of $k$ and $d$ are known only to the in-crowd. Given a message $M$ (most usefully with $\gcd(M,N) = 1$), anyone can encode it by computing $M' = M^e \pmod{N}$.

   a.  Show that the in-crowd can decode by computing $M'^d \pmod{N}$, provided $M < N$. (For large—e.g., 100-digit—primes $p$

---

*cf. Rivest, Shamir and Adleman, "A Method for Obtaining Digital Signatures and Public-Key Cryptosystems," *Communications of the A.C.M.*, **21** (1978), pp. 120-126; Pomerance, "Cryptology and Computational Number Theory—An Introduction," *Proceedings of Symposia in Applied Mathematics* **42** (1990), pp. 1-12.

and $q$ it is virtually impossible to factor $N$, and thereby find $k$ and $d$, in anyone's lifetime. Thus, the code is secure; i.e., even knowing $N$ and $e$, the public cannot decode messages. On the other hand, finding $M^e$ (mod $N$) for large numbers $M$, $e$, and $N$ is not so difficult on computers. See if you can suggest some practical ways to do this.)

b.   Let's agree to convert the letters of the alphabet to two-digit numbers in the obvious way: $A = 01$, $B = 02$, ... , $Z = 26$; we'll denote a space by 00 and a period by 27 (with no necessity to follow a period by a space). The message "HI. FRED IS HERE." would appear in numerical form as "0809270618050400091900080518 0527." To encode this, we need to break it into "words" less than $N$. It will be convenient, then, to have $N$ at least 2726, so that we can use two-letter words. Following Rivest-Shamir-Adleman, we use $p = 47$ and $q = 59$, whence $N = 2773$. (Of course, these numbers are too small to give an effective code, but they illustrate our purposes here.)

Using $d = 101$, check that $e = 317$. We then obtain the following sequence of coded and decoded messages:

| two-letter word | $M$ | $M' = M^e$ (mod 2773) | $M'^d$ (mod 2773) |
|---|---|---|---|
| HI | 0809 | 1663 | 809 |
| . F | 2706 | 363 | 2706 |
| RE | 1805 | 1383 | 1805 |
| D_ | 0400 | 357 | 400 |
| IS | 0919 | 932 | 919 |
| _H | 0008 | 345 | 8 |
| ER | 0518 | 2304 | 518 |
| E. | 0527 | 1210 | 527 |

Try doing this with a different $e$ (using either a good calculator (with at least eight-digit accuracy) or else Mathematica, Matlab, etc.).

c.   Using the same $N$ and $d = 1601$, decode the following message:

$$1914 \quad 1396 \quad 289 \quad 666 \quad 1083 \quad 2414.$$

39.   For any $n \in \mathbb{N}$, $n \geq 2$, let $\varphi(n)$ denote the number of integers $m$ satisfying $1 \leq m < n$ and $\gcd(m, n) = 1$.

a.   Compute $\varphi(n)$, $n = 2, 3, \ldots, 12$.

b.   Prove that for any prime number $p$, $\varphi(p) = p - 1$; and that for distinct prime numbers $p$ and $q$, $\varphi(pq) = (p-1)(q-1)$.
c.   Prove that for any prime number $p$, $\varphi(p^s) = p^{s-1}(p-1)$.
d.   More generally, prove that $\varphi(mn) = \varphi(m)\varphi(n)$ whenever $\gcd(m,n) = 1$. (Hint: Start with $n = p$ (prime) and then consider $n = p^s$.)

40.   Reconsider the postage stamp problem, Exercise 1.2.20. (Hint: If $\gcd(a,b) = 1$, $a > b$, then circle the multiples of $b$ in this table:

| 0     | $a$      | $2a$      | $\ldots$ | $(b-1)a$       |
|-------|----------|-----------|----------|----------------|
| 1     | $a+1$    | $2a+1$    | $\ldots$ | $(b-1)a+1$     |
| 2     | $a+2$    | $2a+2$    | $\ldots$ | $(b-a)a+2$     |
| $\vdots$ | $\vdots$ | $\vdots$ | $\ddots$ | $\vdots$       |
| $a-1$ | $2a-1$   | $3a-1$    | $\ldots$ | $ba-1$         |

Note that the multiples $ka \pmod{b}$, $k = 1, 2, \ldots, b-1$, exhaust $1, \ldots, b-1$.)

41.   Let $p$ be an odd prime number. Prove that

$$\sum_{j=0}^{p} \binom{p}{j}\binom{p+j}{j} \equiv 2^p + 1 \pmod{p^2}.$$

(Hint: see Exercise 1.1.8.)

## 4. $\mathbb{Z}_m$, Rings, Integral Domains, and Fields

In Proposition 3.1 we saw that the rules of modular arithmetic are compatible with the algebraic operations on the set of integers. We now stop to formalize this a bit. We have been working with equivalence classes of integers mod $m$, and we now think of these as the elements (denoted for now by $\bar{0}, \bar{1}, \bar{2}, \ldots, \overline{m-1}$) of a new set $\mathbb{Z}_m$. We assume from here on that $m > 1$. Addition and multiplication are defined modulo $m$, as in the preceding section: if $\bar{a}$ and $\bar{b}$ are elements of $\mathbb{Z}_m$, then

$$\bar{a} + \bar{b} = \overline{a+b}$$

$$\bar{a} \times \bar{b} = \overline{a \times b}.$$

That is, to add (resp., multiply) $\overline{a}$ and $\overline{b}$, we add (resp., multiply) $a$ and $b$ and then take the equivalence class modulo $m$. Proposition 3.1 guarantees that this process is well-defined. For example, here are addition and multiplication tables for $\mathbb{Z}_5$:

| + | $\overline{0}$ | $\overline{1}$ | $\overline{2}$ | $\overline{3}$ | $\overline{4}$ |
|---|---|---|---|---|---|
| $\overline{0}$ | $\overline{0}$ | $\overline{1}$ | $\overline{2}$ | $\overline{3}$ | $\overline{4}$ |
| $\overline{1}$ | $\overline{1}$ | $\overline{2}$ | $\overline{3}$ | $\overline{4}$ | $\overline{0}$ |
| $\overline{2}$ | $\overline{2}$ | $\overline{3}$ | $\overline{4}$ | $\overline{0}$ | $\overline{1}$ |
| $\overline{3}$ | $\overline{3}$ | $\overline{4}$ | $\overline{0}$ | $\overline{1}$ | $\overline{2}$ |
| $\overline{4}$ | $\overline{4}$ | $\overline{0}$ | $\overline{1}$ | $\overline{2}$ | $\overline{3}$ |

| $\times$ | $\overline{0}$ | $\overline{1}$ | $\overline{2}$ | $\overline{3}$ | $\overline{4}$ |
|---|---|---|---|---|---|
| $\overline{0}$ | $\overline{0}$ | $\overline{0}$ | $\overline{0}$ | $\overline{0}$ | $\overline{0}$ |
| $\overline{1}$ | $\overline{0}$ | $\overline{1}$ | $\overline{2}$ | $\overline{3}$ | $\overline{4}$ |
| $\overline{2}$ | $\overline{0}$ | $\overline{2}$ | $\overline{4}$ | $\overline{1}$ | $\overline{3}$ |
| $\overline{3}$ | $\overline{0}$ | $\overline{3}$ | $\overline{1}$ | $\overline{4}$ | $\overline{2}$ |
| $\overline{4}$ | $\overline{0}$ | $\overline{4}$ | $\overline{3}$ | $\overline{2}$ | $\overline{1}$ |

And here are addition and multiplication tables for $\mathbb{Z}_6$:

| + | $\overline{0}$ | $\overline{1}$ | $\overline{2}$ | $\overline{3}$ | $\overline{4}$ | $\overline{5}$ |
|---|---|---|---|---|---|---|
| $\overline{0}$ | $\overline{0}$ | $\overline{1}$ | $\overline{2}$ | $\overline{3}$ | $\overline{4}$ | $\overline{5}$ |
| $\overline{1}$ | $\overline{1}$ | $\overline{2}$ | $\overline{3}$ | $\overline{4}$ | $\overline{5}$ | $\overline{0}$ |
| $\overline{2}$ | $\overline{2}$ | $\overline{3}$ | $\overline{4}$ | $\overline{5}$ | $\overline{0}$ | $\overline{1}$ |
| $\overline{3}$ | $\overline{3}$ | $\overline{4}$ | $\overline{5}$ | $\overline{0}$ | $\overline{1}$ | $\overline{2}$ |
| $\overline{4}$ | $\overline{4}$ | $\overline{5}$ | $\overline{0}$ | $\overline{1}$ | $\overline{2}$ | $\overline{3}$ |
| $\overline{5}$ | $\overline{5}$ | $\overline{0}$ | $\overline{1}$ | $\overline{2}$ | $\overline{3}$ | $\overline{4}$ |

| $\times$ | $\overline{0}$ | $\overline{1}$ | $\overline{2}$ | $\overline{3}$ | $\overline{4}$ | $\overline{5}$ |
|---|---|---|---|---|---|---|
| $\overline{0}$ | $\overline{0}$ | $\overline{0}$ | $\overline{0}$ | $\overline{0}$ | $\overline{0}$ | $\overline{0}$ |
| $\overline{1}$ | $\overline{0}$ | $\overline{1}$ | $\overline{2}$ | $\overline{3}$ | $\overline{4}$ | $\overline{5}$ |
| $\overline{2}$ | $\overline{0}$ | $\overline{2}$ | $\overline{4}$ | $\overline{0}$ | $\overline{2}$ | $\overline{4}$ |
| $\overline{3}$ | $\overline{0}$ | $\overline{3}$ | $\overline{0}$ | $\overline{3}$ | $\overline{0}$ | $\overline{3}$ |
| $\overline{4}$ | $\overline{0}$ | $\overline{4}$ | $\overline{2}$ | $\overline{0}$ | $\overline{4}$ | $\overline{2}$ |
| $\overline{5}$ | $\overline{0}$ | $\overline{5}$ | $\overline{4}$ | $\overline{3}$ | $\overline{2}$ | $\overline{1}$ |

We now want to compare these sets of tables. First, we should check that the same rules of arithmetic hold in $\mathbb{Z}_m$ as hold in $\mathbb{Z}$: for any $\overline{a}, \overline{b}, \overline{c} \in \mathbb{Z}_m$,

(1) $\overline{a} + \overline{b} = \overline{b} + \overline{a}$    (commutative law of addition)
(2) $(\overline{a} + \overline{b}) + \overline{c} = \overline{a} + (\overline{b} + \overline{c})$    (associative law of addition)
(3) $\overline{0} + \overline{a} = \overline{a}$    ($\overline{0}$ is the additive identity)
(4) $\overline{a} + \overline{m - a} = \overline{0}$    ($\overline{m - a}$ is the additive inverse of $\overline{a}$)
(5) $\overline{a} \times \overline{b} = \overline{b} \times \overline{a}$    (commutative law of multiplication)
(6) $(\overline{a} \times \overline{b}) \times \overline{c} = \overline{a} \times (\overline{b} \times \overline{c})$    (associative law of multiplication)
(7) $\overline{a} \times \overline{1} = \overline{a}$    ($\overline{1}$ is the multiplicative identity)
(8) $\overline{a} \times (\overline{b} + \overline{c}) = (\overline{a} \times \overline{b}) + (\overline{a} \times \overline{c})$    (distributive law).

There is nothing particularly exciting in the addition tables. In the two multiplication tables, however, we observe a radical difference. Across each row but the first of the table for $\mathbb{Z}_5$, we find all the elements $\overline{0}$, $\overline{1}$, $\overline{2}$, $\overline{3}$, and $\overline{4}$ of $\mathbb{Z}_5$; that is, for each nonzero $\overline{c} \in \mathbb{Z}_5$, as $x$ ranges through $\mathbb{Z}_5$, $\overline{c}x$ likewise ranges through all of $\mathbb{Z}_5$. In the case of $\mathbb{Z}_6$, however, only in rows 1 and 5 do we find the same phenomenon. What is relevant here, of course, is Proposition 3.5 (actually, Exercise 1.3.25), which says that in order for us to be able to solve the congruence $cx \equiv b \pmod{m}$, $b$ must be divisible by $\gcd(c, m)$.

There is another surprise in the multiplication table of $\mathbb{Z}_6$. We are used to concluding from the hypothesis $ab = 0$ that $a$ or $b$ must equal 0; certainly this is true of arithmetic in $\mathbb{Z}$, but it is not true of arithmetic in $\mathbb{Z}_6$. Nor need it be true that $bx = cx$ implies $b = c$ or $x = 0$; for example, in $\mathbb{Z}_6$, $\overline{2} \times \overline{2} = \overline{5} \times \overline{2} = \overline{4}$.

It is now time to make a general definition.

**Definition.** A set $R$ closed under two operations, addition (denoted by +) and multiplication (denoted by · or, more often, by no symbol at all), is called a **ring** if the following laws hold:

(1) For all $a, b \in R$, $a + b = b + a$   (commutative law of addition).

(2) For all $a, b, c \in R$, $(a + b) + c = a + (b + c)$   (associative law of addition).

(3) $R$ contains an element 0 satisfying $0 + a = a$ for all $a \in R$ (0 is the additive identity).

(4) For each $a \in R$, there is an element $-a$ so that $a + (-a) = 0$ ($-a$ is the additive inverse of $a$).

(5) For all $a, b, c \in R$, $(ab)c = a(bc)$   (associative law of multiplication).

(6) $R$ contains an element 1 satisfying $a \cdot 1 = 1 \cdot a = a$ for all $a \in R$   (1 is the multiplicative identity); we assume further that $1 \neq 0$.

(7) For all $a, b, c \in R$, $a(b + c) = ab + ac$ and $(b + c)a = ba + ca$ (distributive laws).

We say the ring is **commutative** if, moreover,

(8) For all $a, b \in R$, $ab = ba$   (commutative law of multiplication).

Thus, $\mathbb{Z}$ and $\mathbb{Z}_m$ are examples of commutative rings. We have yet to come across an example of a noncommutative ring, but we shall soon. We first want to make a definition to distinguish between the properties we've observed in $\mathbb{Z}_5$ and $\mathbb{Z}_6$.

**Definition.** Let $R$ be a ring. If there are nonzero elements $a, b \in R$ so that $ab = 0$, then we say that $a$ and $b$ are **zero-divisors** in $R$. We say a *commutative* ring $R$ is an **integral domain** if it contains no zero-divisors.

**Example 1.** $\mathbb{Z}_5$ is an integral domain, but $\mathbb{Z}_6$ evidently is not. Indeed, $\bar{2}, \bar{3}$, and $\bar{4}$ are all zero-divisors in $\mathbb{Z}_6$.

Now there is a new distinction to make: $\mathbb{Z}$ and $\mathbb{Z}_5$ are both integral domains; however, the equation $\bar{2}x = \bar{3}$ has a solution in $\mathbb{Z}_5$, but none in $\mathbb{Z}$. (To see why, we note that in $\mathbb{Z}_5$, $\bar{2} \cdot \bar{3} = \bar{1}$, so that, multiplying the equation $\bar{2}x = \bar{3}$ by $\bar{3}$, we obtain $x = \bar{4}$; in $\mathbb{Z}$, we can find no element whose product with 2 is 1.)

**Definition.** Let $R$ be a ring. We say an element $a \in R$ is a **unit** if there is an element $b \in R$ so that $ab = ba = 1$, i.e., if $a$ has a multiplicative inverse in $R$. We denote the multiplicative inverse of $a$ by $a^{-1}$. We say a *commutative* ring $R$ is a **field** if every nonzero element $a \in R$ is a unit.

Note that every field is an integral domain (because if $a$ is a unit and $ab = 0$, then $b = (a^{-1}a)b = a^{-1}(ab) = a \cdot 0 = 0$ by Exercise 6a.). We have already commented that $\mathbb{Z}$ is an integral domain, but it is obviously not a field.

**Theorem 4.1.** *$\mathbb{Z}_m$ is an integral domain if and only if $m$ is prime; when $p$ is prime, $\mathbb{Z}_p$ is a field.*

*Proof.* Obviously, when $m = qr$, $1 < q,r < m$, then both $\bar{q}$ and $\bar{r} \in \mathbb{Z}_m$ are zero-divisors. We now prove that when $p$ is a prime, $\mathbb{Z}_p$ is a field. This is immediate from Corollary 3.6, since, given any nonzero $\bar{a} \in \mathbb{Z}_p$, we can solve the congruence $ax \equiv 1 \pmod{p}$. If $b$ is a solution, then we have $\bar{a} \cdot \bar{b} = 1$ in $\mathbb{Z}_p$, and $\bar{b}$ is the requisite multiplicative inverse of $\bar{a}$.   □

We will now establish the convention that when we write $\mathbb{Z}_p$ in this text, $p$ always denotes a prime number. Working with arithmetic in $\mathbb{Z}_p$, we have the following amusing application (see also Exercises 2 and 16). We'll officially review the definition of the rational numbers $\mathbb{Q}$ in Section 1 of Chapter 2; but meanwhile, recall that $\frac{1}{b} \in \mathbb{Q}$ is the multiplicative inverse $b^{-1}$ of the nonzero number $b \in \mathbb{Z}$, and $\frac{a}{b} = a \cdot \frac{1}{b} = a \cdot b^{-1} \in \mathbb{Q}$. Now consider the arithmetic problem

$(*)$ $\qquad\qquad\qquad \frac{1}{2} + \frac{2}{3} + \frac{3}{4} = \frac{23}{12}$

as a problem in, say $\mathbb{Z}_5$. (Any $\mathbb{Z}_p$ will do here, provided $p$ divides none of our denominators.) What do we mean by $\frac{1}{2}, \frac{2}{3}$, etc.? Since $\frac{1}{2}$ denotes $\bar{1} \cdot \bar{2}^{-1} \in \mathbb{Z}_5$, and since $\bar{2}^{-1} = \bar{3}$, $\frac{1}{2} = \bar{3} \in \mathbb{Z}_5$. Likewise,

$\frac{2}{3} = \overline{2} \cdot \overline{3}^{-1} = \overline{2} \cdot \overline{2} = \overline{4}$, and $\frac{3}{4} = \overline{3} \cdot \overline{4}^{-1} = \overline{3} \cdot \overline{4} = \overline{2}$. Thus, the left-hand side of (∗) in $\mathbb{Z}_5$ is $\overline{3} + \overline{4} + \overline{2} = \overline{4} \in \mathbb{Z}_5$. What about the right-hand side? Here, $\frac{23}{12} = \overline{23} \cdot \overline{12}^{-1} = \overline{3} \cdot \overline{2}^{-1} = \overline{3} \cdot \overline{3} = \overline{4}$, as required. This is, in fact, a rather nifty way of checking arithmetic with rational numbers.

**Example 2.** When $p$ is prime, $\binom{p+k}{k} \equiv 1 \pmod{p}$ for $0 \le k \le p - 1$. We wish to prove that $\overline{\binom{p+k}{k}} = \overline{1} \in \mathbb{Z}_p$. Since $k!$ is not divisible by $p$, we know $\overline{k}!$ is nonzero in $\mathbb{Z}_p$. For $k \ge 1$,

$$\overline{\binom{p+k}{k}} = \overline{\left( \frac{(p + 1)(p + 2) \cdots (p + k)}{k!} \right)} = \frac{\overline{1 \cdot 2 \cdot \ldots \cdot k}}{\overline{k!}} = \frac{\overline{k!}}{\overline{k!}} = \overline{1},$$

as required.

We close this chapter with two examples of noncommutative rings.

**Example 3.** Let

$$M_2(\mathbb{Z}) = \left\{ \begin{bmatrix} a & b \\ c & d \end{bmatrix} : a, b, c, d \in \mathbb{Z} \right\}$$

be the set of $2 \times 2$ matrices with integer entries. Addition and multiplication are defined as follows:

$$\begin{bmatrix} a & b \\ c & d \end{bmatrix} + \begin{bmatrix} e & f \\ g & h \end{bmatrix} = \begin{bmatrix} a + e & b + f \\ c + g & d + h \end{bmatrix}$$

$$\begin{bmatrix} a & b \\ c & d \end{bmatrix} \cdot \begin{bmatrix} e & f \\ g & h \end{bmatrix} = \begin{bmatrix} ae + bg & af + bh \\ ce + dg & cf + dh \end{bmatrix}.$$

It is easy to check that the zero element is $\begin{bmatrix} 0 & 0 \\ 0 & 0 \end{bmatrix}$ and the multiplicative identity is $\begin{bmatrix} 1 & 0 \\ 0 & 1 \end{bmatrix}$. We leave it as an exercise for the reader to check the remainder of properties (1)-(7). To see that multiplication is not commutative, consider $A = \begin{bmatrix} 0 & 1 \\ 0 & 0 \end{bmatrix}$ and $B = \begin{bmatrix} 0 & 0 \\ 1 & 0 \end{bmatrix}$; then $AB = \begin{bmatrix} 1 & 0 \\ 0 & 0 \end{bmatrix}$, whereas $BA = \begin{bmatrix} 0 & 0 \\ 0 & 1 \end{bmatrix}$.

More generally, given any commutative ring $R$, we define $M_n(R)$ to be the ring of $n \times n$ matrices with entries in $R$. Addition and multiplication are defined as usual.

**Example 4.** Consider the following multiplication table for the symbols $1, i, j,$ and $k$:

| · | 1 | $i$ | $j$ | $k$ |
|---|---|---|---|---|
| 1 | 1 | $i$ | $j$ | $k$ |
| $i$ | $i$ | $-1$ | $k$ | $-j$ |
| $j$ | $j$ | $-k$ | $-1$ | $i$ |
| $k$ | $k$ | $j$ | $-i$ | $-1$ |

Let $R = \{a + bi + cj + dk : a,b,c,d \in \mathbb{Z}\}$. Define addition in the obvious way: $(a + bi + cj + dk) + (s + ti + uj + vk) = (a + s) + (b + t)i + (c + u)j + (d + v)k$. Define multiplication by applying the table above and the distributive law:

$$(a + bi + cj + dk) \cdot (s + ti + uj + vk) = (as - bt - cu - dv) +$$
$$(at + bs + cv - du)i + (au - bv + cs + dt)j + (av + bu - ct + ds)k.$$

The additive identity is $0 = 0 + 0i + 0j + 0k$, and the multiplicative identity is $1 = 1 + 0i + 0j + 0k$. Multiplication is obviously not commutative, and we leave it to the reader to check the remaining properties. Are there any zero-divisors in $R$?

## EXERCISES 1.4

1. Construct multiplication tables for $\mathbb{Z}_7$, $\mathbb{Z}_8$, $\mathbb{Z}_9$, and $\mathbb{Z}_{12}$. List the zero-divisors and units in each ring.

2. Compute the following in the indicated $\mathbb{Z}_p$:
   a. $\frac{3}{5} + \frac{4}{3} \in \mathbb{Z}_7$
   b. $\frac{1^2 + 2^2 + 3^2 + 4^2}{7^2 + 8^2 + 9^2 + 10^2} \in \mathbb{Z}_{11}$
   c. $1^2 + 2^2 + \cdots + (\frac{p-1}{2})^2 \in \mathbb{Z}_p$, $p \geq 5$ prime

3. Below are addition and multiplication tables for a set $R$. Is it a commutative ring? What are its zero-divisors and units?

| + | $a$ | $b$ | $c$ | $d$ |
|---|---|---|---|---|
| $a$ | $a$ | $b$ | $c$ | $d$ |
| $b$ | $b$ | $a$ | $d$ | $c$ |
| $c$ | $c$ | $d$ | $a$ | $b$ |
| $d$ | $d$ | $c$ | $b$ | $a$ |

| · | $a$ | $b$ | $c$ | $d$ |
|---|---|---|---|---|
| $a$ | $a$ | $a$ | $a$ | $a$ |
| $b$ | $a$ | $b$ | $a$ | $b$ |
| $c$ | $a$ | $a$ | $c$ | $c$ |
| $d$ | $a$ | $b$ | $c$ | $d$ |

4. Which of the following matrices are units in $M_2(\mathbb{Z})$? Which are zero-divisors? Which are neither?

  a. $\begin{bmatrix} 2 & 5 \\ 1 & 3 \end{bmatrix}$       b. $\begin{bmatrix} 3 & 1 \\ 7 & 2 \end{bmatrix}$       c. $\begin{bmatrix} 2 & 5 \\ 1 & 4 \end{bmatrix}$

  d. $\begin{bmatrix} 2 & 4 \\ 1 & 2 \end{bmatrix}$       e. $\begin{bmatrix} 0 & 0 \\ 1 & -1 \end{bmatrix}$

5. a. Prove that $\gcd(a, m) = 1 \iff \bar{a} \in \mathbb{Z}_m$ is a unit.
   b. Prove that if $\bar{a} \in \mathbb{Z}_m$ is a zero-divisor, then $\gcd(a, m) > 1$, and conversely, provided $m \nmid a$.
   c. Prove that every nonzero element of $\mathbb{Z}_m$ is either a unit or a zero-divisor.
   d. Prove that in any commutative ring $R$, a zero-divisor cannot be a unit, and a unit cannot be a zero-divisor. Do you think c. holds in general?

6. Prove that in any ring $R$:
   a. $0 \cdot a = 0$ for all $a \in R$ (cf. Lemma 1.1);
   b. $(-1)a = -a$ for all $a \in R$ (cf. Lemma 1.2);
   c. $(-a)(-b) = ab$ for all $a, b \in R$;
   d. the multiplicative identity $1 \in R$ is unique.

7. Suppose $R$ is an integral domain, $c, x, y \in R$, and $c \neq 0$. Prove that if $cx = cy$, then $x = y$.

8. Show that if $a, b \in R$ have multiplicative inverses, then $ab$ has a multiplicative inverse, viz., $(ab)^{-1} = b^{-1}a^{-1}$. Give an example to illustrate that the equation $(ab)^{-1} = a^{-1}b^{-1}$ may fail.

9. Show that $R = \{a + b\sqrt{2} : a, b \in \mathbb{Z}\}$ is a commutative ring (where addition and multiplication are defined in the obvious way). Is it an integral domain? What are the units of this ring?

10. a. Prove that the multiplicative inverse of a unit $a$ in a ring $R$ is unique. That is, if $ab = ba = 1$ and $ac = ca = 1$, then $b = c$. (You will need to use associativity of multiplication in $R$.)
    b. Indeed, more is true. If $a \in R$ and there exist $b, c \in R$ so that $ab = 1$ and $ca = 1$, prove that $b = c$ and thus that $a$ is a unit.

11. Show that $\begin{bmatrix} a & b \\ c & d \end{bmatrix}$ is
    a.  a unit in $M_2(\mathbb{Z})$ if and only if $\Delta = ad - bc = \pm 1$;
    b.  a zero-divisor in $M_2(\mathbb{Z})$ if and only if $ad - bc = 0$ and not all of $a$, $b$, $c$, and $d$ are zero.
    (Hints for a.: Either (i) use properties of the determinant (see Section 2 of Appendix B) to show that if $A \in M_2(\mathbb{Z})$ is a unit, then $\Delta$ must be a unit in $\mathbb{Z}$, or (ii) by looking at the entries of $A^{-1} = \dfrac{1}{\Delta}\begin{bmatrix} d & -b \\ -c & a \end{bmatrix}$, show that $\Delta^2 | \Delta$.)

12. Let $R$ be an integral domain having a finite number of elements. Prove $R$ is a field. (Hint: Given any nonzero element $a \in R$, consider the list of elements $ab$ for all $b \in R$.)

13. Let $p$ be a prime number. Use the fact that $\mathbb{Z}_p$ is a field to prove that $(p - 1)! \equiv -1 \pmod{p}$. (Hint: Pair elements of $\mathbb{Z}_p$ with their multiplicative inverses; cf. Exercise 1.3.12.)

14. Let $p$ be a prime number. Prove that $(p - 2)! \equiv 1 \pmod{p}$. (Hint: cf. Exercise 13.)

15. Prove that if $p$ is prime and $p \equiv 1 \pmod 4$, then $x^2 = -\overline{1}$ has a solution in $\mathbb{Z}_p$. (Hint: Consider $x = \overline{1} \cdot \overline{2} \cdot \overline{3} \cdot \ldots \cdot \left(\overline{\frac{p-1}{2}}\right)$, and apply Exercise 13.)

16. Let $p$ be an odd prime. Suppose $a, b \in \mathbb{N}$ and

$$1 + \frac{1}{2} + \frac{1}{3} + \cdots + \frac{1}{p-1} = \frac{a}{b}.$$

    a.  Prove $p | a$. (Hint: Compute the left-hand side in $\mathbb{Z}_p$.)
    b.  If $p > 3$, then prove $p^2 | a$. (Hint: Note that $\frac{1}{k} + \frac{1}{p-k} = \frac{p}{k(p-k)}$. If you factor out one $p$, you must show that $p$ divides what's left; now compute in $\mathbb{Z}_p$ once again. See also Exercise 1.1.4c.)

17. Let $p$ be prime. Prove that
    a.  $\binom{p^2}{p} \equiv p \pmod{p^2}$
    b.  $\binom{2p}{p} \equiv 2 \pmod{p^2}$

18. Let $R$ be the collection $\{ax + b : a, b \in \mathbb{Z}\}$ of affine linear functions. Define the operations of addition and multiplication as

follows:

$$(ax + b) \boxplus (cx + d) = (a + c)x + (b + d)$$
$$(ax + b) \boxdot (cx + d) = (ac)x + (ad + b)$$

Decide whether $R$ is a ring. (Hint: In earlier editions of this text, the author used $R$ as an example of a noncommutative ring.)

19. Let $R$ be the set of all real-valued functions on the closed interval $[0, 1]$.

    a. Prove that $R$ is a commutative ring. (Define addition and multiplication of functions as in calculus: if $f, g \in R$, define $(f + g)(x) = f(x) + g(x)$ and $(fg)(x) = f(x)g(x)$ for all $x \in [0, 1]$.)

    b. Is $R$ an integral domain? What are the units and zero-divisors? Is every nonzero element of $R$ either a unit or a zero-divisor?

    c. Let $R' \subset R$ be the set of all *continuous* functions. Prove that $R'$ is likewise a commutative ring.

    d. Is $R'$ an integral domain? What are the units and zero-divisors? Is every nonzero element of $R'$ either a unit or a zero-divisor?

20. Let $R$ be a (noncommutative) ring. Given that $a$, $b$, and $a + b \in R$ are all units, prove that $a^{-1} + b^{-1}$ is a unit.

21. (To show that Exercise 10b. is not silly) Find a ring $R$ and an element $a \in R$ so that
    (i)   there exists $b \in R$ satisfying $ab = 1$;
    (ii)  there does not exist $c \in R$ satisfying $ca = 1$.
    (Hint: This is not so easy. One possible solution is to consider the "infinite matrix"

$$\begin{bmatrix} 0 & 1 & 0 & 0 & \cdots \\ 0 & 0 & 1 & 0 & \cdots \\ 0 & 0 & 0 & 1 & \cdots \\ \vdots & \vdots & \vdots & \ddots & \ddots \\ \vdots & \vdots & \vdots & \vdots & \ddots \end{bmatrix}.)$$

# CHAPTER 2

# From the Integers to the Complex Numbers

In this chapter we will work our way from the integers through the hierarchy of number systems. The rational numbers are defined as the "fractions" formed by taking ratios of integers. The real numbers are defined by "closing the gaps" in the set of rational numbers (the real number system is conceived so that sequences of rational numbers that ought to be convergent are indeed); interestingly enough, this was historically a tough point for mathematicians, and the real numbers were not satisfactorily defined until late in the nineteenth century. Ironically, people had already been working with the complex numbers, at least in a formal way, since the sixteenth century (e.g., Descartes). They proved necessary, as had irrational numbers centuries earlier, to solve quadratic equations. We discuss the algebra and geometry of the complex number system at some length in §3, and then pursue the geometric aspects in §5 by classifying all isometries of the Euclidean line and plane. For completeness, the standard quadratic formula and Tartaglia's elegant solution (c. 1535) of the arbitrary cubic equation (which was published by Cardano, and therefore is usually credited to him) are included in §4.

## 1. The Rational Numbers

We are all used to working with the rational numbers in their representation as the quotients $\frac{a}{b}$ of two integers, $b \neq 0$. The problem with this is that if all we know is integer arithmetic, we have no context in which to perform this division. Of course, what we are doing is broadening our universe, defining the number $\frac{1}{b}$ in such a way that

$$b \times \frac{1}{b} = \underbrace{\frac{1}{b} + \cdots + \frac{1}{b}}_{b \text{ times}} = 1.$$

Then when $a > 0$, $\frac{a}{b}$ is defined to be $a \times \frac{1}{b} = \underbrace{\frac{1}{b} + \cdots + \frac{1}{b}}_{a \text{ times}}$.

We formally define the larger universe as follows. A rational number will be defined to be an equivalence class of ordered pairs of integers $(a, b)$, $b \neq 0$, where $(a, b) \sim (a', b') \iff ab' = ba'$ (for further discussion, see Section 3 of Appendix A). This is the usual cross-multiplication rule for $\frac{a}{b} = \frac{a'}{b'}$. We denote the equivalence class of $(a, b)$, oddly enough, by $\frac{a}{b}$. We must now define addition and multiplication of these "formal fractions":

Addition: $\dfrac{a}{b} + \dfrac{c}{d} = \dfrac{ad + bc}{bd}$

Multiplication: $\dfrac{a}{b} \times \dfrac{c}{d} = \dfrac{ac}{bd}$

Zero element: $0 = \dfrac{0}{1}$

Additive inverse: $-\dfrac{a}{b} = \dfrac{-a}{b}$

Multiplicative identity: $1 = \dfrac{1}{1}$

Using these definitions, one can check that all the properties of a ring listed in Section 4 of Chapter 1 hold. However, the serious issue is well-definedness: are these definitions independent of the choice of representatives of the respective equivalence classes? Suppose we choose an ordered pair $(a', b')$ equivalent to $(a, b)$; then do we get equivalent answers when we add and multiply? The answer, of course, is yes, as we check here. If $(a', b') \sim (a, b)$, then does $\frac{a'}{b'} + \frac{c}{d} = \frac{a}{b} + \frac{c}{d}$? Well, $\frac{a'}{b'} + \frac{c}{d} = \frac{a'd + b'c}{b'd}$ by definition. But since $a'b = b'a$, we have $a'bd = b'ad$, and so $a'bd + bb'c = b'ad + bb'c$, whence $(a'd + b'c)(bd) = (ad + bc)(b'd)$. This means that $\frac{a'd + b'c}{b'd} = \frac{ad + bc}{bd}$, so $\frac{a'}{b'} + \frac{c}{d} = \frac{a}{b} + \frac{c}{d}$, as we wanted. And $\frac{a'}{b'} \times \frac{c}{d} = \frac{a'c}{b'd} = \frac{ac}{bd}$, since multiplying the equation $a'b = b'a$ by $cd$ gives the needed equality.

The distinguishing feature of this larger universe, however, is that we now have multiplicative inverses: If $a, b \neq 0$, note that

$\frac{a}{b} \times \frac{b}{a} = 1$, so that $\frac{b}{a}$ is the multiplicative inverse of $\frac{a}{b}$. Thus the collection of rational numbers, which we shall call $\mathbb{Q}$, is a **field**.

**Remark.** This ordered-pair construction works, in fact, for any integral domain $R$ (see Exercise 11). The resulting field is called the **field of quotients** of $R$.

Before going on, we need to observe that the integers $\mathbb{Z}$ come with more algebraic structure, namely, an **ordering**. We have a distinguished subset $\mathbb{Z}^+$, the subset of *positive integers*, so that the following are true:

(1) Given $n \in \mathbb{Z}$, either $n \in \mathbb{Z}^+$, $n = 0$, or $-n \in \mathbb{Z}^+$. (This is called the trichotomy principle.)
(2) Given $m, n \in \mathbb{Z}^+$, $m + n \in \mathbb{Z}^+$ and $mn \in \mathbb{Z}^+$ (i.e., the set of positive integers is closed under addition and multiplication).

Of course, $\mathbb{Z}^+ = \mathbb{N}$, and we obtain an ordering $>$ on $\mathbb{Z}$ by defining $m > n$ if and only if $m - n \in \mathbb{Z}^+$. (We likewise say $m \geq n$ provided $m - n > 0$ or $m = n$.)

In general, we may define an **ordered field** to be a field $F$ together with a distinguished subset $F^+$ (the *positive* elements of $F$) satisfying the analogous properties:

(1) Given $x \in F$, either $x \in F^+$, $x = 0$, or $-x \in F^+$.
(2) Given $x, y \in F^+$, $x + y \in F^+$ and $xy \in F^+$.

We write $x > y$ if and only if $x - y \in F^+$, and so on.

**Lemma 1.1.** *The following laws of inequalities are valid in $\mathbb{Z}$ or in any ordered field:*

(i) *If $a > b$, then $a + c > b + c$ for all $c \in \mathbb{Z}$.*
(ii) *If $a > b$ and $c > 0$, then $ac > bc$.*
(iii) *If $a > b$ and $c < 0$, then $ac < bc$.*
(iv) *If $a > b$ and $b > c$, then $a > c$.*
(v) *If $a \neq 0$, then $a^2 > 0$.*

*Proof.* See Exercise 2.  □

The next observation is that the rational numbers $\mathbb{Q}$ inherit an ordering from that on the integers. We define $\frac{a}{b} > 0$ if $ab > 0$, and we let $\mathbb{Q}^+ = \{\frac{a}{b} : \frac{a}{b} > 0\}$ denote the set of positive rational numbers. Note this is well-defined, because if $\frac{a}{b} = \frac{a'}{b'}$, then $ab' = a'b$, so $(ab)b'^2 = (a'b')b^2$. By Lemma 1.1(v), both $b^2$ and $b'^2$ are positive;

by Lemma 1.1(iii), the product of a positive integer and a negative integer is negative, and we know the product of two positive integers is positive. Thus, $ab$ and $a'b'$ have the same sign.

**Lemma 1.2.** $\mathbb{Q}$ *is an ordered field.*

*Proof.* Given $r = \frac{a}{b} \in \mathbb{Q}$, we may first assume $b > 0$ (since $\frac{a}{b} = \frac{-a}{-b}$). Then $r \in \mathbb{Q}^+$ when $a \in \mathbb{Z}^+$, $r = 0$ when $a = 0$, and $-r \in \mathbb{Q}^+$ when $-a \in \mathbb{Z}^+$.

Now it remains to show that $\mathbb{Q}^+$ is closed under addition and multiplication. Let $\frac{a}{b}, \frac{c}{d} \in \mathbb{Q}^+$; we need to show that their product and sum both lie in $\mathbb{Q}^+$. We may assume that $b, d > 0$. Then $a, c > 0$ as well, and now we compute $\frac{a}{b} \times \frac{c}{d} = \frac{ac}{bd}$. Since $\mathbb{Z}^+$ is closed under multiplication, both $ac$ and $bd$ are positive, and therefore so is their quotient. Now $\frac{a}{b} + \frac{c}{d} = \frac{ad+bc}{bd}$ is positive, since $\mathbb{Z}^+$ is closed under both addition and multiplication. $\square$

One property of the rational numbers that will play an important rôle in the next section is the Archimedean Property:

**Proposition 1.3.** *Given any positive rational numbers $r, s$, there is a positive integer $n$ so that $nr > s$.*

*Proof.* Let $r = \frac{a}{b}$, $s = \frac{c}{d}$, where $a, b, c, d \in \mathbb{Z}^+$. We wish to find $n \in \mathbb{Z}^+$ so that $n\frac{a}{b} > \frac{c}{d}$, i.e., so that $nad > bc$. Since all the integers in question are positive, $ad \geq 1$, so it suffices to take $n = 2bc$. $\square$

The rational numbers also have a significant density property.

**Proposition 1.4.** *Given any two distinct rational numbers, there is another rational number between them.*

*Proof.* Given $r > s \in \mathbb{Q}$, consider $z = \frac{r+s}{2}$. We claim that $z \in \mathbb{Q}$ and that $r > z > s$. The former is immediate since $\mathbb{Q}$ is a field. To show $r > z$, we note that $r > s \implies \frac{r}{2} > \frac{s}{2}$. Thus $r = \frac{r}{2} + \frac{r}{2} > \frac{r}{2} + \frac{s}{2} = \frac{r+s}{2} = z$, as required. The remaining inequality is deduced analogously. $\square$

## EXERCISES 2.1

1.  Check the details of the claim that $\mathbb{Q}$ is a field.

2. Prove Lemma 1.1. (Use repeatedly the fact that $\mathbb{Z}^+$ or $F^+$ is closed under addition and multiplication.)

3. Prove or give a counterexample. If the statement is false, modify it to make it correct.
   a. if $a > b$ and $c > d$, then $ac > bd$;
   b. if $c > 0$ and $ac > bc$, then $a > b$.

4. Let $F$ be an ordered field, and let $x, y \in F$. Prove that if $x > 0$ and $y < 0$, then $xy < 0$, and that if $x, y < 0$, then $xy > 0$. Also, from the fact that 1 is the multiplicative identity of $F$, deduce that $1 > 0$.

5. Prove that for $r, s \in \mathbb{Q}$, each of the following inequalities holds. In each case, also determine when equality holds.
   a. $r^2 + s^2 \geq 2rs$;
   b. $r^2 + 2rs + 3s^2 \geq 0$;
   c. $r^2 + rs + s^2 \geq 0$.

6. Prove that if $r > s \geq 0$, then $r^2 > s^2$. (Where do you use the assumption that $r$ and $s$ are nonnegative?) Conversely, prove that if $r, s \geq 0$ and $r^2 > s^2$, then $r > s$.

7. A common mistake made by high school algebra students (and, alas, by college calculus students alike) is the following:

$$\frac{a}{b} + \frac{a}{c} = \frac{a}{b+c}.$$

   a. Show that in any field this equation implies that either $a = 0$ or $b^2 + bc + c^2 = 0$.
   b. Show that in $\mathbb{Q}$ it holds only when $a = 0$.
   c. Give an example of a field where $a \neq 0$ and $b^2 + bc + c^2 = 0$ instead.

8. Prove or give a counterexample: if $F$ is a field and $a, b \in F$, then $a^2 + b^2 = 0 \implies a = b = 0$.

9. Suppose $r = \frac{a}{b}$, $s = \frac{c}{d} \in \mathbb{Q}$, and $r < s$. Prove or give a counterexample: $r < \frac{a+c}{b+d} < s$. If the statement is false, make it correct.

10. Is the following generalization of Proposition 1.4 valid? Between every two distinct rational numbers there are infinitely many rational numbers.

11.  Starting with the ring $\mathbb{Z}$, we constructed the field $\mathbb{Q}$ as a set of equivalence classes of fractions. Show that this construction can be performed generally for any integral domain $R$, yielding the field of quotients of $R$. In order for the construction to work, why is it necessary that $R$ be an integral domain (and not just any commutative ring)?

12.  Let $R = \{\frac{a}{b} \in \mathbb{Q} : b = 2^k$ for some nonnegative integer $k\}$. Show that $R$ is an integral domain, and find its field of quotients.

13.  Let $p$ be a prime number. Let $R = \{\frac{a}{b} \in \mathbb{Q} : \gcd(p, b) = 1\}$. Show that $R$ is an integral domain, and find its field of quotients.

14.  Prove that $1/2 \notin \mathbb{Z}$. (We have taken this fact for granted in Chapter 1.) State and prove an appropriate generalization.

15.  For what values of $x, y \in \mathbb{Q}$ does each of the following equations hold?
     a.  $(x + y)^2 = x^2 + y^2$
     b.  $(x + y)^3 = x^3 + y^3$
     c.  $(x + y)^4 = x^4 + y^4$
     Can you state and prove an appropriate generalization?

16.  Find all functions $f : \mathbb{Q} \to \mathbb{Q}$ satisfying
     (i)  $f(1) = 1$, and
     (ii)  $f(x + y) = f(x) + f(y)$ for all $x, y \in \mathbb{Q}$.

## 2. From the Rational Numbers to the Real Numbers

We have been getting used to the universe of rational numbers. As the Greeks discovered early on, this is a very limited universe: there can be no isosceles right triangle with legs one unit long, because in $\mathbb{Q}$ there is no solution to the equation $x^2 = 2$ (see Exercise 1.2.11). [In case you did not do this exercise, here is the proof. Suppose we had a rational number $x = \frac{a}{b}$ with $x^2 = 2$; we may assume $a$ and $b$ are relatively prime. Then the equation $a^2 = 2b^2$ implies that $a^2$ is even, hence that $a$ is even (by Exercise 1.1.3). Setting $a = 2k$, $a^2 = (2k)^2 = 4k^2 = 2b^2$ implies that $2k^2 = b^2$, and so $b$ is likewise even. Since we began by assuming that $a$ and $b$ are relatively prime, we have arrived at a contradiction.]

Let's try to find rational numbers approximating $\sqrt{2}$. Given integers $a, b$ with $(\frac{a}{b})^2 < 2$, we can find integers $c, d$ so that $\frac{a}{b} < \frac{c}{d}$ and $(\frac{c}{d})^2 < 2$ as well. For example, take $c = 3a + 4b$, $d = 2a + 3b$. Then we leave it as an exercise for the reader to check that $\frac{c}{d}$ has the requisite properties (see Exercise 10). We next establish that the number $\frac{c}{d}$ is significantly closer to $\sqrt{2}$ than $\frac{a}{b}$.

**Lemma 2.1.** *Suppose* $1 < \frac{a}{b}$ *and* $(\frac{a}{b})^2 < 2$. *Let* $c = 3a + 4b$, $d = 2a + 3b$. *If we let* $\delta = 2 - (\frac{a}{b})^2$, *then* $2 - (\frac{c}{d})^2 < \frac{\delta}{25}$.

**Proof.** Given $\delta = 2 - (\frac{a}{b})^2$, $2b^2 - a^2 = \delta b^2$. Now

$$2 - \left(\frac{c}{d}\right)^2 = 2 - \left(\frac{3a + 4b}{2a + 3b}\right)^2 = \frac{2b^2 - a^2}{(2a + 3b)^2} = \frac{\delta}{(2\frac{a}{b} + 3)^2} < \frac{\delta}{25},$$

since $2\frac{a}{b} + 3 > 5$.  □

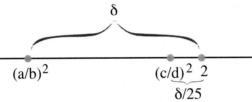

FIGURE 1

By repeatedly applying Lemma 2.1 we can find an infinite number of rational numbers $r = \frac{a}{b}$ whose squares are less than, but arbitrarily close to, 2. Similarly, by Exercise 10, we can find an infinite number of rational numbers whose squares are greater than, but as close as desired to, 2. If we consider the sets $L = \{r \in \mathbb{Q}^+ : r^2 < 2\}$ and $U = \{r \in \mathbb{Q}^+ : r^2 > 2\}$, we observe that every element of $L$ is less than every element of $U$; nevertheless, $L$ has no largest element (by Lemma 2.1) and $U$ has no smallest element.

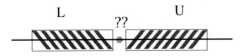

FIGURE 2

The idea behind the real number system is to fill in these gaps in the rational numbers in the most economical way possible (so as still to have a field with a compatible notion of inequality). The details of the construction are quite involved and were not settled carefully until the latter half of the nineteenth century; we will be content to give a precise statement of the properties of the real numbers $\mathbb{R}$.

**Definition.** Let $F$ be an ordered field. If $S \subset F$ and there is an element $x \in F$ so that $x \geq s$ for all $s \in S$, then we say $S$ is **bounded above** and $x$ is called an **upper bound** of $S$.

**Example 1.** In $\mathbb{Q}$, let $S = \{r \in \mathbb{Q} : r^2 < 2\}$. Then $S$ is bounded above by 2 (and by $\frac{3}{2}$). (See Exercise 2 for one way to prove this.)

**Definition.** Let $S \subset F$ be bounded above. We say $x_0 \in F$ is the **least upper bound** (l.u.b.) of $S$ if

(1) $x_0$ is an upper bound of $S$; i.e., $x_0 \geq s$ for all $s \in S$;
(2) for all upper bounds $x$ of $S$, $x_0 \leq x$.

We say $F$ has the **least upper bound property** if every nonempty subset $S \subset F$ that is bounded above has a least upper bound.

**Example 2.** Let $S = \{r \in \mathbb{Q} : r^2 < 2\}$. Then $S$ has no l.u.b. in $\mathbb{Q}$, since the set $U$ discussed above is the set of upper bounds, and $U$ has no least element. Thus, $\mathbb{Q}$ does not have the least upper bound property.

**Example 3.** Let $S = \mathbb{N} \subset \mathbb{Q}$. Then $S$ has no l.u.b., because it is not bounded above (see Exercise 8). Nevertheless, $\mathbb{N}$ does have the least upper bound property (see Exercise 9).

We now come to the following theorem/definition which gives the salient features of the real number system.

**Theorem 2.2.** *There is a (unique) field $\mathbb{R}$ containing $\mathbb{Q}$, having an ordering $>$ so that $\mathbb{Q}^+ \subset \mathbb{R}^+$, and satisfying the least upper bound property.* $\square$

Elements of $\mathbb{R}$ are called **real numbers**, and elements of $\mathbb{R}$ that do not belong to $\mathbb{Q}$ are called **irrational** numbers. Our discussion of "$\sqrt{2}$" above shows that every real number $x$ may be interpreted as the least upper bound of the set of rational numbers less than $x$. This whole discussion may seem somewhat far-fetched, since we have spent most of our lives thinking of and working with real numbers as decimals (e.g., $1.414213562\ldots$ or $3.1415926535\ldots$). On the one hand, the various finite stages of these infinite decimal expansions give a specific set of rational numbers whose least upper bound is the real number in question. On the other hand, you will find that if you try to give a precise definition of addition and multiplication of real numbers using decimal expansions (let alone proving the algebraic properties they must enjoy!), you will be in for a rough time.

**Remark.** The least upper bound property is exactly what's required for the Intermediate Value Theorem (of calculus fame) to hold. If $f$ is a continuous function on an interval $[a, b]$ with $f(a) < 0$ and $f(b) > 0$, how do we find a point $c \in (a, b)$ so that $f(c) = 0$? It is quite natural to define $S = \{x \in [a, b] : f(x) < 0\}$. Then $S$ is nonempty and is bounded above (e.g., by $b$). *Provided S has a l.u.b., c*, then it must be the case that $f(c) = 0$. The salient feature of continuity we need here is this: if $f$ is continuous at $x_o$ and $f(x_o) > 0$ (resp., $< 0$), then $f$ is positive (resp., negative) on some interval $(x_o - \delta, x_o + \delta)$ containing $x_o$. As you can check, we cannot have $f(c) > 0$ (for then $c$ would not be the *least* upper bound of $S$), and we also cannot have $f(c) < 0$ (for then $c$ would not be an upper bound of $S$ at all). Thus, by trichotomy, we must have $f(c) = 0$. In summary, the least upper bound property guarantees that $f(x) = x^2 - 2$ (and lots of other polynomials) will have roots in $\mathbb{R}$. On the other hand, since $f(x) = x^2 - 2$ does not have a root in $\mathbb{Q}$, we infer—once again—that $\mathbb{Q}$ cannot have the least upper bound property.

We next show that the Archimedean and density properties of $\mathbb{Q}$ (Propositions 1.3 and 1.4, respectively) hold in $\mathbb{R}$ as well.

**Proposition 2.3.** *Given any two positive real numbers $x, y$, there is an integer $n$ so that $nx > y$.*

**Proof.** Let $S = \{n \in \mathbb{N} : nx \le y\}$. If $S$ is empty, this means that $x > y$, and we may take $n = 1$. Otherwise, when $S$ is nonempty, note that $S$ is bounded above by, for example, $\frac{y}{x}$. Let $n_o = $ l.u.b. $S$, and observe simply that $(n_o + 1)x > y$, as required. (We really need to check that $n_o \in \mathbb{N}$. See Exercise 9.)  □

**Proposition 2.4.** *Between any two distinct real numbers there is a rational number.*

**Proof.** Let $x > y$ be real numbers. Suppose first that $x - y > 1$; then there must be an integer between $x$ and $y$, and we're done. (If $y \ge 0$, by Proposition 2.3 there is an integer $m$ so that $m(1) >$

$y$. The smallest such integer $m$ (which exists by the Well-Ordering Principle) must satisfy $y < m < x$. The case of $y < 0$ is left to the reader.)

In general, applying Proposition 2.3 again, there is a positive integer $n$ so that $n(x-y) > 1$, and so there is an integer $m$ satisfying $ny < m < nx$, whence $y < \frac{m}{n} < x$, as required.  □

**Remark.** It follows from Proposition 2.4 that between any two distinct real numbers there are infinitely many rational numbers; thus, there ought to be "a lot" of rational numbers. But, in fact, the set of rational numbers is *countable*, i.e., can be put in one-to-one correspondence with $\mathbb{N}$, whereas the set of real numbers is *uncountable*. This remarkable fact was proved by Cantor. (See Exercises 13 and 14.)

We are now in a position to find further examples of fields, all of which lie between $\mathbb{Q}$ and $\mathbb{R}$. We give the prototypical example here and leave others for the reader to check as exercises (see Exercises 4 and 17).

**Example 4.** Let $F = \{a + b\sqrt{2} : a, b \in \mathbb{Q}\} \subset \mathbb{R}$. From here on, $F$ will be denoted as $\mathbb{Q}[\sqrt{2}]$. Addition and multiplication are done in the realm of real arithmetic, although the following algebraic laws are convenient to note:

$$(a + b\sqrt{2}) + (c + d\sqrt{2}) = (a + c) + (b + d)\sqrt{2}$$
$$(a + b\sqrt{2})(c + d\sqrt{2}) = (ac + 2bd) + (ad + bc)\sqrt{2}$$

In particular, $\mathbb{Q}[\sqrt{2}]$ is closed under the operations of addition and multiplication. It is easy to check that $\mathbb{Q}[\sqrt{2}]$ is a commutative ring: it contains the additive and multiplicative identities 0 and 1, and the commutative, associative, and distributive properties are inherited from $\mathbb{R}$. To check that $\mathbb{Q}[\sqrt{2}]$ is a field, we must establish that each nonzero element has a multiplicative inverse. Let $a + b\sqrt{2} \in \mathbb{Q}[\sqrt{2}]$, with $a$ and $b$ not both zero; applying the "conjugate trick" from high school algebra, we obtain

$$\frac{1}{a + b\sqrt{2}} = \frac{1}{a + b\sqrt{2}} \cdot \frac{a - b\sqrt{2}}{a - b\sqrt{2}} = \frac{a - b\sqrt{2}}{a^2 - 2b^2}$$
$$= \left(\frac{a}{a^2 - 2b^2}\right) + \left(\frac{-b}{a^2 - 2b^2}\right)\sqrt{2},$$

which does indeed belong to $\mathbb{Q}[\sqrt{2}]$. Note that since $\sqrt{2}$ is irrational, the denominator $a^2 - 2b^2$ can never vanish so long as $a, b \in \mathbb{Q}$ are not both zero.

## EXERCISES 2.2

1. Show that the various algebraic consequences of ordering in $\mathbb{Z}$ and $\mathbb{Q}$ (e.g., Lemma 1.1, Exercises 2.1.4, 2.1.5, 2.1.6) hold equally well in $\mathbb{R}$.

2. Suppose $x$ and $y$ are nonnegative real numbers and $n \in \mathbb{N}$. Prove that $x \le y \iff x^n \le y^n$.

3. Prove that for $x, y \in \mathbb{R}^+$, $\dfrac{x+y}{2} \ge \sqrt{xy}$.

4. Check explicitly that $\mathbb{Q}[\sqrt{5}]$ is a field. What field is $\mathbb{Q}[\sqrt{8}]$?

5. Do the irrational numbers form a field? In particular, is it true that if $a$ and $b$ are irrational numbers, then $a + b$ and $ab$ are necessarily irrational numbers?

6. Prove that the following numbers are irrational:
   a. $\sqrt{3}$
   b. $\sqrt[3]{2}$
   c. $\log_{10} 3$
   d. $\sqrt{2} + \sqrt{3}$
   e. $e = \sum\limits_{n=0}^{\infty} \frac{1}{n!}$ (Hint: Suppose $e = p/q$, and show that $q! \sum\limits_{n=q+1}^{\infty} \frac{1}{n!}$ cannot be an integer.)

7. Elaborate on the density principle enunciated in Proposition 1.4 as follows:
   a. Using the fact that $0 < \frac{\sqrt{2}}{2} < 1$, prove that between any two distinct rational numbers there is an irrational number.
   b. Deduce that between any two distinct real numbers there is an irrational number.

8. Prove, without using Proposition 2.3, that $\mathbb{N} \subset \mathbb{R}$ is not bounded above. (Hint: If it were, there would be a least upper bound $y$. But then use the Principle of Mathematical Induction to show that $y - 1$ is an upper bound as well.)

9. Let $S \subset \mathbb{N}$ be a nonempty subset that is bounded above. Show that $n_o = \text{l.u.b.}\,S$ is an element of $S$. (Hint: One approach is to apply Exercise 8. Consider $T = \{n \in \mathbb{N} : n \text{ is an upper bound of } S\}$. Show that $T$ has a least element.)

10. Assuming $a, b \in \mathbb{N}$ and $(\frac{a}{b})^2 < 2$, put $c = 3a + 4b$, $d = 2a + 3b$ and $m = a + 2b$, $n = a + b$.
    a. Check that $\frac{a}{b} < \frac{c}{d}$ and that $(\frac{c}{d})^2 < 2$.
    b. Check that $(\frac{m}{n})^2 > 2$ and that $(\frac{m}{n})^2 - 2 < 2 - (\frac{a}{b})^2$.
    c. Conclude (using Lemma 2.1) that there are rational numbers whose squares are arbitrarily close to 2 (both greater than 2 and less than 2).

11. a. Prove that the real number $r$ is rational if and only if its decimal representation is either finite or repeating.
    b. Give an algorithm to express a repeating decimal as a fraction.
    c. We say the decimal $0.142857\overline{142857}\ldots$ has period 6 and the decimal $3.25654\overline{54}\ldots$ has period 2. Find all *prime* numbers $p$ so that $1/p$ has period 3 or 6.

12. Let $K \subset \mathbb{R}$ be a field. Prove that $\mathbb{Q} \subset K$. (Hint: Start by showing that $\mathbb{Z} \subset K$.)

13. Prove that $\mathbb{Q}$ is a countable set, i.e., can be put in a one-to-one correspondence with $\mathbb{N}$. (Hint: Arrange the positive fractions $\frac{m}{n}$ in a two-dimensional array—numerators increasing along rows, denominators increasing down columns. Delete duplicates, and then count along $45°$ lines.)

14. Prove that the set of all real numbers between 0 and 1 is uncountable. Deduce that $\mathbb{R}$ and, in turn, $\mathbb{R} - \mathbb{Q}$ are uncountable. (Hint: Suppose the numbers could be arranged in a sequence $x_1, x_2, \ldots, x_n, \ldots$. Write out a decimal representation of each number: $x_i = 0.a_1^{(i)} a_2^{(i)} \ldots a_i^{(i)} \ldots$, $i = 1, 2, \ldots$. Now consider the real number $y = 0.y_1 y_2 \ldots y_k \ldots$ defined as follows:
$$y_k = \begin{cases} 4, & a_k^{(k)} = 3 \\ 3, & \text{otherwise} \end{cases}.)$$

15. Here is the sketch of a proof, using the l.u.b. axiom, that every positive real number $a$ has a square root. (Query: How is this fact normally proved in a calculus class?)

a.  Let $S = \{x \in \mathbb{R} : x^2 < a\}$. Show that $S$ is nonempty and is bounded above. (Hint: For the latter, it is probably easiest to deal separately with the cases $a < 1$ and $a \geq 1$.)

b.  Let $y = $ l.u.b. $S$. Note $y > 0$. We must show $y^2 = a$. Suppose first that $y^2 > a$, and let $\epsilon = y^2 - a$. Compute $(y - \frac{\epsilon}{2y})^2$ and deduce that $y - \frac{\epsilon}{2y}$ is an upper bound for $S$. Why is this a contradiction?

c.  Suppose now that $y^2 < a$, and let $\epsilon = a - y^2$. Show that there is a small number $\delta$ so that $(y + \delta)^2 < a$, so that $y$ is not an upper bound of $S$ at all. (Hint: Let $\delta$ be a positive number satisfying both $\delta < 1$ and $\delta < \frac{\epsilon}{2y+1}$.)

16.  Find all functions $f \colon \mathbb{R} \to \mathbb{R}$ satisfying the following:
  (i)   $f(1) = 1$,
  (ii)  $f(x + y) = f(x) + f(y)$ for all $x, y \in \mathbb{R}$, and
  (iii) $f(xy) = f(x)f(y)$ for all $x, y \in \mathbb{R}$.
(Hint: Show that $x > 0 \implies f(x) > 0$, and so $x > y \implies f(x) > f(y)$. Cf. Exercise 2.1.16.)

17.  Let $\mathbb{Q}[\sqrt[3]{2}] = \{a + b\sqrt[3]{2} + c\sqrt[3]{4} : a, b, c \in \mathbb{Q}\} \subset \mathbb{R}$. Show that $\mathbb{Q}[\sqrt[3]{2}]$ is a field.

## 3. The Complex Numbers

in Section 2 we were led to the real numbers by trying to solve the innocuous quadratic equation $x^2 = 2$. Here we contemplate the equation $x^2 = -1$. By Lemma 1.1(v), there can be no real number $x$ satisfying this equation. We introduce the "imaginary" number $i$ satisfying the equation $i^2 = -1$ and try to build a field containing $i$ *and* all of $\mathbb{R}$.

We define the complex numbers $\mathbb{C}$ to be the set of all ordered pairs $(a, b)$, $a, b \in \mathbb{R}$, i.e., the set of all vectors in $\mathbb{R}^2$. Addition and multiplication are defined as follows:

$$(a, b) + (c, d) = (a + c, b + d)$$
$$(a, b) \cdot (c, d) = (ac - bd, ad + bc).$$

We denote $(0, 1)$ by the symbol $i$, and $(a, b)$ by $a + bi$. It is easy to remember the definition of multiplication by expanding the product

$(a + bi)(c + di)$ by the distributive property and using the rule $i^2 = -1$.

**Example 1.** $(a + bi)(a - bi) = (a^2 - b(-b)) + (a(-b) + ba)i = a^2 + b^2$.

**Proposition 3.1.** $\mathbb{C}$ *is a field.*

***Proof.*** We must check that $\mathbb{C}$ is a commutative ring in which every nonzero element has a multiplicative inverse. From the definitions of addition and multiplication it is clear that these operations are commutative. Addition is associative because vector addition is associative (which in turn is the case because addition of real numbers is associative). The additive identity is $0 = 0 + 0i$, and additive inverses are given by $-(a + bi) = (-a) + (-b)i$, as expected. The multiplicative identity is $1 = 1 + 0i$. We stop now to verify the associative law of multiplication:

$$
\begin{aligned}
(a + bi)[(c+di)(e + fi)] &= (a + bi)[(ce - df) + (cf + de)i] \\
&= (a(ce - df) - b(cf + de)) + (a(cf + de) + b(ce - df))i \\
&= (ace - adf - bcf - bde) + (acf + ade + bce - bdf)i,
\end{aligned}
$$

whereas

$$
\begin{aligned}
[(a + bi)(c+di)](e + fi) &= [(ac - bd) + (ad + bc)i](e + fi) \\
&= ((ac - bd)e - (ad + bc)f) + ((ac - bd)f + (ad + bc)e)i \\
&= (ace - bde - adf - bcf) + (acf - bdf + ade + bce)i,
\end{aligned}
$$

and so the two agree (whew!). Next comes the distributive law:

$$
\begin{aligned}
(a + bi)[(c + di) + (e + fi)] &= (a + bi)[(c + e) + (d + f)i] \\
&= (a(c + e) - b(d + f)) + (a(d + f) + b(c + e))i \\
&= [(ac - bd) + (ad + bc)i] + [(ae - bf) + (af + be)i] \\
&= (a + bi)(c + di) + (a + bi)(e + fi).
\end{aligned}
$$

This straightforward calculation establishes the fact that $\mathbb{C}$ is a commutative ring. To prove that $\mathbb{C}$ is a field, we try the same trick that we used in Example 4 of Section 2. Formally, write

$$
\begin{aligned}
\frac{1}{a + bi} &= \frac{1}{a + bi} \cdot \frac{a - bi}{a - bi} = \frac{a - bi}{(a + bi)(a - bi)} \\
&= \frac{a - bi}{a^2 + b^2} = \left(\frac{a}{a^2 + b^2}\right) + \left(\frac{-b}{a^2 + b^2}\right)i.
\end{aligned}
$$

Note that this expression makes sense so long as $a + bi$ is nonzero (since $a$ and $b$ are then not both zero, and so $a^2 + b^2 > 0$). The reader can also check directly that $\left(a + bi\right)\left(\left(\frac{a}{a^2+b^2}\right) + \left(\frac{-b}{a^2+b^2}\right)i\right) = 1$.  □

Except for the fact that there is no ordering on $\mathbb{C}$ (see Exercise 1), everything that worked for $\mathbb{Q}$ and $\mathbb{R}$ works for $\mathbb{C}$, and now we are in the position to "solve" the quadratic equation $x^2 + 1 = 0$. Indeed, one benefit of introducing $i$ is that we can now solve *all* quadratic equations. Before coming to this point, we explore some of the geometry behind the algebra of complex numbers.

When we represent $\mathbb{C}$ by ordered pairs of real numbers, we call this the *complex plane*. Addition of complex numbers coincides with addition of vectors in $\mathbb{R}^2$. The horizontal axis is called the real axis; the vertical axis, the imaginary axis. Given a complex number $z = a + bi$, the **real part** of $z$ (denoted $\mathcal{R}e\ z$) is equal to $a$, and the **imaginary part** of $z$ (denoted $\mathcal{I}m\ z$) is equal to $b$. (Note that $\mathcal{I}m\ z$ is itself a *real* number.) The reflection of $z = a + bi$ in the real axis is called the **conjugate** $\bar{z}$ of $z$; thus, $\bar{z} = a - bi$.

**Example 2.** The following formulas are sometimes useful:

$$\mathcal{R}e\ z = \frac{z + \bar{z}}{2}$$

$$\mathcal{I}m\ z = \frac{z - \bar{z}}{2i}.$$

The vector $(a, b) \in \mathbb{R}^2$ has length $\sqrt{a^2 + b^2}$; we also refer to this as the **absolute value** (or **modulus**) of the complex number $a + bi$, denoted $|a + bi|$. Introducing polar coordinates in the plane, as shown in Figure 1, we now write $z = r(\cos\theta + i\sin\theta)$, where $r = |z|$. We note that $\theta$ is only defined up to multiples of $2\pi$ (360°). This is often called the **polar form** of the complex number $z$.

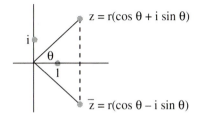

FIGURE 1

**Example 3.** $1 + i = \sqrt{2}(\frac{1}{\sqrt{2}} + \frac{1}{\sqrt{2}}i) = \sqrt{2}(\cos\frac{\pi}{4} + i\sin\frac{\pi}{4})$.

**Example 4.** $-3 - 4i = 5(-\frac{3}{5} - \frac{4}{5}i) = 5(\cos\theta + i\sin\theta)$, where $\theta = \pi + \arctan\frac{4}{3} \approx 4.07$ radians. (Note that it would not help to use arccos or arcsin, since $\arccos(-\frac{3}{5})$ is an angle in the second quadrant and $\arcsin(-\frac{4}{5})$ is an angle in the fourth.)

**Example 5.** Consider the multiplication $(1 - 2i)(-1 + 3i) = 5 + 5i$ from the viewpoint of the moduli and angles of these complex numbers:

$$1 - 2i = \sqrt{5}(\cos\theta_1 + i\sin\theta_1), \quad \theta_1 \approx -63.4°,$$

$$-1 + 3i = \sqrt{10}(\cos\theta_2 + i\sin\theta_2), \quad \theta_2 \approx 108.4°,$$

$$\text{and} \quad 5 + 5i = 5\sqrt{2}(\cos 45° + i\sin 45°).$$

Note that the modulus of $5+5i$ is the product of the moduli of $1-2i$ and $-1+3i$ *and* that the angle 45° is the sum of the corresponding angles $\theta_1$ and $\theta_2$.

The "coincidence" we observed in the last example is, of course, no coincidence and lies at the heart of the geometric interpretation of the algebra of complex numbers.

**Proposition 3.2.** *Let* $z = r(\cos\theta + i\sin\theta)$ *and* $w = \rho(\cos\phi + i\sin\phi)$. *Then*

$$zw = r\rho(\cos(\theta + \phi) + i\sin(\theta + \phi)).$$

*That is, to multiply two complex numbers, we multiply their moduli and add their angles.*

**Proof.** Recall the basic trigonometric formulas

$$\cos(\theta + \phi) = \cos\theta\cos\phi - \sin\theta\sin\phi \quad \text{and}$$
$$\sin(\theta + \phi) = \sin\theta\cos\phi + \cos\theta\sin\phi.$$

Now,

$$zw = (r(\cos\theta + i\sin\theta))(\rho(\cos\phi + i\sin\phi))$$
$$= r\rho(\cos\theta + i\sin\theta)(\cos\phi + i\sin\phi)$$
$$= r\rho((\cos\theta\cos\phi - \sin\theta\sin\phi) + i(\sin\theta\cos\phi + \cos\theta\sin\phi))$$
$$= r\rho(\cos(\theta + \phi) + i\sin(\theta + \phi)),$$

as required. $\square$

As a corollary, we have the famous formula of deMoivre:

**Corollary 3.3** (deMoivre's Theorem). *Let $n$ be any integer, and let $z = r(\cos\theta + i\sin\theta)$. Then $z^n = r^n(\cos n\theta + i\sin n\theta)$.*

**Proof.** See Exercise 5.   □

**Example 6.** We can use Corollary 3.3 to resurrect the double-angle formulas of trigonometry. If $z = \cos\theta + i\sin\theta$, then

$$\cos 2\theta + i\sin 2\theta = z^2 = (\cos\theta + i\sin\theta)^2$$
$$= (\cos^2\theta - \sin^2\theta) + i(2\sin\theta\cos\theta).$$

Comparing real and imaginary parts, we obtain

$$\cos 2\theta = \cos^2\theta - \sin^2\theta \quad \text{and} \quad \sin 2\theta = 2\sin\theta\cos\theta.$$

Using Proposition 3.2, we can give a beautiful geometric description of the reciprocal $\frac{1}{z}$ of the complex number $z$.

**Corollary 3.4.** *Let $z = r(\cos\theta + i\sin\theta)$ be a nonzero complex number. Then*

$$\frac{1}{z} = \frac{1}{r}(\cos\theta - i\sin\theta) = \frac{\bar{z}}{|z|^2}.$$

**Proof.** $z\left(\dfrac{\bar{z}}{|z|^2}\right) = \dfrac{z\bar{z}}{|z|^2} = 1$. See also Exercises 2d. and 3c.   □

As indicated in Figure 2, to obtain $1/z$ from $z$, we first "reflect" $z$ in the unit circle (sending $z$ to the point on the ray $\overrightarrow{0z}$ whose distance from 0 is $1/|z|$), and then conjugate (reflecting in the real axis). In particular, whenever $|z| = 1$, $1/z = \bar{z}$.

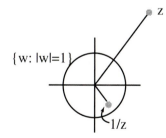

$z$

$\{w: |w|=1\}$

$1/z$

FIGURE 2

One of the important applications of deMoivre's Theorem is to solve the equation $z^n = a$. We begin by finding the $n^{\text{th}}$ roots of 1, commonly called the $n^{\text{th}}$ **roots of unity**. We wish to find the complex numbers $z = r(\cos\theta + i\sin\theta)$ so that $z^n = r^n(\cos n\theta + i\sin n\theta) = 1$;

it follows that $r = 1$ and $n\theta \equiv 0 \pmod{2\pi}$. Thus we conclude that $\theta = \frac{2\pi k}{n}, 0 \le k \le n-1$. Therefore, we have the following proposition.

**Proposition 3.5.** *Let* $\omega = \cos(\frac{2\pi}{n}) + i\sin(\frac{2\pi}{n})$. *Then* $1, \omega, \omega^2, \ldots, \omega^{n-1}$ *are the* $n$ *solutions of* $z^n = 1$.

**Remark.** We shall see in Chapter 3 that there can be no more than $n$ solutions, so we have indeed found them all.

**Example 7.** Here are the sixth roots of unity, shown in Figure 3:

$$\omega = \cos\tfrac{\pi}{3} + i\sin\tfrac{\pi}{3} = \tfrac{1}{2} + \tfrac{\sqrt{3}}{2}i$$

$$\omega^2 = \cos\tfrac{2\pi}{3} + i\sin\tfrac{2\pi}{3} = -\tfrac{1}{2} + \tfrac{\sqrt{3}}{2}i$$

$$\omega^3 = \cos\pi + i\sin\pi = -1$$

$$\omega^4 = \cos\tfrac{4\pi}{3} + i\sin\tfrac{4\pi}{3} = -\tfrac{1}{2} - \tfrac{\sqrt{3}}{2}i$$

$$\omega^5 = \cos\tfrac{5\pi}{3} + i\sin\tfrac{5\pi}{3} = \tfrac{1}{2} - \tfrac{\sqrt{3}}{2}i$$

$$\omega^6 = 1.$$

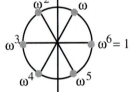

FIGURE 3

**Corollary 3.6.** *Let* $a = \rho(\cos\phi + i\sin\phi)$, *and set* $b = \sqrt[n]{\rho}(\cos\frac{\phi}{n} + i\sin\frac{\phi}{n})$. *Then the* $n$ *solutions of* $z^n = a$ *are* $b, b\omega, b\omega^2, \ldots, b\omega^{n-1}$.

***Proof.*** Clearly $b^n = a$, by deMoivre. On the other hand, for any $k = 0, 1, \ldots, n - 1$, we have $(b\omega^k)^n = b^n(\omega^k)^n = a(\omega^n)^k = a$, as required. □

**Remark.** We may interpret this graphically as follows. Think of the $n^{\text{th}}$ roots of unity as forming the spokes of a wheel. Rotate the wheel so that 1 lines up with a particular $n^{\text{th}}$ root, $b$, of the given complex number $a$. Then the $n$ different spokes of the wheel point towards the $n$ possible $n^{\text{th}}$ roots of $a$. See Figure 4.

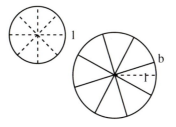

FIGURE 4

**Example 8.** Find the cube roots of $a = -2 + 2i$. Since $|a| = 2\sqrt{2}$, we may write $a = 2\sqrt{2}\left(-\frac{1}{\sqrt{2}} + \frac{1}{\sqrt{2}}i\right)$, and we recognize that $\rho = 2\sqrt{2}$ and $\phi = \frac{3\pi}{4}$. Therefore, one cube root $b$ of $a$ is given by $b = \sqrt[3]{2\sqrt{2}}\left(\cos(\frac{\pi}{4}) + i\sin(\frac{\pi}{4})\right) = \sqrt{2}\left(\frac{1}{\sqrt{2}} + \frac{1}{\sqrt{2}}i\right) = 1 + i$. The other roots are obtained by multiplying by the *cube* roots of unity, $\omega = -\frac{1}{2} + \frac{\sqrt{3}}{2}i$ and $\omega^2 = -\frac{1}{2} - \frac{\sqrt{3}}{2}i$. Thus the cube roots of $a$ are:

$$b = 1 + i$$

$$b\omega = \left(-\tfrac{1}{2} - \tfrac{\sqrt{3}}{2}\right) + \left(\tfrac{\sqrt{3}}{2} - \tfrac{1}{2}\right)i, \text{ and}$$

$$b\omega^2 = \left(-\tfrac{1}{2} + \tfrac{\sqrt{3}}{2}\right) - \left(\tfrac{\sqrt{3}}{2} + \tfrac{1}{2}\right)i.$$

**Example 9.** Find the square roots of $a = -5 + 12i$. Since $|a| = 13$, we may write $a = 13\left(-\frac{5}{13} + \frac{12}{13}i\right)$, so $\rho = 13$ and $\phi = \arccos(-\frac{5}{13})$. Letting $2\theta = \phi$, the square roots of $a$ are given by $b = \pm\sqrt{13}(\cos\theta + i\sin\theta)$. Now we use the double-angle formulas (Example 6) to find $\cos\theta$ and $\sin\theta$, as follows. Since

$$\cos 2\theta = \cos^2\theta - \sin^2\theta = 2\cos^2\theta - 1,$$

we have $2\cos^2\theta - 1 = -\frac{5}{13}$, whence $\cos\theta = \frac{2}{\sqrt{13}}$ (recall that our $\theta$ is an angle in the first quadrant). Since

$$\sin 2\theta = 2\sin\theta\cos\theta,$$

we have $\sin\theta = \dfrac{\frac{12}{13}}{2\frac{2}{\sqrt{13}}} = \frac{3}{\sqrt{13}}$. Thus the two square roots of $a$ are $\pm\sqrt{13}\left(\frac{2}{\sqrt{13}} + \frac{3}{\sqrt{13}}i\right) = \pm(2 + 3i)$, as required.

**Remark.** Euler was the first to observe that the complex numbers provided a link between trigonometric and exponential functions. In particular, you may recall from your calculus course the following Taylor series expansions for $e^x$, $\sin x$, and $\cos x$:

$$e^x = \sum_{j=0}^{\infty} \frac{x^j}{j!} = 1 + x + \frac{x^2}{2!} + \frac{x^3}{3!} + \frac{x^4}{4!} + \cdots$$

$$\sin x = \sum_{j=0}^{\infty} (-1)^j \frac{x^{2j+1}}{(2j+1)!} = x - \frac{x^3}{3!} + \frac{x^5}{5!} - \cdots$$

$$\cos x = \sum_{j=0}^{\infty} (-1)^j \frac{x^{2j}}{(2j)!} = 1 - \frac{x^2}{2!} + \frac{x^4}{4!} - \cdots.$$

As if by magic, when we substitute $x = i\theta$ into the Taylor series for $e^x$, we find

$$e^{i\theta} = \sum_{j=0}^{\infty} \frac{(i\theta)^j}{j!} = (1 - \frac{\theta^2}{2!} + \frac{\theta^4}{4!} - \cdots) + i(\theta - \frac{\theta^3}{3!} + \frac{\theta^5}{5!} - \cdots) = \cos\theta + i\sin\theta.$$

Thus the addition formulas for sin and cos are consequences of the exponential law $e^{a+b} = e^a e^b$ (see Exercise 16).

## EXERCISES 2.3

1. Show that $\mathbb{C}$ is not an ordered field. (Hint: see Lemma 1.1(v).)

2. Recall that the conjugate of the complex number $z = a + bi$ is defined to be $\bar{z} = a - bi$. Prove the following properties of the conjugate:
   a. $\overline{z + w} = \bar{z} + \bar{w}$
   b. $\overline{zw} = \bar{z}\,\bar{w}$
   c. $\bar{z} = z \iff z \in \mathbb{R}$  and  $\bar{z} = -z \iff iz \in \mathbb{R}$.
   d. If $z = r(\cos\theta + i\sin\theta)$, then $\bar{z} = r(\cos\theta - i\sin\theta)$.

3. Recall that the modulus of the complex number $z = a + bi$ is defined to be $|z| = \sqrt{a^2 + b^2}$. Prove the following properties of the modulus:
   a. $|zw| = |z||w|$
   b. $|\bar{z}| = |z|$
   c. $|z|^2 = z\bar{z}$
   d. $|z + w| \le |z| + |w|$ (This is called the triangle inequality; why?)

4.  Let $\alpha = a + bi \in \mathbb{C}$. Express $(z - \alpha)(z - \overline{\alpha})$ in terms of $a = \mathcal{R}e\ \alpha$ and $b = \mathcal{I}m\ \alpha$. Explain why the coefficients of this quadratic polynomial are real.

5.  Prove deMoivre's formula (Corollary 3.3). (Hint: Use Proposition 3.2 and induction on $n$ to establish the formula for positive integers $n$. For negative integers $n$, use Corollary 3.4.)

6.  Use Corollary 3.3 to express the following in terms of $\sin\theta$ and $\cos\theta$ (the binomial theorem may prove helpful):
    a.  $\sin 3\theta$
    b.  $\cos 3\theta$
    c.  $\sin 4\theta$
    d.  $\cos 4\theta$
    e.  $\sin 5\theta$

7.  Prove that the $n^{\text{th}}$ roots of unity are the vertices of a regular $n$-gon inscribed in the unit circle $\{|z| = 1\}$.

8.  Express each of the following complex numbers in polar form, and plot them in the complex plane:
    a.  $1 - i$
    b.  $1 + \sqrt{3}i$
    c.  $2 - 3i$
    d.  $-4 + 3i$
    e.  $-10$
    f.  $-3i$

9.  Find the required roots. Express your answers in the (exact) form $z = a + bi$ without trigonometric functions whenever possible, and plot them in the complex plane.
    a.  the square roots of $-3 + 4i$
    b.  the square roots of $1 - 2\sqrt{2}i$
    c.  the cube roots of $2 - 2i$
    d.  the cube roots of $27i$
    e.  the fourth roots of $2$
    f.  the fourth roots of $-6 - 2\sqrt{3}i$
    g.  the sixth roots of $-3i$

10. a.  Use the answer to Exercise 6d. to express $\cos 4\theta$ as a polynomial in only $\cos\theta$.

b.   Use your answer to a. to prove that $\cos(\pi/8) = \frac{\sqrt{2+\sqrt{2}}}{2}$. (Be sure to remove the $\pm$ ambiguity in your answer.)

11.  Express the following $n^{\text{th}}$ roots of unity in the form $a + bi$ (also see Exercise 2.4.10):

   a.   $n = 8$
   b.   $n = 12$
   c.   $n = 16$    (Answer: $e^{\pi i/8} = \frac{1}{2}(\sqrt{2 + \sqrt{2}} + i\sqrt{2 - \sqrt{2}})$)

12.  Let $c \in \mathbb{C}$ and let $b \in \mathbb{R}$.

   a.   Show that $cz + \overline{c}\overline{z} = b$ is the equation of a line in $\mathbb{C}$.
   b.   For which values of $b$ and $c$ is $|z|^2 + c\overline{z} + \overline{c}z = b$ the equation of a circle in $\mathbb{C}$?

13.  We say a complex number $z$ is a **primitive $n^{\text{th}}$ root of unity** if $z^n = 1$ but $z^m \neq 1$ for $0 < m < n$.

   a.   Show that the number $\omega$ defined in Proposition 3.5 is always a primitive $n^{\text{th}}$ root of unity.
   b.   Show that $\omega^k$ is a primitive $n^{\text{th}}$ root of unity if and only if $\gcd(k, n) = 1$.
   c.   Show that if $z$ is any primitive $n^{\text{th}}$ root of unity, then $1$, $z$, $z^2$, ..., $z^{n-1}$ are distinct and comprise all the $n^{\text{th}}$ roots of unity.

14.  Using the approach of Example 9, show that the square roots of the complex number $a = u + iv$ are

$$\pm \frac{1}{\sqrt{2}} \left( \sqrt{\sqrt{u^2 + v^2} + u} + i\sqrt{\sqrt{u^2 + v^2} - u} \right).$$

15.  Use roots of unity to show that

   a.   $x^6 - 1 = (x - 1)(x + 1)(x^2 - x + 1)(x^2 + x + 1)$
   b.   $x^4 + 1 = (x^2 + \sqrt{2}x + 1)(x^2 - \sqrt{2}x + 1)$

   (Don't just multiply out the polynomials on the right-hand side!)

16.  Verify that from $e^{a+b} = e^a e^b$ and $e^{i\theta} = \cos\theta + i\sin\theta$ you may infer the formulas $\cos(\theta + \phi) = \cos\theta\cos\phi - \sin\theta\sin\phi$ and $\sin(\theta + \phi) = \sin\theta\cos\phi + \cos\theta\sin\phi$.

17.  Use Euler's formula $e^{i\theta} = \cos\theta + i\sin\theta$ to derive the antidifferentiation formula for $\int e^{ax}(\cos bx\, dx + i\sin bx)\, dx$. (Isn't this a lot easier than integrating by parts twice?)

18.  Let $\mathbb{Z}[i] = \{a + bi : a, b \in \mathbb{Z}\} \subset \mathbb{C}$. Show that $\mathbb{Z}[i]$ is an integral domain, and find its field of quotients. What are the units in $\mathbb{Z}[i]$?

19.  Let $\mathbb{Q}[\sqrt{-5}] = \{a + b\sqrt{-5} : a, b \in \mathbb{Q}\} \subset \mathbb{C}$. Show $\mathbb{Q}[\sqrt{-5}]$ is a field.

20.  Let $K \subset \mathbb{C}$ be a field. Prove that $\mathbb{Q} \subset K$. (Hint: cf. Exercise 2.2.12.)

21.  Let $1, \omega, \omega^2, \ldots, \omega^{n-1}$ be the $n^{\text{th}}$ roots of unity.
     a.   Show that the conjugate of any $n^{\text{th}}$ root of unity is another $n^{\text{th}}$ root of unity. In particular, express $\overline{\omega^j}$ in the form $\omega^k$ for the appropriate $k$.
     b.   Find the product of the $n^{\text{th}}$ roots of unity; this is written as
     $$\prod_{k=0}^{n-1} \omega^k.$$
     c.   Find the sum $\sum_{k=0}^{n-1} \omega^k$ of the $n^{\text{th}}$ roots of unity. (Hint: This is a geometric series.)

22.  A precalculus student presents you with the following:
     $$-1 = i \cdot i = \sqrt{-1}\sqrt{-1} = \sqrt{1} = 1.$$
     Is this correct? If not, how do you explain the error?

23.  Let $1, \omega, \omega^2, \ldots, \omega^{n-1}$ be the $n^{\text{th}}$ roots of unity, and let $z \in \mathbb{C}$ be arbitrary. Show that
     $$\sum_{k=0}^{n-1} |z - \omega^k|^2 = n\left(|z|^2 + 1\right).$$
     (Hint: cf. Exercise 21.)

## 4. The Quadratic and Cubic Formulas

We were led to our construction of the complex numbers in order to solve the quadratic equation $x^2 + 1 = 0$. The beauty of the complex number system is that now every quadratic equation can be solved.

**Theorem 4.1** (The Quadratic Formula). *Let $a, b, c \in \mathbb{C}$, $a \neq 0$. Then the quadratic equation $az^2 + bz + c = 0$ has the solutions*
$$z = \frac{-b \pm \sqrt{b^2 - 4ac}}{2a}.$$

***Proof.*** The proof is the usual one, by completing the square.

$$az^2 + bz + c = a(z + \frac{b}{2a})^2 + (c - \frac{b^2}{4a}) = 0$$

$$\Longleftrightarrow \quad a(z + \frac{b}{2a})^2 = -(c - \frac{b^2}{4a}) = \frac{b^2 - 4ac}{4a}$$

$$\Longleftrightarrow \quad (z + \frac{b}{2a})^2 = \frac{b^2 - 4ac}{4a^2}$$

$$\Longleftrightarrow \quad z + \frac{b}{2a} = \pm\frac{\sqrt{b^2 - 4ac}}{2a}$$

$$\Longleftrightarrow \quad z = \frac{-b \pm \sqrt{b^2 - 4ac}}{2a},$$

as required.  □

**Remark.** Here by $\sqrt{w}$ we mean either of the two complex square roots of the number $w \in \mathbb{C}$. Note that by Corollary 3.6, the two solutions to $z^2 = w$ differ by a factor of $-1$ (cf. Exercise 2.3.13). (Although we all agree that when $x$ is a positive *real* number, $\sqrt{x}$ denotes the positive square root of $x$, there is no comparable convention for general complex numbers.)

**Corollary 4.2.** *When $a, b, c \in \mathbb{R}$, the equation $ax^2 + bx + c = 0$ has real roots $\Longleftrightarrow b^2 - 4ac \geq 0$.*  □

More amazingly yet, even all cubic equations can be solved in $\mathbb{C}$. Next we present the so-called Cardano solution of the cubic equation. We first observe that every cubic equation $z^3 + az^2 + bz + c = 0$ can be reduced to the form $z^3 + pz + q = 0$, merely by substituting $z = z' - \frac{a}{3}$ (see Exercise 3).

**Theorem 4.3.** *Suppose $p \neq 0$. Given the cubic equation*

$$(*) \qquad\qquad z^3 + pz + q = 0,$$

*the three roots are given by $z = \sqrt[3]{A} - \frac{p}{3\sqrt[3]{A}}$ for the three possible cube roots of $A = -\frac{q}{2} + \sqrt{\frac{q^2}{4} + \frac{p^3}{27}}$ (with either choice of square root).*

***Proof.*** Substituting $z = v - \frac{p}{3v}$ in the equation $(*)$, we obtain the equation $v^3 - \frac{p^3}{27v^3} + q = 0$. Magically this equation is quadratic in $v^3$; and so, using Theorem 4.1, we may solve for $v^3$:

$$(\dagger) \qquad\qquad v^3 = -\frac{q}{2} \pm \sqrt{\frac{q^2}{4} + \frac{p^3}{27}}.$$

Now, let $A = -\frac{q}{2} + \sqrt{\frac{q^2}{4} + \frac{p^3}{27}}$ and $B = -\frac{q}{2} - \sqrt{\frac{q^2}{4} + \frac{p^3}{27}}$. There are *a priori* six solutions $v$ to (†), and hence ostensibly six solutions $z$ to (∗). But note that if $v$ is a cube root of $A$, then $-\frac{p}{3v}$ is a cube root of $B$, for $(-\frac{p}{3v})^3 = -\frac{p^3}{27A}$ and $AB = -\frac{p^3}{27}$. Thus, the two lists of the values of $v - \frac{p}{3v}$ for $v^3 = A$ and $v^3 = B$, respectively, agree, and there are precisely three roots $z$ of (∗). (Phrased somewhat differently, if we let $v_1, v_2, v_3$ be the three roots of $v^3 = A$ and set $u_j = -\frac{p}{3v_j}$, $j = 1, 2, 3$, then $u_1, u_2, u_3$ are the three roots of $v^3 = B$. The roots of (∗) are $v_j + u_j$, $j = 1, 2, 3$.) □

**Example 1.** Solve $z^3 - 3z + 1 = 0$. In this case we find that $v^3 = -\frac{1}{2} \pm \frac{\sqrt{3}}{2}i$; let $A = -\frac{1}{2} + \frac{\sqrt{3}}{2}i$ and observe that $A$ is the primitive cube root of unity $\cos\frac{2\pi}{3} + i\sin\frac{2\pi}{3} = e^{2\pi i/3}$. Letting $\zeta = e^{2\pi i/9}$, we see that $\zeta, \zeta^4$, and $\zeta^7$ are the cube roots of $A$. Substituting $p = -3$ in the statement of Theorem 4.3, the solutions $z$ to our original equation are therefore $\zeta + 1/\zeta$, $\zeta^4 + 1/\zeta^4$, and $\zeta^7 + 1/\zeta^7$. Since $|\zeta| = 1$, $1/\zeta = \bar{\zeta}$, and so $\zeta + 1/\zeta = \zeta + \bar{\zeta} = 2\cos\frac{2\pi}{9} = 2\cos 40°$. Similarly, the other two roots are $2\cos 80°$ and $2\cos 160° = -2\cos 20°$. (Note, for future reference, that since $2\cos 80° = (2\cos 40°)^2 - 2$ and $2\cos 160° = (2\cos 80°)^2 - 2$, the latter two roots are polynomials in the first.)

**Remark.** This cubic equation has three real roots (a fact which the reader could check directly by graphing the function); yet to find them explicitly required complex numbers. Indeed, as we shall see in Section 6 of Chapter 7, there can be no formula for the roots involving only real numbers. See also Example 3 below.

**Example 2.** Solve $x^3 - 6x^2 + 6x - 2 = 0$. First we eliminate the quadratic term by substituting $z = x - 2$ and obtain the new equation $z^3 - 6z - 6 = 0$. Thus, we have $A = 4$, and the three cube roots of $A$ are $\sqrt[3]{4}, \sqrt[3]{4}\omega, \sqrt[3]{4}\omega^2$, where $\omega$ is a primitive cube root of unity. So the solutions $z$ of $z^3 - 6z - 6 = 0$ are $v + 2/v$, where $v = \sqrt[3]{4}, \sqrt[3]{4}\omega$, and $\sqrt[3]{4}\omega^2$. That is,

$$z = \sqrt[3]{4} + \sqrt[3]{2}, \quad \sqrt[3]{4}\omega + \sqrt[3]{2}\omega^2, \quad \text{and } \sqrt[3]{4}\omega^2 + \sqrt[3]{2}\omega.$$

Then we recover the solutions $x$ of the original equation: $2 + \sqrt[3]{4} + \sqrt[3]{2}$, $2 + \sqrt[3]{4}\omega + \sqrt[3]{2}\omega^2$, and $2 + \sqrt[3]{4}\omega^2 + \sqrt[3]{2}\omega$.

**Example 3.** Solve $z^3 - 7z + 6 = 0$. We substitute $z = v + \frac{7}{3v}$, and obtain the equation $v^3 = A = -3 + \frac{10}{3\sqrt{3}}i$. Writing $A$ in polar form, we get

$$v^3 = \frac{\sqrt{343}}{3\sqrt{3}}\left(-\frac{9\sqrt{3}}{\sqrt{343}} + \frac{10}{\sqrt{343}}i\right) = \left(\frac{7}{3}\right)^{3/2}\left(-\frac{9\sqrt{3}}{\sqrt{343}} + \frac{10}{\sqrt{343}}i\right).$$

If $v = r(\cos\theta + i\sin\theta)$, then, by deMoivre's Theorem, Corollary 3.3, $v^3 = r^3(\cos 3\theta + i\sin 3\theta)$; so in this case we can read off $r = \sqrt{\frac{7}{3}}$, $\cos 3\theta = -\frac{9\sqrt{3}}{\sqrt{343}}$, and $\sin 3\theta = \frac{10}{\sqrt{343}}$. Benefiting from divine inspiration, we try $\cos\theta = \sqrt{\frac{3}{7}}$ and $\sin\theta = \sqrt{1 - \frac{3}{7}} = \frac{2}{\sqrt{7}}$; we leave it to the reader to check that these in fact work. This gives us our cube root:

$$v_1 = \sqrt{\tfrac{7}{3}}\left(\sqrt{\tfrac{3}{7}} + \tfrac{2}{\sqrt{7}}i\right) = 1 + \tfrac{2}{\sqrt{3}}i.$$

We obtain the other roots by multiplying $v_1$ by the cube roots of unity $\omega$ and $\overline{\omega}$:

$$v_2 = \left(1 + \tfrac{2}{\sqrt{3}}i\right)\left(-\tfrac{1}{2} + \tfrac{\sqrt{3}}{2}i\right) = -\tfrac{3}{2} + \tfrac{1}{2\sqrt{3}}i$$

$$v_3 = \left(1 + \tfrac{2}{\sqrt{3}}i\right)\left(-\tfrac{1}{2} - \tfrac{\sqrt{3}}{2}i\right) = \tfrac{1}{2} + \tfrac{-5}{2\sqrt{3}}i$$

Finally, the roots $z_1, z_2, z_3$ of our original equation are:

$$z_1 = v_1 + \tfrac{7}{3v_1} = \left(1 + \tfrac{2}{\sqrt{3}}i\right) + \tfrac{7}{3}\left(\frac{1 - \tfrac{2}{\sqrt{3}}i}{\tfrac{7}{3}}\right) = 2,$$

$$z_2 = v_2 + \tfrac{7}{3v_2} = \left(-\tfrac{3}{2} + \tfrac{1}{2\sqrt{3}}i\right) + \tfrac{7}{3}\left(\frac{-\tfrac{3}{2} - \tfrac{1}{2\sqrt{3}}i}{\tfrac{7}{3}}\right) = -3, \text{ and}$$

$$z_3 = v_3 + \tfrac{7}{3v_3} = \left(\tfrac{1}{2} + \tfrac{-5}{2\sqrt{3}}i\right) + \tfrac{7}{3}\left(\frac{\tfrac{1}{2} + \tfrac{5}{2\sqrt{3}}i}{\tfrac{7}{3}}\right) = 1.$$

Once again, look at the amount of effort expended to obtain these amazingly simple roots (which one might have unearthed easily by applying Proposition 3.1 of Chapter 3)!

We finish this discussion of the cubic equation with a brief word about the number of real roots of the cubic $x^3 + px + q = 0$ when $p$ and $q$ are real numbers. We define $\Delta = q^2/4 + p^3/27$; $-108\Delta$ is called the **discriminant** of the cubic polynomial $x^3 + px + q$. (The rationale for the weird factor $-108$ will be explained in Section 6 of Chapter 7.) The first observation is that the equation will have two equal roots when $\Delta = 0$ (see Exercise 5). Now we make the following claim.

**Proposition 4.4.** *When $p, q \in \mathbb{R}$, the equation $x^3 + px + q = 0$ has only real roots if and only if $\Delta \le 0$.*

*Proof.* Although this result can be proved by purely algebraic methods (see Exercise 7.6.7), we give a sketch of a proof using basic differential calculus. By Rolle's Theorem, the function $f(x) =$

$x^3 + px + q$ has three real roots (counting multiplicities) only if its derivative $f'(x) = 3x^2 + p$ has two real roots (counting multiplicities). Note that this means that $p \le 0$.

The two critical points, then, are $\pm\sqrt{-p/3}$ (see Figure 1). Let $x_1 = -\sqrt{-p/3}$ and $x_2 = \sqrt{-p/3}$; then $f(x)$ has three real roots if and only if $f(x_1) \ge 0$ and $f(x_2) \le 0$. Now $f(x_1) \ge 0 \iff \sqrt{-p^3/27} + q/2 \ge 0$ and $f(x_2) \le 0 \iff \sqrt{-p^3/27} - q/2 \ge 0$. Thus, $f(x)$ has three real roots if and only if $(-p/3)^{3/2} \ge |q/2| \iff \Delta \le 0$. $\quad\square$

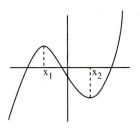

FIGURE 1

**Remark.** There is a similar formula for the solutions of a quartic (fourth-degree) equation; it involves square roots and cube roots, and was found by Italian mathematicians during the sixteenth century. One might hope to find similar formulas—involving roots— for the solutions of polynomial equations of degree five and greater. This problem plagued mathematicians for centuries until Abel and Galois proved in the early nineteenth century that no such formulas could exist. We will discuss the ideas behind their proof later in Section 6 of Chapter 7.

### EXERCISES 2.4

1.  Solve the following equations using Theorem 4.1:
    a.  $z^2 - (6 + i)z + (8 + 2i) = 0$
    b.  $z^2 - iz + 2 = 0$
    c.  $z^4 + 2z^2 + 4 = 0$
    d.  $(x^2 + 3x + 4)(x^2 + 3x + 5) = 6$
    e.  $2z^4 - 3z^3 + 4z^2 - 3z + 2 = 0$ (Hint: Divide by $z^2$.)
    f.  $16x^4 - 20x^2 + 5 = 0$

2.  a.  A rectangle with area 2 sq. units is inscribed in a 3-4-5 right triangle, as pictured in Figure 2. Find its dimensions.

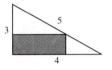

FIGURE 2

   b.  The corners of a square with side 2 in. are cut off in such a way as to form a regular octagon. What is the length of a side of the octagon?

   c.  A jogger ran 10 miles in a certain amount of time. To cover the same distance in 1/4 hour less, she must run 2 m.p.h. faster. How fast did she run, and for how long?

3.  Show that the substitution of $z = z' - \frac{a}{3}$ converts the cubic equation $z^3 + az^2 + bz + c = 0$ to one of the form $z'^3 + pz' + q = 0$.

4.  Note that in Example 1 above, $B = -\frac{1}{2} - \frac{\sqrt{3}}{2}i$. Find the three cube roots of $B$, and verify that the resulting solutions $z$ to the original cubic equation are the ones already obtained.

5.  a.  Show that the quadratic equation $az^2 + bz + c = 0$ has two equal roots $\iff b^2 - 4ac = 0$.

   b.  Show that the cubic equation $z^3 + pz + q = 0$ has two equal roots $\iff \Delta = q^2/4 + p^3/27 = 0$.
   (Hint: By Exercise 3.1.15, the number $r$ is a double root of $f(z)$ if and only if $f(r) = f'(r) = 0$.)

6.  Solve the following cubic equations:
   a.  $z^3 - 9z - 28 = 0$   (Answer: $4, -2 \pm \sqrt{3}i$)
   b.  $z^3 - 9z^2 + 9z - 8 = 0$   (Answer: $8, \frac{1 \pm \sqrt{3}i}{2}$)
   c.  $z^3 - 3z - 1 = 0$   (Answer: $2\cos 20°, 2\cos 140°, 2\cos 260°$)
   Compute $\Delta$ and verify Proposition 4.4 in each case.

7.  a.  A prism with a square base is inscribed in a sphere with diameter $3\sqrt{3}$ units. If the volume of the prism is 27 cubic units, what is its altitude?

   b.  The altitude of a right circular cone is 6 and the radius of its base is 4. A right circular cylinder is inscribed in this cone and has volume 4/9 that of the cone. Find the altitude of the cylinder.

8. Let $\alpha \in \mathbb{C}$ be a root of $z^2 + z + 1 = 0$. Show that $\mathbb{Q}[\alpha] = \{a + b\alpha : a, b \in \mathbb{Q}\} \subset \mathbb{C}$ is a field.

9. Make addition and multiplication tables for
$$\mathbb{Z}_2[\alpha] = \{\overline{0}, \overline{1}, \alpha, \alpha + \overline{1}\},$$
where by definition arithmetic is done in $\mathbb{Z}_2$ according to each of the following rules:
  a. $\alpha^2 = \alpha + \overline{1}$;
  b. $\alpha^2 = \overline{1}$;
Decide in each case whether or not $\mathbb{Z}_2[\alpha]$ is a field.

10. Show that the primitive fifth root of unity $e^{2\pi i/5} = \frac{\sqrt{5}-1}{4} + i\sqrt{\frac{5+\sqrt{5}}{8}}$. (Hint: see Exercises 1f. and 2.3.6e.)

11. Find all cubic polynomials $f(x)$ so that all the *roots* of $f(x)$ and all the *critical points* of $f(x)$ are integers.

## 5. The Isometries of $\mathbb{R}$ and $\mathbb{C}$

There will be a great deal of discussion in this course of symmetries of geometric objects—where algebra and geometry intermingle. We begin with the rather less complicated discussion of the "symmetries" of the real line and complex plane, i.e., the motions that preserve distance.

Since we measure distance using the absolute value, we begin by making the following definition.

**Definition.** An **isometry** of $\mathbb{R}$ is a distance-preserving function from $\mathbb{R}$ to $\mathbb{R}$, i.e., a function $f: \mathbb{R} \to \mathbb{R}$ so that for all $x, y \in \mathbb{R}$,

(†) $$|f(x) - f(y)| = |x - y|.$$

(Recall the definition of the absolute value:

$$|x| = \begin{cases} x, & \text{if } x \geq 0 \\ -x, & \text{if } x < 0 \end{cases}.$$

In particular, $|x| = |-x|$ represents the distance along the real axis from 0 to $\pm x$, and $|x - y|$ is the distance between $x$ and $y$.)

We begin with some examples of isometries. Of course, the identity function $\iota(x) = x$ is an isometry, since each point is mapped precisely back to itself. The function $\rho(x) = -x$ is also an isometry, since the distance from $-x$ to $-y$ is the same as the distance from

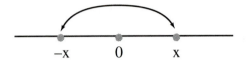

FIGURE 1

$x$ to $y$. Geometrically, $\rho$ flips $\mathbb{R}$ across the origin and is called a **reflection**. Another simple example is the function $\tau(x) = x + 1$, which pushes all the points one unit to the right. $\tau$ is called a **translation** of $\mathbb{R}$. Our goal now is to find all possible isometries. We start

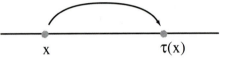

FIGURE 2

by finding all isometries $f: \mathbb{R} \to \mathbb{R}$ that leave 0 fixed.

**Lemma 5.1.** *Suppose $f$ is an isometry of $\mathbb{R}$ and $f(0) = 0$. Then either $f(x) = x$ for all $x$ or $f(x) = -x$ for all $x$.*

**Proof.** First, $|f(x)| = |f(x) - f(0)| = |x - 0| = |x|$; so for each individual real number $x$, $f(x)$ equals either $x$ or $-x$. Now suppose there were nonzero numbers $x$ and $y$ with $f(x) = x$ and $f(y) = -y$. From (†) it follows that $|x + y| = |x - y|$. We claim that this implies $x = 0$ or $y = 0$, establishing the fact that $f(x)$ either equals $x$ for all $x$ or equals $-x$ for all $x$. The claim follows either from an algebraic verification (see Exercise 1) or from the following geometric argument: $|x + y| = |x - y|$ implies that $x$ is equidistant from $y$ and $-y$; but unless $y = -y$, this means that $x = 0$. $\square$

**Theorem 5.2.** *Suppose $f$ is an isometry of $\mathbb{R}$. Then there is $a \in \mathbb{R}$ so that either $f(x) = x + a$ (for all $x$) or $f(x) = -x + a$ (for all $x$).*

**Proof.** Suppose $f(0) = a$. Then the function $g(x) = f(x) - a$ is also an isometry (since $|g(x) - g(y)| = |(f(x) - a) - (f(y) - a)| = |f(x) - f(y)| = |x - y|$), and $g(0) = f(0) - a = 0$. Applying Lemma 5.1 to $g$, we infer that either $g(x) = x$ for all $x \in \mathbb{R}$ or $g(x) = -x$

for all $x \in \mathbb{R}$. The result follows by adding $a$ to both sides of the respective equations. $\square$

**Remark.** We now see that our original two examples are typical. If $f(x) = x + a$, then $f$ is a translation by $a$. If $f(x) = -x + a$, then observe that $f(\frac{a}{2}) = \frac{a}{2}$. Let $\tau(x) = x + \frac{a}{2}$ be the translation by $a/2$ units to the right, and let $\rho(x) = -x$ be the reflection of $\mathbb{R}$ about the origin. Then we can write

$$f(x) = -x + a = \tfrac{a}{2} - (x - \tfrac{a}{2}) = \left(\tau \circ \rho \circ \tau^{-1}\right)(x);$$

that is, $f$ is the composition of a translation by $-a/2$ units, a reflection across the origin, and a translation by $a/2$ units. By doing this interesting composition of functions, we have "transported" the fixed point of the reflection from the origin to the point $a/2$, and we infer that $f$ is a reflection of $\mathbb{R}$ about the point $a/2$. We shall encounter this idea many times in the future.

We say $y$ is a **fixed point** of $f$ if $f(y) = y$. Our classification of the isometries of $\mathbb{R}$ shows that there are three possibilities:

| fixed point set | geometric description of isometry |
|:---:|:---:|
| none | translation |
| point | reflection |
| all | identity |

We next turn to the isometries (or "symmetries") of the complex plane $\mathbb{C}$. The definition that follows is completely analogous.

**Definition.** An **isometry** of $\mathbb{C}$ is a distance-preserving function from $\mathbb{C}$ to $\mathbb{C}$, i.e., a function $f \colon \mathbb{C} \to \mathbb{C}$ so that for all $z, w \in \mathbb{C}$,

$$(\ddagger) \qquad\qquad |f(z) - f(w)| = |z - w|.$$

This is akin to the notion of a congruence in Euclidean geometry, insofar as isometries preserve angles as well as distances. (This follows on general grounds, as we shall see when we treat Euclidean motions of three-dimensional space in Section 4 of Chapter 7.)

We begin with three examples. As in the real case, we have translations $\tau(z) = z + c$, $c \in \mathbb{C}$. Conjugation, $\gamma(z) = \bar{z}$, is also an isometry, since $|\gamma(z) - \gamma(w)| = |\bar{z} - \bar{w}| = |\overline{z - w}| = |z - w|$ (see Exercises 2.3.2 and 2.3.3). Lastly, rotation $R_\theta$ about the origin through an angle $\theta$ is an isometry: let $\zeta = e^{i\theta} = \cos\theta + i\sin\theta$; then

$R_\theta(z) = \zeta z$ gives the desired rotation. But now $|R_\theta(z) - R_\theta(w)| = |\zeta z - \zeta w| = |\zeta(z - w)| = |\zeta||z - w| = |z - w|$.

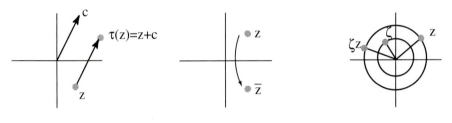

FIGURE 3

**Lemma 5.3.** *Suppose $f: \mathbb{C} \to \mathbb{C}$ is an isometry. Then $f$ maps a circle of radius $r$ centered at $a$ to a circle of the same radius centered at $f(a)$.*

**Proof.** This is immediate from (‡). See Exercise 3.   □

We begin with the analogue of Lemma 5.1.

**Lemma 5.4.** *Suppose $f: \mathbb{C} \to \mathbb{C}$ is an isometry satisfying $f(0) = 0$ and $f(1) = 1$ (i.e., 0 and 1 are fixed points). Then either $f(z) = z$ for all $z \in \mathbb{C}$ or $f(z) = \overline{z}$ for all $z \in \mathbb{C}$.*

**Proof.** Since $f(0) = 0$ and $f(1) = 1$, $f$ preserves the circles centered at 0 and 1 (by Lemma 5.3). Given an arbitrary point $z \in \mathbb{C}$, it lies on a unique circle centered at 0 and on a unique circle centered at 1. Therefore $f(z)$ lies on both these circles as well; since the circles intersect in only the points $z$ and $\overline{z}$ (see Figure 4), it follows that $f(z) = z$ or $f(z) = \overline{z}$. Suppose now that $z, w \in \mathbb{C}$ and $f(z) = z \neq \overline{z}$,

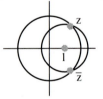

FIGURE 4

whereas $f(w) = \overline{w} \neq w$. Then by (‡), $|z - \overline{w}| = |z - w|$, so that $z$ is equidistant from $w$ and $\overline{w}$. This can happen in two ways: either $w = \overline{w}$, which is ruled out; or $z$ lies on the perpendicular bisector of the line joining $w$ and $\overline{w}$—the real axis—and this is likewise not allowed.   □

**Theorem 5.5.** *Let $f: \mathbb{C} \to \mathbb{C}$ be an isometry. Then there are $c \in \mathbb{C}$ and $\zeta = \cos \theta + i \sin \theta$ so that either $f(z) = \zeta z + c$ for all $z \in \mathbb{C}$ or $f(z) = \zeta \bar{z} + c$ for all $z \in \mathbb{C}$.*

**Proof.** Given an isometry $f$, let $f(0) = c$, and consider $g(z) = f(z) - c$; then 0 is a fixed point of $g$. Now since $g$ is an isometry, the circle of radius 1 centered at 0 is preserved by $g$; let $g(1) = \zeta$, and finally define $h(z) = \zeta^{-1} g(z)$. Also, $h$ is an isometry fixing both 0 and 1; so, by Lemma 5.4, $h(z) = z$ for all $z$ or $h(z) = \bar{z}$ for all $z$. In either event, we now retrace our tracks to solve for $f$. This concludes the proof. $\square$

We'd like to interpret the possible isometries of $\mathbb{C}$ a bit more geometrically. Let's start with functions of the form $f(z) = \zeta z + c$, which are called **proper isometries**. If $\zeta = 1$, this is simply a translation, and there are no fixed points. But if $\zeta \neq 1$, we see that $p = c/(1 - \zeta)$ is a fixed point and that $f$ is a rotation through angle $\theta$ about this point (see Exercise 5b.).

To understand isometries of the type $f(z) = \zeta \bar{z} + c$, called **improper isometries**, we first assume $c = 0$. When $\zeta = 1$, the function $y(z) = \bar{z}$ is a reflection in the real axis. In general, set $\zeta = e^{i\theta}$ and $\omega = e^{i\theta/2}$, so that $\zeta = \omega^2$. Now

$$f(z) = \zeta \bar{z} = \omega^2 \bar{z} = \omega(\overline{\bar{\omega} z}) = (R_{\theta/2} \circ y \circ R_{-\theta/2})(z).$$

This formula shows that $f$ is a reflection of $\mathbb{C}$ in the line making angle $\theta/2$ with the positive real axis (note, for example, that a point lying on this line rotates onto the real axis, is left fixed by $y$, and then rotates back to its original location).

FIGURE 5

**Example.** Let $f(z) = i\bar{z} + \frac{1+i}{2}$. We know that $z \leadsto i\bar{z}$ is reflection in the line through the origin with slope 1. On the other hand, the point $\frac{1+i}{2}$ lies on this line, so $f(z)$ is the reflection of $z$ in this line, followed by a translation parallel to the line. Such an isometry is called a **glide reflection**. See the diagram at the right in Figure 5.

**Proposition 5.6.** *Suppose $f(z) = \zeta \bar{z} + c$. Then $f$ is of two possible geometric types:*

(1) *If $\zeta \bar{c} + c = 0$, then $f$ is a reflection in the line passing through $z = c/2$ with slope $\tan \frac{\theta}{2}$.*

(2) *If $\zeta \bar{c} + c \neq 0$, then $f$ is a glide reflection with this same axis, with translation vector $\frac{1}{2}(\zeta \bar{c} + c)$.*

**Proof.** The main issue is to decide when $f$ has a fixed point; if it does, $f$ must be a reflection in a line passing through that fixed point. So we begin by showing $f$ has a fixed point $\iff \zeta \bar{c} + c = 0$. Suppose $f(a) = a$ for some $a \in \mathbb{C}$; then $\zeta \bar{a} + c = a \implies \bar{\zeta} a + \bar{c} = \bar{a} \implies \zeta(\bar{\zeta} a + \bar{c}) + c = a \implies \zeta \bar{c} + c = 0$. On the other hand, if $\zeta \bar{c} + c = 0$, then it is easy to check that $f(c/2) = c/2$, so $c/2$ is a fixed point. But now we claim that every point on the line through $c/2$ with slope $\tan \frac{\theta}{2}$ is fixed by $f$. Let $\omega = \sqrt{\zeta}$, i.e., $\omega = \cos \frac{\theta}{2} + i \sin \frac{\theta}{2}$; then note that $\zeta \bar{\omega} = \omega^2 \bar{\omega} = \omega(\omega \bar{\omega}) = \omega$. Now for any $t \in \mathbb{R}$, we compute $f(\frac{c}{2} + t\omega) = \zeta(\overline{\frac{c}{2} + t\omega}) + c = \zeta \frac{\bar{c}}{2} + t\omega + c = -\frac{c}{2} + t\omega + c = \frac{c}{2} + t\omega$, as required. It follows that $f$ is reflection in this line (see Exercise 13).

From a more geometric standpoint, $\zeta \bar{c} + c = 0$ if and only if $c$ is perpendicular to $\sqrt{\zeta}$ (the vector whose real multiples give the axis of reflection). (See Exercise 4.) If $c$ is perpendicular to the line $\ell$ passing through the origin, reflecting in $\ell$ and then translating by $c$ is the same as reflecting through the line $\ell'$ parallel to $\ell$ and passing through the point $c/2$.

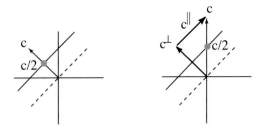

FIGURE 6

Now if $\zeta \bar{c} + c \neq 0$, then we wish to show $f$ is a glide reflection. The complex number $\zeta \bar{c} + c$ is a real multiple of $\omega$, as $\zeta \bar{c} + c = \omega(\omega \bar{c} + \bar{\omega} c) = \omega(\omega \bar{c} + \overline{\omega \bar{c}})$, and similarly the complex number $-\zeta \bar{c} +$

$c$ is an imaginary multiple of $\omega$. Since $c = \frac{1}{2}(\zeta\bar{c} + c) + \frac{1}{2}(-\zeta\bar{c} + c)$, we let $c^{\|} = \frac{1}{2}(\zeta\bar{c} + c)$ and $c^{\perp} = \frac{1}{2}(-\zeta\bar{c} + c)$, and we have expressed $c$ in the form $c = c^{\perp} + c^{\|}$, where $c^{\|}$ is parallel to $\sqrt{\zeta}$ and $c^{\perp}$ is perpendicular to $\sqrt{\zeta}$. By the first part of the proof, $g(z) = \zeta\bar{z} + c^{\perp}$ represents a reflection in the line parallel to $\sqrt{\zeta}$ passing through the midpoint of $c^{\perp}$—or, what is the same, passing through the midpoint of $c$. We then translate by $c^{\|}$, which is parallel to the axis of reflection. This establishes that $f$ is the desired glide reflection.   □

In summary, we have established the following theorem.

**Theorem 5.7.** *An isometry of* ℂ *is a translation, a rotation, a reflection, or a glide reflection.*   □

| fixed point set | proper isometry | improper isometry |
|:---:|:---:|:---:|
| none | translation | glide reflection |
| point | rotation through angle $\theta \neq 0$ | ———— |
| line | ———— | reflection |
| all | identity | ———— |

We will continue in this vein when we study the isometries of $\mathbb{R}^3$ in Section 4 of Chapter 7.

## EXERCISES 2.5

1. Give an algebraic verification that if $|x - y| = |x + y|$, $x, y \in \mathbb{R}$, then either $x = 0$ or $y = 0$. (Hint: Square both sides.)

2. Give an algebraic verification that if $|z - w| = |z - \overline{w}|$, $z, w \in \mathbb{C}$, then either $z \in \mathbb{R}$ or $w \in \mathbb{R}$. (Hint: Square both sides and show that $(z - \bar{z})(w - \overline{w}) = 0$.)

3. Prove Lemma 5.3.

4. Viewing $\mathbb{C} = \mathbb{R}^2$, we can identify the complex numbers $z = a + bi$ and $w = c + di$ with the vectors $(a, b)$ and $(c, d) \in \mathbb{R}^2$, respectively. Then we can form their dot product, $(a, b) \cdot (c, d) = ac + bd$.
   a. Prove that $\mathcal{R}e\,(z\overline{w}) = ac + bd$.
   b. Prove that $\zeta\bar{c} + c = 0 \iff c$ is orthogonal to $\sqrt{\zeta}$.

5.  Let $f(z) = \zeta z + c$, $\zeta = \cos \theta + i \sin \theta \neq 1$.
    a.  Show that $p = c/(1 - \zeta)$ is a fixed point of $f$.
    b.  Show that $f(z) - p = \zeta(z - p)$, and conclude that $f$ is a rotation of $\mathbb{C}$ about $p$ through an angle $\theta$.

6.  Show that the rotation of $\mathbb{C}$ about a point $p$ through angle $\theta$ is obtained by doing two successive reflections in lines passing through $p$. (Hint: If $p = 0$, let the first be conjugation.)

7.  Find formulas for the following isometries $f: \mathbb{C} \to \mathbb{C}$.
    a.  rotation about $2 + 3i$ through angle $\pi/3$
    b.  reflection in the line $5z + (3 - 4i)\bar{z} = -4 + 2i$
    c.  glide reflection with axis $z + \bar{z} = 2$ and translation vector $3i$

8.  Consider the triangle $\triangle_1$ with vertices $1$, $4 + 4i$, and $2 + 5i$, and the triangle $\triangle_2$ with vertices $3i$, $-5 + 2i$, and $-4$.
    a.  Check that the triangles are congruent.
    b.  Find an isometry $f: \mathbb{C} \to \mathbb{C}$ carrying $\triangle_1$ to $\triangle_2$. Is it unique?

9.  Given two congruent triangles in the plane, prove that there is an isometry of the plane carrying one to the other. Furthermore,
    a.  When will this isometry be unique?
    b.  Under what circumstances will the isometry be improper? proper?

10. Analyze the following isometries:
    a.  $f(z) = -iz + 2$
    b.  $f(z) = \bar{z} + 1$
    c.  $f(z) = i\bar{z} + 1$
    d.  $f(z) = (\frac{1+i}{\sqrt{2}})\bar{z} + (\sqrt{2} - 1 + i)$

11. Suppose $f: \mathbb{C} \to \mathbb{C}$ is an isometry. Show that for any $z, w \in \mathbb{C}$, $f$ carries the line joining $z$ and $w$ to the line joining $f(z)$ and $f(w)$. (Hint: For a geometric proof, let $y$ lie on the line joining $z$ and $w$, and consider circles centered at $z$ and $w$, respectively, and passing through $y$. For a more algebraic approach, when does equality hold in the triangle inequality (Exercise 2.3.3d.)?)

12. Use Exercise 11 to prove that isometries $f: \mathbb{C} \to \mathbb{C}$ preserve the (unsigned) measure of angles.

13. If $f$ is an isometry of $\mathbb{C}$ that fixes each point on a line, then prove that $f$ is either the identity or reflection in that line.

14.   Suppose $f: \mathbb{C} \to \mathbb{C}$ is an isometry and $|f(z) - z| \le 5$ for all $z \in \mathbb{C}$. Prove $f$ must be a translation or the identity map.

15.   List the possible isometries $f: \mathbb{C} \to \mathbb{C}$ satisfying
       (i)   $f(0) = 0$, and
       (ii)   $f$ maps the real axis to the real axis.
       Prove your list is complete. (Hint: What can $f(1)$ be?) Query: What is the case if we remove the requirement that $f(0) = 0$?

16.   a.   Prove that a translation is the composition of two reflections. (Hint: Consider reflections in parallel lines.)
       b.   Using the results of Theorem 5.7 and Exercise 6, prove that *every* isometry of $\mathbb{C}$ is the composition of (some number of) reflections.

17.   Prove that if $f: \mathbb{C} \to \mathbb{C}$ is an isometry, then $f$ is a bijection: i.e., $f$ is one-to-one and onto. (Hint: One-to-one is immediate from the definition. Here is an idea for a geometric proof that an isometry is onto: show that $f$ maps a circle $C$ of radius $r$ centered at 0 *onto* the circle of radius $r$ centered at $f(0)$ by considering in turn the circles centered at a particular point $P$ of $C$.)

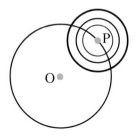

FIGURE 7

# CHAPTER 3

# Polynomials

We pursue our study of algebraic structures by turning to polynomials. As was the case with Euclid, Descartes developed a good deal of the structure of polynomials and so-called theory of equations in concert with his development of analytic geometry in *La Géométrie*, published in 1637. He was one of the first to come to terms with irrational, negative, and even complex numbers. He discovered a way to estimate the number of real roots of a real polynomial (now called Descartes' Rule of Signs), but a proof came only in the eighteenth century.

We explore next how the algebraic structures we studied for the integers are paralleled for polynomials, including the wealth of applications of the Euclidean algorithm. Here that algorithm is based on the "long division" of polynomials we all learn in high school algebra. The analogue of prime numbers is irreducible polynomials; we prove that any polynomial (with coefficients in a field) can be factored uniquely as a product of irreducible polynomials.

Gauss was the first to give a correct proof of the Fundamental Theorem of Algebra: Every nonconstant polynomial can be factored over the complex numbers as the product of linear polynomials. Given a polynomial $f(x)$ with rational coefficients, it is now natural to ask for the smallest field (containing the rational numbers and contained in the complex numbers) in which $f(x)$ can be factored as the product of linear polynomials; this is the splitting field of $f(x)$. In §3 we concentrate on tests for irreducibility of polynomials with integer coefficients, as this will be one of the crucial ingredients we need for a deeper study of polynomials and splitting fields.

## 1. The Euclidean Algorithm

Let $R$ be a commutative ring (e.g., $\mathbb{Z}$, $\mathbb{Z}_m$, or a field $F$ such as $\mathbb{Q}$, $\mathbb{R}$, $\mathbb{C}$, or $\mathbb{Z}_p$). Let $n$ be a nonnegative integer; a **polynomial** of **degree** $n$ with coefficients in $R$ is an expression of the form

$$f(x) = a_n x^n + a_{n-1} x^{n-1} + \cdots + a_1 x + a_0, \qquad a_j \in R,\ a_n \neq 0;$$

we write $\deg(f(x)) = n$. (An individual term of the form $a_j x^j$ is called a **monomial**.) We denote by $R[x]$ the set of all polynomials with coefficients in $R$. We refer to $a_j$ as the **coefficient** of $x^j$; and if $j$ is larger than the degree of $f(x)$, we say for convenience that the coefficient is automatically zero. We say two polynomials $f(x) = a_n x^n + a_{n-1} x^{n-1} + \cdots + a_1 x + a_0$ and $g(x) = b_m x^m + b_{m-1} x^{m-1} + \cdots + b_1 x + b_0$ are equal if all their corresponding coefficients are equal. By convention, the zero polynomial (all of whose coefficients are zero) has no degree. When $f(x)$ is a polynomial of degree $n$, we call $a_n$ its **leading coefficient**, and we say $f(x)$ is a **monic** polynomial if its leading coefficient is 1.

Given two polynomials $f(x) = a_n x^n + a_{n-1} x^{n-1} + \cdots + a_1 x + a_0$ and $g(x) = b_m x^m + b_{m-1} x^{m-1} + \cdots + b_1 x + b_0$, we define their sum and product as follows. For convenience, we suppose $n \geq m$ (and formally take $a_j = 0$ for $j > n$ and $b_\ell = 0$ for $\ell > m$).

$$f(x) + g(x) = \sum_{j=0}^{n} (a_j + b_j) x^j;$$

$$f(x) \cdot g(x) = \sum_{j=0}^{n+m} \left( \sum_{k=0}^{j} a_k b_{j-k} \right) x^j.$$

The latter is the usual formula obtained by imposing the distributive law and then "collecting terms."

**Example 1.** Let $R = \mathbb{Z}$, $f(x) = 3x^2 - x + 5$, and $g(x) = -x^3 + 4x^2 - 6x + 1$. Then $f(x) + g(x) = -x^3 + 7x^2 - 7x + 6$ and $f(x) \cdot g(x) = -3x^5 + 13x^4 - 27x^3 + 29x^2 - 31x + 5$.

**Proposition 1.1.** *$R[x]$ is a commutative ring.*

*Proof.* First, the zero polynomial is the additive identity, and the polynomial $1 = 1x^0$ is the multiplicative identity, as the reader may easily check. The commutative and associative properties of addition are immediate consequences of the respective properties for $R$. We give the proof that multiplication is commutative.

$$f(x) \cdot g(x) = \sum_{j=0}^{n+m} \left( \sum_{k=0}^{j} a_k b_{j-k} \right) x^j \quad \text{and}$$

$$g(x) \cdot f(x) = \sum_{j=0}^{m+n} \left( \sum_{k=0}^{j} b_k a_{j-k} \right) x^j .$$

$\sum_{k=0}^{j} b_k a_{j-k} = \sum_{k=0}^{j} a_{j-k} b_k$ by the commutative law of multiplication

in $R$; but $\sum_{k=0}^{j} a_{j-k} b_k = \sum_{\ell=0}^{j} a_\ell b_{j-\ell} = \sum_{k=0}^{j} a_k b_{j-k}$, and so we're done. The associative law of multiplication and the distributive law are similarly verified (see Exercise 17).  $\square$

**Proposition 1.2.** *Let $R$ be an integral domain. If $f(x), g(x) \in R[x]$ are nonzero polynomials, then $\deg(f(x) \cdot g(x)) = \deg(f(x)) + \deg(g(x))$. Moreover, $R[x]$ is an integral domain.*

*Proof.* Let $\deg(f(x)) = n$ and $\deg(g(x)) = m$; note that so long as $R$ is an integral domain, $a_n b_m \neq 0$ since neither $a_n$ nor $b_m$ is 0. Thus, $\deg(f(x) \cdot g(x)) = m + n$. In particular, if neither $f(x)$ nor $g(x)$ is the zero polynomial, the product $f(x) \cdot g(x)$ cannot be the zero polynomial.  $\square$

**Example 2.** Let $R = \mathbb{Z}_6$, $f(x) = \bar{2}x + \bar{4}$, and $g(x) = \bar{3}x + \bar{3} \in \mathbb{Z}_6[x]$. Then $f(x) \cdot g(x) = \bar{0}$. As a result, when $R$ has zero divisors, bizarre things can occur. (See also Example 7 and the remark following Theorem 1.8.)

We will be particularly interested in the ring $F[x]$ of polynomials with coefficients in a field $F$, since in this case there is a division algorithm. We say that $g(x)$ divides $f(x)$ (written $g(x)|f(x)$) if there is a polynomial $h(x)$ so that $f(x) = g(x)h(x)$ (from this point on, we delete the "$\cdot$").

**Proposition 1.3** (Division Algorithm). *Let $f(x), g(x) \in F[x]$ be nonzero polynomials. Then there are unique polynomials $q(x), r(x) \in F[x]$ so that*

$$f(x) = q(x)g(x) + r(x), \quad \text{with} \quad \deg(r(x)) < \deg(g(x)) \quad \text{or} \quad r(x) = 0.$$

**Proof.** Note first that whenever $\deg(f(x)) < \deg(g(x))$, we have $q(x) = 0$ and $r(x) = f(x)$. Second, if $\deg(g(x)) = 0$, then $g(x) = b_0 \in F$; in this case, $q(x) = f(x)/b_0$ and $r(x) = 0$.

Fix a polynomial $g(x) = b_m x^m + b_{m-1} x^{m-1} + \cdots + b_1 x + b_0$ of degree $m \geq 1$. We prove the proposition by complete induction on the degree of $f(x)$. If $\deg(f(x)) = 1$ and $m > 1$ there is nothing to prove, so suppose $m = 1$ as well. Then $f(x) = a_1 x + a_0$, $g(x) = b_1 x + b_0$, so $f(x) = \frac{a_1}{b_1} g(x) + (a_0 - \frac{a_1 b_0}{b_1})$. Thus $q(x) = \frac{a_1}{b_1}$ and $r(x) = (a_0 - \frac{a_1 b_0}{b_1})$, as required; note that if $r(x) \neq 0$, $\deg(r(x)) = 0 < 1$.

Now suppose the result is known whenever $\deg(f(x)) \leq k$, and we are given a polynomial $F(x) = a_{k+1} x^{k+1} + a_k x^k + \cdots + a_1 x + a_0$ of degree $k + 1$. (Once again the problem is only interesting if $m \leq k + 1$.) The coefficients of $x^{k+1}$ in $F(x)$ and in the polynomial $\frac{a_{k+1}}{b_m} x^{k+1-m} g(x)$ are the same, so the polynomial $f(x) = F(x) - \frac{a_{k+1}}{b_m} x^{k+1-m} g(x)$ has degree at most $k$. Thus, by induction hypothesis, we can write $f(x) = q(x)g(x) + r(x)$, where $\deg(r(x)) < m$ or $r(x) = 0$. Set $Q(x) = q(x) + \frac{a_{k+1}}{b_m} x^{k+1-m}$ and $R(x) = r(x)$; then $F(x) = Q(x)g(x) + R(x)$, as desired.

To establish uniqueness, suppose we had $f(x) = q(x)g(x) + r(x) = \tilde{q}(x)g(x) + \tilde{r}(x)$, where $\deg(r(x)) < \deg(g(x))$ or $r(x) = 0$ and the same is true of $\tilde{r}(x)$. Then we have $(q(x) - \tilde{q}(x))g(x) = \tilde{r}(x) - r(x)$. If $q(x) \neq \tilde{q}(x)$, then the degree of $(q(x) - \tilde{q}(x))g(x)$ is at least $\deg(g(x))$, while the degree of $\tilde{r}(x) - r(x)$ must be strictly less. Since this cannot happen, we must have $q(x) = \tilde{q}(x)$, and so $r(x) = \tilde{r}(x)$ as well.   □

**Remark.** Of course, this is the usual "long division" of polynomials from high school algebra, as we shall see in the examples below.

**Example 3.** Let $F = \mathbb{Z}_5$, and suppose $f(x) = x^3 + \overline{2}$ and $g(x) = \overline{2}x^2 + x + \overline{1}$. Then we perform the long division:

$$
\begin{array}{r}
\overline{3}x + \quad \overline{1} \\
\overline{2}x^2 + x + \overline{1} \overline{\smash{\big)}\, x^3 \qquad\qquad + \overline{2}} \\
\underline{x^3 + \overline{3}x^2 + \overline{3}x} \\
\overline{2}x^2 + \overline{2}x + \overline{2} \\
\underline{\overline{2}x^2 + \ x + \overline{1}} \\
x + \overline{1}
\end{array}
$$

We find that $x^3 + \overline{2} = (\overline{2}x^2 + x + \overline{1})(\overline{3}x + \overline{1}) + (x + \overline{1})$, so $q(x) = \overline{3}x + \overline{1}$ and $r(x) = x + \overline{1}$.

**Example 4.** Let $F = \mathbb{Q}$, and suppose $f(x) = x^3 + 4x^2 - x + 7$ and $g(x) = x - 2$. We perform the long division:

$$
\begin{array}{r}
x^2 + 6x + 11 \\
x - 2 \overline{)\, x^3 + 4x^2 - \phantom{1}x + 7} \\
\underline{x^3 - 2x^2} \phantom{aaaaaaaaa} \\
6x^2 - \phantom{1}x \phantom{aaa} \\
\underline{6x^2 - 12x} \phantom{aaa} \\
11x + 7 \\
\underline{11x - 22} \\
29
\end{array}
$$

Thus $f(x) = (x^2 + 6x + 11)(x - 2) + 29$, so that $q(x) = x^2 + 6x + 11$ and $r(x) = 29$.

**Example 5.** Let $F = \mathbb{Q}$, and suppose $f(x) = x^3 + 1$ and $g(x) = 2x^2 + x + 1$. Then once again long division shows that $f(x) = (2x^2 + x + 1)(\frac{1}{2}x - \frac{1}{4}) + (-\frac{1}{4}x + \frac{5}{4})$, so that $q(x) = \frac{1}{2}x - \frac{1}{4}$ and $r(x) = -\frac{1}{4}x + \frac{5}{4}$.

The following corollary of Proposition 1.3 is so important that we give it a name.

**Corollary 1.4** (Remainder Theorem). *Let $c \in F$ and $f(x) \in F[x]$. When we divide $f(x)$ by $x - c$, the remainder is $f(c)$.*

**Proof.** We have $f(x) = (x - c)q(x) + r(x)$, where $\deg(r(x)) < 1$ or $r(x) = 0$. Thus $r(x)$ is a constant, $r$. Evaluating both sides at $x = c$, $f(c) = r$. □

And this result in turn gives rise to the following definition.

**Definition.** We say $c \in F$ is a **root** of $f(x) \in F[x]$ if $f(c) = 0$.

**Corollary 1.5** (Root-Factor Theorem). *Let $f(x) \in F[x]$. Then $(x - c)$ is a factor of $f(x)$ if and only if $c$ is a root of $f(x)$.*

**Example 6.** $x - \overline{2}$ is a factor of $f(x) = x^4 + x^3 + \overline{2}x^2 + \overline{1} \in \mathbb{Z}_3[x]$, for $f(\overline{2}) = f(-\overline{1}) = (-\overline{1})^4 + (-\overline{1})^3 + (-\overline{1})(-\overline{1})^2 + \overline{1} = \overline{0}$.

As a consequence of Corollary 1.5, a polynomial of degree $n$ with coefficients in a field $F$ has at most $n$ roots in $F$ (see Exercise

6). The main idea is this: each root gives rise to a linear factor, and by Proposition 1.2 we cannot have more than $n$ linear factors.

**Example 7.** As usual, odd things can occur in $R[x]$ when $R$ is not an integral domain. Consider $f(x) = x^2 - \overline{1} \in \mathbb{Z}_8[x]$. Then it is easy to check that $\overline{1}, \overline{3}, \overline{5}$, and $\overline{7}$ are all roots of $f(x)$.

In Section 2 of Chapter 1 we deduced the existence of the greatest common divisor and the unique factorization property of the integers from the Euclidean algorithm. These results will follow identically for the polynomial ring $F[x]$ once we have the appropriate notions of prime and greatest common divisor.

**Definition.** Let $F$ be a field. A nonconstant polynomial $f(x) \in F[x]$ is called **irreducible** in $F[x]$ if it cannot be expressed as a product of nonconstant polynomials in $F[x]$. (In other words, if $f(x)$ is irreducible and there are $g(x), h(x) \in F[x]$ so that $f(x) = g(x)h(x)$, then either $g(x)$ or $h(x)$—but not both—must be a constant polynomial.)

**Examples 8.** Every polynomial of degree 1 is irreducible. A polynomial of degree 2 or 3 is irreducible in $F[x]$ if and only if it has no root in $F$ (see Exercise 8). However, the analogous statement is not in general true for polynomials of degree $\geq 4$: consider $f(x) = (x^2 + 1)(x^2 + x + 1) \in \mathbb{Q}[x]$. The polynomial $f(x)$ is patently not irreducible, and yet it has no root in $\mathbb{Q}$.

We next explore the existence of the greatest common divisor of two polynomials by mimicking the proof of Theorem 2.3 of Chapter 1.

**Theorem 1.6.** *Given nonzero polynomials $f(x), g(x) \in F[x]$, let*

$$S = \{h(x) \in F[x] : h(x) = a(x)f(x) + b(x)g(x)$$
$$\textit{for some } a(x), b(x) \in F[x]\}.$$

*Then there is some polynomial $d(x) \in S$ of smallest degree, and every $h(x) \in S$ is divisible by $d(x)$.*

**Proof.** We consider the set $S'$ of all *nonzero* polynomials in $S$. Choose a polynomial $d(x) = s(x)f(x) + t(x)g(x) \in S'$ of smallest degree (which must exist by the Well-Ordering Principle, Proposition 1.3 of Chapter 1).

First we claim that $d(x)|f(x)$ (and likewise $g(x)$). Apply the division algorithm, Proposition 1.3, to write $f(x) = q(x)d(x) +$

$r(x)$ with $\deg(r(x)) < \deg(d(x))$ or $r(x) = 0$. If $r(x) \neq 0$, then $r(x) = f(x) - (s(x)f(x) + t(x)g(x))q(x) = (1 - s(x)q(x))f(x) - (t(x)q(x))g(x) \in S'$; and $r(x)$ has degree less than that of $d(x)$, contradicting the way $d(x)$ was chosen. Thus we must have $r(x) = 0$ and $d(x)|f(x)$, as desired (and likewise for $g(x)$). On the other hand, if $d(x)|f(x)$ and $d(x)|g(x)$, then for every $a(x), b(x) \in F[x]$, $d(x)|(a(x)f(x) + b(x)g(x))$; and so $d(x)$ is a factor of every polynomial in $S$. $\square$

**Remark.** To what extent is the $d(x)$ we've just constructed unique? If $d(x)$ and $\tilde{d}(x)$ both are factors of every polynomial in $S$, then in particular $d(x)|\tilde{d}(x)$ and $\tilde{d}(x)|d(x)$, so each must be a constant multiple of the other.

Armed with Theorem 1.6, we proceed as in Chapter 1. Given two nonzero polynomials $f(x), g(x) \in F[x]$, we define their **greatest common divisor** (g.c.d.) $d(x) = \gcd(f(x), g(x))$ to be the *monic* polynomial satisfying

(i) $d(x)|f(x)$ and $d(x)|g(x)$; and
(ii) for any $e(x) \in F[x]$, if $e(x)|f(x)$ and $e(x)|g(x)$, then $e(x)|d(x)$.

The greatest common divisor $d(x)$ and the polynomials $s(x)$ and $t(x)$ may be constructed explicitly, as in the case of the integers (see Section 2 of Chapter 1), by applying the division algorithm repeatedly. We call this process the **Euclidean algorithm** for polynomials.

**Examples 9.**

(a) Let $F = \mathbb{Q}$, $f(x) = x^3 - 8$, and $g(x) = x^2 - x - 2$. Then $d(x) = x - 2$, and $d(x) = \frac{1}{3}f(x) + (-\frac{1}{3})(x + 1)g(x)$.
(b) Let $F = \mathbb{Z}_3$, $f(x) = x^3 - \bar{1}$, and $g(x) = x^3 - x^2 - x + \bar{1}$. Now $f(x) = (x - \bar{1})^3$ (cf. Exercise 1.3.29) and $g(x) = (x^2 - \bar{1})(x - \bar{1}) = (x - \bar{1})^2(x + \bar{1})$, so $d(x) = (x - \bar{1})^2$. And $d(x) = f(x) - g(x)$!

We say $f(x)$ and $g(x)$ are **relatively prime** if $\gcd(f(x), g(x)) = 1$. Now the analogue of Proposition 2.5 of Chapter 1 will follow, and, from this, the fact that any polynomial in $F[x]$ can be factored as a product of irreducible polynomials.

**Proposition 1.7.** *Suppose $f(x)$ is irreducible and $f(x)|g(x)h(x)$. Then $f(x)|g(x)$ or $f(x)|h(x)$.*

**Proof.** Suppose $f(x)$ divides $g(x)h(x)$ and does not divide $h(x)$. Then $\gcd(f(x), h(x)) = 1$, and so there are polynomials $s(x)$ and $t(x)$ so that $1 = s(x)f(x) + t(x)h(x)$. Thus $g(x) = s(x)g(x)f(x) + t(x)g(x)h(x)$; since $f(x)|t(x)g(x)h(x)$ and $f(x)|s(x)g(x)f(x)$, we infer that $f(x)|g(x)$. □

**Theorem 1.8** (Unique Factorization in $F[x]$). *Let $F$ be a field. Then every nonconstant polynomial $f(x) \in F[x]$ can be written as a product of irreducible polynomials in $F[x]$; the resulting expression is unique, except for rearrangement and nonzero constant factors.*

**Proof.** As in the proof of Theorem 2.7 of Chapter 1, we would proceed by complete induction on the degree of $f(x)$. We leave this as an exercise for the reader (see Exercise 18). □

**Remark.** Lest the reader take this theorem as being totally obvious, consider the polynomial $f(x) = x^2 - \overline{1} \in \mathbb{Z}_8[x]$. It has four roots, *viz.*, $\pm\overline{1}$ and $\pm\overline{3}$ (see Example 7). Indeed, we can write $f(x) = (x - \overline{1})(x + \overline{1})$ or $f(x) = (x - \overline{3})(x + \overline{3})$, so factorization is most definitely not unique. Are there any other possible factorizations?

We end this section by showing how the partial fractions decomposition (which one learns in integral calculus) can be deduced from the Euclidean algorithm. We first begin with a definition.

**Definition.** The field of quotients (see p. 47 and Exercise 2.1.11) of the polynomial ring $F[x]$ is called the field of **rational functions** with coefficients in $F$, denoted $F(x)$. Clearly,

$$F(x) = \left\{ \frac{f(x)}{g(x)} : f(x), g(x) \in F[x], \ g(x) \neq 0 \right\}.$$

When $\deg(g(x)) \leq \deg(f(x))$, we may apply the division algorithm to write $f(x) = q(x)g(x) + r(x)$, as usual, and so $\frac{f(x)}{g(x)} = q(x) + \frac{r(x)}{g(x)}$, where $\deg(r(x)) < \deg(g(x))$. So it suffices to treat the case of rational functions $\frac{f(x)}{g(x)}$ when $\deg(f(x)) < \deg(g(x))$.

**Theorem 1.9** (Partial Fractions Decomposition). *Let $f(x)$, $g(x)$ $\in F[x]$, where $\deg(f(x)) < \deg(g(x))$. If $g(x) = u(x)v(x)$, where $u(x)$ and $v(x)$ are relatively prime, then there are polynomials $a(x)$ and $b(x)$, with $\deg(a(x)) < \deg(u(x))$ and $\deg(b(x)) < \deg(v(x))$, so that*

$$\frac{f(x)}{g(x)} = \frac{a(x)}{u(x)} + \frac{b(x)}{v(x)}.$$

**Proof.** By Theorem 1.6, there are polynomials $s(x)$ and $t(x)$ so that $1 = s(x)u(x) + t(x)v(x)$. Thus

$$\frac{f(x)}{u(x)v(x)} = \frac{f(x)t(x)}{u(x)} + \frac{f(x)s(x)}{v(x)}.$$

Now use the division algorithm to write

$$\frac{f(x)t(x)}{u(x)} = q(x) + \frac{a(x)}{u(x)}, \quad \deg(a(x)) < \deg(u(x)),$$

$$\frac{f(x)s(x)}{v(x)} = Q(x) + \frac{b(x)}{v(x)}, \quad \deg(b(x)) < \deg(v(x)), \quad \text{whence}$$

$$\frac{f(x)}{u(x)v(x)} = \frac{a(x)}{u(x)} + \frac{b(x)}{v(x)}.$$

Note that the sum of the polynomials $q(x)$ and $Q(x)$ must be 0 since the left-hand side has no polynomial term. $\square$

For a change, this is a constructive proof! That is, the proof actually gives an *algorithm* for calculating the partial fractions decomposition.

**Example 10.** Consider the rational function $\dfrac{1}{(x-a)(x-b)}$, where $a \neq b$. Since $u(x) = x - a$ and $v(x) = x - b$ are relatively prime, we write $1 = s(x)u(x) + t(x)v(x)$ for appropriate polynomials $s(x), t(x)$. It isn't difficult:

$$1 = \frac{1}{b-a}((x-a) - (x-b)), \quad \text{whence}$$

$$\frac{1}{(x-a)(x-b)} = \frac{-\frac{1}{b-a}}{x-a} + \frac{\frac{1}{b-a}}{x-b}.$$

**Example 11.** Consider the rational function $\dfrac{x+3}{x^3(x-1)^2}$. Applying the Euclidean algorithm, we find that

$$1 = (3x^2 + 2x + 1)(x-1)^2 - (3x-4)x^3.$$

Therefore,

$$\frac{x+3}{x^3(x-1)^2} = \frac{(x+3)(3x^2 + 2x + 1)}{x^3} - \frac{(x+3)(3x-4)}{(x-1)^2}$$

$$= \frac{11x^2 + 7x + 3}{x^3} + \frac{-11x + 15}{(x-1)^2}.$$

(Note that the polynomial parts—3 and −3, respectively—of the middle two fractions cancel.)

**Remark.** More generally, let $p_1(x)$, ..., $p_m(x)$ be irreducible polynomials; and let $g(x) = p_1(x)^{v_1} p_2(x)^{v_2} \cdots p_m(x)^{v_m}$, for $v_1$, ..., $v_m \in \mathbb{N}$. Then there are polynomials $r_1(x)$, ..., $r_m(x)$, with $\deg(r_j(x)) < \deg(p_j(x))$, $j = 1,\ldots,m$, so that

$$\frac{f(x)}{g(x)} = \frac{r_1(x)}{p_1(x)^{v_1}} + \frac{r_2(x)}{p_2(x)^{v_2}} + \cdots + \frac{r_m(x)}{p_m(x)^{v_m}}.$$

This can be proved easily by induction using Theorem 1.9.

In calculus books the result is often stated somewhat differently. When $v_j > 1$, i.e., when $p_j(x)$ is a "repeated factor" of $g(x)$, then the appropriate term $\dfrac{r_j(x)}{p_j(x)^{v_j}}$ can be rewritten as

$$\sum_{\ell=1}^{v_j} \frac{q_\ell(x)}{p_j(x)^\ell}, \quad \deg(q_\ell(x)) < \deg(p_j(x)), \ \ell = 1,\ldots,v_j.$$

For example, $\dfrac{x^3 + 2x^2 + 3x + 1}{(x^2 + 1)^2} = \dfrac{x + 2}{x^2 + 1} + \dfrac{2x - 1}{(x^2 + 1)^2}$. See Exercise 21.

## EXERCISES 3.1

1.  Apply the division algorithm to the following polynomials $f(x)$, $g(x) \in F[x]$:
    a.  $f(x) = 4x^3 - x^2 - 3x + 5$, $g(x) = 2x + 5$, $F = \mathbb{Q}$
    b.  $f(x) = (x - 3)^3$, $g(x) = x - 1$, $F = \mathbb{Q}$
    c.  $f(x) = x^6 + \overline{3}x^5 + \overline{4}x^2 - \overline{3}x + \overline{2}$, $g(x) = x^2 + \overline{2}x - \overline{3}$, $F = \mathbb{Z}_7$
    d.  $f(x) = x^6 + \overline{3}x^5 + \overline{4}x^2 - \overline{3}x + \overline{2}$, $g(x) = \overline{3}x^2 + \overline{2}x - \overline{3}$, $F = \mathbb{Z}_7$
    e.  $f(x) = x^7 + x^6 + x^4 + x + \overline{1}$, $g(x) = x^3 + x + \overline{1}$, $F = \mathbb{Z}_2$

2.  Find the greatest common divisors $d(x)$ of the following polynomials $f(x), g(x) \in F[x]$, and express $d(x) = s(x)f(x) + t(x)g(x)$ for appropriate $s(x), t(x) \in F[x]$:
    a.  $f(x) = x^3 - 1$, $g(x) = x^4 + x^3 - x^2 - 2x - 2$, $F = \mathbb{Q}$
    b.  $f(x) = x^2 + (1 - \sqrt{2})x - \sqrt{2}$, $g(x) = x^2 - 2$, $F = \mathbb{R}$
    c.  $f(x) = x^2 + 1$, $g(x) = x^2 - i + 2$, $F = \mathbb{C}$
    d.  $f(x) = x^2 + 2x + 2$, $g(x) = x^2 + 1$, $F = \mathbb{Q}$
    e.  $f(x) = x^2 + 2x + 2$, $g(x) = x^2 + 1$, $F = \mathbb{C}$

3.  Follow the proof of Theorem 1.9 to give the partial fractions decompositions of the following rational functions in $\mathbb{Q}(x)$:

    a.  $\frac{5x+7}{x^2+2x-3}$

    b.  $\frac{2x^3+x^2+2x-1}{x^4-1}$

    c.  $\frac{7x^2+x-3}{x^3-x^2}$

4.  Follow the proof of Theorem 1.9 to give the partial fractions decomposition of these rational functions:

    a.  $\frac{4x+2}{x^3+2x^2+4x+8} \in \mathbb{Q}(x)$

    b.  $\frac{\overline{4}x+\overline{2}}{x^3+\overline{2}x^2+\overline{4}x+\overline{8}} \in \mathbb{Z}_5(x)$

5.  Suppose $\deg(f(x)) = n$, $\deg(g(x)) = m$, and $m \geq n$. Prove or give a counterexample:

    a.  $\deg(f(x) + g(x)) = m$.

    b.  $\deg(f(x) \cdot g(x)) = m + n$.

6.  Prove that if $F$ is a field, $f(x) \in F[x]$, and $\deg(f(x)) = n$, then $f(x)$ has at most $n$ roots in $F$. (Hint: Use Corollary 1.5 and induction.)

7.  Show that Corollaries 1.4 and 1.5 are valid even when the coefficients of $f(x)$ lie in an integral domain $R$, but that Theorem 1.6 may fail. (Hint: The g.c.d. may no longer be monic. Consider $f(x) = 2x$, $g(x) = 4x + 6 \in \mathbb{Z}[x]$.)

8.  Let $F$ be a field. Prove that if $f(x) \in F[x]$ is a polynomial of degree 2 or 3, then $f(x)$ is irreducible in $F[x]$ if and only if $f(x)$ has no root in $F$.

9.  a.  Show that unique factorization fails horribly in $R[x]$ when $R$ is not an integral domain. Consider, for example, $(\overline{2}x + \overline{4})(\overline{3}x^2 + \overline{3})$ and $(\overline{2}x + \overline{3})(\overline{3}x + \overline{2})$ in $\mathbb{Z}_6[x]$.

    b.  Show that Proposition 1.7 also fails when $R$ is not an integral domain.

    c.  How many roots does $f(x) = \overline{2}x - \overline{4} \in \mathbb{Z}_6[x]$ have?

10. Decide whether each of the following polynomials $f(x)$ is irreducible in $F[x]$ for the given field $F$:

    a.  $f(x) = x^2 + \overline{1}$, $F = \mathbb{Z}_5$

    b.  $f(x) = x^2 + \overline{1}$, $F = \mathbb{Z}_7$

    c.  $f(x) = x^2 + \overline{1}$, $F = \mathbb{Z}_{19}$

    d.  $f(x) = x^3 - \overline{9}$, $F = \mathbb{Z}_{11}$

e.   $f(x) = x^3 + x + \bar{1}, \ F = \mathbb{Z}_2$
f.   $f(x) = x^4 + x^2 + \bar{1}, \ F = \mathbb{Z}_2$

11.  Find all odd prime numbers $p$ so that $x + \bar{2}$ is a factor of $f(x) = x^4 + x^3 + x^2 - x + \bar{1} \in \mathbb{Z}_p[x]$.

12.  Let $f(x) = a_n x^n + a_{n-1} x^{n-1} + \cdots + a_1 x + a_0 \in F[x]$, and let $b \in F$. The substitution of $x + b$ in $f(x)$ yields another polynomial $g(x) = f(x + b)$.
     a.   Prove that $c$ is a root of $f(x) \iff c - b$ is a root of $g(x)$.
     b.   Prove that $f(x)$ is irreducible $\iff g(x)$ is irreducible.

13.  List all the irreducible polynomials in $\mathbb{Z}_2[x]$ of degree $\leq 4$. Factor $f(x) = x^7 + \bar{1}$ as a product of irreducible polynomials in $\mathbb{Z}_2[x]$.

14.  For each of the following numbers $c$, find an irreducible polynomial in $\mathbb{Q}[x]$ that has the number $c \in \mathbb{C}$ as a root:
     a.   $1 + \sqrt{3}$
     b.   $2 + \sqrt[3]{2}$
     c.   $2 + i$
     d.   $\sqrt{1 + \sqrt{3}}$

15.  Let $f(x) \in \mathbb{R}[x]$, and let $f'(x)$ be its derivative. (If $f(x) = a_n x^n + a_{n-1} x^{n-1} + \cdots + a_1 x + a_0$, then, of course, $f'(x) = n a_n x^{n-1} + (n - 1) a_{n-1} x^{n-2} + \cdots + a_1$.) If $\mu \in \mathbb{N}$ and $(x - c)^{\mu}$ divides $f(x)$, but $(x - c)^{\mu+1}$ does not, we say that $c$ is a root of $f(x)$ of **multiplicity** $\mu$.
     a.   Suppose that $c$ is a root of $f(x)$ of multiplicity $\mu > 1$. Prove that $c$ is a root of $f'(x)$ as well.
     b.   Conversely, suppose that $c$ is a root of both $f(x)$ and $f'(x)$. Prove that $c$ is a root of $f(x)$ of multiplicity $\mu > 1$.

16.  Let $R$ be any commutative ring. If $f(x), g(x) \in R[x]$ and the leading coefficient of $g(x)$ is a *unit*, prove that there are polynomials $q(x), r(x) \in R[x]$ so that $f(x) = q(x)g(x) + r(x)$, with $\deg(r(x)) < \deg(g(x))$ or $r(x) = 0$. Thus, Proposition 1.3 holds in this situation.

17.  Finish the verification that $R[x]$ is a commutative ring (Proposition 1.1).

18.  Prove Theorem 1.8.

19. Prove that if $F$ is a field, there is no rational function $\frac{f(x)}{g(x)} \in F(x)$ whose square is $x$. (Where do you use the fact that $F$ is a field? Does it hold for $R(x)$ when $R$ is an integral domain?)

20. a. Let $F$ be a field, and let $f(x), g(x), h(x) \in F[x]$. We say
$$f(x) \equiv g(x) \pmod{h(x)} \quad \text{if} \quad h(x) \mid (f(x) - g(x)).$$
Show that this is an equivalence relation: i.e.,
    (i) $f(x) \equiv f(x) \pmod{h(x)}$,
    (ii) if $f(x) \equiv g(x) \pmod{h(x)}$, then $g(x) \equiv f(x)$ $\pmod{h(x)}$, and
    (iii) if $f(x) \equiv g(x)$ and $g(x) \equiv j(x) \pmod{h(x)}$, then $f(x) \equiv j(x) \pmod{h(x)}$.

    b. Give all polynomials $f(x) \in \mathbb{Q}[x]$ solving the simultaneous congruences
$$f(x) \equiv x + 3 \pmod{x^2}$$
$$f(x) \equiv 4 \pmod{x + 1}.$$
    (Cf. the Chinese Remainder Theorem, Theorem 3.7 of Chapter 1.)

    c. Give all polynomials $f(x) \in \mathbb{Z}_3[x]$ solving the simultaneous congruences
$$f(x) \equiv x^2 + \overline{1} \pmod{x^3 + x + \overline{2}}$$
$$f(x) \equiv \overline{2}x + \overline{1} \pmod{x^2 + x + \overline{2}}.$$

21. Let $p(x)$ be an irreducible polynomial of degree $m$, and let $r(x)$ be a polynomial of degree $< m\nu$. Show that there are polynomials $r_1(x), \ldots, r_\nu(x)$ with $\deg(r_j(x)) < m$ so that
$$\frac{r(x)}{p(x)^\nu} = \frac{r_1(x)}{p(x)} + \frac{r_2(x)}{p(x)^2} + \cdots + \frac{r_\nu(x)}{p(x)^\nu}.$$
Note that we can rewrite this $r(x) = \sum_{j=1}^{\nu} r_j(x) p(x)^{\nu - j}$. (Hint: $r_1(x)$ is the quotient when $r(x)$ is divided by $p(x)^{\nu - 1}$.)

22. Exercise 21 may be generalized as follows. Given $f(x) \in F[x]$ and any nonconstant polynomial $g(x) \in F[x]$, show there are an integer $m$ and (unique) polynomials $f_0(x), f_1(x), \ldots, f_m(x)$ with $\deg(f_j(x)) < \deg(g(x))$ for all $j = 0, \ldots, m$, so that $f(x) = \sum_{j=0}^{m} f_j(x) g(x)^{m - j}$.

23. Let $F$ be a finite field (i.e., a field with finitely many elements).
    a. Compute the sum of all the elements of $F$. (Be careful if $1 + 1 = 0$ in $F$.)
    b. Prove that $a = a^{-1} \in F \iff a = \pm 1$.
    c. Compute the product of the nonzero elements of $F$ (cf. Exercise 1.4.13).

24. Let $F = \mathbb{Q}(x)$, the field of rational functions with coefficients in $\mathbb{Q}$.
    a. Let $F^+ = \left\{ \frac{f(x)}{g(x)} \in F : \text{the leading coefficients of } f(x) \text{ and } g(x) \right.$ have the same sign$\left. \right\}$. Check that $F^+$ is well-defined, and deduce that $F$ is an ordered field.
    b. $\mathbb{Q}$ is a subset of $F$ in a natural way. Show that $\mathbb{N} \subset F$ is bounded above.
    c. (With thanks to Dino Lorenzini) Does $F$ have the least upper bound property? (Hint: see Exercise 2.2.8.)

## 2. Roots of Polynomials

We saw in Section 4 of Chapter 2 that every quadratic and cubic polynomial can be "solved" in $\mathbb{C}$; by this we meant that we could give explicit formulas expressing all the roots of the polynomials in terms of (square and cube) roots. But far more is true:

**Theorem 2.1** (The Fundamental Theorem of Algebra). *Suppose $f(x) \in \mathbb{C}[x]$ is a polynomial of degree $n \geq 1$. Then $f(x)$ has a root in $\mathbb{C}$.*

Once we show that $f(x)$ has a root in $\mathbb{C}$, it will follow by induction (using Corollary 1.5) that $f(x)$ can be factored as a product of linear polynomials in $\mathbb{C}[x]$. Thus, $f(x)$ has $n$ roots (counting multiplicities) in $\mathbb{C}$. Gauss was the first to give a correct proof of this theorem, in 1799, and his proof hinged on the consequences of the fact that a polynomial is a continuous function. It is instructive to see roughly how one proof (not Gauss' original) might go. (The reader might want to look at some other proofs: cf. Birkhoff and MacLane, *A Survey of Modern Algebra*, 4th ed., pp. 113-116; McCoy and Janusz, *Introduction to Modern Algebra*, 4th ed., pp. 182-185;

Spivak, *Calculus*, 3rd ed., pp. 539-541; Guillemin and Pollack, *Differential Topology*, p. 82, p. 110; or numerous books on complex analysis.)

**Proof.** We use $z \in \mathbb{C}$ as a variable, and we may assume that $f(z)$ is monic: $f(z) = z^n + a_{n-1}z^{n-1} + \cdots + a_1z + a_0$. First, note that $f(z) = z^n\left(1 + \frac{a_{n-1}}{z} + \frac{a_{n-2}}{z^2} + \cdots + \frac{a_0}{z^n}\right)$; since $\lim_{|z| \to \infty}\left[\frac{a_{n-1}}{z} + \frac{a_{n-2}}{z^2} + \cdots + \frac{a_0}{z^n}\right] = 0$, there is a positive real number $R$ so that whenever $|z| > R$, $\left|\frac{a_{n-1}}{z} + \frac{a_{n-2}}{z^2} + \cdots + \frac{a_0}{z^n}\right| < \frac{1}{2}$. Thus whenever $|z| > R$, $|f(z)| > |z|^n(1 - \frac{1}{2}) > \frac{R^n}{2}$. Certainly, then, $f(z)$ has no root in this region.

On the other hand, it follows from the continuity of the polynomial $f(z)$ that on the (closed and bounded, or *compact*) set $\{z \in \mathbb{C} : |z| \le R\}$, $|f(z)|$ achieves a minimum value, say at the point $z_o$. This minimum value will turn out to be 0; otherwise, arbitrarily close to the point $z_o$ there will be points where $|f(z)|$ is yet smaller.

For convenience, we assume $z_o = 0$. Then $f(0) = a_0$; and if $a_0 \ne 0$, let $j$ be the smallest *positive* integer $k$ so that $a_k \ne 0$. We put $g(z) = \frac{a_{j+1}}{a_j}z + \cdots + \frac{a_n}{a_j}z^{n-j}$ and then $f(z) = a_0 + a_jz^j(1 + g(z))$. By Corollary 3.6 of Chapter 2, there is a complex number $y$ satisfying $y^j = -\frac{a_0}{a_j}$. Consider now the values of $f(ty) = a_0 + a_j(ty)^j(1 + g(ty)) = a_0 - a_0t^j(1 + g(ty)) = a_0\left(1 - t^j(1 + g(ty))\right)$ for $t$ a small positive (real) number. For $t$ sufficiently small, $|g(ty)| = \left|\left[\frac{a_{j+1}}{a_j}(ty) + \cdots + \frac{a_n}{a_j}(ty)^{n-j}\right]\right| < \frac{1}{2}$; for such $t$, $|f(ty)| = |a_0||1 - t^j(1 + g(ty))| \le |a_0||1 - t^j/2| < |a_0|$. This shows that $z_o$ could not have been the minimum point of $|f(z)|$, and from this contradiction we conclude that the minimum value must in fact be 0. Thus $f(z_o) = 0$, as desired. $\square$

Given a polynomial $f(x) \in \mathbb{Q}[x]$, Theorem 2.1 guarantees that it will have all its roots in $\mathbb{C}$. Our immediate goal is to find the *smallest possible* field $K$ containing them all. (Note that any field $K \subset \mathbb{C}$ necessarily contains $\mathbb{Q}$, by Exercise 2.3.20.) We need first to introduce officially the notation $F[\alpha]$, which we've used so far in an *ad hoc* manner. Let $F$ be any field, let $K$ be a field containing $F$, and suppose $\alpha \in K$. Then we define

$$F[\alpha] = \{p(\alpha) \in K : p(x) \in F[x]\}.$$

Note that $F[\alpha]$ is a **subring** of $K$; this means that $F[\alpha]$ is a subset of $K$ and is itself a ring. (All the ring properties are inherited from the polynomial ring $F[x]$.) The reader should check that this is consistent with our use of this notation in Chapter 2. Similarly, if $\alpha, \beta \in K$, we define $F[\alpha, \beta] = (F[\alpha])[\beta]$. (See Exercise 13.)

**Example 1.** In Example 4 of Section 2 of Chapter 2, we introduced the ring $\{a + b\sqrt{2} : a, b \in \mathbb{Q}\} \subset \mathbb{R}$. This is consistent with our present definition of $\mathbb{Q}[\sqrt{2}]$: if $p(x) = a_n x^n + a_{n-1} x^{n-1} + \cdots + a_1 x + a_0 \in \mathbb{Q}[x]$, then $p(\sqrt{2}) = a_n (\sqrt{2})^n + a_{n-1} (\sqrt{2})^{n-1} + \cdots + a_2 (\sqrt{2})^2 + \alpha_1 \sqrt{2} + a_0$ can be written, collecting terms, in the form $a + b\sqrt{2}$, where $a = a_0 + 2a_2 + 4a_4 + \ldots$ and $b = a_1 + 2a_3 + 4a_5 + \ldots$. Indeed, if we set $q(x) = a + bx$, then $p(\sqrt{2}) = q(\sqrt{2})$. Now the addition and multiplication rules defined in that example are consistent with those for polynomials: if $\alpha = a + b\sqrt{2}$ and $\beta = c + d\sqrt{2}$, let $q(x) = a + bx$ and $r(x) = c + dx$; then $\alpha + \beta = (a + c) + (b + d)\sqrt{2} = q(\sqrt{2}) + r(\sqrt{2})$ and $\alpha\beta = (ac + 2bd) + (ad + bc)\sqrt{2} = q(\sqrt{2}) \cdot r(\sqrt{2})$.

Consider now $\mathbb{Q}[\sqrt{2}, i] = (\mathbb{Q}[\sqrt{2}])[i]$. We have established that elements of $\mathbb{Q}[\sqrt{2}]$ can be written in the form $a + b\sqrt{2}$; similarly, since $i^2 = -1$, elements of $(\mathbb{Q}[\sqrt{2}])[i]$ can be written in the form $(a + b\sqrt{2}) + (c + d\sqrt{2})i = a + b\sqrt{2} + ci + d(\sqrt{2}i)$, where $a, b, c, d \in \mathbb{Q}$.

**Examples 2.**

(a) Consider $\mathbb{Q}[\sqrt{3}i]$ and $\mathbb{Q}[\sqrt{3}, i]$. Note that $\sqrt{3}$ and $i$ are obviously elements of the latter, and so their product $\sqrt{3}i$ is as well. It follows, then, that $\mathbb{Q}[\sqrt{3}i] \subset \mathbb{Q}[\sqrt{3}, i]$. We claim that the inclusion is proper. In fact, we shall now argue that $i \notin \mathbb{Q}[\sqrt{3}i]$. Since $(\sqrt{3}i)^2 = -3$, $\mathbb{Q}[\sqrt{3}i] = \{a + b\sqrt{3}i : a, b \in \mathbb{Q}\}$. Suppose $i = a + b\sqrt{3}i$ for some $a, b \in \mathbb{Q}$. First, we cannot have $b = 0$, since $i \notin \mathbb{Q}$. Then, by elementary algebra, $i(1 - b\sqrt{3}) = a \implies (1 - b\sqrt{3})^2 = -a^2 \implies \sqrt{3} = \frac{1 + a^2 + 3b^2}{2b} \in \mathbb{Q}$, contradicting the irrationality of $\sqrt{3}$.

(b) Now we modify this a bit. Consider $\mathbb{Q}[\sqrt{3} + i] \subset \mathbb{Q}[\sqrt{3}, i]$. In this case, we claim the two rings are equal. Let $\alpha = \sqrt{3} + i$. A straightforward computation shows that $\alpha^3 = 8i$, whence $i = \frac{1}{8}\alpha^3 \in \mathbb{Q}[\alpha]$. Since $\sqrt{3} = \alpha - i$, $\sqrt{3} = \alpha - \frac{1}{8}\alpha^3 \in \mathbb{Q}[\alpha]$ as well, and so any polynomial in $\sqrt{3}$ and $i$ can be written as a polynomial in $\alpha$; this shows that $\mathbb{Q}[\sqrt{3}, i] \subset \mathbb{Q}[\sqrt{3} + i]$, and so the two fields are equal.

We went to some pains in Chapter 2 to check that $F[\alpha]$ was again a field for certain examples of fields $F$ and numbers $\alpha$. We now settle this issue once and for all. The proof uses—what else?—the Euclidean Algorithm and is one that we shall see numerous times.

**Proposition 2.2.** *Let $f(x) \in F[x]$, and let $K \supset F$ be a field containing a root $\alpha$ of $f(x)$. Then $F[\alpha] \subset K$ is a field (containing both $\alpha$ and the original field $F$).*

**Proof.** Since we know that $F[\alpha]$ is a subring of $K$, we need only prove that any nonzero $\beta \in F[\alpha]$ has a multiplicative inverse in $F[\alpha]$. By definition, $\beta = p(\alpha)$ for some polynomial $p(x) \in F[x]$. Factor $f(x)$ in $F[x]$ as a product of irreducible polynomials. Then $\alpha$ must be a root of one of these irreducible factors, say $g(x)$. Note that $g(x)$ cannot be a factor of $p(x)$, as $p(\alpha) \neq 0$. From the fact that $g(x)$ is irreducible, we infer that $\gcd(g(x), p(x)) = 1$; therefore, there are $s(x), t(x) \in F[x]$ so that $1 = s(x)p(x) + t(x)g(x)$. Evaluating at $x = \alpha$, doing arithmetic in the field $K$, we obtain $1 = s(\alpha)p(\alpha)$ (since $g(\alpha) = 0$). Thus, the multiplicative inverse of $\beta$ is $s(\alpha) \in F[\alpha]$, and we are done. □

**Remark.** We often refer to $F[\alpha]$ as the field obtained by **adjoining** $\alpha$ to $F$. Note that it is in fact the *smallest* field containing both $F$ and $\alpha$.

**Example 3.** We can apply the proof of Proposition 2.2 to find the multiplicative inverse of $\beta = \alpha^2 + \alpha - 1 \in \mathbb{Q}[\alpha]$, where $\alpha$ satisfies the equation $\alpha^3 + \alpha + 1 = 0$. Let $f(x) = x^3 + x + 1 \in \mathbb{Q}[x]$ and $p(x) = x^2 + x - 1$. Then $\alpha$ is a root of the irreducible polynomial $f(x) \in \mathbb{Q}[x]$, and $\beta = p(\alpha)$. Applying the Euclidean algorithm, we find that $1 = \frac{1}{3}((x+1)f(x) - (x^2+2)p(x))$. Thus $1 = -\frac{1}{3}(\alpha^2+2)p(\alpha)$, so that $(\alpha^2+\alpha-1)^{-1} = -\frac{1}{3}(\alpha^2+2)$. As a check, $-\frac{1}{3}(\alpha^2+2)(\alpha^2+\alpha-1) = -\frac{1}{3}(\alpha^4 + \alpha^3 + \alpha^2 + 2\alpha - 2) = 1$.

The same sort of process takes place when we work with finite fields such as $\mathbb{Z}_p$. In Chapters 4 and 5 we shall discuss the rigorous process by which fields containing $\mathbb{Z}_p$ are created, but meanwhile, we can formally consider the ring $\mathbb{Z}_2[\alpha]$ with the rule $\alpha^2 + \alpha + \overline{1} = 0$, and check that it is indeed a field. To get a feel for this ring, note the following: By the division algorithm, given any polynomial $p(x) \in \mathbb{Z}_2[x]$, there are a polynomial $q(x) \in \mathbb{Z}_2[x]$ and a polynomial $r(x) \in \mathbb{Z}_2[x]$ of degree *at most* 1 so that $p(\alpha) = q(\alpha)(\alpha^2 + \alpha + \overline{1}) + r(\alpha) = r(\alpha)$. Thus $p(\alpha)$ must be equal to either $\overline{0}$, $\overline{1}$, $\alpha$, or $\alpha + \overline{1}$. So, this ring consists of the four elements $\overline{0}$, $\overline{1}$, $\alpha$, and $\alpha + \overline{1}$, with the multiplication table given below.

| · | $\overline{0}$ | $\overline{1}$ | $\alpha$ | $\alpha + \overline{1}$ |
|---|---|---|---|---|
| $\overline{0}$ | $\overline{0}$ | $\overline{0}$ | $\overline{0}$ | $\overline{0}$ |
| $\overline{1}$ | $\overline{0}$ | $\overline{1}$ | $\alpha$ | $\alpha + \overline{1}$ |
| $\alpha$ | $\overline{0}$ | $\alpha$ | $\alpha + \overline{1}$ | $\overline{1}$ |
| $\alpha + \overline{1}$ | $\overline{0}$ | $\alpha + \overline{1}$ | $\overline{1}$ | $\alpha$ |

Note in particular that $\alpha$ and $\alpha + \overline{1}$ are multiplicative inverses. More-over, the proof given of Proposition 2.2 applies *verbatim*, except for the somewhat murky problem of knowing what $K$ might be in this instance.

Given $f(x) \in \mathbb{Q}[x]$, we now return to the issue of finding the *smallest* possible field $K \subset \mathbb{C}$ so that $f(x)$ can be factored as a product of *linear* polynomials in $K[x]$.

**Definition.** Let $F$ be a field and let $f(x) \in F[x]$. Let $K$ be a field so that $F \subset K$; we call $K$ a **field extension** of $F$. We say $f(x)$ **splits** in $K$ if $f(x)$ can be written as a product of linear polynomials in $K[x]$. If $f(x)$ splits in $K$, but in no field $E$ satisfying $F \subsetneq E \subsetneq K$, then $K$ is called a **splitting field** of $f(x)$.

We say more casually that $K$ is a splitting field of $f(x) \in F[x]$ if $K$ is the "smallest" field extension of $F$ in which $f(x)$ can be written as a product of linear polynomials. Here are some straightforward examples of splitting fields. (As we shall verify the uniqueness of splitting fields in Section 6 of Chapter 7, we will often take the liberty of referring to "the" splitting field of a polynomial.)

**Examples 4.**

(a) Let $f(x) = x^2 + 2x + 2 \in \mathbb{Q}[x]$. Then by the quadratic formula, the roots of $f(x)$ are $-1 \pm i$, so

$$f(x) = (x - (-1 + i))(x - (-1 - i)).$$

If we set $K = \mathbb{Q}[i]$, this factorization makes sense in $K[x]$. Note that any field containing both roots $-1 + i$ and $-1 - i$ must contain their difference and therefore $i$, and so $\mathbb{Q}[i]$ is the smallest field containing $\mathbb{Q}$ in which the polynomial splits. That is, $\mathbb{Q}[i]$ is the splitting field of $f(x)$.

(b) Let $f(x) = x^2 - 2x - 1 \in \mathbb{Q}[x]$. Then by the quadratic formula, the roots of $f(x)$ are $1 \pm \sqrt{2}$, so

$$f(x) = (x - (1 + \sqrt{2}))(x - (1 - \sqrt{2})).$$

In analogy with the example just completed, $\mathbb{Q}[\sqrt{2}]$ is the splitting field of $f(x) \in \mathbb{Q}[x]$.

(c) Let $f(x) = x^4 - 2 \in \mathbb{Q}[x]$. From Corollary 3.6 of Chapter 2 it follows that the roots of $f(x)$ are $\sqrt[4]{2}, \sqrt[4]{2}i, -\sqrt[4]{2}$, and $-\sqrt[4]{2}i$. Any field containing $\mathbb{Q}$ and these four complex numbers must contain $\sqrt[4]{2}$ and $\frac{1}{\sqrt[4]{2}}(\sqrt[4]{2}i) = i$, so the splitting field of $f(x)$ is $K = \mathbb{Q}[\sqrt[4]{2}, i]$.

(d) Let $f(x) = x^6 - 1 \in \mathbb{Q}[x]$. Since we have the factorization

$$f(x) = (x^3 - 1)(x^3 + 1)$$
$$= (x - 1)(x + 1)(x^2 + x + 1)(x^2 - x + 1),$$

the non-rational roots of $f(x)$ are $\frac{\pm 1 \pm \sqrt{3}i}{2}$. Thus, the splitting field of $f(x)$ is $K = \mathbb{Q}[\sqrt{3}i]$. (Cf. the discussion in Example (a) above.) The reader should check that we obtain the same result upon realizing that the roots of $f(x)$ are the sixth roots of unity.

(e) Let $f(x) = x^3 - 6x^2 + 6x - 2 \in \mathbb{Q}[x]$. Glancing back at Example 2 in Section 4 of Chapter 2, we find that the roots of $f(x)$ are $2 + \sqrt[3]{4} + \sqrt[3]{2}, 2 + \sqrt[3]{4}\omega + \sqrt[3]{2}\omega^2$, and $2 + \sqrt[3]{4}\omega^2 + \sqrt[3]{2}\omega$, where $\omega = -\frac{1}{2} + \frac{\sqrt{3}}{2}i$ is a primitive cube root of unity. Then the splitting field of $f(x)$ is $K = \mathbb{Q}[\sqrt[3]{2}, \sqrt{3}i]$.

We conclude this section with a brief (and optional) discussion of Descartes' rule of signs, a theorem that used to be a standard topic in high school and introductory college algebra. It can be useful in estimating the number of *real* roots of a polynomial with real coefficients.

**Theorem 2.3** (Descartes' Rule of Signs). *Let $f(x) = x^n + a_{n-1}x^{n-1} + \cdots + a_1x + a_0 \in \mathbb{R}[x]$. Let $C_+$ be the number of times the coefficients of $f(x)$ change signs (here we ignore the zero coefficients); let $Z_+$ be the number of positive roots of $f(x)$, counting multiplicities. Then $Z_+ \leq C_+$ and $Z_+ \equiv C_+$ (mod 2). Moreover, if we set $g(x) = f(-x)$, let $C_-$ be the number of times the coefficients of $g(x)$ change signs, and $Z_-$ the number of negative roots of $f(x)$. Then $Z_- \leq C_-$ and $Z_- \equiv C_-$ (mod 2).*

*Proof.* Let $C_+(f)$ denote the number of sign changes in the coefficients of $f(x)$, and let $C_+(f')$ denote the number of sign changes in the coefficients of $f'(x)$, the derivative of $f(x)$. Similarly, let $Z_+(f)$ denote the number of positive roots of $f(x)$, and $Z_+(f')$ the number of positive roots of $f'(x)$ (counting multiplicities in both cases). Now, Rolle's Theorem from calculus tells us that between every two roots of a differentiable function $f(x)$ there must be at least one root of $f'(x)$. Thus, if $f(x)$ has $n$ roots, $f'(x)$ must have at least $n - 1$ roots; it follows that

(A) $$Z_+(f) \le Z_+(f') + 1.$$

Differentiating $f(x) = x^n + a_{n-1}x^{n-1} + \cdots + a_1 x + a_0$, we obtain $f'(x) = nx^{n-1} + (n-1)a_{n-1}x^{n-2} + \cdots + 2a_2 x + a_1$. Since $f(x) \to \infty$ as $x \to \infty$, if $f(0) > 0$ (resp., $< 0$), then $f(x)$ must have an even (resp., odd) number of positive roots (counting multiplicities). The same holds for $f'(x)$. Since $f(0) = a_0$ and $f'(0) = a_1$, we draw the following conclusions:

(B) If $a_0$ and $a_1$ have the same sign, then
$$Z_+(f) \equiv Z_+(f') \;(\text{mod } 2) \text{ and, in fact, } Z_+(f) \le Z_+(f'),$$

by virtue of (A).

(C) If $a_0$ and $a_1$ have opposite signs, then
$$Z_+(f) \equiv Z_+(f') + 1 \;(\text{mod } 2).$$

Our last preparatory observation is that

(D) $$C_+(f) = \begin{cases} C_+(f'), & \text{if } a_1 \text{ and } a_0 \text{ have the same sign} \\ C_+(f') + 1, & \text{if } a_1 \text{ and } a_0 \text{ have opposite signs} \end{cases}.$$

The proof now proceeds by induction on the degree of the polynomial. When $\deg(f(x)) = 1$, we have $f(x) = x + a_0$, and so $f(x)$ has a positive root $\iff a_0 < 0 \iff C_+ = 1$. Now let $\deg(f(x)) = k \ge 2$, and suppose we have proved the result for all polynomials of degree $k - 1$. Given a polynomial $f(x)$ of degree $k$, its derivative $f'(x)$ is a polynomial of degree $k - 1$. Therefore, we have (by inductive hypothesis) $Z_+(f') \le C_+(f')$ and $Z_+(f') \equiv C_+(f')$ (mod 2). We now must consider different cases: if $a_0$ and $a_1$ have the same sign, we conclude that

$$Z_+(f) \underset{(B)}{\le} Z_+(f') \le C_+(f') \underset{(D)}{=} C_+(f), \text{ and}$$

$$Z_+(f) \underset{(B)}{\equiv} Z_+(f') \equiv C_+(f') \underset{(D)}{=} C_+(f) \;(\text{mod } 2),$$

as required. Likewise, if $a_0$ and $a_1$ have opposite signs, we have

$$Z_+(f) \underset{(A)}{\leq} Z_+(f') + 1 \leq C_+(f') + 1 \underset{(D)}{=} C_+(f), \text{ and}$$

$$Z_+(f) \underset{(C)}{\equiv} Z_+(f') + 1 \equiv C_+(f') + 1 \underset{(D)}{\equiv} C_+(f) \pmod{2}.$$

Now we stop to tie up a few technical loose ends. If $a_0 = 0$, then $f(x) = xg(x)$, where $\deg(g(x)) = k - 1$. By inductive hypothesis, $Z_+(g) \leq C_+(g)$ and $Z_+(g) \equiv C_+(g) \pmod{2}$. But $Z_+(f) = Z_+(g)$, and likewise $C_+(f) = C_+(g)$, so the desired result holds for $f(x)$ by transitivity. Lastly, if $a_0 \neq 0$ but $a_1 = 0$, then assume $a_1 = a_2 = \cdots = a_{\ell-1} = 0$ but $a_\ell \neq 0$. Since an analysis of the graph of $f'(x)$ shows that $a_\ell > 0 \implies Z_+(f') \equiv 0 \pmod{2}$ and $a_\ell < 0 \implies Z_+(f') \equiv 1 \pmod{2}$, the argument proceeds exactly as before. We leave the details to the punctilious reader.

The results for $C_-$ and $Z_-$ follow when we observe that positive roots of $g(x)$ correspond to negative roots of $f(x)$. $\square$

**Example 5.** Let $f(x) = x^5 + x^4 - x^2 - 1$. Then $C_+ = 1$ and $C_- = 2$, so $Z_+ = 1$ and $Z_-$ is either 0 or 2. This is the best we can expect without considering the actual values of the coefficients, since it turns out that this polynomial has no negative roots. On the other hand, $g(x) = x^5 + 3x^4 - x^2 - 1$ has two negative roots.

## EXERCISES 3.2

1.  Suppose $f(x) \in \mathbb{C}[x]$ is a monic polynomial of degree $n$ with roots $c_1, c_2, \ldots, c_n$. Prove that the sum of the roots is $-a_{n-1}$ and their product is $(-1)^n a_0$.

2.  Prove that
    a.  $\mathbb{Q}[\sqrt{2}, i] = \mathbb{Q}[\sqrt{2} + i]$, but $\mathbb{Q}[\sqrt{2}i] \subsetneq \mathbb{Q}[\sqrt{2}, i]$
    b.  $\mathbb{Q}[\sqrt{2}, \sqrt{3}] = \mathbb{Q}[\sqrt{2} + \sqrt{3}]$, but $\mathbb{Q}[\sqrt{6}] \subsetneq \mathbb{Q}[\sqrt{2}, \sqrt{3}]$
    c.  $\mathbb{Q}[\sqrt[3]{2} + i] = \mathbb{Q}[\sqrt[3]{2}, i]$; what about $\mathbb{Q}[\sqrt[3]{2}i] \subset \mathbb{Q}[\sqrt[3]{2}, i]$?

3.  Find splitting fields of the following polynomials in $\mathbb{Q}[x]$:
    a.  $f(x) = x^6 + 1$
    b.  $f(x) = (x^2 - 3)(x^3 + 1)$
    c.  $f(x) = x^4 - 9$
    d.  $f(x) = x^8 - 1$
    e.  $f(x) = x^6 - 2x^4 + x^2 - 2$

   f.   $f(x) = x^4 - 2x^2 + 9$   (Hint: Find $\sqrt{1 + 2\sqrt{2}i}$ explicitly.)

   g.   $f(x) = x^6 + 81$

   h.   $f(x) = x^5 - 1$

   i.   $f(x) = x^4 - 10x^2 + 1$   (Hint: $\sqrt{2} + \sqrt{3}$ is one root.)

4.   Decide whether each of the following subsets of $\mathbb{R}$ is a ring, a field, or neither.

   a.   $\{a + b\sqrt[3]{2} : a, b \in \mathbb{Q}\}$

   b.   $\{a + b\sqrt[3]{2} + c\sqrt[3]{4} : a, b, c \in \mathbb{Q}\}$

   c.   $\{a + b\sqrt{2} + c\sqrt{3} : a, b, c \in \mathbb{Q}\}$

5.   Construct multiplication tables for each of the following rings, and decide whether or not each is a field.

   a.   $\mathbb{Z}_2[\alpha]$, $\alpha^2 + \bar{1} = \bar{0}$

   b.   $\mathbb{Z}_2[\alpha]$, $\alpha^3 + \alpha + \bar{1} = \bar{0}$

   c.   $\mathbb{Z}_3[\alpha]$, $\alpha^2 - \alpha + \bar{1} = \bar{0}$

   d.   $\mathbb{Z}_3[\alpha]$, $\alpha^2 + \bar{1} = \bar{0}$

6.   Suppose $\alpha \in \mathbb{C}$ is a root of the given irreducible polynomial $f(x) \in \mathbb{Q}[x]$. Find the multiplicative inverse of $\beta \in \mathbb{Q}[\alpha]$.

   a.   $f(x) = x^2 + 3x - 3$, $\beta = \alpha - 1$         (Answer: $-\alpha - 4$)

   b.   $f(x) = x^3 + x^2 - 2x - 1$, $\beta = \alpha + 1$     (Answer: $-\alpha^2 + 2$)

   c.   $f(x) = x^3 + x^2 + 2x + 1$, $\beta = \alpha^2 + 1$

   d.   $f(x) = x^3 - 2$, $\beta = \alpha + 1$

   e.   $f(x) = x^3 + x^2 - x + 1$, $\beta = \alpha + 2$

   f.   $f(x) = x^3 - 2$, $\beta = r + s\alpha + t\alpha^2$

                                (Answer: $\frac{(r^2 - 2st) + (2t^2 - rs)\alpha + (s^2 - rt)\alpha^2}{r^3 + 2s^3 - 6rst + 4t^3}$)

   g.   $f(x) = x^4 + x^2 - 1$, $\beta = \alpha^3 + \alpha - 1$

7.   Let $f(x) \in \mathbb{R}[x]$.

   a.   Prove that the complex roots of $f(x)$ come in "conjugate pairs"; i.e., $\alpha \in \mathbb{C}$ is a root of $f(x)$ if and only if $\bar{\alpha}$ is also a root.

   b.   Prove that the only irreducible polynomials in $\mathbb{R}[x]$ are linear polynomials and quadratic polynomials $ax^2 + bx + c$ with $b^2 - 4ac < 0$.

8.   Use Descartes' rule of signs, Theorem 2.3, to determine the possible numbers of positive, negative, and complex roots of each of the following real polynomials.

   a.   $f(x) = 3x^4 + x^2 - 2x - 5$

   b.   $f(x) = 4x^3 - 6x^2 + x - 1$

c.  $f(x) = x^6 - 3x^4 + x^3 + 3x^2 + 5$

d.  $f(x) = x^3 + px + q$ (cf. also Proposition 4.4 of Chapter 2)

9.  Give proofs or counterexamples:

    a.  If $C_+$ or $C_-$ is odd, then $f(x)$ must have a real root.

    b.  $Z_+ = C_+$ and $Z_- = C_-$ if and only if $f(x)$ has all real roots.

    c.  If $\deg(f(x)) = n$, then $f$ has at least $n - (C_+ + C_-)$ nonreal roots.

10. Decide whether $f(x) = x^3 - 2$ is irreducible in $\mathbb{Q}[\sqrt{2}][x]$.

11. Let $f(x) \in F[x]$, and suppose $\alpha$ is a root of $f(x)$ in some field extension of $F$. Show that there is an irreducible polynomial $g(x) \in F[x]$ having the property that if $h(x) \in F[x]$ also has $\alpha$ as a root, then $g(x)|h(x)$ in $F[x]$. (Hint: Consider $S = \{h(x) \in F[x] : h(\alpha) = 0\} \subset F[x]$.)

12. Prove that any element of $\mathbb{C}(z)$ can be written as the sum of a polynomial and rational functions in which each numerator is a constant and each denominator is a power of a linear function. (Hint: Theorem 1.9.)

13. Let $K$ be a field extension of $F$, and suppose $\alpha, \beta \in K$. Show that $(F[\alpha])[\beta] = (F[\beta])[\alpha]$, so that $F[\alpha, \beta]$ makes good sense. (Remark: One way to do this is to think about the ring of polynomials in two variables. The other way is just to show directly that every element of one ring belongs to the other.)

14. Decide whether the following polynomials have the same splitting fields:

$$f(x) = x^5 - 6x^3 - x^2 + 6$$
$$g(x) = x^4 + 5x^2 + 6$$
$$h(x) = x^6 + 8$$
$$j(x) = x^6 - 8$$

15. Suppose $f(x), g(x) \in F[x]$ and $K$ is a field extension of $F$. We can consider $f(x)$ and $g(x)$ as polynomials with coefficients in $K$ and suppose we find their g.c.d. in $K[x]$ to be $d(x)$. Prove that $d(x) \in F[x]$. (Hint: Compare degrees of the g.c.d. in $K[x]$ and the g.c.d. in $F[x]$.)

16. Is $\mathbb{Q}[\pi]$ a field?

17.   Let $P_0, \ldots, P_{n-1}$ be $n$ equally spaced points on the unit circle. Compute the product of the distances from $P_0$ to all the remaining points. (Hint: see Exercise 2.3.7.)

18.   Decide whether the fields $\mathbb{Q}[\sqrt[4]{5}(1+i)]$ and $\mathbb{Q}[\sqrt[4]{5}(1-i)] \subset \mathbb{C}$ are equal.

19.   Give an alternative proof of Theorem 2.3 along the following lines.
      a.   Let $h(x) \in \mathbb{R}[x]$ be a monic polynomial of degree $m$, and let $\alpha > 0$. Suppose the coefficients of $h(x)$ change sign $k$ times. Prove that the coefficients of $g(x) = (x - \alpha)h(x)$ change sign at least $k + 1$ times. (Hint: Reduce to the case $\alpha = 1$ by considering $g(\alpha x)$. Then use induction to show that there is at least one more sign change in the sequence $-a_0, a_0 - a_1, a_1 - a_2, \ldots, a_{m-1} - 1, 1$ than in the sequence $a_0, a_1, \ldots, a_{m-1}, 1$.)
      b.   Conclude that $f(x)$ must have at least $Z_+$ sign changes in its coefficients.
      c.   Check the parity statement as follows: if $a_0 > 0$, then $C_+ \equiv 0 \pmod 2$, and if $a_0 < 0$, then $C_+ \equiv 1 \pmod 2$. But $a_0 > 0 \implies Z_+ \equiv 0 \pmod 2$ and $a_0 < 0 \implies Z_+ \equiv 1 \pmod 2$.

## 3. Polynomials with Integer Coefficients

In this section we develop different methods to establish the irreducibility of polynomials in $\mathbb{Q}[x]$. Of course, by clearing denominators, we can assume from the start that our polynomials lie in $\mathbb{Z}[x]$. The first result should be quite familiar.

**Proposition 3.1** (Rational Roots Theorem). *Let* $f(x) = a_n x^n + a_{n-1}x^{n-1} + \cdots + a_1 x + a_0 \in \mathbb{Z}[x]$. *If* $r/s$ *is a rational root of* $f(x)$ *expressed in lowest terms (i.e.,* $\gcd(r,s) = 1$*), then* $r \mid a_0$ *and* $s \mid a_n$.

**Proof.** If $f(\frac{r}{s}) = 0$, then $a_n(\frac{r}{s})^n + a_{n-1}(\frac{r}{s})^{n-1} + \cdots + a_1(\frac{r}{s}) + a_0 = 0$. We clear denominators by multiplying by $s^n$, and obtain the equation $a_n r^n + a_{n-1}r^{n-1}s + \cdots + a_1 r s^{n-1} + a_0 s^n = 0$. Rewriting this in the form $-r(a_n r^{n-1} + a_{n-1}r^{n-2}s + \cdots + a_1 s^{n-1}) = a_0 s^n$, we infer that $r \mid a_0 s^n$. Since $\gcd(r,s) = 1$, we conclude (see Exercise 1.2.7) that $r \mid a_0$, as desired. We leave it to the reader to complete the proof (see Exercise 1). $\square$

With diligence we could apply this result to find all possible rational roots of a given polynomial. (Of course, having found one root $c$ of the polynomial $f(x)$, we can then consider the polynomial $f(x)/(x - c)$, which has lesser degree.)

**Corollary 3.2.** *If $f(x) = x^n + a_{n-1}x^{n-1} + \cdots + a_1x + a_0 \in \mathbb{Z}[x]$ is monic, any rational root must be an integer $r$ dividing $a_0$.*

**Examples 1.** Find the rational roots of the following polynomials.

(a) $f(x) = x^3 - x^2 - 10x - 8$. Since $f(x)$ is monic, any rational root must in fact be an integer $r$ that divides $a_0 = -8$. Thus, the possibilities are 1, $-1$, 2, $-2$, 4, $-4$, 8, and $-8$. Evaluating $f(x)$ at each of these eight values, we find that, indeed, $f(-1) = f(-2) = f(4) = 0$, and so $-1$, $-2$, and 4 are the roots of $f(x)$.

(b) $f(x) = 15x^4 + 4x^3 + 11x^2 + 4x - 4$. Here there are 24 possible rational roots $r/s$: $r \in \{\pm1, \pm2, \pm4\}$ and $s \in \{1, 3, 5, 15\}$. Only $-2/3$ and $2/5$ are roots. (We can tell from a rough curve-sketch—or by examining the remaining quadratic factor—that the other two roots are complex.)

(c) $f(x) = x^5 - 3$ has no rational root (since the only possibilities are $\pm1$ and $\pm3$, and these are easily checked). As a consequence, $\sqrt[5]{3}$ is irrational.

We remind the reader to be careful, however; just because a polynomial has no rational root, it need not be irreducible (see Section 1). We now turn to two very useful techniques for establishing irreducibility of polynomials in $\mathbb{Q}[x]$, both based on transforming the question to $\mathbb{Z}_p[x]$. We begin with a simple observation: Given a polynomial $f(x) = a_nx^n + a_{n-1}x^{n-1} + \cdots + a_1x + a_0 \in \mathbb{Z}[x]$, we obtain a polynomial $\overline{f}(x) \in \mathbb{Z}_p[x]$ naturally by "reducing" each of the coefficients mod $p$ (i.e., by replacing each coefficient $a_j$ by its mod $p$ equivalence class $\overline{a}_j$). Moreover, if $\deg(f(x)) = n$ and $a_n \not\equiv 0 \pmod{p}$, then $\deg(\overline{f}(x)) = n$ as well. Before going any further, we need the following fact:

> Given a polynomial $f(x)$ with *integer* coefficients, $f(x)$ factors as a product of nonconstant polynomials in $\mathbb{Q}[x]$ if and only if $f(x)$ so factors in $\mathbb{Z}[x]$.

More precisely, we have the following result, which is more subtle than it first appears.

**Proposition 3.3** (Gauss' Lemma). *If $f(x) \in \mathbb{Z}[x]$ and $g(x), h(x) \in \mathbb{Q}[x]$ are nonconstant polynomials so that $f(x) = g(x)h(x)$, then there are nonconstant polynomials $\tilde{g}(x), \tilde{h}(x) \in \mathbb{Z}[x]$ so that $f(x) = \tilde{g}(x)\tilde{h}(x)$.*

**Proof.** There are integers $k, \ell$ so that $g_1(x) = kg(x)$ and $h_1(x) = \ell h(x) \in \mathbb{Z}[x]$ (clear denominators). Therefore,

$$(*) \qquad k\ell f(x) = g_1(x)h_1(x).$$

Now let $p \in \mathbb{N}$ be a prime factor of $k\ell$; then $p \mid g_1(x)h_1(x)$—by which we mean that $p$ divides every coefficient of the polynomial $g_1(x)h_1(x) \in \mathbb{Z}[x]$—and so we claim that $p \mid g_1(x)$ or $p \mid h_1(x)$. Thus $p$ divides one of the two factors of the right-hand side, and so we may cancel a factor of $p$ from each side of $(*)$. We continue this process until we are left with the desired equation $f(x) = \tilde{g}(x)\tilde{h}(x)$.

Now we turn to a proof of the claim. We reduce mod $p$: since $p$ divides $g_1(x)h_1(x)$, $\overline{g_1(x)h_1(x)} = \overline{0}$. Here is the crucial point: By the definition of polynomial multiplication and Proposition 3.1 of Chapter 1, $\overline{g_1(x)h_1(x)} = \overline{g}_1(x)\overline{h}_1(x)$. Insofar as $\mathbb{Z}_p[x]$ is an integral domain (by Proposition 1.2), if $\overline{g}_1(x)\overline{h}_1(x) = \overline{0}$, then one of the factors $\overline{g}_1(x), \overline{h}_1(x)$ must be $\overline{0}$. Therefore, $p$ divides either $g_1(x)$ or $h_1(x)$. $\square$

**Example 2.** The polynomial $f(x) = 2x^2 - 3x - 2$ can be factored as $(x + \frac{1}{2})(2x - 4)$; but of course we can transfer the 2 from the second factor to the first, obtaining $f(x) = (2x + 1)(x - 2) \in \mathbb{Z}[x]$.

**Example 3.** Given $f(x) = x^4 - x^3 + 2$, we are to decide whether $f(x)$ is irreducible in $\mathbb{Q}[x]$. It has no rational roots, but it may factor as a product of two irreducible quadratic polynomials. By virtue of Proposition 3.3, if it factors, it does so as the product of two polynomials with *integer* coefficients. So, suppose $f(x) = x^4 - x^3 + 2 = (x^2 + ax + b)(x^2 + cx + d)$, where $a, b, c, d \in \mathbb{Z}$ (note that we may assume both polynomials are monic; why?). Then comparing coefficients on the left- and right-hand sides of this equation, we obtain these equations:

$$a + c = -1$$
$$b + ac + d = 0$$
$$(†) \qquad ad + bc = 0$$
$$bd = 2 \, .$$

From the last equation, we see there are four possible pairs of values for $b$ and $d$: $(b, d)$ equals either $(1, 2)$, $(-1, -2)$, $(2, 1)$ or $(-2, -1)$. We give the rest of the reasoning in the first case, and leave it to the reader to complete the argument. If $b = 1$ and $d = 2$, then we now obtain these equations:

$$a + c = -1$$
$$ac = -3$$
$$2a + c = 0 \, .$$

From this, we obtain the contradictory equations $a = 1$, $c = -2$, $ac = -3$. The upshot (after checking out the remaining three cases similarly) is that there can be no integers $a$, $b$, $c$, $d$ so that the original equation holds. Thus, $f(x)$ is irreducible in $\mathbb{Q}[x]$.

**Remark.** This approach is often called the **method of undetermined coefficients**, and, although tedious, is often quite useful. It was apparently discovered by Descartes.

**Proposition 3.4.** *Given $f(x) = a_n x^n + a_{n-1} x^{n-1} + \cdots + a_1 x + a_0 \in \mathbb{Z}[x]$, if for some prime number $p$, $a_n \not\equiv 0$ (mod $p$) and the corresponding polynomial $\overline{f}(x) \in \mathbb{Z}_p[x]$ is irreducible, then $f(x)$ is irreducible in $\mathbb{Q}[x]$.*

**Proof.** We prove the contrapositive. Suppose $f(x)$ is not irreducible in $\mathbb{Q}[x]$. By Proposition 3.3, $f(x)$ can be written as a product of nonconstant polynomials in $\mathbb{Z}[x]$. If $f(x) = g(x)h(x) \in \mathbb{Z}[x]$, with $\deg(g(x)), \deg(h(x)) \geq 1$, then reducing mod $p$, we obtain $\overline{f}(x) = \overline{g}(x)\overline{h}(x)$ in $\mathbb{Z}_p[x]$. We must now use the hypothesis that $a_n \not\equiv 0$ (mod $p$) to conclude that $\deg(\overline{f}(x)) = \deg(f(x))$ and therefore $\deg(\overline{g}(x)) = \deg(g(x))$, $\deg(\overline{h}(x)) = \deg(h(x))$, whence $\overline{f}(x)$ is factored in $\mathbb{Z}_p[x]$ as a product of polynomials of degree $\geq 1$. $\square$

**Examples 4.** The polynomial $x^2 + x + \overline{1}$ is irreducible in $\mathbb{Z}_2[x]$ (why?); and so $f(x) = 3x^2 - x + 7$, $f(x) = x^2 + 27x - 15$, $f(x) = 9x^2 - 21x + 325, \ldots$ are all irreducible in $\mathbb{Q}[x]$. Likewise, $x^3 + x + \overline{1}$ is irreducible in $\mathbb{Z}_2[x]$, and so $f(x) = x^3 + 4x^2 + 5x + 5879$ is irreducible in $\mathbb{Q}[x]$. Moving to higher degree, $x^4 + x + \overline{1}$ is irreducible in $\mathbb{Z}_2[x]$ (see Exercise 6b.); therefore, $f(x) = x^4 + 2x^3 - 6x^2 + 7x - 13$ is irreducible in $\mathbb{Q}[x]$.

**Example 5.** Let's consider $f(x) = x^4 - x^3 + 2$ from Example 3 above. Working in $\mathbb{Z}_3[x]$, $\overline{f}(x) = x^4 - x^3 - \overline{1}$ obviously has no root. If it factors as $\overline{f}(x) = (x^2 + ax + b)(x^2 + cx + d)$, we obtain the same equations (†) as before. But working in $\mathbb{Z}_3$, it is quite easy to see that there is no solution: $b = \overline{1}$ and $d = -\overline{1}$ (or *vice versa*), so $a - c = \overline{0}$; since $a + c = \overline{2}$, we must have $a = c = \overline{1}$, and so $b + ac + d = \overline{1}$, contradicting the second equation. Thus $\overline{f}(x) \in \mathbb{Z}_3[x]$ is irreducible, and so $f(x)$ is irreducible in $\mathbb{Q}[x]$.

**Example 6.** Consider $f(x) = 2x^2 + 3x + 1 \in \mathbb{Z}[x]$. Reducing mod 2, we get $\overline{f}(x) = x + \overline{1} \in \mathbb{Z}_2[x]$, which is certainly irreducible. But $f(x) = (2x + 1)(x + 1)$ is not.

**Remark.** Be warned: the converse of Proposition 3.4 is definitely false! The polynomial $f(x) = x^2 + x + 2$ is irreducible in $\mathbb{Q}[x]$, yet $\overline{f}(x) \in \mathbb{Z}_2[x]$ is certainly not irreducible, as $\overline{f}(x) = x^2 + x = x(x + \overline{1}) \in \mathbb{Z}_2[x]$.

Query. Is it conceivable that this test is powerful enough to detect all irreducible polynomials in $\mathbb{Q}[x]$, or might there be a polynomial $f(x) \in \mathbb{Z}[x]$ that is irreducible in $\mathbb{Q}[x]$ and yet fails to be irreducible (mod $p$) in $\mathbb{Z}_p[x]$ for every prime number $p$? (See Exercise 10.)

We now give the last test for irreducibility in $\mathbb{Q}[x]$.

**Theorem 3.5** (Eisenstein Criterion). *Let* $f(x) = a_n x^n + a_{n-1} x^{n-1} + \cdots + a_1 x + a_0 \in \mathbb{Z}[x]$. *Let* $p$ *be a prime so that* $a_j \equiv 0 \pmod{p}$, *for all* $j = 0, 1, \ldots, n - 1$, *but* $a_n \not\equiv 0 \pmod{p}$, *and* $a_0 \not\equiv 0 \pmod{p^2}$. *Then* $f(x)$ *is irreducible in* $\mathbb{Q}[x]$.

***Proof.*** Suppose $f(x) = g(x)h(x)$, where $g(x), h(x) \in \mathbb{Z}[x]$. Put $g(x) = b_k x^k + \cdots + b_0$ and $h(x) = c_{n-k} x^{n-k} + \cdots + c_0$. Since $a_0 = b_0 c_0$, we have $b_0 c_0 \equiv 0 \pmod{p}$ and $\not\equiv 0 \pmod{p^2}$. Thus, $p$ divides either $b_0$ or $c_0$, but not both. To be specific, let's say that $p$ divides $c_0$ and does not divide $b_0$.

Now consider reduction mod $p$: since all but the leading coefficient of $f(x)$ are divisible by $p$, $\overline{f}(x) = \overline{a}_n x^n$, and $\overline{a}_n \neq \overline{0}$. As in the proof of Proposition 3.3, $\overline{f}(x) = \overline{g}(x)\overline{h}(x)$. Since $\mathbb{Z}_p$ is a field, by

Theorem 1.8 there is unique factorization in $\mathbb{Z}_p[x]$. Therefore, the only way the product of $\overline{g}(x)$ and $\overline{h}(x)$ can be a (nonzero) monomial is for each of them in turn to be a nonzero monomial. On the other hand, $\overline{b}_k \overline{c}_{n-k} = \overline{a}_n \neq \overline{0}$, so, in particular, $\overline{b}_k \neq \overline{0}$. Since $\overline{g}(x) = \overline{b}_k x^k + \cdots + \overline{b}_0$ and $\overline{b}_0 \neq \overline{0}$ (remember that $p \nmid b_0$), $g(x)$ must be a constant polynomial. Therefore, $f(x)$ is irreducible in $\mathbb{Q}[x]$. $\square$

### Examples 7.

(a) $f(x) = x^3 + 2x + 6$ is irreducible in $\mathbb{Q}[x]$, since we may apply Theorem 3.5 with $p = 2$.

(b) $f(x) = 5x^3 + 6x - 12$ is irreducible in $\mathbb{Q}[x]$, since we may apply Theorem 3.5 with $p = 3$. Note that we may not use $p = 2$, since $12 \equiv 0 \pmod 4$.

(c) $f(x) = x^4 + x^3 + x^2 + x + 1$ seems to be impervious to this approach. However, we consider $g(x) = f(x+1) = (x+1)^4 + (x+1)^3 + (x+1)^2 + (x+1) + 1 = x^4 + 5x^3 + 10x^2 + 10x + 5$. Applying Theorem 3.5 with $p = 5$, we infer that $g(x)$ is irreducible, and hence by Exercise 3.1.12, $f(x)$ is irreducible as well. (See also Exercise 7.)

Query. The trick in Example (c) showed that we could make an appropriate substitution of $x + n$ for $x$ and apply Theorem 3.5 successfully. Will this always be possible? (See Exercise 9 below.)

Remark. It may be useful at this point to summarize this section by developing some strategy for testing polynomials $f(x) \in \mathbb{Q}[x]$ for irreducibility. Obviously, if we are fortunate enough to be able to apply the Eisenstein criterion, this is the easiest possible test. Failing this, if we are given a quadratic or cubic polynomial, we should look for possible roots (reducing mod $p$ for a convenient $p$ will most often be easier than checking all possible rational roots). In general, if we have handy a list of irreducible polynomials of sufficiently high degree in $\mathbb{Z}_p[x]$ for some convenient prime $p$, then reduction mod $p$ is a good test to try (cf. Exercise 6b.). As a last resort for polynomials of degree $\geq 4$, we can always try the method of undetermined coefficients.

## EXERCISES 3.3

1. Finish the proof of Proposition 3.1; i.e., prove that $s|a_n$.

2. Decide which of the following polynomials are irreducible in $\mathbb{Q}[x]$.
   a. $f(x) = x^3 + 4x^2 - 3x + 5$
   b. $f(x) = 4x^4 - 6x^2 + 6x - 12$
   c. $f(x) = x^3 + x^2 + x + 1$
   d. $f(x) = x^4 - 180$
   e. $f(x) = x^4 + x^2 - 6$
   f. $f(x) = x^4 - 2x^3 + x^2 + 1$
   g. $f(x) = x^3 + 17x + 36$
   h. $f(x) = x^4 + x + 1$
   i. $f(x) = x^5 + x^3 + x^2 + 1$
   j. $f(x) = x^5 + x^3 + x + 1$

3. Find the rational roots of the following polynomials:
   a. $f(x) = 3x^3 + 5x^2 + 5x + 2$
   b. $f(x) = x^5 - x^4 - x^3 - x^2 - x - 2$
   c. $f(x) = 2x^4 + 3x^3 + 4x + 6$
   d. $f(x) = x^3 + x^2 - 2x - 3$
   e. $f(x) = x^3 - \frac{1}{5}x^2 - 4x + \frac{4}{5}$

4. Show that each of the following polynomials has no rational root:
   a. $x^{200} - x^{41} + 4x + 1$
   b. $x^8 - 54$
   c. $x^{2k} + 3x^{k+1} - 12, \ k \geq 1$

5. Prove that for any prime number $p$ and any $n \in \mathbb{N}$, $f(x) = x^n - p$ is irreducible in $\mathbb{Q}[x]$.

6. a. Prove that $f(x) \in \mathbb{Z}_2[x]$ has $x + \overline{1}$ as a factor if and only if it has an even number of nonzero coefficients.
   b. List the irreducible polynomials in $\mathbb{Z}_2[x]$ of degrees $2, 3, 4$, and $5$.

   **Remark.** For the reader's interest and convenience, we provide on the next page a list of the monic irreducible polynomials in $\mathbb{Z}_3[x]$ of degrees 2, 3, and 4. (For legibility, we omit the equivalence class "bars" on coefficients.)

7.  Prove that for any prime number $p$, $f(x) = x^{p-1} + x^{p-2} + \cdots + x + 1$ is irreducible in $\mathbb{Q}[x]$. (Hint: Consider $f(x + 1)$.)

8.  a.  Let $p$ be a prime. Prove that $x^p - x$ has $p$ distinct roots in $\mathbb{Z}_p$. (For one approach, see Proposition 3.3 of Chapter 1.)
    b.  Deduce that $x^{p-1} - \overline{1} = (x - \overline{1})(x - \overline{2}) \ldots (x - \overline{p - 1})$ in $\mathbb{Z}_p[x]$.
    c.  Conclude that $(p - 1)! \equiv -1 \pmod{p}$ (see Exercise 1.4.13).

9.  Find a quadratic polynomial $f(x) \in \mathbb{Z}[x]$ that is irreducible in $\mathbb{Q}[x]$, but has the property that for every $n \in \mathbb{Z}$ the Eisenstein criterion cannot be applied successfully to prove irreducibility of $f(x + n)$.

10.  Show that $f(x) = x^4 - 10x^2 + 1$ is irreducible in $\mathbb{Q}[x]$, yet reducible in $\mathbb{Z}_p[x]$ for every prime $p$. You may need the following result, whose proof is postponed to Exercise 6.3.33.

     **Lemma.** *If neither* 2 *nor* 3 *is a square mod* $p$, *then* 6 *is a square mod* $p$.

     (Hint: Try to write $f(x)$ as a difference of squares in several different ways.)

---

MONIC IRREDUCIBLE POLYNOMIALS IN $\mathbb{Z}_3[x]$
OF DEGREES 2, 3, AND 4

| | |
|---|---|
| $x^2 + 1$ | $x^4 + x^2 + x + 1$ |
| $x^2 + x + 2$ | $x^4 + x^2 + 2x + 1$ |
| $x^2 + 2x + 2$ | $x^4 + 2x^2 + 2$ |
| | $x^4 + x^3 + 2$ |
| $x^3 + 2x + 1$ | $x^4 + x^3 + 2x + 1$ |
| $x^3 + 2x + 2$ | $x^4 + x^3 + x^2 + 1$ |
| $x^3 + x^2 + 2$ | $x^4 + x^3 + x^2 + x + 1$ |
| $x^3 + x^2 + x + 2$ | $x^4 + x^3 + x^2 + 2x + 2$ |
| $x^3 + x^2 + 2x + 1$ | $x^4 + x^3 + 2x^2 + 2x + 2$ |
| $x^3 + 2x^2 + 1$ | $x^4 + 2x^3 + 2$ |
| $x^3 + 2x^2 + x + 1$ | $x^4 + 2x^3 + x + 1$ |
| $x^3 + 2x^2 + 2x + 2$ | $x^4 + 2x^3 + x^2 + 1$ |
| | $x^4 + 2x^3 + x^2 + x + 2$ |
| $x^4 + x + 2$ | $x^4 + 2x^3 + x^2 + 2x + 1$ |
| $x^4 + 2x + 2$ | $x^4 + 2x^3 + 2x^2 + x + 2$ |
| $x^4 + x^2 + 2$ | |

# CHAPTER 4

# Homomorphisms and Quotient Rings

In this chapter we develop more of the structure of rings and the fundamental notion of a function enters. To study rings more carefully, we must be able to tell when two are "the same": we must be able to give a bijective function from one to the other that is compatible with addition and multiplication in the respective rings; such a function is called a (ring) isomorphism. More generally, we study functions, called homomorphisms, from one ring to another that are compatible with the algebraic structures. The arithmetic properties of prime and composite integers generalize naturally to the study of certain subsets of rings, called ideals. And there is a comparable generalization of the ring of equivalence classes of integers mod $m$, called the quotient ring construction: We now consider equivalence classes of elements of our ring defined by an ideal. We are able to construct splitting fields of polynomials in greater generality, and to understand the "complexity" of a complex number by the size of the ring it generates.

In an attempt to prove Fermat's infamous last theorem (which states that there are no positive integral solutions of $x^n + y^n = z^n$ for $n \geq 3$, and which has just been proved by A. Wiles—in an incredible *tour de force* of algebraic number theory—as I write at the beginning of 1995), Kummer in the middle of the nineteenth century was led to consider rings in which there is not unique factorization (e.g., in $\mathbb{Z}[\sqrt{-13}]$, we have $14 = 2 \cdot 7 = (1 + \sqrt{-13})(1 - \sqrt{-13})$). In order to "force" unique factorization, he factored the numbers as products of *ideal* numbers that were not necessarily in his original ring. These ultimately developed into the *ideals* we shall soon study. We return to the number-theoretic origins of rings in §3, where complex numbers, geometry, and ring theory combine in a study of the ring of Gaussian integers. In particular, we will use complex arithmetic to determine a necessary and sufficient condition for a prime number to be a sum of two squares.

## 1. Ring Homomorphisms and Ideals

> **Caveat:** From this point on, we drop the "bar" notation for elements of $\mathbb{Z}_m$. We assume that the reader is now fully familiar with $\mathbb{Z}_m$, and the notation will be used in a more general context from now on.

In order to unify the material we've studied in the first three chapters, we next come to functions in an algebraic setting. We have already seen this idea informally when associating to each integer its equivalence class mod $m$ (thus giving a function from $\mathbb{Z}$ to $\mathbb{Z}_m$), and when reducing a polynomial in $\mathbb{Z}[x]$ mod $p$ (thus giving a function from $\mathbb{Z}[x]$ to $\mathbb{Z}_p[x]$).

**Definition.** Let $R$ and $S$ be rings. A function $\phi: R \to S$ is called a **ring homomorphism** if for all $a, b \in R$,

(1) $\phi(a + b) = \phi(a) + \phi(b)$,
(2) $\phi(ab) = \phi(a)\phi(b)$, and
(3) $\phi(1_R) = 1_S$.

(Note that addition and multiplication on the left-hand side are in $R$, whereas those on the right-hand side are in $S$.)

Thus a ring homomorphism is compatible with the algebraic structures of $R$ and $S$. We begin with a few technical observations.

**Lemma 1.1.** *Let* $\phi: R \to S$ *be a ring homomorphism. Then*

(i) $\phi(0_R) = 0_S$.
(ii) *The image of* $\phi$ *is a subring of S. (Recall that* image $(\phi) = \phi(R) = \{s \in S : s = \phi(r) \text{ for some } r \in R\}$.)

***Proof.*** To prove (i), note that $\phi(0_R) = \phi(0_R + 0_R) = \phi(0_R) + \phi(0_R)$; so, adding the additive inverse of $\phi(0_R)$ (in $S$, of course) to both sides, we obtain $\phi(0_R) = 0_S$.

To prove (ii), we must show that $\phi(R)$ is closed under addition and multiplication, and contains additive inverses and the multiplicative identity $1_S$. All the remaining properties of a ring are inherited from $S$. First note that given $a, b \in R$, $\phi(a) + \phi(b) = \phi(a + b) \in \phi(R)$, and $\phi(a)\phi(b) = \phi(ab) \in \phi(R)$; so $\phi(R)$ is indeed closed under addition and multiplication. The additive inverse of $\phi(a)$ is $-\phi(a) = \phi(-a) \in \phi(R)$ (see Exercise 1). Lastly, since $\phi(1_R) = 1_S$, the multiplicative identity $1_S$ belongs to $\phi(R)$.  □

We now give various examples of ring homomorphisms; the reader should check that the relevant properties hold.

**Examples 1.**

(a) First, we have the unimaginative example: the identity homomorphism $\iota: R \to R$ is given by $\iota(a) = a$ for all $a \in R$.

(b) Closely related to (a), let $\phi: \mathbb{Z} \to \mathbb{Q}$ be given by $\phi(n) = n$ for all $n \in \mathbb{Z}$. This is obviously a homomorphism, with image $\mathbb{Z} \subset \mathbb{Q}$.

(c) Let $\phi: \mathbb{Z} \to \mathbb{Z}_m$ be given by $\phi(a) = a \pmod{m}$. $\phi$ assigns to each integer (the equivalence class of) its remainder upon division by $m$. As a consequence of Proposition 3.1 of Chapter 1, $\phi$ is a homomorphism.

(d) Let $\psi: \mathbb{Z}[x] \to \mathbb{Z}_m[x]$, $\psi(f(x)) = \overline{f}(x)$, be defined in terms of the homomorphism $\phi$ in (c) as follows:

$$\psi\left(\sum_{j=0}^{n} a_j x^j\right) = \sum_{j=0}^{n} \phi(a_j)x^j = \sum_{j=0}^{n} (a_j \pmod{m})x^j.$$

The fact that $\psi$ is a ring homomorphism was crucial in the proof of Proposition 3.3 and Theorem 3.5 of Chapter 3.

(e) Given a commutative ring $R$ and an element $a \in R$, define the evaluation homomorphism $\mathrm{ev}_a: R[x] \to R$ by $\mathrm{ev}_a(f(x)) = f(a)$. This will prove to be one of our most important examples.

(f) Let $\phi: \mathbb{C} \to \mathbb{C}$ be complex conjugation, $\phi(z) = \overline{z}$.

(g) Define $\phi: \mathbb{Z}_6 \to \mathbb{Z}_3$ by $\phi(a \pmod{6}) = a \pmod{3}$. To see that $\phi$ is well-defined, note that if $a \equiv b \pmod{6}$, then $a \equiv b \pmod{3}$ since $3|6$.

(h) Given a ring (most interestingly, not commutative) $R$ and a unit $a \in R$, define $\psi_a: R \to R$ by $\psi_a(x) = axa^{-1}$. (Cf. Theorem 1.1 of Appendix B for a relation with linear algebra.) To prove that $\psi_a$ is a homomorphism, let $x, y \in R$.

(1) $\psi_a(x) + \psi_a(y) = axa^{-1} + aya^{-1} = a(x + y)a^{-1} = \psi_a(x + y)$;

(2) $\psi_a(x)\psi_a(y) = (axa^{-1})(aya^{-1}) = (ax)(a^{-1}a)(ya) = a(xy)a^{-1} = \psi_a(xy)$;

(3) $\psi_a(1_R) = a1_R a^{-1} = 1_R$, as required.

We assume from here on that $R$ and $S$ are commutative rings, and we proceed to study a special sort of subset of a commutative ring. For motivation, consider the **kernel** of the ring homomorphism $\phi: R \to S$, defined as follows:

$$\ker \phi = \{a \in R : \phi(a) = 0_S\}.$$

Let's see what algebraic properties $\ker \phi$ has. First, $\ker \phi$ is closed under addition, since if $a, b \in \ker \phi$, then $\phi(a + b) = \phi(a) + \phi(b) = 0 + 0 = 0$, whence $a + b \in \ker \phi$. Similarly, $\ker \phi$ is closed under multiplication, but in fact more is true. Given $a \in \ker \phi$ and $r \in R$ arbitrary, $\phi(ra) = \phi(r)\phi(a) = \phi(r)0 = 0$ (see Exercise 1.4.6a.), so $ra \in \ker \phi$. This leads us to make the following general definition.

**Definition.** Let $R$ be a commutative ring. We say a nonempty subset $I \subset R$ is an **ideal** if

(1) $a, b \in I \implies a + b \in I$, and

(2) $a \in I, r \in R \implies ra \in I$.

### Examples 2.

(a) Let $R$ be any commutative ring, and let $I = \{0\}$; this is called the *zero ideal*.

(b) Let $R = \mathbb{Z}$, and let $I = \{3n : n \in \mathbb{Z}\}$ be the subset consisting of all the multiples of 3.

(c) Let $R = \mathbb{Z}$, let $a, b \in \mathbb{Z}$ be fixed, not both zero, and let $I = \{ma + nb : m, n \in \mathbb{Z}\}$. It is easy to check that $I$ is an ideal, and by Theorem 2.3 of Chapter 1, $I = \{kd : k \in \mathbb{Z}\}$, where $d = \gcd(a, b)$.

(d) Let $R = F[x]$, and let $f(x) \in F[x]$. Let $I = \{g(x)f(x) : g(x) \in F[x]\}$ be the set of all polynomials that are divisible by $f(x)$.

(e) Let $R = \mathbb{Q}[x]$, and let $I = \{f(x) \in \mathbb{Q}[x] : f(2) = 0\}$. Then $I$ is the ideal of all polynomials having 2 as a root. By Corollary 1.5 of Chapter 3, $I$ consists of all polynomials in $\mathbb{Q}[x]$ that are divisible by $x - 2$.

(f) Let $R = \mathbb{R}$, and let $I = \mathbb{Q}$. Then $I$ is *not* an ideal: although $\mathbb{Q}$ is closed under addition, the second criterion fails (consider multiplication by any irrational $r \in \mathbb{R}$).

We must introduce some notation. Given a commutative ring $R$, we often consider the ideal consisting of all multiples of a given element $a \in R$. We call this the **ideal generated by** $a$, and denote it by $\langle a \rangle$. That is,

$$\langle a \rangle = \{ra : r \in R\}.$$

More generally, if $a_1, \ldots, a_n \in R$, we may consider the ideal generated by them all:

$$\langle a_1, \ldots, a_n \rangle = \{r_1 a_1 + r_2 a_2 + \cdots + r_n a_n : r_1, \ldots, r_n \in R\}.$$

In Examples 2, the ideals would be written in our new notation, respectively, as follows: $\langle 0 \rangle \subset R$, $\langle 3 \rangle \subset \mathbb{Z}$, $\langle a, b \rangle \subset \mathbb{Z}$, $\langle f(x) \rangle \subset F[x]$, and $\langle x - 2 \rangle \subset \mathbb{Q}[x]$.

**Example 3.** Suppose $a, b \in R$ and $\langle b \rangle \subset \langle a \rangle$. This means in particular that $b \in \langle a \rangle$, so $b = ra$ for some $r \in R$. Now suppose $\langle b \rangle = \langle a \rangle$. This means that $b = ra$ for some $r \in R$ *and* $a = sb$ for some $s \in R$. But be careful! We cannot conclude that $a = b$. Substituting, we find $b = (rs)b$. Now if $R$ is an integral domain, we can conclude (see Exercise 1.4.7) that $rs = 1$, and so $r$ and $s$ are units. Therefore, $\langle a \rangle = \langle b \rangle \iff a = sb$ for some unit $s \in R$. However, if $R$ is not an integral domain, this need not be true. Consider the example $\langle 2 \rangle = \langle 4 \rangle \subset \mathbb{Z}_6$.

**Definition.** We say an ideal $I \subset R$ is **principal** if $I = \langle a \rangle$ for some $a \in R$, i.e., if it is generated by a single element. If $R$ is an integral domain and every ideal of $R$ is principal, we say $R$ is a **principal ideal domain**.

Any ring in which there is a division algorithm (such as $\mathbb{Z}$ and $F[x]$, and we shall see another—quite different—example in Section 3) must be a principal ideal domain. Indeed, given integers $a$ and $b$ or polynomials $f(x)$ and $g(x)$, we have found their g.c.d.'s exactly by finding the element that generates the respective ideals $\langle a, b \rangle \subset \mathbb{Z}$ and $\langle f(x), g(x) \rangle \subset F[x]$. For the record, we reformulate this officially.

**Proposition 1.2.** $\mathbb{Z}$ *and the polynomial ring* $F[x]$ *(for any field $F$) are principal ideal domains.*

***Proof.*** Let $I \subset \mathbb{Z}$ be an ideal. If $I = \langle 0 \rangle$, we're done. If not, let $a$ be the *smallest* positive integer in $I$. We claim $I = \langle a \rangle$. Certainly, $\langle a \rangle \subset I$, but why is every element of $I$ a multiple of $a$? Suppose $b \in I$; then there are integers $q$ and $r$ so that $b = qa + r$ with $0 \le r < a$. Since $b \in I$ and $a \in I$, it follows that $r \in I$. Because $a$ is the smallest positive element of $I$, we must have $r = 0$, and so $b$ is a multiple of $a$, as desired. (The reader should compare this with the proof of Theorem 2.3 of Chapter 1.)

The proof for the polynomial ring is analogous, following the ideas of Theorem 1.6 of Chapter 3. The reader should complete the proof. □

We obtain from Proposition 1.2 an amazingly simple, yet powerful, result we shall use numerous times.

**Corollary 1.3.** *If $I \subsetneq \mathbb{Z}$ is an ideal and a prime number $p \in I$, then $I = \langle p \rangle$. Similarly, if $I \subsetneq F[x]$ is an ideal, $f(x) \in I$, and $f(x)$ is irreducible in $F[x]$, then $I = \langle f(x) \rangle$.*

***Proof.*** Since $\mathbb{Z}$ is a principal ideal domain, $I = \langle a \rangle$ for some nonzero $a \in \mathbb{Z}$; we may assume $a$ is positive, and, insofar as $\langle 1 \rangle = \mathbb{Z}$, $a$ must be at least 2. Because $p \in I$, we have $p = ra$ for some $r \in \mathbb{Z}$; but since $p$ is prime, $r = 1$ and $a = p$.

Similarly, since $F[x]$ is a principal ideal domain, $I = \langle g(x) \rangle$ for some $g(x) \in F[x]$, and $f(x) = r(x)g(x)$ for some $r(x) \in F[x]$. Because $f(x)$ is irreducible in $F[x]$, either $r(x)$ or $g(x)$ must be a (nonzero) constant. If $g(x)$ were a constant, we'd have $\langle g(x) \rangle = F[x]$ (see Exercise 3). As a result, $r(x)$ must be a constant, and so $\langle f(x) \rangle = \langle g(x) \rangle = I$, as desired. □

And from this we deduce the following result (which the reader should compare with Proposition 2.2 of Chapter 2).

**Corollary 1.4.** *Suppose $f(x) \in F[x]$ is irreducible in $F[x]$, and $K \supset F$ is a field containing a root $\alpha$ of $f(x)$. Then the ideal of all polynomials in $F[x]$ vanishing at $\alpha$ is generated by $f(x)$. In other words, letting $\operatorname{ev}_\alpha \colon F[x] \to K$ denote the evaluation homomorphism, $\ker \operatorname{ev}_\alpha = \langle f(x) \rangle$.*

***Proof.*** This is immediate from Corollary 1.3, since $\ker \operatorname{ev}_\alpha$ is not all of $F[x]$ (the constant polynomial 1 does not vanish at $\alpha$). □

Here is an example of an integral domain that is *not* a principal ideal domain.

**Example 4.** Consider $R = \mathbb{Z}[x]$ and $I = \langle 3, x+2 \rangle$. We sketch two proofs that $I$ is not principal. The first is the straightforward argument: suppose $I$ were principal and were generated by the polynomial $f(x) \in \mathbb{Z}[x]$. Then both 3 and $x + 2$ must be divisible by $f(x)$; from the former we infer that $\deg(f(x)) = 0$: i.e, $f(x) = c$ is a constant. Since 3 is prime, either $c = \pm 1$ or $c = \pm 3$. As we'll establish in a moment, $1 \notin I$, so $f(x) = \pm 3$. But $3 \nmid (x + 2)$. Thus, $I$ cannot be principal.

Here is an argument that $1 \notin I$: Suppose there were polynomials $p(x)$ and $q(x) \in \mathbb{Z}[x]$ so that $3p(x) + q(x)(x + 2) = 1$. Evaluating at $x = -2$, we obtain $3p(-2) = 1$. Since $p(-2) \in \mathbb{Z}$, we have arrived at a contradiction; thus, $1 \notin I$.

The second proof is equivalent, but a bit more elegant. Let $\psi \colon \mathbb{Z}[x] \to \mathbb{Z}_3[x]$ be the homomorphism that reduces polynomials mod 3. By Exercise 16b., $\psi(I) = J$ is an ideal in $\mathbb{Z}_3[x]$; obviously, $J = \langle x + 2 \rangle$. Now if $I$ were principal, say $I = \langle f(x) \rangle$, then we would have $J = \langle \psi(f(x)) \rangle = \langle \overline{f}(x) \rangle$ (see Exercise 16c.). Therefore, $\overline{f}(x) = \pm(x + 2)$, and so $\deg(f(x)) \geq 1$, contradicting the fact that $f(x) | 3$ in $\mathbb{Z}[x]$.

We now come to one of the most important constructions in mathematics. The example you should keep in mind as you read this is $\mathbb{Z}_m$. Given a commutative ring $R$ and an ideal $I \subset R$, we say that two elements $a, b \in R$ are equivalent (mod $I$), written $a \equiv b$ (mod $I$), if and only if $a - b \in I$. For example, in $\mathbb{Z}$,

$$a \equiv b \ (\text{mod } \langle m \rangle) \iff a - b \in \langle m \rangle \iff m | (a - b)$$
$$\iff a \equiv b \ (\text{mod } m),$$

as before. The analogue of Proposition 3.1 of Chapter 1 is the following proposition.

**Proposition 1.5.** *Let $I \subset R$ be an ideal, and suppose $a \equiv a'$ and $b \equiv b'$ (mod $I$). Then*

(1) $a + b \equiv a' + b'$ (mod $I$), *and*

(2) $ab \equiv a'b'$ (mod $I$).

***Proof.*** By definition, there are elements $x, y \in I$ so that $a = a' + x$ and $b = b' + y$. Then $a + b = (a' + b') + (x + y)$; since $x + y \in I$ (as ideals are closed under addition), it follows that $a + b \equiv a' + b'$ (mod $I$). On the other hand, $ab = (a' + x)(b' + y) = a'b' + b'x + a'y + xy$. Now $b'x$, $a'y$, and $xy$ are all elements of

$I$ (since the product of an element of an ideal by an arbitrary ring element is still in the ideal), and so by closure under addition their sum is an element of $I$. Therefore, $ab \equiv a'b' \pmod{I}$, as desired.  □

We denote the equivalence class of $a$ by either $a \pmod{I}$ or $\overline{a}$, when no confusion will arise. We denote by $R \bmod I$, or, more commonly, $R/I$, the set of equivalence classes. Note that $\mathbb{Z}_m$ and $\mathbb{Z}/\langle m \rangle$ coincide. Because of the following result, we call $R/I$ a **quotient ring**.

**Proposition 1.6.** *Let $R$ be a commutative ring and let $I \subsetneq R$ be an ideal. Then $R/I$ is a commutative ring.*

***Proof.*** We define addition and multiplication of equivalence classes in the obvious way:

$$\overline{a} + \overline{b} = \overline{a + b}$$
$$\overline{a} \cdot \overline{b} = \overline{ab}.$$

The crucial point is that these are well-defined because of Proposition 1.5. For example, if $\overline{a} = \overline{a'}$ (i.e., $a$ and $a'$ represent the same equivalence class) and $\overline{b} = \overline{b'}$, this means that $a \equiv a'$ and $b \equiv b' \pmod{I}$, so $a + b \equiv a' + b' \pmod{I}$, whence $\overline{a + b} = \overline{a' + b'}$. So we get the same answer for $\overline{a} + \overline{b}$ independent of the choice of representatives of the equivalence classes.

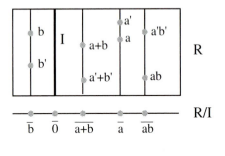

FIGURE 1

The commutative, associative, and distributive laws are inherited from those in $R$. Of course, the additive identity of $R/I$ is $\overline{0}$, and the multiplicative identity is $\overline{1}$; note that $\overline{1} \neq \overline{0}$ because $1 \notin I$ (see Exercise 3).  □

**Remark.** We must be careful not to assume that all properties of $R$ are inherited by $R/I$. For example, $\mathbb{Z}$ is an integral domain, but $\mathbb{Z}/\langle 6 \rangle$ is not.

**Examples 5.**

(a) Consider the ideal $I = \langle 2 \rangle \subset \mathbb{Z}_6$. When we list the elements 0, 1, 2, 3, 4, 5 of $\mathbb{Z}_6$, we see that $0 \equiv 2 \equiv 4 \pmod{I}$ and $1 \equiv 3 \equiv 5 \pmod{I}$. So the quotient ring $\mathbb{Z}_6/I$ consists of the two equivalence classes $\overline{0}, \overline{1}$ with the obvious addition and multiplication laws.

(b) Consider $\mathbb{Q}[x]/\langle x^2 - 2 \rangle$, i.e., $R = \mathbb{Q}[x]$ and $I = \langle x^2 - 2 \rangle$. The first observation is that any polynomial $f(x) \in \mathbb{Q}[x]$ is equivalent mod $I$ to a linear polynomial, for mod $I$ we may replace $x^2$ by 2, $x^3$ by $2x$, $x^4$ by $2^2$, etc. Alternatively, apply the division algorithm to replace $f(x)$ by its remainder upon division by $x^2 - 2$. Thus, in the quotient ring, $\overline{x}$ plays the rôle of $\sqrt{2}$, and we strongly suspect that the ring $R/I$ "is" $\mathbb{Q}[\sqrt{2}]$.

(c) Consider $\mathbb{Z}_2[x]/\langle x^2 + x + 1 \rangle$. As in the previous example, working mod $I = \langle x^2 + x + 1 \rangle$, any polynomial in $\mathbb{Z}_2[x]$ is equivalent to a linear polynomial, since $x^2 \equiv x + 1 \pmod{I}$ (recall $-1 = 1$ in $\mathbb{Z}_2$). Thus, the elements of the quotient ring are the equivalence classes $\overline{0}, \overline{1}, \overline{x}$, and $\overline{x+1}$ with the addition and multiplication tables below.

| + | $\overline{0}$ | $\overline{1}$ | $\overline{x}$ | $\overline{x+1}$ |
|---|---|---|---|---|
| $\overline{0}$ | $\overline{0}$ | $\overline{1}$ | $\overline{x}$ | $\overline{x+1}$ |
| $\overline{1}$ | $\overline{1}$ | $\overline{0}$ | $\overline{x+1}$ | $\overline{x}$ |
| $\overline{x}$ | $\overline{x}$ | $\overline{x+1}$ | $\overline{0}$ | $\overline{1}$ |
| $\overline{x+1}$ | $\overline{x+1}$ | $\overline{x}$ | $\overline{1}$ | $\overline{0}$ |

| $\cdot$ | $\overline{0}$ | $\overline{1}$ | $\overline{x}$ | $\overline{x+1}$ |
|---|---|---|---|---|
| $\overline{0}$ | $\overline{0}$ | $\overline{0}$ | $\overline{0}$ | $\overline{0}$ |
| $\overline{1}$ | $\overline{0}$ | $\overline{1}$ | $\overline{x}$ | $\overline{x+1}$ |
| $\overline{x}$ | $\overline{0}$ | $\overline{x}$ | $\overline{x+1}$ | $\overline{1}$ |
| $\overline{x+1}$ | $\overline{0}$ | $\overline{x+1}$ | $\overline{1}$ | $\overline{x}$ |

(For example, $(x + 1)^2 = x^2 + 1 \equiv (x + 1) + 1 = x \pmod{I}$.)

## EXERCISES 4.1

1. a. Prove that for any ring homomorphism $\phi: R \to S$, $\phi(-a) = -\phi(a)$ for all $a \in R$.

   b. Given two rings $R$ and $S$, is the zero function $\phi: R \to S$ ($\phi(a) = 0_S$ for all $a \in R$) a ring homomorphism? Explain.

2. Consider the function $\phi: \mathbb{Z}_2 \to \mathbb{Z}_6$ defined by $\phi(0 \text{ (mod 2)}) = 0 \text{ (mod 6)}$ and $\phi(1 \text{ (mod 2)}) = 3 \text{ (mod 6)}$. Show that $\phi$ satisfies conditions (1) and (2) for a ring homomorphism, but not (3). This shows that the third condition is not a consequence of the first two.

3. a. Prove that if $I \subset R$ is an ideal and $1 \in I$, then $I = R$.

   b. Prove that $a \in R$ is a unit if and only if $\langle a \rangle = R$.

   c. Prove that the only ideals in a (commutative) ring $R$ are $\langle 0 \rangle$ and $R$ if and only if $R$ is a field.

4. Find all the ideals in the following rings:

   a. $\mathbb{Z}$

   b. $\mathbb{Z}_7$

   c. $\mathbb{Z}_6$

   d. $\mathbb{Z}_{12}$

   e. $\mathbb{Z}_{36}$

   f. $\mathbb{Q}$

   g. $\mathbb{Z}[i]$ (see Exercise 2.3.18)

5. a. Let $I = \langle f(x) \rangle$, $J = \langle g(x) \rangle$ be ideals in $F[x]$. Prove that $I \subset J \iff g(x) | f(x)$.

   b. List all the ideals of $\mathbb{Q}[x]$ containing the element

$$f(x) = (x^2 + x - 1)^3 (x - 3)^2.$$

6. Prove that $\phi: \mathbb{Z}_p \to \mathbb{Z}_p$, $\phi(a) = a^p$, is a ring homomorphism, and find $\ker \phi$.

7. Find all ring homomorphisms $\phi: \mathbb{Z} \to \mathbb{Z}$.

8. Prove that a ring homomorphism $\phi: R \to S$ is one-to-one if and only if $\ker \phi = \langle 0 \rangle$.

9. Construct the multiplication table for the ring $\mathbb{Z}_3[x]/\langle x^2 \rangle$. What do you observe?

10. Prove that if $f(x) \in F[x]$ is not irreducible, then $F[x]/\langle f(x) \rangle$ contains zero-divisors.

11. Let $R = \left\{ \begin{bmatrix} a & -b \\ b & a \end{bmatrix} : a, b \in \mathbb{R} \right\} \subset M_2(\mathbb{R})$. Define $\phi: R \to \mathbb{C}$ by $\phi\left( \begin{bmatrix} a & -b \\ b & a \end{bmatrix} \right) = a + bi$. Prove that $\phi$ is a homomorphism. Does it map onto $\mathbb{C}$? What is $\ker \phi$?

12. Show that the equation $y^2 = 4$ has at least four solutions in the ring $\mathbb{Z}_5[x]/\langle x^2 + 1 \rangle$. What do you conclude?

13. Use the ring homomorphism $\phi: \mathbb{Z} \to \mathbb{Z}_m$ (for appropriate values of $m$) to prove the following:
    a. The equation $x^2 - 5y^2 = 2$ has no solution for $x, y \in \mathbb{Z}$.
    b. The equation $x^3 - 7y^2 = 17$ has no solution for $x, y \in \mathbb{Z}$.

14. Mimicking Example 5(c), give the addition and multiplication tables of
    a. $\mathbb{Z}_2[x]/\langle x^2 + x \rangle$
    b. $\mathbb{Z}_3[x]/\langle x^2 + x - 1 \rangle$
    c. $\mathbb{Z}_2[x]/\langle x^3 + x + 1 \rangle$
    In each case, is the quotient ring an integral domain? a field?

15. Find all ring homomorphisms
    a. $\phi: \mathbb{Z}_2 \to \mathbb{Z}$
    b. $\phi: \mathbb{Z}_2 \to \mathbb{Z}_6$
    c. $\phi: \mathbb{Z}_6 \to \mathbb{Z}_2$
    d. $\phi: \mathbb{Z}_m \to \mathbb{Z}_n$ (Your answer will depend on the relation between $m$ and $n$.)
    e. $\phi: \mathbb{Q} \to \mathbb{Q}$ (cf. Exercise 2.1.16)

16. Let $R$ and $S$ be commutative rings, and let $\phi: R \to S$ be a ring homomorphism.
    a. Given an ideal $J \subset S$, define

    $$= \{a \in R : \phi(a) \in J\} \subset R.$$

    (This is usually denoted by $\phi^{-1}(J)$.) Prove that     is an ideal.

    b. Given an ideal $I \subset R$, define

    $$= \{\phi(a) : a \in I\} \subset S.$$

(This is usually denoted by $\phi(I)$.) Prove that   is an ideal, provided $\phi$ maps onto $S$. Give an example to demonstrate that the latter hypothesis is necessary.

  c.  As a particular case of b., prove that if $\phi$ maps onto $S$, then $\phi(\langle a \rangle) = \langle \phi(a) \rangle$.

17.  Let $R$ be a commutative ring and let $I, J \subset R$ be ideals. Define

$$I \cap J = \{a \in R : a \in I \text{ and } a \in J\}$$
$$I + J = \{a + b \in R : a \in I, b \in J\}.$$

  a.  Prove that $I \cap J$ and $I + J$ are ideals.

  b.  Suppose $R = \mathbb{Z}$ or $F[x]$, $I = \langle a \rangle$, and $J = \langle b \rangle$. Identify $I \cap J$ and $I + J$.

  c.  Let $a_1, \ldots, a_n \in R$. Prove that $\langle a_1, \ldots, a_n \rangle = \langle a_1 \rangle + \cdots + \langle a_n \rangle$.

18.  An element $a$ of a commutative ring $R$ is called **nilpotent** if $a^n = 0$ for some $n \in \mathbb{N}$.

  a.  Find the nilpotent elements in $\mathbb{Z}_6$, $\mathbb{Z}_{12}$, $\mathbb{Z}_8$, $\mathbb{Z}_{36}$.

  b.  Find the nilpotent elements in $\mathbb{Q}[x]/\langle x^2 \rangle$.

  c.  Show that the collection $N$ of all nilpotent elements in $R$ is an ideal. (Hint: To show closure under addition, the binomial theorem, Theorem 1.4 of Chapter 1, is needed. Your job will be to pick an appropriate exponent.)

  d.  Show that the quotient ring $R/N$ has no nonzero nilpotent elements.

19.  Prove that every ideal in $\mathbb{Z}_m$ is principal. (Hint: Apply Exercise 16 to the obvious homomorphism $\phi : \mathbb{Z} \to \mathbb{Z}_m$.)

20.  We motivated the definition of an ideal by considering the kernel of a ring homomorphism. Let $R$ be a commutative ring. Is it true that *every* ideal $I \subsetneq R$ is the kernel of a ring homomorphism $\phi : R \to S$ for some ring $S$?

21.  Show that there is a one-to-one correspondence

$$\{\text{ideals of } R/I\} \leftrightarrow \{\text{ideals of } R \text{ containing } I\}.$$

More generally, if $\phi : R \to S$ is a ring homomorphism, mapping onto $S$, then there is a one-to-one correspondence

$$\{\text{ideals of } S\} \leftrightarrow \{\text{ideals of } R \underline{\hspace{2cm}}\}.$$

(Part of your job here is to fill in the blank.)

22. Generalizing Exercise 6, let $f(x) \in \mathbb{Z}_p[x]$ be any polynomial and let $R = \mathbb{Z}_p[x]/\langle f(x) \rangle$. Define $\phi: R \to R$ by $\phi(a) = a^p$ for all $a \in R$. Prove that $\phi$ is a homomorphism. Can you determine ker $\phi$?

## 2. Isomorphisms and the Fundamental Homomorphism Theorem

In the previous section we defined quotient rings; in this section we will explore them further and will give a powerful application to the study of splitting fields initiated in Chapter 3. We begin by saying what it means for two rings to be "the same."

**Definition.** A ring homomorphism $\phi: R \to S$ is an **isomorphism** if it maps one-to-one and onto $S$. We say the rings $R$ and $S$ are **isomorphic** (denoted by $R \cong S$) if there exists an isomorphism between them.

Recall that a function $f: R \to S$ is **one-to-one** if whenever $x, y \in R$, $f(x) = f(y) \implies x = y$. It maps **onto** $S$ provided its image is all of $S$; i.e., given any $s \in S$, there is an $x \in R$ so that $f(x) = s$. Here is a useful criterion for a homomorphism to be one-to-one.

**Lemma 2.1.** *A ring homomorphism $\phi: R \to S$ is one-to-one if and only if* ker $\phi = \langle 0 \rangle$.

**Proof.** Since, by Lemma 1.1(i), $\phi(0) = 0$, if $\phi$ is one-to-one, then ker $\phi = \langle 0 \rangle$. On the other hand, suppose ker $\phi = \langle 0 \rangle$ and $\phi(x) = \phi(y)$. Then $\phi(x - y) = 0 \implies x - y \in$ ker $\phi$. Thus, by hypothesis, $x - y = 0$ and $x = y$, as required.  $\square$

### Examples 1.

(a) The conjugation homomorphism $\phi: \mathbb{C} \to \mathbb{C}$, $\phi(z) = \overline{z}$, is an isomorphism: $\phi(z) = 0 \implies \overline{z} = 0 \implies z = 0$, so ker $\phi = \langle 0 \rangle$; $\phi$ maps onto $\mathbb{C}$ (given $a \in \mathbb{C}$, we have $\phi(\overline{a}) = a$).

(b) Define $\phi: \mathbb{R}[x]/\langle x^2 + 1 \rangle \to \mathbb{C}$ by

$$\phi(f(x) \pmod{\langle x^2 + 1 \rangle}) = f(i) \in \mathbb{C}.$$

Note that if $f(x) \equiv g(x) \pmod{\langle x^2 + 1 \rangle}$, then $x^2 + 1$ divides $f(x) - g(x)$; since $i^2 + 1 = 0$, $f(i) = g(i)$, and so $\phi$ is well-defined. By the division algorithm, every polynomial is equivalent mod $\langle x^2 + 1 \rangle$ to a linear polynomial $bx + a$, $a, b \in \mathbb{R}$, and

$$\phi(bx + a \ (\mathrm{mod} \ \langle x^2 + 1 \rangle)) = a + bi.$$

We can check directly from this formula that $\phi$ is in fact a homomorphism:

$$\phi((bx + a) + (dx + c) \ (\mathrm{mod} \ \langle x^2 + 1 \rangle)) = (a + c) + (b + d)i,$$

and $\phi((bx + a) \cdot (dx + c) \ (\mathrm{mod} \ \langle x^2 + 1 \rangle))$

$$= \phi(bdx^2 + (ad + bc)x + ac \ (\mathrm{mod} \ \langle x^2 + 1 \rangle))$$
$$= \phi((ad + bc)x + (ac - bd) \ (\mathrm{mod} \ \langle x^2 + 1 \rangle))$$
$$= (ac - bd) + (ad + bc)i = (a + bi)(c + di),$$

as desired. It is also easy to see from this formula that $\phi$ is both one-to-one and onto, and so $\phi$ is an isomorphism.

(c) For a different interpretation of $\mathbb{C}$, let

$$R = \left\{ \begin{bmatrix} a & -b \\ b & a \end{bmatrix} : a, b \in \mathbb{R} \right\} \subset M_2(\mathbb{R}).$$

Define $\phi \colon R \to \mathbb{C}$ by $\phi\left( \begin{bmatrix} a & -b \\ b & a \end{bmatrix} \right) = a + bi$. As Exercise 4.1.11 established, $\phi$ is an isomorphism.

**Example 2.** Define a ring $R = \{(a, b) : a \in \mathbb{Z}_2, b \in \mathbb{Z}_3\}$, where the ring operations are performed component by component. We leave it to the reader to check this is actually a ring, but we'd better point out that $(0, 0)$ is the additive identity and $(1, 1)$ is the multiplicative identity (see Exercise 2a.). It is easy to see that $R$ has six elements, and so we'll try to show that $R$ is isomorphic to the only ring with six elements that we know, $\mathbb{Z}_6$. We need only write down a ring homomorphism $\phi \colon \mathbb{Z}_6 \to R$ that is one-to-one (why?), and there's only one way to do so. $\phi(0_{\mathbb{Z}_6})$ must be $0_R$ and $\phi(1_{\mathbb{Z}_6})$ must be $1_R$, and the additive structure determines the rest:

$$\phi(0) = (0, 0)$$
$$\phi(1) = (1, 1)$$
$$\phi(2) = (0, 2)$$
$$\phi(3) = (1, 0)$$
$$\phi(4) = (0, 1)$$
$$\phi(5) = (1, 2)$$

Note the happenstance (or is it?) that

$$\phi(k \ (\text{mod } 6)) = (k \ (\text{mod } 2), \ k \ (\text{mod } 3)),$$

and so, indeed,

$$\phi(k\ell \ (\text{mod } 6)) = (k\ell \ (\text{mod } 2), k\ell \ (\text{mod } 3))$$
$$= \phi(k \ (\text{mod } 6)) \cdot \phi(\ell \ (\text{mod } 6)).$$

Therefore, $\phi$ is an isomorphism.

We may as well give an official definition here. Given two rings $R$ and $S$, their **direct product** is $R \times S = \{(r,s) : r \in R, \ s \in S\}$. Addition and multiplication are defined component by component:

$$(r,s) + (r',s') = (r + r', s + s')$$
$$(r,s) \cdot (r',s') = (rr', ss').$$

We set $0_{R \times S} = (0_R, 0_S)$ and $1_{R \times S} = (1_R, 1_S)$. We leave it to the reader to check that $R \times S$ enjoys all the requisite properties of a ring (see Exercise 2a.).

It is useful to notice that if $\phi: R \to S$ is an isomorphism, then for any $a \in R$, $a$ and $\phi(a)$ will share any distinctive algebraic property. For example, if $a$ is a unit in $R$, $\phi(a)$ must be a unit in $S$. Suppose there exists $b \in R$ so that $ab = ba = 1_R$; then $\phi(a)\phi(b) = \phi(b)\phi(a) = 1_S$. If $a$ is a zero-divisor in $R$, then $\phi(a)$ must be a zero-divisor in $S$. If there are nonzero elements $a, b \in R$ so that $ab = 0_R$, then $\phi(a)\phi(b) = 0_S$; now, $\phi(a)$ and $\phi(b)$ are nonzero elements of $S$ since $\ker \phi = \langle 0 \rangle$, so $\phi(a)$ is a zero-divisor in $S$. This sort of observation can be quite useful.

**Example 3.** Are the rings $\mathbb{Z}_4$ and $\mathbb{Z}_2 \times \mathbb{Z}_2$ isomorphic? Since each of them has four elements, there is certainly a one-to-one correspondence between them. However, from the remarks we've just made, it follows that there must be a one-to-one correspondence between the zero-divisors in isomorphic rings. There is one zero-divisor in $\mathbb{Z}_4$ (namely 2), whereas there are two zero-divisors in $\mathbb{Z}_2 \times \mathbb{Z}_2$ (namely $(1,0)$ and $(0,1)$). Thus the rings cannot be isomorphic. (See Exercise 11 for some related examples.) Be warned that for $R$ to be isomorphic to $S$, it is necessary, but rarely sufficient, that there be equal numbers of zero-divisors in each.

We now come to one of the central results in modern algebra, one that will provide many striking applications. The theorem is easy to state and not hard to prove, but this is because we've worked hard to lay the groundwork.

**Theorem 2.2** (Fundamental Homomorphism Theorem). *Let $R$ and $S$ be commutative rings, and suppose $\phi \colon R \to S$ is a ring homomorphism* onto *$S$. Then $R/\ker\phi \cong S$.*

**Proof.** Note, first, that if $a \equiv b \pmod{\ker\phi}$, then $a - b = r$ for some $r \in \ker\phi$; and so $\phi(a - b) = \phi(r) = 0$, whence $\phi(a) = \phi(b)$. Thus, denoting the equivalence class of $a$ in $R/\ker\phi$ by $\overline{a}$, we obtain a well-defined function $\overline{\phi} \colon R/\ker\phi \to S$ by setting $\overline{\phi}(\overline{a}) = \phi(a)$, for all $\overline{a} \in R/\ker\phi$.

Next, $\overline{\phi}$ is a homomorphism, simply because $\phi$ is:

$$\overline{\phi}(\overline{a} + \overline{b}) = \overline{\phi}(\overline{a + b}) \qquad \text{by definition of addition in } R/\ker\phi,$$
$$= \phi(a + b) \qquad \text{by definition of } \overline{\phi},$$
$$= \phi(a) + \phi(b) \qquad \text{since } \phi \text{ is a homomorphism,}$$
$$= \overline{\phi}(\overline{a}) + \overline{\phi}(\overline{b}) \qquad \text{by definition of } \overline{\phi} \text{ again;,}$$

and similarly for products. The homomorphism $\overline{\phi}$ maps onto $S$ since $\phi$ does so. What remains to be checked is that $\overline{\phi}$ is one-to-one, i.e., that $\ker\overline{\phi} = \langle \overline{0} \rangle$. So suppose $\overline{\phi}(\overline{a}) = 0$; this means that $\phi(a) = 0$, whence $a \in \ker\phi$ and $\overline{a} = \overline{0}$. This concludes the proof. $\square$

**Examples 4.**

(a) By applying Theorem 2.2, we can rephrase Example 1(b). Consider the evaluation homomorphism

$$\mathrm{ev}_i \colon \mathbb{R}[x] \to \mathbb{C}, \quad \mathrm{ev}_i(f(x)) = f(i).$$

Since $\mathrm{ev}_i(bx + a) = a + bi$, the homomorphism $\mathrm{ev}_i$ maps onto $\mathbb{C}$, and so $\mathbb{R}[x]/\ker\mathrm{ev}_i \cong \mathbb{C}$. The kernel of $\mathrm{ev}_i$ contains the irreducible polynomial $x^2 + 1 \in \mathbb{R}[x]$, and so, by Corollary 1.4, $\ker\mathrm{ev}_i = \langle x^2 + 1 \rangle$.

(b) Similarly, the evaluation homomorphism

$$\mathrm{ev}_{\sqrt{2}} \colon \mathbb{Q}[x] \to \mathbb{Q}[\sqrt{2}]$$

has kernel $\langle x^2 - 2 \rangle$ (why?), and so $\mathbb{Q}[x]/\langle x^2 - 2 \rangle \cong \mathbb{Q}[\sqrt{2}]$.

(c) Consider the homomorphism

$$\phi \colon \mathbb{Z}_{12} \to \mathbb{Z}_3, \quad \phi(m \pmod{12}) = m \pmod{3}.$$

The homomorphism is clearly onto, and its kernel is generated by 3 (mod 12). Thus $\mathbb{Z}_{12}/\langle \overline{3} \rangle \cong \mathbb{Z}_3$.

(d) Let $\mathbb{Z}[\frac{1}{2}] = \{\frac{r}{s} : r \in \mathbb{Z}, \ s = 2^m$ for some integer $m \geq 0\}$. We claim that $\mathbb{Z}[x]/\langle 2x - 1 \rangle \cong \mathbb{Z}[\frac{1}{2}]$. Define a homomorphism $\phi \colon \mathbb{Z}[x] \to \mathbb{Z}[\frac{1}{2}]$, as usual, by the evaluation map: Let $\phi(f(x)) = f(\frac{1}{2})$; note first that this number is in $\mathbb{Z}[\frac{1}{2}]$. Now $\phi$ maps onto $\mathbb{Z}[\frac{1}{2}]$. Given $r/2^m \in \mathbb{Z}[\frac{1}{2}]$, let $f(x) = rx^m \in \mathbb{Z}[x]$; then $\phi(f(x)) = r/2^m$, as required.

It is a bit more difficult, however, to see that $\ker \phi = \langle 2x - 1 \rangle$. Note, first, that $\langle 2x - 1 \rangle \subset \ker \phi$. Now, think temporarily of polynomials with integer coefficients as having rational coefficients. If $f(x) \in \mathbb{Z}[x]$ and $f(\frac{1}{2}) = 0$, then $x - \frac{1}{2}$ is a factor (in $\mathbb{Q}[x]$) of $f(x)$; i.e., $f(x) = \left(x - \frac{1}{2}\right) g(x)$ for some $g(x) \in \mathbb{Q}[x]$. By Gauss' Lemma (Proposition 3.3 of Chapter 3) and unique factorization in $\mathbb{Q}[x]$, it follows that $f(x) = (2x - 1)\tilde{g}(x)$, where $\tilde{g}(x) \in \mathbb{Z}[x]$. Therefore, $f(x) \in \langle 2x - 1 \rangle$, and $\ker \phi = \langle 2x - 1 \rangle$.

By the Fundamental Homomorphism Theorem, it follows that $\phi$ induces an isomorphism $\mathbb{Z}[x]/\langle 2x - 1 \rangle \xrightarrow{\cong} \mathbb{Z}[\frac{1}{2}]$.

We have seen numerous examples of fields (e.g., $\mathbb{Q}[\sqrt{2}]$, $\mathbb{Q}[i]$, $\mathbb{Z}_2[\alpha]$ where $\alpha^2 + \alpha + 1 = 0$) built by adjoining some number to a base field subject to some polynomial relation. On the other hand, we have seen other examples of such a construction that failed to be a field (see Exercises 2.4.9 and 3.2.5). We now can interpret this construction as a quotient ring ($\mathbb{Q}[x]/\langle x^2 - 2 \rangle$, $\mathbb{Q}[x]/\langle x^2 + 1 \rangle$, $\mathbb{Z}_2[x]/\langle x^2 + x + 1 \rangle$ respectively), and so, once and for all, we should decide when the resulting ring will be a field.

**Example 5.** The polynomial $f(x) = x^2 \in \mathbb{Q}[x]$ is not irreducible, and the quotient ring $\mathbb{Q}[x]/\langle x^2 \rangle$ is not even an integral domain, since $\overline{x}$ is a zero-divisor.

**Theorem 2.3.** *Let $F$ be a field, $f(x) \in F[x]$. Then $F[x]/\langle f(x) \rangle$ is a field if and only if $f(x)$ is irreducible in $F[x]$.*

**Proof.** Suppose $f(x) = g(x)h(x)$, where $\deg(g(x)), \deg(h(x)) \geq 1$. Then in $F[x]/\langle f(x) \rangle$, we have $\overline{g(x)h(x)} = \overline{g(x)} \cdot \overline{h(x)} = \overline{0}$. Yet neither $\overline{g(x)}$ nor $\overline{h(x)}$ can be $\overline{0}$, since $g(x), h(x) \notin \langle f(x) \rangle$. Thus, we've proved that when $f(x)$ fails to be irreducible, $F[x]/\langle f(x) \rangle$ fails to be an integral domain, and so it is certainly not a field.

Conversely, suppose $f(x)$ is irreducible, and let $g(x) \in F[x]$ with $\overline{g(x)} \neq \overline{0}$. Since $f(x)$ is irreducible, $f(x)$ and $g(x)$ are relatively prime; and so by Theorem 1.6 of Chapter 3, there are polynomials

$s(x)$ and $t(x)$ so that $1 = s(x)f(x) + t(x)g(x)$. Therefore, in the quotient ring $F[x]/\langle f(x) \rangle$, we find that $\overline{t(x)}$ is the multiplicative inverse of $\overline{g(x)}$, establishing the fact that $F[x]/\langle f(x) \rangle$ is a field.  □

**Example 6.** In $\mathbb{Z}_3[x]$, $f(x) = x^3 - x^2 + 1$ is irreducible (why?). To compute the multiplicative inverse of $\overline{x^2 + x + 1}$ in $\mathbb{Z}_3[x]/\langle f(x) \rangle$, we apply the Euclidean algorithm to find that

$$x^3 - x^2 + 1 = (x + 1)(x^2 + x + 1) + x$$

$$x^2 + x + 1 = (x + 1)x + 1.$$

And so,

$$1 = -(x + 1)(x^3 - x^2 + 1) + (x^2 - x - 1)(x^2 + x + 1).$$

Thus, in $\mathbb{Z}_3[x]/\langle f(x) \rangle$ the multiplicative inverse of $\overline{x^2 + x + 1}$ is $\overline{x^2 - x - 1}$. The reader can check this directly.

There is a wonderful application of this theorem to the construction of splitting fields of polynomials (first discussed in Section 2 of Chapter 3).

**Theorem 2.4.** *Given an irreducible polynomial $f(x) \in F[x]$, the field $K = F[x]/\langle f(x) \rangle$ contains a root $\alpha$ of $f(x)$, and, indeed, $K \cong F[\alpha]$.*

**Proof.** Let $\alpha = \overline{x} \in F[x]/\langle f(x) \rangle$; then $f(\alpha) = \overline{f(x)} = 0$, so $\alpha$ is indeed a root of $f(x)$ in $K$. Now the evaluation homomorphism $\text{ev}_\alpha \colon F[x] \to K$ maps onto $F[\alpha]$ and has kernel $\langle f(x) \rangle$ (by Corollary 1.4). Therefore, by Theorem 2.2, we have $K \cong F[\alpha]$.  □

The reader should compare this result with Proposition 2.2 of Chapter 3. Now we are in a position to prove the following corollary.

**Corollary 2.5.** *Any polynomial has a splitting field.*

**Proof.** Given a polynomial $f(x) \in F[x]$ of degree $n$ with roots $c_1, \ldots, c_k \in F$, write $f(x) = (x - c_1)(x - c_2) \cdots (x - c_k)g(x)$, where $g(x)$ has no roots in $F$. Now factor $g(x)$ as a product of irreducible polynomials in $F[x]$; let $h(x)$ be one of these. Pass to the field $\tilde{F} = F[x]/\langle h(x) \rangle$, in which $h(x)$ has at least one root $c_{k+1}$. Now $f(x)$ has roots $c_1, \ldots, c_k, c_{k+1}, \ldots, c_\ell \in \tilde{F}$. Since $f(x)$ has at most $n$ roots in any field extension of $F$ (see Exercise 3.1.6), we may repeat this procedure until we have just reached a field $K$ containing all the roots of $f(x)$.  □

## Examples 7.

(a) Let $a \in F$ and suppose $f(x) = x^2 - a$ is irreducible. Then $K = F[x]/\langle f(x) \rangle$ is the splitting field of $f(x)$, because once $K$ contains one root $\sqrt{a}$ of $f(x)$, it contains both roots. Indeed, the same is true for any quadratic polynomial.

(b) Let $f(x) = x^3 - 2 \in \mathbb{Q}[x]$. Then we pass, by Theorem 2.4, to $K = \mathbb{Q}[x]/\langle f(x) \rangle \cong \mathbb{Q}[\sqrt[3]{2}]$. Is $K$ the splitting field of $f(x)$? Of course, it cannot be, since $\mathbb{Q}[\sqrt[3]{2}] \subset \mathbb{R}$, and we know that $f(x)$ has two complex (non-real) roots. But we can attempt a direct argument by brute force: let $\alpha$ be the root of $f(x)$ in $K$. By long division we obtain $\dfrac{x^3 - 2}{x - \alpha} = x^2 + \alpha x + \alpha^2$. This quadratic polynomial factors in $K[x]$ if and only if (by the quadratic formula) $\sqrt{-3\alpha^2} \in K$, and the latter holds if and only if $\sqrt{-3\alpha^2} = A + B\alpha + C\alpha^2$ for some $A, B, C \in \mathbb{Q}$. Squaring both sides of this equation, we obtain

$$-3\alpha^2 = (A^2 + 4BC) + 2(AB + C^2)\alpha + (B^2 + 2AC)\alpha^2.$$

Since $1, \alpha$, and $\alpha^2$ are linearly independent over $\mathbb{Q}$ (by Lemma 1.6 of Chapter 5), $A^2 + 4BC = AB + C^2 = 0$ and $B^2 + 2AC = -3$. The first equations imply that $A^3 - 4C^3 = 0$, and so $A = C = 0$ (since $\sqrt[3]{4} \notin \mathbb{Q}$); then $B^2 = -3$, which is impossible. Therefore, $x^2 + \alpha x + \alpha^2$ is irreducible in $K[x]$.

Thus, we must pass to the next field

$$L = K[y]/\langle y^2 + \alpha y + \alpha^2 \rangle \cong K[\sqrt{3}i],$$

and the quadratic splits. In conclusion, the splitting field of the original polynomial $f(x) = x^3 - 2$ is given by $L \cong K[\sqrt{3}i] \cong \mathbb{Q}[\sqrt[3]{2}, \sqrt{3}i]$, as, of course, we knew earlier.

(c) The analysis in the preceding example suggests that *any* irreducible cubic polynomial in $\mathbb{Q}[x]$ splits upon adjoining either a single root or that root and a square root as well. We shall see in Section 6 of Chapter 7 that the splitting field $K$ is of the form $\mathbb{Q}[\alpha]$ (where $\alpha$ is one root) if and only if $\sqrt{-108\Delta} \in \mathbb{Q}$, where the discriminant $-108\Delta$ is as defined in Section 4 of Chapter 2 (see p. 70). Note that in the case of the cubic polynomial $z^3 - 3z + 1$ analyzed in Section 4 of Chapter 2, $-108\Delta = 81$. Once we have the root $\alpha = \zeta + 1/\zeta$, the two remaining roots $\zeta^4 + 1/\zeta^4$ and $\zeta^7 + 1/\zeta^7$ are in $\mathbb{Q}[\alpha]$ (see Exercise 7).

(d) Let $f(x) = x^2 + x + 1 \in \mathbb{Z}_2[x]$; $f(x)$ is irreducible and so $K = \mathbb{Z}_2[x]/\langle f(x) \rangle$ is a field containing a root $\alpha$ of $f(x)$ (cf. the discussion in Section 2 of Chapter 3). Note that $f(\alpha + 1) = (\alpha + 1)^2 + (\alpha + 1) + 1 = 0$, so $\alpha$ and $\alpha + 1$ are the two roots of $f(x)$, both lying in $K$; thus $K$ is the splitting field of $f(x)$. But this was too easy ...

(e) Let $f(x) = x^3 + x + 1 \in \mathbb{Z}_2[x]$; since $f(x)$ is irreducible, we consider the field $K = \mathbb{Z}_2[x]/\langle f(x) \rangle$. Let $\alpha \in K$ be one root of $f(x)$, and calculate $\dfrac{x^3 + x + 1}{x - \alpha} = x^2 + \alpha x + (\alpha^2 + 1) = (x - \alpha^2)(x - [\alpha^2 + \alpha])$; so $f(x)$ splits in $K$.

We asked in Exercise 3.2.16 whether $\mathbb{Q}[\pi] \subset \mathbb{R}$ is a field. What is known about $\pi$ is that it is a **transcendental** number, i.e., it is not the root of any (nonzero) polynomial in $\mathbb{Q}[x]$. This was proved by Lindemann in 1882 and is a rather difficult result. (See C. R. Hadlock's *Field Theory and its Classical Problems* for the most elementary exposition I know of. The proof that $e$ is transcendental is similar and only slightly more elementary, but can be found, for example, in Spivak's *Calculus*.) Conversely, we say a complex number is **algebraic** if it is the root of a nonzero polynomial in $\mathbb{Q}[x]$. We can now rephrase Theorem 2.3 as follows.

**Corollary 2.6.** *Let $c \in \mathbb{C}$. Then $\mathbb{Q}[c] \subset \mathbb{C}$ is a field if and only if $c$ is an algebraic number.*

***Proof.*** By definition, $\mathbb{Q}[c]$ is the image of the evaluation homomorphism $\operatorname{ev}_c \colon \mathbb{Q}[x] \to \mathbb{C}$. By Theorem 2.2, $\mathbb{Q}[c] \cong \mathbb{Q}[x]/\ker \operatorname{ev}_c$. Moreover, since $\mathbb{Q}[x]$ is a principal ideal domain, $\ker \operatorname{ev}_c = \langle f(x) \rangle$ for some $f(x) \in \mathbb{Q}[x]$.

If $c$ is transcendental, the only polynomial $f(x)$ having $c$ as a root is the zero polynomial, and so in this case $\ker \operatorname{ev}_c = \langle 0 \rangle$ and $\mathbb{Q}[c] \cong \mathbb{Q}[x]$, which, of course, is not a field. On the other hand, if $c$ is algebraic, $\ker \operatorname{ev}_c \neq \langle 0 \rangle$, and so by Corollary 1.4, $\ker \operatorname{ev}_c = \langle f(x) \rangle$ for some irreducible polynomial $f(x) \in \mathbb{Q}[x]$. Now it follows from Theorem 2.3 that $\mathbb{Q}[c]$ is a field.  $\square$

We conclude this section by discussing a criterion (credited to Liouville in 1851) for a real number to be algebraic.

**Theorem 2.7** (Liouville). *If $\alpha \in \mathbb{R}$ is the root of an irreducible polynomial $f(x) \in \mathbb{Z}[x]$ of degree $n > 1$, then there is a constant*

$c > 0$ *so that*

$$|\alpha - \tfrac{p}{q}| \geq \tfrac{c}{q^n}$$

*for all rational numbers* $\tfrac{p}{q}$, $p \in \mathbb{Z}, q \in \mathbb{N}$.

**Remark.** This says that when we approximate an algebraic number by rational numbers $p/q$, the error *must* be of the order of $1/q^n$, so that it is impossible to get a better rational approximation than this accidentally. The contrapositive is more useful:

If for every $n > 1$ there are $p_n \in \mathbb{Z}$ and $q_n \in \mathbb{N}$ so that

$$(*) \qquad \lim_{n \to \infty} q_n^n \left| \alpha - \tfrac{p_n}{q_n} \right| = 0,$$

then $\alpha$ is transcendental.

It was for exactly this purpose that Liouville discovered the result, and he was the first to exhibit explicitly a transcendental number.

**Corollary 2.8.** *The number* $\alpha = \sum\limits_{k=0}^{\infty} \dfrac{1}{10^{k!}}$ *is transcendental.*

**Proof.** We apply $(*)$ above. Let $q_n = 10^{n!}$, and let $\tfrac{p_n}{q_n} = 1 + \tfrac{1}{10} + \tfrac{1}{10^2} + \tfrac{1}{10^6} + \cdots + \tfrac{1}{10^{n!}}$. Then

$$\left| \alpha - \frac{p_n}{q_n} \right| = \frac{1}{10^{(n+1)!}} + \frac{1}{10^{(n+2)!}} + \cdots < \frac{1}{10^{(n+1)!}} \sum_{\ell=0}^{\infty} \frac{1}{10^\ell} < \frac{2}{10^{(n+1)!}},$$

and so

$$q_n^n \left| \alpha - \frac{p_n}{q_n} \right| < (10^{n!})^n \frac{2}{10^{(n+1)!}} = \frac{2}{10^{n!}},$$

which goes to 0 as $n \to \infty$. $\square$

**Proof of Theorem 2.7.** Let $f(x) = a_n x^n + a_{n-1} x^{n-1} + \cdots + a_1 x + a_0$, $a_j \in \mathbb{Z}$, $j = 0, 1, \ldots, n$. Recall the Mean Value Theorem from differential calculus: since $f$ is everywhere differentiable, given any points $a, b \in \mathbb{R}$, there is a point $\xi$ between $a$ and $b$ so that

$$\frac{f(a) - f(b)}{a - b} = f'(\xi).$$

Let the number $M$ be the maximum value of $|f'(x)|$ for $x \in [\alpha - 1, \alpha + 1]$, and let $c$ be the smaller of the two numbers 1 and $1/M$. We claim this is the $c$ stipulated by the theorem.

Suppose first that $|\alpha - \tfrac{p}{q}| > 1$. Then since $q \geq 1$, $1 \geq \tfrac{c}{q^n}$; and so $|\alpha - \tfrac{p}{q}| > \tfrac{c}{q^n}$, as required. Now suppose

$|\alpha - \frac{p}{q}| \leq 1$. Then $\frac{p}{q}$ lies in the interval $[\alpha - 1, \alpha + 1]$, and so it follows from the Mean Value Theorem that

$$\frac{f(\alpha) - f(\frac{p}{q})}{\alpha - \frac{p}{q}} = f'(\xi)$$

for some $\xi$ between $\alpha$ and $\frac{p}{q}$. Since $\deg(f(x)) > 1$, we know $\alpha \notin \mathbb{Q}$. Because $f(\alpha) = 0$ and $|f'(\xi)| \leq M$,

$$\left| \frac{f(\frac{p}{q})}{\alpha - \frac{p}{q}} \right| \leq M,$$

whence $|\alpha - \frac{p}{q}| \geq \frac{1}{M} |f(\frac{p}{q})| \geq c|f(\frac{p}{q})|$. But

$$\left| f(\frac{p}{q}) \right| = \left| a_n (\frac{p}{q})^n + \cdots + a_1 \frac{p}{q} + a_0 \right|$$

$$= \frac{1}{q^n} |a_n p^n + a_{n-1} p^{n-1} q + \cdots + a_1 p q^{n-1} + a_0 q^n|$$

$$\geq \frac{1}{q^n},$$

insofar as $|a_n p^n + a_{n-1} p^{n-1} q + \cdots + a_1 p q^{n-1} + a_0 q^n|$ is a nonnegative integer and cannot equal zero (why?). This completes the proof. □

## EXERCISES 4.2

1.  a.  Prove that the function $\phi \colon \mathbb{Q}[\sqrt{2}] \to \mathbb{Q}[\sqrt{2}]$ defined by $\phi(a + b\sqrt{2}) = a - b\sqrt{2}$ is an isomorphism.
    b.  Define $\phi \colon \mathbb{Q}[\sqrt{3}] \to \mathbb{Q}[\sqrt{7}]$ by $\phi(a + b\sqrt{3}) = a + b\sqrt{7}$. Is $\phi$ an isomorphism? Is there any isomorphism?

2.  a.  Check that $R \times S$ is indeed a ring.
    b.  Check that if $R$ and $S$ are commutative rings, then $R \times S$ is a commutative ring.
    c.  Is $R \times S$ ever an integral domain?

3. Establish the following isomorphisms (preferably, using Theorem 2.2):
   a.  $\mathbb{R}[x]/\langle x^2 + 6 \rangle \cong \mathbb{C}$
   b.  $\mathbb{Z}_{18}/\langle \overline{6} \rangle \cong \mathbb{Z}_6$
   c.  $\mathbb{Q}[x]/\langle x^2 + x + 1 \rangle \cong \mathbb{Q}[\sqrt{3}i]$
   d.  $\mathbb{Z}[x]/\langle 2x - 3 \rangle \cong \mathbb{Z}[\frac{1}{2}] = \{\frac{a}{b} \in \mathbb{Q} : b = 2^j \text{ for some } j \geq 0\} \subset \mathbb{Q}$
   e.  $F[x]/\langle x \rangle \cong F$
   f.  $\mathbb{Z}_3 \times \mathbb{Z}_4 \cong \mathbb{Z}_{12}$

4. a.  Prove that the composition of two ring isomorphisms is a ring isomorphism.
   b.  Suppose $\phi: R \to S$ is a ring isomorphism. Prove that the inverse function $\phi^{-1}: S \to R$ is a homomorphism (therefore also an isomorphism).

5. Let $R = \left\{ \begin{bmatrix} a & b \\ 0 & a \end{bmatrix} : a, b \in \mathbb{R} \right\} \subset M_2(\mathbb{R})$ and $I = \left\{ \begin{bmatrix} 0 & b \\ 0 & 0 \end{bmatrix} : b \in \mathbb{R} \right\}$.
   a.  Prove $I$ is an ideal in $R$.
   b.  Identify the quotient ring $R/I$ by exhibiting a homomorphism whose kernel is $I$.

6. Let $f(x) = x^2 + x - 1$. Find the multiplicative inverse of the element $\overline{x^3 + x + 2}$ in the following quotient rings:
   a.  $\mathbb{Q}[x]/\langle f(x) \rangle$
   b.  $\mathbb{Z}_3[x]/\langle f(x) \rangle$

7. Let $\zeta = e^{2\pi i/9}$. Check that $\mathbb{Q}[\zeta + \zeta^{-1}]$ is a splitting field of $f(x) = x^3 - 3x + 1 \in \mathbb{Q}[x]$.

8. Let $d \in \mathbb{Z}$ be an integer that is not a perfect square. Show that

$$\mathbb{Q}[\sqrt{d}\,] \cong \left\{ \begin{bmatrix} a & db \\ b & a \end{bmatrix} : a, b \in \mathbb{Q} \right\} \subset M_2(\mathbb{Q}).$$

9. Implement the proof of Corollary 2.5 to find the splitting fields of the following polynomials:
   a.  $f(x) = x^4 + 5x^2 + 4$
   b.  $f(x) = x^4 - x^2 - 2$
   c.  $f(x) = x^6 + x^4 - 4x^2 - 4$
   d.  $f(x) = x^6 + 2x^4 - 5x^2 - 6$

10. Let $f(x) \in F[x]$, and let $K$ be a field extension of $F$ containing the root $\alpha$ of $f(x)$. If $\sigma: K \to K$ is a ring isomorphism with the

property that $\sigma(a) = a$ for all $a \in F$, show that $\sigma(\alpha)$ is likewise a root of $f(x)$. Apply this to show the following.

a.  The complex roots of a real polynomial occur in conjugate pairs.

b.  If $n \in \mathbb{N}$ is not a perfect square, and $\sqrt{n}$ is a root of $f(x) \in \mathbb{Q}[x]$, then $-\sqrt{n}$ is a root as well.

c.  $\sqrt{2} + \sqrt{3}$ is a root of $f(x) = x^4 - 10x^2 + 1$; show that $\mathbb{Q}[\sqrt{2}, \sqrt{3}]$ is the splitting field of $f(x)$.

11. True or false? (Give proofs or disproofs.)

a.  $\mathbb{Z}_2[x]/\langle x^2 \rangle \cong \mathbb{Z}_4$, or $\mathbb{Z}_2[x]/\langle x^2 \rangle \cong \mathbb{Z}_2 \times \mathbb{Z}_2$?

b.  Same questions for $\mathbb{Z}_2[x]/\langle x^2 + x \rangle$.

c.  Same questions for $\mathbb{Z}_2[x]/\langle x^2 + 1 \rangle$.

d.  $\mathbb{Z}_3[x]/\langle x^2 - 1 \rangle \cong \mathbb{Z}_3 \times \mathbb{Z}_3$?

e.  $\mathbb{Q}[x]/\langle x^2 - 1 \rangle \cong \mathbb{Q} \times \mathbb{Q}$?

12. Let $R$ be a commutative ring, $I \subset R$ an ideal. Suppose $a \in R$, $a \notin I$, and $I + \langle a \rangle = R$ (see Exercise 4.1.17 for the notion of the sum of two ideals). Prove that $\bar{a} \in R/I$ is a unit.

13. Let $R$ be a commutative ring, and let $I \subsetneq R$ be an ideal. We say $I$ is a **prime ideal** if $ab \in I \implies a \in I$ or $b \in I$. We say $I$ is a **maximal ideal** if the only ideal properly containing $I$ is $R$ itself (i.e., if $J$ is an ideal and $I \subsetneq J$, then $J = R$).

a.  Find the prime and maximal ideals in $\mathbb{Z}$. Is $\langle x \rangle \subset \mathbb{Z}[x]$ prime? maximal?

b.  Prove that $R$ is an integral domain if and only if $\langle 0 \rangle$ is a prime ideal.

c.  Prove that $I$ is a prime ideal if and only if $R/I$ is an integral domain.

d.  Prove that $I$ is a maximal ideal if and only if $R/I$ is a field (see Exercises 12 and 4.1.3).

e.  Prove that every maximal ideal is a prime ideal.

14. Let $I = \{f(x) : f(1) = 0\} \subset \mathbb{Q}[x]$ and $J = \{f(x) : f(1) = f(-1) = 0\} \subset \mathbb{Q}[x]$.

a.  Prove that $I$ is a maximal ideal in $\mathbb{Q}[x]$.

b.  Prove that $J$ is an ideal in $\mathbb{Q}[x]$. Is it maximal? Is it prime?

15. Let $R = \mathbb{Z}[i] = \{a + bi : a, b \in \mathbb{Z}\}$. Let $I = \langle 1 + i \rangle$, $J = \langle 2 \rangle$.

a.  Decide whether each of $I$ and $J$ is a prime ideal, maximal ideal, or neither.

     b.    To what rings are $R/I$ and $R/J$ isomorphic? Is either a field? (Hint: You might start by counting the number of elements in each.)

16.    Decide (with proofs) whether each of the following rings is an integral domain, a principal ideal domain, or a field. (Hint: Use Exercise 13.)

    a.    $\mathbb{Q}[x]/\langle x^2 - x - 6 \rangle$
    b.    $\mathbb{Z}[i]/\langle 3 + i \rangle$
    c.    $\mathbb{Q}[x]/\langle x^4 + x^3 + 2x^2 - 3 \rangle$
    d.    $\mathbb{Q}[x, y]/\langle y - x^2 \rangle$

17.    a.    Suppose $I \subsetneq R$ is an ideal with the property that every element $a \notin I$ is a unit. Prove that $I$ is a maximal ideal. Are there any others?

    b.    Let $R = \{\frac{r}{s} : r, s \in \mathbb{Z}, s \text{ odd}\} \subset \mathbb{Q}$. Find the units in $R$ and find the maximal ideals in $R$.

18.    Let $R$ be a principal ideal domain. Prove that every prime ideal $I \neq \langle 0 \rangle$ is a maximal ideal. (Hint: Let $I = \langle a \rangle$, and suppose $I \subsetneq J = \langle b \rangle$. Prove that $b$ is a unit.)

19.    Prove that the maximal ideals of $\mathbb{C}[x]$ are in one-to-one correspondence with points of $\mathbb{C}$, whereas the maximal ideals of $\mathbb{R}[x]$ are in one-to-one correspondence with points of $\{z \in \mathbb{C} : Im\, z \geq 0\}$.

20.    Let $\gcd(m, n) = 1$. Prove that $\mathbb{Z}_{mn} \cong \mathbb{Z}_m \times \mathbb{Z}_n$. (Hint: This is a "strong" restatement of the Chinese Remainder Theorem, Theorem 3.7 of Chapter 1.)

21.    Prove that $\mathbb{Z}_2[x]/\langle x^3 + x + 1 \rangle$ and $\mathbb{Z}_2[x]/\langle x^3 + x^2 + 1 \rangle$ are both fields with eight elements. Decide whether or not they are isomorphic. (Hint: Note that a ring homomorphism

$$\phi: \mathbb{Z}_2[x]/\langle x^3 + x + 1 \rangle \to \mathbb{Z}_2[x]/\langle x^3 + x^2 + 1 \rangle$$

is completely determined by the value $\phi(\overline{x})$. But $\beta = \phi(\overline{x}) \in \mathbb{Z}_2[x]/\langle x^3 + x^2 + 1 \rangle$ must satisfy the equation $\beta^3 + \beta + 1 = 0$.)

22.    Suppose $f(x) \in F[x]$ is irreducible, and let $K$ be a field containing $F$. Suppose $\alpha$ and $\beta$ are two roots of $f(x)$ in $K$. Prove that there is an isomorphism $\phi: F[\alpha] \cong F[\beta]$ satisfying $\phi(a) = a$ for all $a \in F$.

23. Find all ring homomorphisms
    a. $\phi: \mathbb{Z} \times \mathbb{Z} \to \mathbb{Z} \times \mathbb{Z}$
    b. $\phi: \mathbb{Z} \times \mathbb{Z}_6 \to \mathbb{Z} \times \mathbb{Z}_3$
    c. $\phi: \mathbb{Z}_6 \times \mathbb{Z}_6 \to \mathbb{Z}_2 \times \mathbb{Z}_3$
    d. $\phi: \mathbb{Z}_6 \times \mathbb{Z}_6 \to \mathbb{Z}_6 \times \mathbb{Z}_6$
    (Warning: Be sure to check that your candidates for homomorphism are compatible with multiplication!)

24. Describe as completely as you can the following quotient rings:
    a. $\mathbb{Z}[x]/\langle 2x \rangle$
    b. $F[x,y]/\langle xy - 1 \rangle$

25. Proposition 2.2 of Chapter 3 and Theorem 2.3 of this chapter seem to be slightly at odds. Resolve the following paradox. Let $\alpha = \sqrt{2}$, $\beta = \sqrt{3}$. Is $\mathbb{Q}[x]/\langle (x^2 - 2)(x^2 - 3) \rangle$ isomorphic to $\mathbb{Q}[\alpha]$, $\mathbb{Q}[\beta]$, or $\mathbb{Q}[\alpha, \beta]$? What would Proposition 2.2 of Chapter 3 imply? (Hint: When we consider $\mathbb{Q}[x]/\langle x^2 - 2 \rangle$, $\bar{x}$ "knows" it must play the rôle of $\sqrt{2}$. What happens, then, in the case of $\mathbb{Q}[x]/\langle x^2 - 1 \rangle$? What is the implication of Proposition 2.2 here?)

26. Is $\alpha = \sum_{n=0}^{\infty} \frac{1}{3^{n!}}$ transcendental? $\alpha = \sum_{n=1}^{\infty} \frac{1}{10^{n^n}}$? Query: What about $\alpha = \sum_{n=0}^{\infty} \frac{1}{10^{2^n}}$?

27. To what common ring is $\mathbb{Z}_6[x]/\langle 2x - 3 \rangle$ isomorphic? What about $\mathbb{Z}_6[x]/\langle 2x - 1 \rangle$? Give proofs.

28. Let $R$ be an integral domain. We say a nonzero element $a \in R$ is *irreducible* if $a$ is not a unit and $a = bc \implies b$ or $c$ is a unit. (How does this relate to concepts we studied earlier?) We also say $b$ *divides* $a$ (written $b|a$, as usual) if $a = bc$ for some $c \in R$.
    a. Prove that if $p \in R$ and $\langle p \rangle$ is a prime ideal, then $p$ is irreducible.
    b. Prove that when $R$ is a principal ideal domain, the converse of a. holds (see Exercise 18). Give a counterexample when $R$ isn't a principal ideal domain.
    c. Prove that if $R$ is a principal ideal domain, then every element of $R$ can be written as a (finite) product of irreducible elements of $R$. (Hint: Here you must use the tricky observation that if $\langle a_1 \rangle \subset \langle a_2 \rangle \subset \langle a_3 \rangle \subset \cdots \subset \langle a_n \rangle \subset \ldots$, then

$$\bigcup_{i=1}^{\infty} \langle a_i \rangle$$ will be an ideal. Now use the fact that $R$ is a principal ideal domain to conclude that this "chain" of ideals cannot be infinite, so that the factorization process must terminate.)

d.   Now use b. to prove that when $R$ is a principal ideal domain, the product you've obtained in c. is unique up to multiplication by units. The upshot is that every principal ideal domain has the unique factorization property.

## 3. The Gaussian Integers

In this section we will focus on a particular example, the ring of **Gaussian integers** $\mathbb{Z}[i] \cong \mathbb{Z}[x]/\langle x^2 + 1 \rangle$. It is a beautiful example for (at least) two reasons—it admits a division algorithm, and the question of factoring in $\mathbb{Z}[i]$ is intricately tied in with some basic questions in number theory, as we shall soon see. Recall that $\mathbb{Z}[i] = \{a + bi : a, b \in \mathbb{Z}\} \subset \mathbb{C}$. Graphically, $\mathbb{Z}[i]$ is the set of all points in the complex plane with integer coordinates, as pictured in Figure 1.

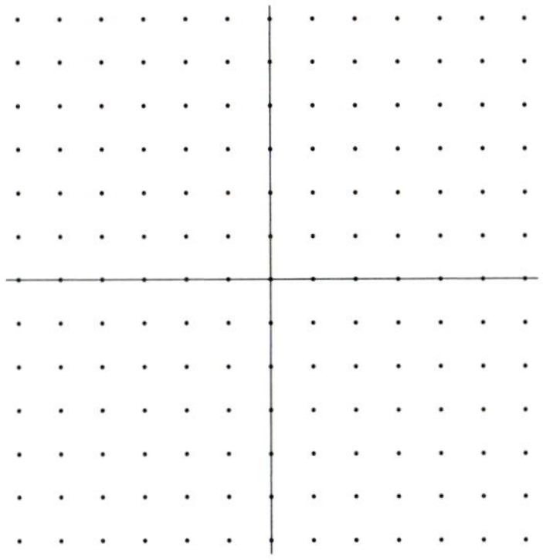

FIGURE 1

We begin by pointing out that, as in the case of $\mathbb{Z}$ and $F[x]$, there is a division algorithm in $\mathbb{Z}[i]$. First we need the notion of "size" of elements of $\mathbb{Z}[i]$; for $\mathbb{Z}$ we used absolute value, and for $F[x]$ we used the degree of the polynomial. For $\mathbb{Z}[i]$, so as to have an integer, we use the *square* of the length of the complex number.

**Proposition 3.1** (Division Algorithm). *Let $z = a + bi$ and $w = c + di \in \mathbb{Z}[i]$, $w \neq 0$. Then there are Gaussian integers $q$ and $r$ so that*

$$z = qw + r, \qquad \text{with} \qquad |r|^2 < |w|^2.$$

**Proof.** Consider the complex number $\frac{z}{w} = \frac{a+bi}{c+di} = x + yi \in \mathbb{Q}[i]$, and let $m$ and $n$ be integers so that $|m - x| \leq \frac{1}{2}$ and $|n - y| \leq \frac{1}{2}$. Set $q = m + ni$ and $r = z - qw$. Now we only need to check that $|r|^2 < |w|^2$. Well, $|r|^2 < |w|^2 \iff |\frac{r}{w}| < 1 \iff |\frac{z}{w} - q| < 1$. And $|\frac{z}{w} - q| = |(x + yi) - (m + ni)| = |(x - m) + (y - n)i| \leq \frac{1}{2}\sqrt{2} < 1$, as required. $\square$

**Example 1.** Apply the division algorithm to $z = -4 + 2i$ and $w = 3 + 2i$:

$$\frac{z}{w} = \frac{-4 + 2i}{3 + 2i} = \frac{-8 + 14i}{13}.$$

So we take $q = -1 + i$ and $r = (-4 + 2i) - (-1 + i)(3 + 2i) = 1 + i$.

**Remark.** Geometrically, we can interpret the division algorithm as follows. Given $w \in \mathbb{Z}[i]$, we consider the ideal $\langle w \rangle \subset \mathbb{Z}[i]$ it generates. We obtain a "tiling" of $\mathbb{Z}[i]$ by squares whose vertices are the elements of this ideal, as shown in Figure 2. Then any Gaussian integer $z$ lies in one of these squares, and its distance from the nearest vertex of that square can be at most $\frac{\sqrt{2}}{2}|w| < |w|$. The nearest vertex, of course, is $qw$.

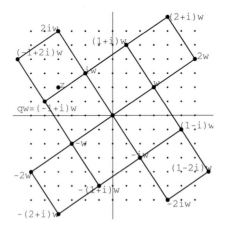

FIGURE 2

**Example 2.** Apply the division algorithm to $z = 7 + 4i$, $w = 1 - 5i$:

$$\frac{z}{w} = \frac{7 + 4i}{1 - 5i} = \frac{(7 + 4i)(1 + 5i)}{26} = \frac{-13 + 39i}{26} = \frac{-1 + 3i}{2} .$$

Thus, one of the nearest points in $\mathbb{Z}[i]$ is $q = i$, and $\frac{z}{w} - q = -\frac{1}{2} + \frac{1}{2}i$. Therefore, $r = (-\frac{1}{2} + \frac{1}{2}i)w = 2 + 3i$, as we can check directly: $7 + 4i = (i)(1 - 5i) + (2 + 3i)$. Note that in this example the "nearest" vertex is not unique, so the Gaussian integers $q$ and $r$ in Proposition 3.1 need not be unique. Interpret this non-uniqueness geometrically in terms of the "tiling" discussed above.

Once we have a division algorithm, we know how to find the g.c.d. of two Gaussian integers; and, as usual, there is a unique factorization theorem analogous to our results in Chapters 1 and 3 (see Exercise 7). The additional concern here is that we have more units than in $\mathbb{Z}$. If $z, u, v \in \mathbb{Z}[i]$ and $z = uv$, then $|z|^2 = |u|^2|v|^2$. As a result, $u$ is a unit if and only if $|u| = 1$, whence $u = 1, -1, i$, or $-i$. We say a Gaussian integer $z$ is **irreducible** if $z$ is not a unit and $z$ cannot be written as a product of non-units.

**Lemma 3.2.** *If $|z|^2$ is a prime number, then $z \in \mathbb{Z}[i]$ is irreducible.*

**Proof.** If $z = uv$ and neither $u$ nor $v$ is a unit, then (since $|u|^2, |v|^2 > 1$) we obtain a nontrivial factorization $|z|^2 = |u|^2|v|^2$.  $\square$

For example, $2 + 3i$ is irreducible, since $|2 + 3i|^2 = 13$. But the converse of Lemma 3.2 is false, since $3 \in \mathbb{Z}[i]$ is irreducible, and yet $|3|^2$ certainly is no prime number. On the other hand, the prime number 5 can be factored in $\mathbb{Z}[i]$ as $5 = (2 + i)(2 - i)$.

Before going any further, we illustrate the application of the Euclidean algorithm to find the g.c.d. of two Gaussian integers.

**Example 3.** Continuing with Example 2, let's find the g.c.d. of $1 - 5i$ and $7 + 4i$. As usual,

$$7 + 4i = (i)(1 - 5i) + (2 + 3i)$$
$$1 - 5i = (-1 - i)(2 + 3i) .$$

Therefore, $\gcd(1 - 5i, 7 + 4i) = 2 + 3i$, and it is unique up to units.

We now stop to illustrate some examples of quotient rings. The reader may notice a striking resemblance to Exercise 4.2.15.

**Example 4.** Consider the ring $R = \mathbb{Z}[i]/\langle 2 + i \rangle$. We observe first from the diagram that $R$ consists of five elements; that is, every $z \in \mathbb{Z}[i]$ is equivalent mod the ideal $\langle 2 + i \rangle$ to one of the elements $0$, $i$, $2i$, $1 + i$, or $1 + 2i$. By Lemma 3.2, $2 + i$ is irreducible, and so we expect $R$ to be a field. Since the only ring with five elements we know is $\mathbb{Z}_5$, we set about trying to prove that $R \cong \mathbb{Z}_5$.

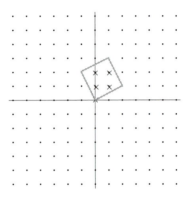

FIGURE 3

The pedestrian approach is to observe that a homomorphism must carry multiplicative identity to multiplicative identity, and so the only possible homomorphism $\phi: \mathbb{Z}_5 \to R$ is given by the formula:

$$\phi(k \ (\mathrm{mod}\ 5)) = k \ (\mathrm{mod}\ 2 + i).$$

Specifically, we have

$$\phi(0 \ (\mathrm{mod}\ 5)) = 0 \ (\mathrm{mod}\ 2 + i)$$
$$\phi(1 \ (\mathrm{mod}\ 5)) = 2i \ (\mathrm{mod}\ 2 + i)$$
$$\phi(2 \ (\mathrm{mod}\ 5)) = 1 + 2i \ (\mathrm{mod}\ 2 + i)$$
$$\phi(3 \ (\mathrm{mod}\ 5)) = i \ (\mathrm{mod}\ 2 + i)$$
$$\phi(4 \ (\mathrm{mod}\ 5)) = 1 + i \ (\mathrm{mod}\ 2 + i).$$

The original formula makes it clear that $\phi$ will be a ring homomorphism, *provided* it is well-defined. Suppose $k, \ell \in \mathbb{Z}$ and $k \equiv \ell \ (\mathrm{mod}\ 5)$; does it follow that $k \equiv \ell \ (\mathrm{mod}\ 2 + i)$? Well, yes, since $5 = (2 + i)(2 - i)$, if $5 | (k - \ell)$, then $(2 + i) | (k - \ell)$ as well. So, $\phi$ is an isomorphism, being a one-to-one correspondence between two five-element sets.

A more inspired and more interesting approach is to apply Theorem 2.2. We seek a homomorphism $\psi$ from $\mathbb{Z}[i]$ onto $\mathbb{Z}_5$ whose kernel is $\langle 2 + i \rangle$. Since a homomorphism must carry the multiplicative identity to the multiplicative identity, once again we are forced to define $\psi(1) = 1 \in \mathbb{Z}_5$. Suppose $\psi(i) = \beta \in \mathbb{Z}_5$. Since $i^2 = -1$ in $\mathbb{Z}[i]$, we must have $\beta^2 = -1$ in $\mathbb{Z}_5$, so $\beta$ must be 2 or 3. We assign $\beta = 3$, so that, as a consequence, $\psi(2 + i) = \psi(2) + \psi(i) = 2\psi(1) + \psi(i) = 2 + \beta = 0$, as desired. We thus arrive at the rule:

$$\psi: \mathbb{Z}[i] \to \mathbb{Z}_5$$
$$\psi(a + bi) = a + 3b.$$

It clearly is compatible with addition, and we check multiplication here:

$$\psi((a + bi)(c + di)) = \psi((ac - bd) + (ad + bc)i)$$
$$= (ac - bd) + 3(ad + bc) = (a + 3b)(c + 3d) \in \mathbb{Z}_5$$

because $9 \equiv -1 \pmod 5$. We leave it to the reader to check that $\psi$ maps onto $\mathbb{Z}_5$ and that $\ker \psi = \langle 2 + i \rangle$.

**Example 5.** Next, we wish to examine the ring $R = \mathbb{Z}[i]/\langle 5 \rangle$. Guided once again by the picture, we surmise that $R$ contains 25 elements. How many rings with 25 elements do we actually know? Well, at least $\mathbb{Z}_{25}$ and $\mathbb{Z}_5 \times \mathbb{Z}_5$.

If $R$ were to be isomorphic to $\mathbb{Z}_{25}$, we'd have to find an element $a \in R$, all of whose *integral* multiples give the entire ring $R$. This contradicts our intuitive notion that $R$ looks set-theoretically like $\mathbb{Z}_5 \times \mathbb{Z}_5$ (see Figure 4). Recall that we defined in Section 2 a ring structure on $\mathbb{Z}_5 \times \mathbb{Z}_5$ by adding and multiplying component by component. (It follows immediately that the "obvious" function $\phi: R \to \mathbb{Z}_5 \times \mathbb{Z}_5$ given by $\phi(a + bi \mod \langle 5 \rangle) = (a, b)$ cannot even be a homomorphism. Why?)

We proceed to define a homomorphism $\phi$ from $\mathbb{Z}[i]$ onto $\mathbb{Z}_5 \times \mathbb{Z}_5$; we will see that its kernel is $\langle 5 \rangle$, and we'll then infer from the Fundamental Homomorphism Theorem that $\mathbb{Z}[i]/\langle 5 \rangle \cong \mathbb{Z}_5 \times \mathbb{Z}_5$. Define $\phi(1) = (1, 1)$ (as is automatic) and $\phi(i) = (2, 3)$. (Where did this come from? As we argued in Example 4, $\phi(i) \in \mathbb{Z}_5 \times \mathbb{Z}_5$ must be a square root of $-1$, so the possibilities are $(\pm 2, \pm 2)$. But if we choose $\phi(i) = \pm(2, 2)$, then $\phi(2 \mp i) = 0$, and the kernel is too large.) The additive structure then dictates that

$$\phi(m + ni) = (m + 2n, m + 3n) \pmod 5 \in \mathbb{Z}_5 \times \mathbb{Z}_5 .$$

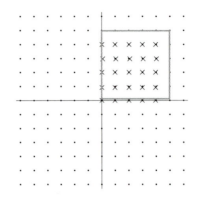

FIGURE 4

Is this in fact a homomorphism—i.e., does the multiplicative structure check out? Well,

$$\phi\left((m + ni)(p + qi)\right) = \phi\left((mp - nq) + (np + mq)i\right)$$
$$= (mp - nq + 2np + 2mq, mp - nq + 3np + 3mq)$$
$$= \phi(m + ni)\phi(p + qi),$$

all because $4 \equiv -1 \pmod 5$ and $9 \equiv -1 \pmod 5$.

Now we have to check that $\phi$ maps onto $\mathbb{Z}_5 \times \mathbb{Z}_5$: given $(a, b) \in \mathbb{Z}_5 \times \mathbb{Z}_5$, solve the system of linear equations

$$(*) \qquad \begin{aligned} m + 2n &= a \\ m + 3n &= b, \end{aligned}$$

with all arithmetic in the field $\mathbb{Z}_5$. The coefficient matrix $\begin{bmatrix} 1 & 2 \\ 1 & 3 \end{bmatrix}$ has determinant 1, and so there's a unique solution for every $a, b \in \mathbb{Z}_5$.

Lastly, what is $\ker \phi$? Suppose $\phi(m + ni) = (0, 0) \in \mathbb{Z}_5 \times \mathbb{Z}_5$. The same analysis we just completed shows that in $\mathbb{Z}_5$ the linear system $(*)$ has only the solution $m = n = 0 \in \mathbb{Z}_5$ when $a = b = 0 \in \mathbb{Z}_5$. This shows that the integers $m$ and $n$ must be multiples of 5, and, indeed, $\ker \phi = \langle 5 \rangle$. Whew!!

When can a prime number $p$ be expressed as the sum of two squares? The answer to this question is related to the irreducibility of $p$ in $\mathbb{Z}[i]$. We begin with the following lemma, which gives an obvious necessary criterion for $p$ to be a sum of squares.

**Lemma 3.3.** *If the odd prime number $p$ can be written in the form $p = a^2 + b^2$, $a, b \in \mathbb{Z}$, then $p \equiv 1 \pmod 4$.*

**Proof.** Since $p$ is odd, one of the two integers $a$ and $b$ must be even and the other odd. But the square of an even number is congruent to 0 (mod 4), and the square of an odd number is congruent to 1 (mod 4). □

**Proposition 3.4.** *If $p$ is a prime number and $p \equiv 1$ (mod 4), then the polynomial $x^2 + 1$ has a root in $\mathbb{Z}_p$.*

**Proof.** This was the gist of Exercise 1.4.15, but we give the proof here for completeness. Let $x = 1 \cdot 2 \cdot \ldots \cdot \frac{p-1}{2} \in \mathbb{Z}_p$. We claim that $x^2 = -1$. Note that $x$ is the product of an even number of terms (since $p \equiv 1$ (mod 4)), and so

$$x^2 = (1 \cdot 2 \cdot \ldots \cdot \tfrac{p-1}{2})((-1) \cdot (-2) \cdot \ldots \cdot (-\tfrac{p-1}{2})).$$

Note that this is the product of all the nonzero elements of $\mathbb{Z}_p$. We compute this product by pairing elements with their *multiplicative* inverses (with the exception of 1 and $-1$, which are their own multiplicative inverses). Thus the net result is $-1$, and we are done. □

**Proposition 3.5.** *The prime number $p$ fails to be irreducible in $\mathbb{Z}[i] \iff p = a^2 + b^2$ for some $a, b \in \mathbb{Z}$.*

**Proof.** If $p = a^2 + b^2$, then $p = (a + bi)(a - bi)$, so $p$ is not irreducible in $\mathbb{Z}[i]$. Conversely, if $p$ is not irreducible in $\mathbb{Z}[i]$, then there is an irreducible factor $z = a + bi$. Since $p$ is real, $\bar{z} = a - bi$ also divides $p$. If $p = 2$, then $2 = (1 + i)(1 - i)$. If $p$ is odd, then by Exercise 5, $\gcd(a + bi, a - bi) = 1$, and so $(a + bi)(a - bi)$ divides $p$ (why?). In either event, $a^2 + b^2$ divides $p$. Since $a^2 + b^2 > 1$ and $p$ is prime, it follows that $p = a^2 + b^2$. □

We now come to the most sophisticated of this series of results, one relating Propositions 3.4 and 3.5. It is here that the notion of quotient rings proves its worth (but read the proof carefully and see Exercise 16 for a point that is sloughed over).

**Proposition 3.6.** *$p$ is irreducible in $\mathbb{Z}[i]$ if and only if the polynomial $x^2 + 1$ is irreducible in $\mathbb{Z}_p[x]$.*

**Proof.** $p$ is irreducible in $\mathbb{Z}[i]$ if and only if $\mathbb{Z}[i]/\langle p \rangle$ is an integral domain (see Exercise 4.2.13 and Exercise 3). Recall that we can interpret $\mathbb{Z}[i]$ itself as a quotient ring, and so

$$\mathbb{Z}[i]/\langle p \rangle \cong (\mathbb{Z}[x]/\langle x^2 + 1 \rangle)/\langle p \rangle \cong \mathbb{Z}[x]/\langle x^2 + 1, p \rangle \cong \mathbb{Z}_p[x]/\langle x^2 + 1 \rangle.$$

Thus $\mathbb{Z}[i]/\langle p \rangle$ is an integral domain if and only if $\mathbb{Z}_p[x]/\langle x^2 + 1 \rangle$ is an integral domain, and the latter occurs if and only if $x^2 + 1$ is irreducible in $\mathbb{Z}_p[x]$, as promised.  $\square$

Now we can assemble all the pieces to prove our main result.

**Theorem 3.7.** *Let $p$ be an odd prime. The following are equivalent:*

(1) $p \equiv 1 \pmod 4$;
(2) $p = a^2 + b^2$ *for some* $a, b \in \mathbb{Z}$;
(3) $x^2 + 1 \equiv 0 \pmod p$ *has a solution.*

**Proof.** $(2) \Rightarrow (1)$ follows from Lemma 3.3. $(1) \Rightarrow (3)$ is the content of Proposition 3.4. And $(3) \Rightarrow (2)$ follows immediately from Propositions 3.6 and 3.5.  $\square$

As a corollary of this result, we are able to prove that there are infinitely many prime numbers congruent to 1 (mod 4), the result parallel to Proposition 3.4 of Chapter 1.

**Theorem 3.8.** *There are infinitely many primes of the form $p = 4k + 1$, $k \in \mathbb{N}$.*

**Proof.** Suppose, as usual, that there were only finitely many such primes, and list them all: $p_1, p_2, \ldots, p_m$. Consider the integer $N = 4(p_1 p_2 \cdots p_m)^2 + 1$. $N$ can be factored as a product of primes, but none of the primes $p_1, p_2, \ldots, p_m$ is a divisor of $N$. Let $p$ be a prime factor of $N$ (note that $p$ is necessarily odd). The congruence $x^2 + 1 \equiv 0 \pmod p$ has a solution, since, letting $x = 2p_1 p_2 \cdots p_m$, we have $x^2 + 1 \equiv 0 \pmod N$, and therefore mod $p$ as well. So, by Theorem 3.7, $p \equiv 1 \pmod 4$, contradicting the fact that $p_j \nmid N$, $j = 1, \ldots, m$.  $\square$

## EXERCISES 4.3

1. Apply the Euclidean algorithm to find $\gcd(z, w)$:
   a.  $z = 8 + 6i$, $w = 5 - 15i$
   b.  $z = 4 - i$, $w = 1 + i$
   c.  $z = 16 + 7i$, $w = 10 - 5i$

2. Factor each of the following as a product of irreducible elements in $\mathbb{Z}[i]$:

     a.   6
     b.   11+7i
     c.   7
     d.   4+3i

3. Prove that $z \in \mathbb{Z}[i]$ is irreducible if and only if $\langle z \rangle$ is a prime ideal in $\mathbb{Z}[i]$. (See Exercise 4.2.13.)

4. Prove there are infinitely many irreducible elements in $\mathbb{Z}[i]$.

5. Prove that if $a, b \in \mathbb{Z}$ are relatively prime and $a^2 + b^2$ is odd, then $\gcd(a + bi, a - bi) = 1$.

6. Prove that $\mathbb{Z}[i]$ is a principal ideal domain.

7. Prove that the ring $\mathbb{Z}[i]$ enjoys unique factorization; i.e., every nonzero element can be written as a product of irreducible elements, and this expression is unique up to multiplication by units. (Hint: Mimic the proof of Theorem 2.7 of Chapter 1, doing complete induction on $|z|^2$.)

8. Prove or give a counterexample: if $p \equiv 1 \pmod 4$, then $p$ can be written *uniquely* as a sum of squares (i.e., $p = a^2 + b^2$, and there is only one pair of numbers $a^2$, $b^2$).

9. Suppose $a, b \in \mathbb{Z}$ are integers. Show that $\gcd(a, b)$ is the same if we view them both as Gaussian integers. (Cf. Exercise 3.2.15.)

10. Establish the following isomorphisms:
     a.   $\mathbb{Z}[i]/\langle 1 + i \rangle \cong \mathbb{Z}_2$
     b.   $\mathbb{Z}[i]/\langle 3 + i \rangle \cong \mathbb{Z}_{10}$
     c.   $\mathbb{Z}[i]/\langle 13 \rangle \cong \mathbb{Z}_{13} \times \mathbb{Z}_{13}$
     d.   $\mathbb{Z}[i]/\langle 2 + 3i \rangle \cong \mathbb{Z}_{13}$
    What about
     e.   $\mathbb{Z}[i]/\langle 2 \rangle$
     f.   $\mathbb{Z}[i]/\langle 3 \rangle$
     g.   $\mathbb{Z}[i]/\langle 3 + 4i \rangle$
    Query. Can you generalize these results?

11. Let $\omega = -\frac{1}{2} + \frac{\sqrt{3}}{2}i$. Show that there is a division algorithm for the ring $\mathbb{Z}[\omega]$. (Hint: Proceed pictorially.)

12. Show that unique factorization fails in the ring $\mathbb{Z}[2i]$; show that the ring is not a principal ideal domain, either.

13. Show that unique factorization fails in the ring $\mathbb{Z}[\sqrt{-5}\,]$; show that the ring is not a principal ideal domain, either.

14. Prove that the irreducible elements of $\mathbb{Z}[i]$ are precisely the primes $p \equiv 3 \pmod 4$ (along with $-p$, $pi$, and $-pi$) and the elements $a + bi$ where $a^2 + b^2$ is prime.

15. Generalizing Theorem 3.7, we can classify all natural numbers $n$ that can be written as a sum of two squares. Prove that $n = a^2 + b^2$ for some $a, b \geq 0$ if and only if any prime $p \equiv 3 \pmod 4$ appearing in the prime factorization of $n$ has an even exponent. (Hint: Factor $n$ as a product of irreducible elements in $\mathbb{Z}[i]$.)

16. Let $R$ be a commutative ring, and let $I$ and $J$ be ideals in $R$. Let $\phi\colon R \to R/I$ be the obvious homomorphism. Write $R/I = \overline{R}$ and let $\phi(I + J) = \overline{J}$. Prove that $R/(I + J) \cong \overline{R}/\overline{J}$. Apply this to make the relevant justifications in the proof of Proposition 3.6. (See Exercises 4.1.16b. and 4.1.17.)

17. Find the units in $\mathbb{Z}[\sqrt{-3}]$ and $\mathbb{Z}[\sqrt{2}]$.

18. a. If $z = a + bi$, $w = c + di \in \mathbb{Z}[i]$, let $N(z, w)$ denote the number of points in $\mathbb{Z}[i]$ lying either inside the parallelogram spanned by the vectors $z$ and $w$ or along either of these two vectors, not including the points $a + bi$ and $c + di$ themselves. For example, see Figure 5. Prove that $N(z, w) = \left| \det \begin{bmatrix} a & c \\ b & d \end{bmatrix} \right|$. (Hint: Show that $N(z, w + nz) = N(z, w)$ for any $n \in \mathbb{Z}$, and cf. Section 2 of Appendix B.)

    b. Given a nonzero element $z$ of $\mathbb{Z}[i]$, prove that the quotient ring $\mathbb{Z}[i]/\langle z \rangle$ has $|z|^2$ elements.

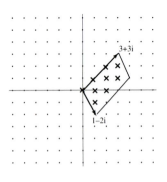

FIGURE 5

# CHAPTER 5

# Field Extensions

As Descartes had developed the subject we now call *analytic geometry*, giving an underlying algebraic structure to the geometry of the plane, Laguerre and Clairaut did so for three-dimensional space in the early part of the eighteenth century. In 1788 Lagrange introduced vectors in his *Analytic Mechanics* to deal with forces, velocities, and accelerations. It was only at the end of the nineteenth century, with the development of the theory of electricity and magnetism, that vector algebra and vector calculus began to develop into the powerful tools that they are today. Cayley gave a formal definition of $n$-dimensional space and then, in 1855 (when he was 34 years old, after he'd been a lawyer for several years), he defined matrices and linear transformations, as he and Sylvester pursued the study of invariant theory (eigenvalues being an elementary example).

It's time to bring the power of linear algebra to bear on our considerations of splitting fields. In §1 we give a self-contained review of vector spaces: linear independence, spanning, bases, and dimension. We then give an application to field extensions, obtaining a precise way to measure the complexity of algebraic numbers and a powerful counting tool. In §2, we use this approach to address the classic question of compass and straightedge constructions: one cannot "square the circle," "duplicate the cube," or trisect the general angle using only a compass and straightedge. (On the other hand, Gauss understood how to obtain *positive* construction results: at the age of nineteen he showed how to construct the regular 17-gon.) In §3, we give a brief and optional treatment of finite fields. This material has fascinating modern applications to coding, computer design, and experiment design (see the Supplementary Reading).

## 1. Vector Spaces and Dimension

We begin with a review of the notion of **dimension** of a vector space. Our treatment will be self-contained, but quite brief. In elementary linear algebra courses, vector spaces arise as the solution sets to systems of homogeneous linear equations and dimension, as the number of "degrees of freedom" in these solutions. For our immediate applications, vector spaces will arise as field extensions $K \supset F$.

**Definition.** Let $F$ be a field. A **vector space** $V$ over $F$ is a set closed under addition (+) and under multiplication by elements of $F$ (called *scalar multiplication*), subject to the following laws:

(1) For all $\mathbf{v}, \mathbf{w} \in V$, $\mathbf{v} + \mathbf{w} = \mathbf{w} + \mathbf{v}$.
(2) For all $\mathbf{u}, \mathbf{v}, \mathbf{w} \in V$, $\mathbf{u} + (\mathbf{v} + \mathbf{w}) = (\mathbf{u} + \mathbf{v}) + \mathbf{w}$.
(3) There is an element $\mathbf{0} \in V$, called the zero vector, so that $\mathbf{0} + \mathbf{v} = \mathbf{v}$ for all $\mathbf{v} \in V$.
(4) For each $\mathbf{v} \in V$, there is an element $-\mathbf{v}$ so that $\mathbf{v} + (-\mathbf{v}) = \mathbf{0}$.
(5) For all $c, d \in F$ and $\mathbf{v} \in V$, $c(d\mathbf{v}) = (cd)\mathbf{v}$.
(6) For all $c \in F$ and $\mathbf{v}, \mathbf{w} \in V$, $c(\mathbf{v} + \mathbf{w}) = c\mathbf{v} + c\mathbf{w}$.
(7) For all $c, d \in F$ and $\mathbf{v} \in V$, $(c + d)\mathbf{v} = c\mathbf{v} + d\mathbf{v}$.
(8) For all $\mathbf{v} \in V$, $1\mathbf{v} = \mathbf{v}$ (where 1 is the multiplicative identity of $F$).

## Examples 1.

(a) The familiar example is $\mathbb{R}^n$, the set of all ordered $n$-tuples $(x_1, \ldots, x_n)$, $x_1, \ldots, x_n \in \mathbb{R}$. This is a vector space over $\mathbb{R}$, and the same construction works over any field $F$.
(b) For any field $F$, the polynomial ring $F[x]$ is a vector space over $F$ (addition is addition of polynomials, and scalar multiplication is the obvious thing).
(c) Let $F = \mathbb{Q}$, $V = \mathbb{Q}[\sqrt[3]{2}] = \{a + b\sqrt[3]{2} + c(\sqrt[3]{2})^2 : a, b, c \in \mathbb{Q}\}$. Then $V$ is a vector space over $\mathbb{Q}$, and this example is typical of what we shall be encountering.
(d) $\mathbb{C}$ is a vector space over $\mathbb{R}$ (and over $\mathbb{Q}$ as well).

(e) We can generalize both the preceding examples by taking any two fields $F$ and $K$ with $F \subset K$. (Recall from Chapter 4 that $K$ is called a field extension of $F$.) Then $K$ is a vector space over $F$.

We say a nonempty subset $W \subset V$ is a **subspace** of $V$ if it is closed under addition and scalar multiplication, i.e., if for all $\mathbf{v}, \mathbf{w} \in W$ and $c \in F$, $\mathbf{v} + \mathbf{w} \in W$ and $c\mathbf{v} \in W$. Given vectors $\mathbf{v}_1, \dots, \mathbf{v}_k \in V$ and scalars $c_1, \dots, c_k \in F$, the vector $c_1\mathbf{v}_1 + \cdots + c_k\mathbf{v}_k$ is called a **linear combination** of the given vectors. The set of all linear combinations of $\mathbf{v}_1, \dots, \mathbf{v}_k$ is denoted by $\mathrm{Span}(\mathbf{v}_1, \dots, \mathbf{v}_k)$.

**Lemma 1.1.** $\mathrm{Span}(\mathbf{v}_1, \dots, \mathbf{v}_k)$ *is a subspace of $V$, called the subspace spanned by* $\mathbf{v}_1, \dots, \mathbf{v}_k$.

**Proof.** See Exercise 2.  □

**Example 2.** Let $F$ be a field, $f(x) \in F[x]$, and suppose $\alpha$ is a root of $f(x)$ in some field $K$ containing $F$. Then $K$ is a vector space over $F$; $F[\alpha]$ is a subspace of $K$ and is spanned by $1$, $\alpha$, $\alpha^2$, ..., $\alpha^{\deg(f(x))-1}$.

We say the vectors $\mathbf{v}_1, \dots, \mathbf{v}_k \in V$ are **linearly independent** if

$$c_1\mathbf{v}_1 + \cdots + c_k\mathbf{v}_k = \mathbf{0} \implies c_1 = \cdots = c_k = 0,$$

i.e., if the only relation among $\mathbf{v}_1, \dots, \mathbf{v}_k$ is the trivial relation. The vectors $\mathbf{v}_1, \dots, \mathbf{v}_k$ are **linearly dependent** if they are not linearly independent.

**Examples 3.**
(a) The vectors $\mathbf{v}_1 = (1,0,0,0)$, $\mathbf{v}_2 = (0,1,0,0)$, $\mathbf{v}_3 = (0,0,1,0) \in \mathbb{R}^4$ are linearly independent.
(b) The vectors $\mathbf{v}_1 = (2,3,-1)$, $\mathbf{v}_2 = (1,1,-2)$, $\mathbf{v}_3 = (1,2,1) \in \mathbb{R}^3$ are linearly dependent, as $-\mathbf{v}_1 + \mathbf{v}_2 + \mathbf{v}_3 = \mathbf{0}$.
(c) The vectors $\mathbf{v}_1 = 1$, $\mathbf{v}_2 = 2i \in \mathbb{C}$ are linearly independent when we view $\mathbb{C}$ as a vector space over $\mathbb{R}$, although they are certainly linearly dependent when we view $\mathbb{C}$ as a vector space over itself.

**Definition.** We say $\mathbf{v}_1, \dots, \mathbf{v}_k \in V$ are a **basis** for $V$ if
   (i) $\mathbf{v}_1, \dots, \mathbf{v}_k$ are linearly independent; and
   (ii) $\mathbf{v}_1, \dots, \mathbf{v}_k$ span $V$, i.e., $\mathrm{Span}(\mathbf{v}_1, \dots, \mathbf{v}_k) = V$.

For convenience, we call a vector space $V$ **finite-dimensional** if $V$ is spanned by a finite number of vectors. So $\mathbb{R}^n$ is a finite-dimensional vector space over $\mathbb{R}$, but $F[x]$ is not a finite-dimensional vector space over $F$. Is $\mathbb{R}$ a finite-dimensional vector space over $\mathbb{Q}$? Our main goal is to show that if $V$ is a finite-dimensional vector space, then $V$ has a basis, and that any two bases for $V$ consist of the same number of elements. This number is called the **dimension** of $V$, abbreviated $\dim(V)$ (or $\dim_F(V)$ to remove any ambiguity). N.B.: We adopt the convention that $\dim\{\mathbf{0}\} = 0$, and assume from now on that we are dealing with vector spaces having more than just the zero element.

**Proposition 1.2.** *Let $V$ be a finite-dimensional vector space, and suppose $V = \text{Span}(\mathbf{v}_1, \dots, \mathbf{v}_k)$. Then some subset of $\{\mathbf{v}_1, \dots, \mathbf{v}_k\}$ forms a basis for $V$.*

**Proof.** We proceed by induction on $k$. When $k = 1$ and $V \neq \{\mathbf{0}\}$, $\mathbf{v}_1$ must itself be a basis for $V$. Assume now that whenever vectors $\mathbf{v}_1, \dots, \mathbf{v}_m$ span a vector space $W$, some subset of $\{\mathbf{v}_1, \dots, \mathbf{v}_m\}$ forms a basis for $W$. Given a vector space $V$ spanned by $\mathbf{v}_1, \dots, \mathbf{v}_m, \mathbf{v}_{m+1}$, consider the subspace $W = \text{Span}(\mathbf{v}_1, \dots, \mathbf{v}_m)$. By induction hypothesis, there is a subset $\mathcal{S} \subset \{\mathbf{v}_1, \dots, \mathbf{v}_m\}$ that forms a basis for $W$. If $\mathbf{v}_{m+1} \in W$, then $\mathcal{S}$ forms a basis for all of $V$. If not, then by Exercise 5, $\mathcal{S} \cup \{\mathbf{v}_{m+1}\}$ forms a basis for $V$.  □

It follows from Proposition 1.2 that every finite-dimensional vector space has a basis.

**Theorem 1.3.** *Let $V$ be a finite-dimensional vector space. Suppose $\mathbf{v}_1, \dots, \mathbf{v}_k$ span $V$ and $\mathbf{w}_1, \dots, \mathbf{w}_\ell \in V$ are linearly independent. Then $k \geq \ell$.*

**Proof.** The proof we give here is somewhat abstract, although conceptually appealing. For a more concrete proof in terms of the theory of systems of linear equations, see Exercise 24. Let $\mathcal{L} = \{\mathbf{w}_1, \dots, \mathbf{w}_\ell\}$ ($\mathcal{L}$ for "linearly independent") and $\mathcal{S} = \{\mathbf{v}_1, \dots, \mathbf{v}_k\}$ ($\mathcal{S}$ for "spanning"). If, by some miracle, $\mathcal{L} \subset \mathcal{S}$, then the result is immediate (since if $\mathcal{L}$ and $\mathcal{S}$ are finite sets and $\mathcal{L} \subset \mathcal{S}$, then $\mathcal{L}$ has at most as many elements as $\mathcal{S}$). If not, here is our game plan: Suppose $\mathbf{w}_1 \notin \mathcal{S}$; then we will replace $\mathbf{w}_1$ by some $\mathbf{v}_j \in \mathcal{S}$, say $\mathbf{v}_1$, with the property that $\mathbf{v}_1, \mathbf{w}_2, \dots, \mathbf{w}_\ell$ are still linearly independent. We let $\mathcal{L}_{(1)} = \{\mathbf{v}_1, \mathbf{w}_2, \dots, \mathbf{w}_\ell\}$, and observe that $\mathcal{L}_{(1)}$ has the same number of elements as $\mathcal{L}$. If $\mathcal{L}_{(1)} \subset \mathcal{S}$, we are done. If not, we continue this process, replacing $\mathbf{w}$'s with $\mathbf{v}$'s until the final set $\mathcal{L}_{(\text{final})}$ of linearly independent vectors *is* a subset of $\mathcal{S}$; and then we are finished.

It remains only to see that given $\mathbf{w}_1, \ldots, \mathbf{w}_\ell$ linearly independent with $\mathbf{w}_1 \notin \mathcal{S}$, there must be some $\mathbf{v}_j \in \mathcal{S}$ so that $\mathbf{v}_j, \mathbf{w}_2, \ldots, \mathbf{w}_\ell$ are still linearly independent. If not, by Exercise 5, every vector $\mathbf{v}_1, \ldots, \mathbf{v}_k \in \text{Span}(\mathbf{w}_2, \ldots, \mathbf{w}_\ell)$; since $\mathbf{v}_1, \ldots, \mathbf{v}_k$ span $V$, this implies in particular that $\mathbf{w}_1 \in \text{Span}(\mathbf{w}_2, \ldots, \mathbf{w}_\ell)$; and this contradicts the linear independence of $\mathbf{w}_1, \ldots, \mathbf{w}_\ell$.   □

**Corollary 1.4.** *Any two bases for a finite-dimensional vector space V have the same number of elements.*

**Proof.** Let $\mathbf{v}_1, \ldots, \mathbf{v}_k$ and $\mathbf{w}_1, \ldots, \mathbf{w}_\ell$ be two bases for $V$. Then $\mathbf{v}_1, \ldots, \mathbf{v}_k$ span $V$, and $\mathbf{w}_1, \ldots, \mathbf{w}_\ell$ are linearly independent, so $k \geq \ell$. Reversing the rôles of the $\mathbf{v}$'s and $\mathbf{w}$'s, $\mathbf{v}_1, \ldots, \mathbf{v}_k$ are linearly independent, and $\mathbf{w}_1, \ldots, \mathbf{w}_\ell$ span $V$; so $\ell \geq k$. Thus $k = \ell$, as required.   □

Dimension of a finite-dimensional vector space, as we've seen, does make sense. We shall exploit this important notion in the remainder of this chapter: let $F$ and $K$ be fields with $F \subset K$. If $K$ is a finite-dimensional vector space over $F$, we call $K$ a **field extension** of $F$ of **degree** $\dim_F(K)$; we denote the degree of $K$ over $F$ by $[K : F]$. The following computational tool will prove quite useful.

**Proposition 1.5.** *Suppose E is a field extension of F of degree $[E : F]$ and K is a field extension of E of degree $[K : E]$. Then K is a field extension of F of degree $[K : F] = [K : E][E : F]$.*

**Proof.** Let $\alpha_1, \ldots, \alpha_m$ be a basis for $E$ over $F$, and let $\beta_1, \ldots, \beta_n$ be a basis for $K$ over $E$. We claim that $\alpha_i \beta_j$, $1 \leq i \leq m$, $1 \leq j \leq n$, are a basis for $K$ over $F$. (Here the multiplication is done in the field $K$.) We must check that these vectors span $K$ and are linearly independent over $F$.

Let $\beta \in K$ be arbitrary. Then there are elements $x_1, \ldots, x_n \in E$ so that

(A)
$$\beta = \sum_{j=1}^{n} x_j \beta_j,$$

insofar as $\beta_1, \ldots, \beta_n$ are a basis for $K$ over $E$. On the other hand, since $\alpha_1, \ldots, \alpha_m$ are a basis for $E$ over $F$, there are elements $y_{1j}, \ldots,$

$y_{mj} \in F$, $j = 1, \ldots, n$, so that

(B) $$x_j = \sum_{i=1}^{m} y_{ij}\alpha_i, \quad j = 1, \ldots, n.$$

Substituting (B) in (A), we obtain

(C) $$\beta = \sum_{j=1}^{n} \sum_{i=1}^{m} y_{ij}(\alpha_i\beta_j), \quad y_{ij} \in F,$$

whence the vectors $\alpha_i\beta_j$ span $K$ as a vector space over $F$.

To prove linear independence, suppose $\beta = 0$ in (C). Then, reversing the process we just completed, we have

$$0 = \sum_{j=1}^{n} \sum_{i=1}^{m} y_{ij}(\alpha_i\beta_j) = \sum_{j=1}^{n} \left(\sum_{i=1}^{m} y_{ij}\alpha_i\right)\beta_j.$$

Since $\beta_1, \ldots, \beta_n$ are linearly independent, we must have $\sum_{i=1}^{m} y_{ij}\alpha_i = 0$ for all $j = 1, \ldots, n$. But since $\alpha_1, \ldots, \alpha_m$ are linearly independent, this means that $y_{ij} = 0$ for all $i = 1, \ldots, m$ and $j = 1, \ldots, n$, as needed. $\square$

**Lemma 1.6.** *Suppose $K$ is a field extension of $F$ and $\alpha \in K$ is the root of an irreducible polynomial $f(x) \in F[x]$ of degree $n$. Then $[F[\alpha] : F] = n$.*

**Proof.** Recall that $F[\alpha] = \{p(\alpha) : p(x) \in F[x]\}$, so $F[\alpha]$ is spanned by $1, \alpha, \alpha^2, \ldots$. Given any polynomial $p(x) \in F[x]$, by the division algorithm, there is $r(x) \in F[x]$ so that $p(x) \equiv r(x) \pmod{f(x)}$ with $\deg(r(x)) < n$ or $r(x) = 0$. Since $f(\alpha) = 0$, $p(\alpha) = r(\alpha)$ is a linear combination of $1, \alpha, \alpha^2, \ldots, \alpha^{n-1}$. Thus, these $n$ elements span $F[\alpha]$ as a vector space over $F$. On the other hand, they are linearly independent; if $c_0(1) + c_1\alpha + c_2\alpha^2 + \cdots + c_{n-1}\alpha^{n-1} = 0$, with $c_j \in F$ not all zero, then $g(x) = c_{n-1}x^{n-1} + \cdots + c_2x^2 + c_1x + c_0 \in F[x]$ is a polynomial of degree less than $n$ having $\alpha$ as a root. This contradicts the irreducibility of $f(x)$ (see Corollary 1.4 of Chapter 4). $\square$

**Corollary 1.7.** *Suppose $[K : F] = n$ and $\alpha \in K$ is the root of an irreducible polynomial $f(x) \in F[x]$. Then $\deg(f(x)) \mid n$.*

**Proof.** We have $F \subset F[\alpha] \subset K$. Since $[K : F[\alpha]][F[\alpha] : F] = n$ (see Exercise 18), $[F[\alpha] : F] \mid n$. Now the conclusion follows from Lemma 1.6. $\square$

**Example 4.** Suppose $[K : F] = 4$ and $f(x) \in F[x]$ is an irreducible polynomial of degree 3. Then $f(x)$ is likewise irreducible in $K[x]$. This follows immediately from Corollary 1.7: in order for a cubic polynomial not to be irreducible in $K[x]$, it must have a root in $K$; since $3 \nmid 4$, this cannot happen. For example, this shows that $f(x) = x^3 - 2$ is irreducible in $\mathbb{Q}[\sqrt[4]{2}][x]$ or in $\mathbb{Q}[\sqrt{2}, i][x]$.

**Example 5.** Here is a more powerful application of Proposition 1.5 and the divisibility properties of integers. Suppose $m$ and $n$ are relatively prime, $[F[\alpha] : F] = m$, and $[F[\beta] : F] = n$. Let $K = F[\alpha, \beta]$; then what can we say about $[K : F]$? From the "tower" $F \subset F[\alpha] \subset K$, we infer that $m \mid [K : F]$. Similarly, from $F \subset F[\beta] \subset K$ we infer that $n \mid [K : F]$. Since $m$ and $n$ are relatively prime, we now know that

$$(\dagger) \qquad\qquad mn \mid [K : F].$$

Let $f(x) \in F[x]$ be the irreducible polynomial (of degree $n$) satisfied by $\beta$ (see Exercise 19). Now we may consider $f(x)$ as a polynomial with coefficients in $F[\alpha]$, but conceivably $f(x)$ might not be irreducible in $F[\alpha][x]$. Nevertheless, $\beta$ must be a root of an irreducible factor of $f(x)$, and so we know that $[F[\alpha][\beta] : F[\alpha]] \leq n$. Since $K = F[\alpha, \beta] = F[\alpha][\beta]$,

$$(\ddagger) \qquad\qquad [K : F] = [K : F[\alpha]] [F[\alpha] : F] \leq mn.$$

The upshot of $(\dagger)$ and $(\ddagger)$, then, is that $[K : F] = mn$.

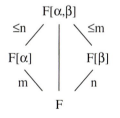

FIGURE 1

**EXERCISES 5.1**

1.  Let $V$ be a vector space over $F$. Prove the following:
    a.  $c\mathbf{0}_V = \mathbf{0}_V$ for all scalars $c \in F$ (here $\mathbf{0}_V$ denotes the zero vector).

    b.   $0_F \mathbf{v} = \mathbf{0}_V$ for all $\mathbf{v} \in V$ (here $0_F$ denotes the zero element of $F$)

    c.   $(-c)\mathbf{v} = c(-\mathbf{v}) = -(c\mathbf{v})$ for all $c \in F$ and $\mathbf{v} \in V$.

2. Prove Lemma 1.1.

3. Let $V$ be a vector space over $\mathbb{Q}$. Prove that if $\mathbf{v}, \mathbf{w} \in V$ are linearly independent, then so are $\mathbf{v} + \mathbf{w}, 2\mathbf{v} - \mathbf{w}$.

4. Prove that the real numbers $1$ and $\sqrt{3}$ are linearly independent over $\mathbb{Q}$. Do the same for $1$, $\sqrt{3}$, and $\sqrt{5}$.

5. Suppose $\mathbf{v}_1, \ldots, \mathbf{v}_k \in V$ are linearly independent and $\mathbf{v} \in V$. Prove that $\mathbf{v}_1, \ldots, \mathbf{v}_k, \mathbf{v}$ are linearly independent if and only if $\mathbf{v} \notin \mathrm{Span}(\mathbf{v}_1, \ldots, \mathbf{v}_k)$. (Hint: "Contrapositive.")

6. Prove: $\mathbf{v}_1, \ldots, \mathbf{v}_k \in V$ are linearly dependent if and only if for some $j$, $1 \le j \le k$, $\mathbf{v}_j$ is a linear combination of $\mathbf{v}_1, \ldots, \mathbf{v}_{j-1}$, $\mathbf{v}_{j+1}, \ldots, \mathbf{v}_k$.

7. Suppose $\mathbf{v} \in \mathrm{Span}(\mathbf{v}_1, \ldots, \mathbf{v}_k)$. Prove that $\mathrm{Span}(\mathbf{v}_1, \ldots, \mathbf{v}_k, \mathbf{v}) = \mathrm{Span}(\mathbf{v}_1, \ldots, \mathbf{v}_k)$.

8. Prove that $\mathbf{v}_1, \ldots, \mathbf{v}_k$ are a basis for $V$ if and only if every vector in $V$ can be written *uniquely* as a linear combination of $\mathbf{v}_1, \ldots, \mathbf{v}_k$.

9. Let $\mathbf{v}_1, \ldots, \mathbf{v}_k \in V$ be linearly independent vectors in a finite-dimensional vector space. Show that there are vectors $\mathbf{v}_{k+1}, \ldots, \mathbf{v}_n \in V$ so that $\mathbf{v}_1, \ldots, \mathbf{v}_k, \mathbf{v}_{k+1}, \ldots, \mathbf{v}_n$ form a basis for $V$. (Hint: see Exercise 5.)

10. Let $V$ be an $n$-dimensional vector space over $F$.

    a.   Prove that if $\mathbf{v}_1, \ldots, \mathbf{v}_n \in V$ are linearly independent, then they span $V$.

    b.   Prove that if $\mathbf{v}_1, \ldots, \mathbf{v}_n \in V$ span $V$, then they are linearly independent.

11. Give a basis for each of the given vector spaces over the given field. What is the degree of each field extension?

    a.   $\mathbb{Q}[\sqrt{2}]$ over $\mathbb{Q}$

    b.   $\mathbb{Q}[i\sqrt{3}]$ over $\mathbb{Q}$

    c.   $\mathbb{Q}[\sqrt{3}, i]$ over $\mathbb{Q}$

    d.   $\mathbb{Q}[\sqrt{3}, i]$ over $\mathbb{Q}[i\sqrt{3}]$

    e.   $\mathbb{Q}[\sqrt{3} + i]$ over $\mathbb{Q}[\sqrt{3}]$

    f.   $\mathbb{Q}[e^{\pi i/3}]$ over $\mathbb{Q}$

    g.  $\mathbb{Z}_2[x]/\langle x^3 + x + 1 \rangle$ over $\mathbb{Z}_2$

    h.  $\mathbb{Q}[\sqrt[5]{8}]$ over $\mathbb{Q}$

    i.   $\mathbb{C}$ over $\mathbb{R}$

12.  Show that the subset of $F[x]$ consisting of polynomials of degree $\leq n$ is a finite-dimensional subspace of $F[x]$, and find its dimension.

13.  Suppose $[K : F]$ is prime. Show that if $\alpha \in K$ and $\alpha \notin F$, then $K = F[\alpha]$. Does this result hold in greater generality?

14.  If $F$ is a finite field with $q$ elements and $V$ is an $n$-dimensional vector space over $F$, show that $V$ contains $q^n$ elements.

15.  a.  Prove that $\mathbb{Q}[\sqrt[3]{2}, \sqrt[40]{2}] \subset \mathbb{Q}[\sqrt[120]{2}]$.

     b.  Use Corollary 1.7 to prove that $\sqrt[3]{2} \notin \mathbb{Q}[\sqrt[40]{2}]$.

     c.  Use degree to prove that in fact $\mathbb{Q}[\sqrt[3]{2}, \sqrt[40]{2}] = \mathbb{Q}[\sqrt[120]{2}]$.

16.  Use the technique of Example 5 to establish these results:

     a.  $\mathbb{Q}[\sqrt[4]{5}, \sqrt[9]{5}] = \mathbb{Q}[\sqrt[36]{5}]$

     b.  $\mathbb{Q}[\sqrt[4]{5}, \sqrt[6]{5}] = \mathbb{Q}[\sqrt[12]{5}]$

17.  Is $\mathbb{R}$ a finite-dimensional vector space over $\mathbb{Q}$? Why or why not?

18.  Suppose $F \subset E \subset K$ are fields and $[K : F]$ is finite. Prove that $[K : E]$ and $[E : F]$ are both finite.

19.  Suppose $K$ is a field extension of $F$ of finite degree. Prove that if $\alpha \in K$, then there is an irreducible polynomial $f(x) \in F[x]$ having $\alpha$ as a root. (Hint: If $[K : F] = n$, consider $1, \alpha, \alpha^2, \ldots, \alpha^n$.)

20.  Let $W$ and $Z$ be subspaces of $V$. Define

$$W \cap Z = \{\mathbf{v} \in V : \mathbf{v} \in W \text{ and } \mathbf{v} \in Z\} \quad \text{and}$$

$$W + Z = \{\mathbf{v} \in V : \mathbf{v} = \mathbf{w} + \mathbf{z} \text{ for some } \mathbf{w} \in W \text{ and } \mathbf{z} \in Z\}.$$

     a.  Prove that $W \cap Z$ and $W + Z$ are subspaces of $V$.

     b.  Prove that if $W$ and $Z$ are finite-dimensional, then so are $W + Z$ and $W \cap Z$.

     c.  Prove that $\dim(W + Z) = \dim(W) + \dim(Z) - \dim(W \cap Z)$. (Hint: Start with a basis for $W \cap Z$, and apply Exercise 9 twice.)

21.  Let $V$ and $W$ be vector spaces over a field $F$. Recall that a **linear map** $T: V \to W$ is a function satisfying the properties
     (i)   $T(\mathbf{v} + \mathbf{v}') = T(\mathbf{v}) + T(\mathbf{v}')$ for all $\mathbf{v}, \mathbf{v}' \in V$;
     (ii)  $T(c\mathbf{v}) = cT(\mathbf{v})$ for all $\mathbf{v} \in V$ and $c \in F$.
     Prove that
     a.  $\ker T = \{\mathbf{v} \in V : T(\mathbf{v}) = \mathbf{0}\}$ is a subspace of $V$;
     b.  $\text{image } T = \{\mathbf{w} \in W : \mathbf{w} = T(\mathbf{v}) \text{ for some } \mathbf{v} \in V\}$ is a subspace of $W$;
     c.  For any subspace $U \subset V$,

     $$T(U) = \{\mathbf{w} \in W : \mathbf{w} = T(\mathbf{u}) \text{ for some } \mathbf{u} \in U\}$$

     is a subspace of $W$.
     d.  For any subspace $Z \subset W$,

     $$T^{-1}(Z) = \{\mathbf{v} \in V : T(\mathbf{v}) \in Z\}$$

     is a subspace of $V$.

22.  Let $T: V \to W$ be a linear map. If $T$ maps $V$ *onto* $W$ and $\mathbf{v}_1, \ldots, \mathbf{v}_k$ span $V$, prove that $T(\mathbf{v}_1), \ldots, T(\mathbf{v}_k)$ span $W$.

23.  Recall the relevant definitions from Exercise 21. Let $V$ be a finite-dimensional vector space, and let $T: V \to W$ be a linear map.
     a.  Suppose $\ker T = \{\mathbf{0}\}$. Prove that if $\mathbf{v}_1, \ldots, \mathbf{v}_k \in V$ are linearly independent, then $T(\mathbf{v}_1), \ldots, T(\mathbf{v}_k) \in W$ are linearly independent.
     b.  More generally, let $\mathbf{u}_1, \ldots, \mathbf{u}_\ell \in V$ be a basis for $\ker T$, and extend to a basis $\mathbf{u}_1, \ldots, \mathbf{u}_\ell, \mathbf{v}_1, \ldots, \mathbf{v}_k$ for all of $V$ (see Exercise 9). Now show that $T(\mathbf{v}_1), \ldots, T(\mathbf{v}_k)$ give a basis for image $T$.
     c.  Deduce from b. that

     $$\dim V = \dim(\ker T) + \dim(\text{image } T).$$

     This important result is usually called the *nullity-rank theorem*.

24.  Prove Theorem 1.3 as follows. Since $\mathbf{v}_1, \ldots, \mathbf{v}_k$ span $V$, there are scalars $a_{ij}$ so that $\mathbf{w}_j = \sum_{i=1}^{k} a_{ij}\mathbf{v}_i$, $j = 1, \ldots, \ell$. Let $A$ be the $k \times \ell$ matrix with entries $a_{ij}$. Show that the only solution to the equation $A\mathbf{x} = \mathbf{0}$, $\mathbf{x} \in F^\ell$, is the zero solution $\mathbf{x} = \mathbf{0}$; conclude that $\ell \leq k$.

25.  Suppose $f(x) \in F[x]$ is a polynomial of degree $n$. Let $K$ be its splitting field. Prove that $[K : F] \leq n!$. (Hint: Use Lemma 1.6.)

## 2. Constructions with Compass and Straightedge

This section concerns a geometric application of field extensions and the multiplicativity formula, Proposition 1.5. The most famous result is that "one cannot trisect an angle using compass and straightedge." Lest the reader believe that all the results are negative, Gauss figured out exactly which regular polygons could be constructed using compass and straightedge.

We begin by stating the rules of the game for compass and straightedge constructions. Initially, we are given two points in the plane a unit distance apart; these points are considered constructed, and we fix coordinates by placing them at $(0,0)$ and $(1,0)$, respectively. Given any two constructed points, we are allowed to draw the line passing through them or a circle centered at one and passing through the other. Points of intersection of lines and circles that have already been constructed are constructed points. We say the real number $\alpha$ is **constructible** if we can (in a finite number of steps) construct two points a distance $|\alpha|$ apart. Now, using these rules, what constructions can we perform?

(1) Given constructed point $P$ and line $\ell$, we can construct a line perpendicular to $\ell$ passing through $P$.

(2) Given constructed point $P$ and line $\ell$, $P$ not lying on $\ell$, we can construct a line parallel to $\ell$ passing through $P$.

(3) Given constructed point $P$ and line $\ell$ with $P$ on $\ell$, and given a constructible number $\alpha$, we can construct $Q$ on $\ell$ so that the length of line segment $\overline{PQ}$ is $\alpha$.

In Exercise 4, the reader is asked to perform these constructions explicitly.

**Proposition 2.1.** *The constructible numbers form a subfield of* $\mathbb{R}$.

*Proof.* We must show that if $\alpha$ and $\beta$ are constructible numbers, then so are $\alpha \pm \beta$, $1/\alpha$ (assuming, of course, that $\alpha \neq 0$), and $\alpha\beta$. Addition and subtraction are easy, using construction (3). For the product we use similar triangles, as shown in the diagram at the left in Figure 1; for the reciprocal, we likewise use similar triangles as shown in the middle diagram. □

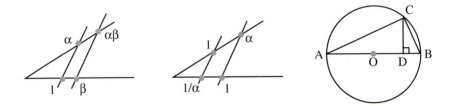

FIGURE 1

**Proposition 2.2.** *If $\alpha > 0$ is a constructible number, then so is $\sqrt{\alpha}$.*

*Proof.* It is again a matter of using the right similar triangles. Construct a circle of radius $\frac{\alpha+1}{2}$ centered at $O$, with $DB = 1$. Construct a perpendicular to the diameter $\overline{AB}$ passing through $D$, and let its intersection with the circle be $C$. Then $CD = \sqrt{\alpha}$, because $\frac{DB}{CD} = \frac{CD}{AD}$. See the rightmost diagram in Figure 1. □

We next wish to give an algebraic criterion for a real number to be constructible. Note first that a point in the plane is constructible if and only if its coordinates are constructible numbers (see Exercise 2). Suppose $F \subset \mathbb{R}$ is an arbitrary subfield, $(x_i, y_i)$, $i = 1, 2, 3, 4$, are given, and $x_i, y_i \in F$, $i = 1, 2, 3, 4$. Then if we draw lines through pairs of these points, the point of intersection of the lines again has coordinates in $F$ (since we must just solve a pair of linear equations with coefficients in $F$). Suppose we consider the circle centered at $(x_1, y_1)$ and passing through $(x_2, y_2)$. Its equation is

$$(*) \qquad (x - x_1)^2 + (y - y_1)^2 = (x_2 - x_1)^2 + (y_2 - y_1)^2.$$

The line passing through $(x_3, y_3)$ and $(x_4, y_4)$ has an equation of the form $ax + by + c = 0$, $a, b, c \in F$. Solving these equations simultaneously, we obtain a quadratic equation (either in $x$ or in $y$), all of whose coefficients lie in $F$. Thus the solutions will lie in a field of the form $F[\sqrt{\Delta}]$, where $\Delta \in F$ (see Theorem 4.1 of Chapter 2). Therefore, the coordinates of the point(s) of intersection of the line and the circle lie in $F[\sqrt{\Delta}]$. A similar argument applies to find the point(s) of intersection of two circles, one with equation $(*)$ and the other with equation

$$(**) \qquad (x - x_3)^2 + (y - y_3)^2 = (x_4 - x_3)^2 + (y_4 - y_3)^2.$$

In this case, subtracting equation $(**)$ from equation $(*)$ gives a linear equation; and solving this linear equation simultaneously with $(*)$, we are back in the previous case. Once again, the coordinates of the points of intersection lie in a field of the form $F[\sqrt{\Delta}]$, $\Delta \in F$. As a result, we have the following theorem.

**Theorem 2.3.** *Let $\alpha$ be a constructible number. There is a sequence of fields $\mathbb{Q} = F_0 \subset F_1 \subset \cdots \subset F_n = K$ so that*

(i) $K \subset \mathbb{R}$,

(ii) $\alpha \in K$, *and*

(iii) *for each $i = 0, 1, \ldots, n - 1$, the field $F_{i+1}$ is of the form $F_{i+1} = F_i[\sqrt{r_i}]$ for some positive number $r_i \in F_i$.*

*Conversely, given a sequence of fields satisfying (i) and (iii), every element of $K$ is constructible.*

**Proof.** To construct the number $\alpha$ we must draw lines and circles using points already constructed and consider intersections of such objects. By our earlier remarks, each such construction involves a field extension of degree one or two, and so the result follows by induction on the number of constructions required to reach $\alpha$. The converse follows similarly, provided we use Propositions 2.1 and 2.2. □

Perhaps the most useful result is the following corollary.

**Corollary 2.4.** *If $\alpha$ is a constructible number, then $[\mathbb{Q}[\alpha] : \mathbb{Q}]$ is a power of 2.*

**Proof.** By Theorem 2.3, there is a sequence of fields $\mathbb{Q} = F_0 \subset F_1 \subset \cdots \subset F_n = K$ so that $\alpha \in K$ and $[F_{i+1} : F_i] = 2$ for each $i = 0, 1, \ldots, n - 1$. Thus, $[K : \mathbb{Q}] = 2^n$, by Proposition 1.5. Since $\mathbb{Q} \subset \mathbb{Q}[\alpha] \subset K$, we have $[\mathbb{Q}[\alpha] : \mathbb{Q}] \mid 2^n$; and so, $[\mathbb{Q}[\alpha] : \mathbb{Q}]$ is itself a power of 2. □

**Remark.** A word of warning! The converse of Corollary 2.4 is false. There is, for example, an irreducible polynomial of degree 4 in $\mathbb{Q}[x]$ whose (real) roots fail to be constructible. See Exercise 7.6.25.

We now deduce the classic non-constructibility results.

**Proposition 2.5.** *Using compass and straightedge, it is impossible to construct an edge of a cube with volume* 2.

**Proof.** The edges of such a cube would have to have length $\alpha = \sqrt[3]{2}$; since $[\mathbb{Q}[\sqrt[3]{2}] : \mathbb{Q}] = 3$, $\alpha$ is not constructible.   $\square$

**Proposition 2.6.** *It is impossible to trisect a* 60° *angle using compass and straightedge.*

**Proof.** To construct a 20° angle, we must construct $\sin 20°$ and $\cos 20°$ (see Exercise 9). We show that $\alpha = \cos 20°$ is not constructible by finding the irreducible polynomial in $\mathbb{Q}[x]$ whose root it is. Since $\cos 60° = \frac{1}{2}$, we use the formula $\cos 3\theta = \cos^3 \theta - 3\cos\theta\sin^2\theta = 4\cos^3\theta - 3\cos\theta$ (see Exercise 2.3.6b.). Thus, $\alpha$ is a root of the polynomial $f(x) = 8x^3 - 6x - 1$. We leave it to the reader to check that this polynomial is irreducible (see Exercise 8). Thus, $[\mathbb{Q}[\alpha] : \mathbb{Q}] = 3$, and so by Corollary 2.4, $\alpha$ cannot be constructible.   $\square$

**Remark.** Amazingly, Archimedes knew how to trisect the general angle using *ruler* and compass! Here is his construction: Given $\angle AOB = \phi$, draw a circle with center $O$ and radius 1, say; let $P$ and $Q$ be the points of intersection of the circle with $\overline{OA}$ and $\overline{OB}$ respectively. Mark two points, $X$ and $Y$, on the ruler a distance 1 apart. Keeping $X$ on the line $\overline{OA}$ and the edge of the ruler passing through $Q$, adjust the ruler until $Y$ lies on the circle. Then $\angle OXY = \theta = \phi/3$. Now consider Figure 2. Since $\triangle OXY$ and $\triangle OYQ$ are isosceles, we

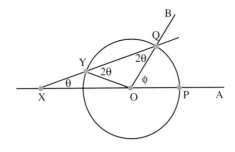

FIGURE 2

infer that $\angle OYQ = 2\theta$; and so $\phi = \angle OQX + \angle OXQ = 3\theta$, as required.

Which regular $n$-gons are constructible? Well, the regular $n$-gon is constructible if and only if $\cos(\frac{2\pi}{n})$ is constructible, and so we can conclude immediately that the regular 3-, 4-, 5-, and 6-gons are all constructible (for the pentagon, see Exercise 2.4.10). The regular 9-gon is not, however, since a 40° angle cannot be constructed (why?). Gauss proved that the regular $n$-gon is constructible if and only if every odd prime factor $p$ is of the form $p = 2^{2^r} + 1$ for some $r \in \mathbb{N}$ (see Exercise 1.2.17), and each such prime factor appears only once in the factorization of $n$. Thus, the regular 17-gon and the regular 60-gon are constructible.

### EXERCISES 5.2

1.  Which of the following real numbers are constructible?
    a.  $\sqrt[4]{5 + \sqrt{2}}$
    b.  $\sqrt[6]{2}$
    c.  $\dfrac{3}{4+\sqrt{13}}$
    d.  $3 + \sqrt[5]{8}$

2.  Prove that the point $P = (a, b) \in \mathbb{R}^2$ is constructible if and only if $a, b \in \mathbb{R}$ are constructible numbers.

3.  a.  Prove that every constructible number is algebraic, i.e., is the root of some polynomial with rational coefficients (see Section 4.2).
    b.  Prove that one cannot "square the circle," i.e., construct a square whose area is that of the unit circle. (See the discussion of $\pi$ in Section 4.2.)

4.  Show explicitly how to perform constructions (1), (2), and (3) using compass and straightedge.

5.  a.  Can one trisect the general line segment?
    b.  Show that one can bisect a (constructed) angle using compass and straightedge.

6.  Show how you would construct the circles inscribed in and circumscribed about a given (constructed) triangle. What do you conclude about the centers and radii of these respective circles?

7.  Prove or give a counterexample.
    a.  There is always a constructible circle passing through two constructed points.

    b.    There is always a non-constructible circle passing through two constructed points.

    c.    There is always a constructible circle passing through two non-constructible points.

8.    Prove that $f(x) = 8x^3 - 6x - 1 \in \mathbb{Q}[x]$ is irreducible.

9.    An angle $\alpha$ is constructible if one can construct lines intersecting at angle $\alpha$.

    a.    Show that if angles $\alpha$ and $\beta$ are constructible, then so are angles $\alpha + \beta$ and $\alpha - \beta$.

    b.    Show that the angle $\alpha$ is constructible if and only if $\cos \alpha$ and $\sin \alpha$ are both constructible real numbers (see Exercise 2).

10.    a.    Prove that if the regular $mn$-gon is constructible, then the regular $m$- and $n$-gons are constructible as well.

    b.    Prove that if $\gcd(m, n) = 1$ and the regular $m$- and $n$-gons are both constructible, then the regular $mn$-gon is constructible.

11.    Show that it is possible to trisect $2\pi/5$ and construct a regular 15-gon. (Hint: The angle $2\pi/5$ is constructible (why?). Apply Exercise 9.)

12.    Can one trisect the angle $\arccos 6/7$ using compass and straightedge?

13.    Show that an angle of $3°$ is constructible, whereas an angle of $1°$ is not. (Do this without quoting Gauss's result.) Now decide what angles $n°$, $n \in \mathbb{N}$, are constructible.

14.    Decide whether one can construct with compass and straightedge

    a.    an equilateral triangle with area 2 square units;

    b.    an isosceles triangle with area 2 square units and perimeter 8 units.

15.    Prove that the regular heptagon (7-gon) is not constructible. (Hint: Show that $\sin(\frac{2\pi}{7})$ is not constructible. Recall Corollary 3.3 of Chapter 2.)

16. Let $\zeta = \cos(\frac{2\pi}{p}) + i\sin(\frac{2\pi}{p})$, $p$ an odd prime.

    a. Prove that $\zeta$ is a root of the irreducible polynomial $f(x) = x^{p-1} + x^{p-2} + \cdots + x + 1 \in \mathbb{Q}[x]$. (See Exercise 3.3.7.)

    b. Prove that if the regular $p$-gon is constructible, then $p = 2^s + 1$ for some $s \in \mathbb{N}$. It follows from Exercise 1.2.17 that $p$ must be a Fermat prime $2^{2^q} + 1$. (Hint: If $\cos(\frac{2\pi}{p})$ is constructible, what can you say about $[\mathbb{Q}[\zeta] : \mathbb{Q}]$?)

17. Can an arbitrary constructible angle be divided into five equal parts ("quinquisected"?) with compass and straightedge? Give a proof of your answer. (Hint: see Exercise 2.3.6e.)

18. Construct a regular pentagon. (Hint: see Exercise 2.4.10; also, Figure 3 may be suggestive.)

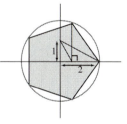

FIGURE 3

## 3. An Introduction to Finite Fields

Finite fields are extremely important in computer applications (particularly in coding theory) and in the design of experiments (Latin squares and block designs); cf. the books by Gilbert and Lidl and Pilz in the Supplementary Reading. They are also important in number theory and algebraic geometry. We illustrate the power of the machinery we have developed by showing that given any prime number $p$ and $n \in \mathbb{N}$, there is a field consisting of $p^n$ elements. We explain first why we must start with a prime number $p$ here.

We say a field $F$ has **characteristic** $m$ if $m$ is the smallest positive integer so that $m \cdot 1_F = \underbrace{1_F + \cdots + 1_F}_{m \text{ times}} = 0$. We say it has characteristic 0 if no such $m$ exists. For example, $\mathbb{Q}$ has characteristic 0; and, for any prime $p$, $\mathbb{Z}_p$ has characteristic $p$.

**Lemma 3.1.** *The characteristic of a field F is either 0 or a prime number p.*

**Proof.** Suppose $F$ has characteristic $m > 0$. If $m$ is composite, $m = rs$ for some integers $1 < r, s < m$. Then $m \cdot 1_F = (r \cdot 1_F)(s \cdot 1_F) = 0$; since $F$ is an integral domain, either $r \cdot 1_F = 0$ or $s \cdot 1_F = 0$. This contradicts the assumption that $m$ is the smallest positive integer with $m \cdot 1_F = 0$. □

Note that if $F$ is a field of characteristic $p$, then it naturally contains as a subfield a copy of $\mathbb{Z}_p$, namely, the set of all integral multiples of $1_F$. But then $F$ is a vector space over $\mathbb{Z}_p$.

**Proposition 3.2.** *Let F be a finite field of characteristic p. Then F has $p^n$ elements, for some $n \in \mathbb{N}$.*

**Proof.** $F$ is a vector space over $\mathbb{Z}_p$ of dimension $n \in \mathbb{N}$. Therefore, by Exercise 5.1.14, $F$ contains $p^n$ elements. □

The interesting point is that given any prime power $q = p^n$, there is a field $\mathbb{F}_q$ with $q$ elements. (Using Theorem 2.4 of Chapter 4 and Lemma 1.6, it would be sufficient to show that there are irreducible polynomials of every degree $n \geq 2$ over $\mathbb{Z}_p$. We leave this approach—which has its difficulties—to the reader in Exercise 8.)

**Theorem 3.3.** *Given a prime number p and $n \in \mathbb{N}$, there is a field $\mathbb{F}_q$ with $q = p^n$ elements.*

**Proof.** Let $f(x) = x^q - x \in \mathbb{Z}_p[x]$. By Corollary 2.5 of Chapter 4, this polynomial has a splitting field $K \supset \mathbb{Z}_p$. Now we claim the roots of $f(x)$ in $K$ are distinct and form a subfield $\mathbb{F}_q$ of $K$. This will complete the proof of the theorem. Indeed, by the definition of a splitting field, it follows that $K = \mathbb{F}_q$.

First, if $\alpha, \beta \in K$ are roots of $f(x)$, then we check that $\alpha\beta$, $1/\alpha$ (assuming $\alpha \neq 0$), and $\alpha + \beta$ are likewise. If $\alpha^q = \alpha$, then $(1/\alpha)^q = 1/\alpha$, and if $\beta^q = \beta$, then $(\alpha\beta)^q = \alpha^q\beta^q = \alpha\beta$, as required. Since $(\alpha+\beta)^q = \alpha^q+\beta^q$ by the characteristic $p$ binomial theorem (Exercise 1.3.29b.), we have $(\alpha + \beta)^q = \alpha + \beta$. Thus the roots of $f(x)$ form a field inside $K$.

Next, we maintain that $f(x)$ has distinct roots in $K$. Certainly 0 is a simple root, since $f(x) = x(x^{q-1} - 1)$. Now suppose $\alpha \in K$, $\alpha \neq 0$, is a root of $f(x)$ of multiplicity $m$. Then $f(x) = (x-\alpha)^m g(x)$. Substituting $x = y + \alpha$, we obtain $f(y + \alpha) = y^m g(y + \alpha)$; on the other hand, $f(y + \alpha) = (y + \alpha)^q - (y + \alpha) = (y^q + \alpha^q) - (y + \alpha) =$

$y^q - y = f(y)$, again by the characteristic $p$ binomial theorem. Putting these together, $f(y) = y^m g(y + \alpha)$, or, if you prefer, $f(x) = x^m g(x + \alpha)$. Since $K[x]$ has the unique factorization property, it follows that $m = 1$ and that every root $\alpha$ is a simple root. This concludes the proof. $\square$

**Remark.** It will follow from Proposition 1.2 and Corollary 3.4 of Chapter 6 that the polynomial $f(x) = x^q - x$ will split in *any* field with $q$ elements. As we shall see in Corollary 6.7 of Chapter 7, any splitting fields of a polynomial are isomorphic. Thus, any two fields with $q = p^n$ elements are isomorphic, and so we may denote "the" field $\mathbb{F}_q$ with impunity.

The symmetries of $\mathbb{F}_q$ (i.e., the isomorphisms $\mathbb{F}_q \to \mathbb{F}_q$ leaving $\mathbb{Z}_p \subset \mathbb{F}_q$ fixed) are of great importance in applications. Consider $\sigma: \mathbb{F}_q \to \mathbb{F}_q$, $\sigma(\alpha) = \alpha^p$. $\sigma$ is called the *Frobenius automorphism* of $\mathbb{F}_q$. Fermat's Theorem (Proposition 3.3 of Chapter 1) implies that $\sigma(\alpha) = \alpha$ for all $\alpha \in \mathbb{Z}_p$; once again the characteristic $p$ binomial theorem implies that $\sigma(\alpha + \beta) = \sigma(\alpha) + \sigma(\beta)$, and so $\sigma$ is a ring homomorphism (that $\sigma(\alpha\beta) = \sigma(\alpha)\sigma(\beta)$ is immediate). More interestingly, we can compose $\sigma$ with itself $j$ times, obtaining $\sigma^j(\alpha) = \alpha^{p^j}$. Since all the elements of $\mathbb{F}_q$ are roots of $x^q - x$, it follows that $\sigma^n$ is the identity, and so we have a list of $n$ symmetries: $1, \sigma, \sigma^2, \ldots, \sigma^{n-1}$. Is this it? Yes, one of the beautiful results of Galois theory (see Section 6 of Chapter 7) says that there can be no more symmetries than the degree of the field extension ($n$ in this case).

We close with one last observation.

**Proposition 3.4.** *Let* $q = p^n$, $q' = p^{n'}$. *Then* $\mathbb{F}_q \subset \mathbb{F}_{q'} \iff n \mid n'$.

**Proof.** If $\mathbb{F}_{q'}$ is a field extension of $\mathbb{F}_q$, then it is a vector space over $\mathbb{F}_q$; therefore, $q' = q^s$ for some $s \in \mathbb{N}$. This says that $n' = ns$, and so $n \mid n'$. Conversely, suppose $n' = ns$. Then $q' - 1 = q^s - 1 = (q-1)(q^{s-1} + \cdots + q + 1)$, and so $x^{q'-1} - 1 = (x^{q-1})^{(q^{s-1} + \cdots + q + 1)} - 1 = (x^{q-1} - 1)(\ldots)$, whence $(x^{q-1} - 1) \mid (x^{q'-1} - 1)$. Since $x^{q'-1} - 1$ splits in $\mathbb{F}_{q'}$, so must its factor $x^{q-1} - 1$. This implies (by uniqueness of $\mathbb{F}_q$) that $\mathbb{F}_q$ is a subfield of $\mathbb{F}_{q'}$, as required. $\square$

**Example.** We have the following diagram of the fields with 2, 4, 8, 16, 32, and 64 elements.

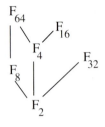

FIGURE 1

**EXERCISES 5.3**

1. Find all the subfields of $\mathbb{F}_{2^{12}}$.

2. Construct explicitly a field with 32 elements.

3. The polynomial $f(x) = x^2 + 1$ is irreducible in $\mathbb{Z}_3[x]$, and so
   $K = \mathbb{Z}_3[x]/\langle x^2 + 1\rangle$ is a field with nine elements. Let $\alpha \in K$ be
   a root of $f(x)$. Find irreducible polynomials in $\mathbb{Z}_3[x]$ having as
   roots, respectively,
   a. $\alpha + 1$
   b. $\alpha - 1$.

4. Construct explicitly an isomorphism

$$\mathbb{Z}_2[x]/\langle x^3 + x + 1\rangle \to \mathbb{Z}_2[x]/\langle x^3 + x^2 + 1\rangle .$$

5. Let $F$ be a finite field of characteristic $p$. Show that every element $a \in F$ can be written in the form $a = b^p$ for some $b \in F$.
   (Hint: Consider the Frobenius automorphism.)

6. Let $f(x) \in \mathbb{Z}_p[x]$ be an irreducible polynomial of degree $n$. Let
   $K = \mathbb{Z}_p[x]/\langle f(x)\rangle$.
   a. Prove directly that if $\alpha \in K$ is a root of $f(x)$, then $f(x)$ splits
      in $K$ with roots $\alpha, \alpha^p, \alpha^{p^2}, \ldots, \alpha^{p^{n-1}}$. (Hint: see Exercise
      4.2.10.)
   b. Check the result of a. for $f(x) = x^3 + x^2 + 1 \in \mathbb{Z}_2[x]$.
   c. Check the result of a. for $f(x) = x^2 + 1 \in \mathbb{Z}_3[x]$.

7. Let $q = p^n$, and let $f(x) = x^q - x$.
   a. Prove that if $g(x)$ is an irreducible polynomial of degree $d$
      in $\mathbb{Z}_p[x]$, then $g(x)$ divides $f(x)$ if and only if $d|n$.

  b.    Prove that $f(x)$ is the product of all monic, irreducible poly-
        nomials in $\mathbb{Z}_p[x]$ whose degrees divide $n$.

8.    Prove directly that there are irreducible polynomials of every
      degree $n \geq 2$ in $\mathbb{Z}_p[x]$. (In particular, you might start by show-
      ing there are $\frac{p^2-p}{2}$ monic irreducible quadratics and $\frac{p^3-p}{3}$ monic
      irreducible cubics in $\mathbb{Z}_p[x]$. After this, it gets harder, although
      Exercise 7 may be of some use.)

# CHAPTER 6

# Groups

Groups first arose naturally in the study of solutions of polynomials (work of Galois, Lagrange, Cauchy, and Abel); Cayley proposed the first definition of an abstract group in 1849. Although we shall study this beautiful application of group theory in Section 6 of Chapter 7, we begin with more concrete and geometric considerations. We start with a number of examples and then, in parallel with our treatment of rings in Chapter 4, discuss homomorphisms, cosets, quotient groups, and the fundamental homomorphism theorem. We'll see that matters are somewhat more subtle here than with rings: If $H$ is a subgroup of $G$, then we can use $H$ to define an equivalence relation on $G$ (as we did with an ideal $I$ in a commutative ring $R$); but the set of equivalence classes will not always form a group (whereas $R/I$ always has a natural ring structure). The class of subgroups for which the quotient carries a natural group structure is that of *normal* subgroups.

In §4 we discuss the symmetric group (which is usually emphasized more in traditional texts), even and odd permutations, and the fact that the group of even permutations (of at least five objects) has no nontrivial normal subgroup. As we shall see in Chapter 7, this has both geometric and algebraic implications—one case is the symmetry group of the regular icosahedron, and simplicity here tells us that there can be no formula to solve a general quintic polynomial by roots. As an amusing application, we discuss the 15-puzzle and its solution.

## 1. The Basic Definitions

By a **symmetry** of a set $S$ we mean a bijection (one-to-one and onto map) from $S$ to itself. When the set $S$ is a geometric object, i.e., a subset of Euclidean space, a symmetry of $S$ is a motion (isometry) of the Euclidean space that maps $S$ to itself. We begin by examining the symmetries of an equilateral triangle in the Euclidean plane. Any such symmetry must move the triangle so that it occupies the same "space" as it did originally; in particular, if we mark the initial locations of the three vertices, after the symmetry has been performed, the vertices must be redistributed somehow among these marked positions. To fix ideas, we can label the vertices 1, 2, and 3 counterclockwise as shown in Figure 1; then we can keep track of the symmetries by noting which vertices occupy positions 1, 2, and 3. (We should point out that if a symmetry leaves all three vertices fixed, the edges joining the pairs of vertices must likewise be fixed, and so the symmetry must be the identity map.) Since there are six permutations of the set $\{1, 2, 3\}$, there are (at most) six symmetries of the equilateral triangle.

We may rotate the triangle 120° about its centroid (say, counterclockwise), so that vertex 3 has replaced vertex 1, 1 has replaced 2, and 2 has replaced 3. We may similarly rotate once again through 120°, and then once again, returning to our original position.

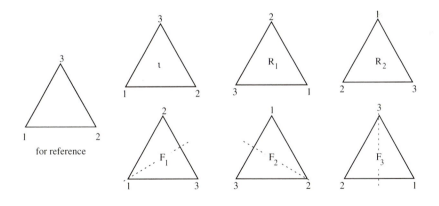

FIGURE 1

There are three other possible symmetries of the triangle: namely, the "flips" (or reflections) about the perpendicular bisectors of each of the three edges. Each of these fixes the opposite vertex, as shown.

The crucial observation now is this: we have exhibited six symmetries of the equilateral triangle, and there can be no more. As a consequence, the *composition* of any two symmetries on our list must be on our list, too. If we label our symmetries $\iota$ (identity), $R_1$ (120° rotation counterclockwise), $R_2$ (240° rotation counterclockwise), $F_1$ (flip fixing vertex 1), $F_2$ (flip fixing vertex 2), and $F_3$ (flip fixing vertex 3), respectively, then we have the following "multiplication table" (here we do composition of functions, reading right to left: $R \cdot F = RF$ means that we first do symmetry $F$, then symmetry $R$):

| $\cdot$ | $\iota$ | $R_1$ | $R_2$ | $F_1$ | $F_2$ | $F_3$ |
|---|---|---|---|---|---|---|
| $\iota$ | $\iota$ | $R_1$ | $R_2$ | $F_1$ | $F_2$ | $F_3$ |
| $R_1$ | $R_1$ | $R_2$ | $\iota$ | $F_3$ | $F_1$ | $F_2$ |
| $R_2$ | $R_2$ | $\iota$ | $R_1$ | $F_2$ | $F_3$ | $F_1$ |
| $F_1$ | $F_1$ | $F_2$ | $F_3$ | $\iota$ | $R_1$ | $R_2$ |
| $F_2$ | $F_2$ | $F_3$ | $F_1$ | $R_2$ | $\iota$ | $R_1$ |
| $F_3$ | $F_3$ | $F_1$ | $F_2$ | $R_1$ | $R_2$ | $\iota$ |

This is the basic example of a group, and so now we make the official definition. Following general custom, we label the identity element $e$ rather than $\iota$.

**Definition.** A **group** is a set $G$ closed under an operation (usually denoted by $\cdot$ or by no symbol at all, and called "multiplication"), so that

(1) If $a, b, c \in G$, then $(a \cdot b) \cdot c = a \cdot (b \cdot c)$    (associative law).
(2) There is an identity element $e \in G$ satisfying $a \cdot e = e \cdot a = a$ for all $a \in G$.
(3) If $a \in G$, there is an element $a^{-1} \in G$ (called the inverse of $a$) satisfying $a \cdot a^{-1} = a^{-1} \cdot a = e$.

We say $G$ is **abelian** if $a \cdot b = b \cdot a$ for all $a, b \in G$. (In this event we often write the operation with a +.)

**Examples 1.** We begin with some examples of abelian groups, the first few based on our earlier experience with rings and fields.

(a) $\mathbb{Z}$, $\mathbb{Z}_m$, $\mathbb{Q}$, $\mathbb{R}$, and $\mathbb{C}$ are all abelian groups with respect to the usual addition operation; indeed, it follows from properties (1), (2), (3), and (4) of a ring (see p. 38) that any ring $R$ forms an abelian group with respect to its addition operation. The group operation is denoted by +, the identity element by 0, and the inverse of $a$ by $-a$.

(b) Let $\mathbb{R}^{\times} = \mathbb{R} - \{0\}$ with the operation of multiplication. Since multiplication of real numbers is commutative and associative, this is again an abelian group (1 is the identity, and $1/a$ is the inverse element of $a \in \mathbb{R} - \{0\}$). The same works for $\mathbb{C}^{\times}$, and, indeed, for the nonzero elements of any field (cf. Proposition 1.2 below).

(c) Let the abelian group $\mathcal{V}$ be given by the following table:

| · | $e$ | $a$ | $b$ | $c$ |
|---|---|---|---|---|
| $e$ | $e$ | $a$ | $b$ | $c$ |
| $a$ | $a$ | $e$ | $c$ | $b$ |
| $b$ | $b$ | $c$ | $e$ | $a$ |
| $c$ | $c$ | $b$ | $a$ | $e$ |

We shall encounter $\mathcal{V}$—named the Klein four-group—often.

**Examples 2.** Here now are some non-abelian examples.

(a) The group of symmetries of the equilateral triangle, which we discussed above is an important example, denoted $\mathcal{T}$.

(b) The multiplication table of an interesting group with eight elements, called $\mathcal{Q}$ (the *quaternion group*), is given below.

| · | $1$ | $-1$ | $i$ | $-i$ | $j$ | $-j$ | $k$ | $-k$ |
|---|---|---|---|---|---|---|---|---|
| $1$ | $1$ | $-1$ | $i$ | $-i$ | $j$ | $-j$ | $k$ | $-k$ |
| $-1$ | $-1$ | $1$ | $-i$ | $i$ | $-j$ | $j$ | $-k$ | $k$ |
| $i$ | $i$ | $-i$ | $-1$ | $1$ | $k$ | $-k$ | $-j$ | $j$ |
| $-i$ | $-i$ | $i$ | $1$ | $-1$ | $-k$ | $k$ | $j$ | $-j$ |
| $j$ | $j$ | $-j$ | $-k$ | $k$ | $-1$ | $1$ | $i$ | $-i$ |
| $-j$ | $-j$ | $j$ | $k$ | $-k$ | $1$ | $-1$ | $-i$ | $i$ |
| $k$ | $k$ | $-k$ | $j$ | $-j$ | $-i$ | $i$ | $-1$ | $1$ |
| $-k$ | $-k$ | $k$ | $-j$ | $j$ | $i$ | $-i$ | $1$ | $-1$ |

Note that every element $a$ of this group other than $\pm 1$ satisfies the equation $a^2 = -1$.

(c) Let $GL(2, \mathbb{R})$ be the set of invertible $2 \times 2$ real matrices with operation matrix multiplication. If $A$ and $B$ are invertible, then so is $AB$, since $(AB)^{-1} = B^{-1}A^{-1}$ (see Exercise 1.4.8). Matrix multiplication, although not commutative, *is* associative, so $GL(2, \mathbb{R})$ is a group, called the *general linear group*.

(d) Here is a more subtle example: let $SL(2, \mathbb{R}) \subset GL(2, \mathbb{R})$ be the set of $2 \times 2$ matrices with determinant 1. Since $\det(AB) = \det(A)\det(B)$ (see Theorem 2.9 of Appendix B), the product and inverse of matrices with determinant 1 again have determinant 1. This is called the *special linear group*.

Here are some basic properties of groups that everyone should know.

**Lemma 1.1.** *Let $G$ be a group with identity element $e$.*

(i) *The identity element is unique; i.e., if $e'a = ae' = a$ for all $a \in G$, then $e' = e$.*

(ii) *If $ab = e$, then $b = a^{-1}$; as a consequence, each element $a$ has a unique inverse.*

(iii) *If $a, b, c \in G$ and $ac = bc$ (or $ca = cb$), then $a = b$.*

(iv) *If $a \in G$, then $(a^{-1})^{-1} = a$.*

(v) *If $a, b \in G$, then $(ab)^{-1} = b^{-1}a^{-1}$.*

**Proof.** We'll prove (ii) and (v) here, and leave the rest to the reader (see Exercise 3). Assume $ab = e$; by the associative law, $b = (a^{-1}a)b = a^{-1}(ab) = a^{-1}e = a^{-1}$, establishing (ii). For (v), consider the product $(ab)(b^{-1}a^{-1}) = a((b \cdot b^{-1})a^{-1}) = a((e)a^{-1}) = a \cdot a^{-1} = e$. Thus, by (ii), $b^{-1}a^{-1} = (ab)^{-1}$, as required. $\square$

The next result provides a rather pretty link with our earlier study of rings.

**Proposition 1.2.** *Let $R$ be any ring. Then $R^{\times} = \{$units in $R\}$ forms a group with respect to the multiplication operation in $R$.*

**Proof.** We must first and foremost check that $R^{\times}$ is closed under multiplication: if $a$ and $b$ are units in $R$, then they have multiplicative inverses $a^{-1}, b^{-1} \in R$. By Exercise 1.4.8, $(ab)^{-1} = b^{-1}a^{-1}$, so $ab$ is a unit in $R$ as well. (Why can we not just apply Lemma 1.1(v)?) Associativity follows from the associative law of multiplication in the ring $R$. The multiplicative identity of $R$ will obviously be the identity element of our group. Lastly, by the very definition of a unit, if $a \in R^{\times}$, then there is a multiplicative inverse $a^{-1} \in R$, and $a^{-1} \in R^{\times}$ (since $a$ is the inverse of $a^{-1}$). $\square$

Given a group $G$, we say that $H \subset G$ is a **subgroup** if $H$ is nonempty and closed under multiplication and inverse, i.e., if $a, b \in H \implies ab \in H$ and $a^{-1} \in H$. The point is that these conditions are sufficient to guarantee that $H$ (with the multiplication operation of $G$) will be a group unto itself (see Exercise 6). We say $H$ is a **proper** (or nontrivial) **subgroup** of $G$ if $H$ is neither all of $G$ nor just $\{e\}$.

Given an element $a \in G$, we observe that $\langle a \rangle = \{a^n : n \in \mathbb{Z}\}$ is a subgroup of $G$, called the **cyclic subgroup** generated by $a$. If there is an element $a \in G$ such that the cyclic subgroup generated by $a$ is *all* of $G$, we say $G$ is **cyclic**. Note that the subgroup $\langle a \rangle$ may happen to be finite; if so, we say the **order** of $a$ is the smallest integer $n \in \mathbb{N}$ so that $a^n = e$. Just to confuse matters, we say a group $G$ is finite if it contains finitely many elements, and we call the number of elements the **order** of $G$, written $|G|$. But life is not all that confusing, because now the order of $a$ is equal to the order of the cyclic subgroup $\langle a \rangle$ it generates.

**Examples 3.**

(a) Given any $m \in \mathbb{Z}$, $\langle m \rangle \subset \mathbb{Z}$ is a subgroup. Indeed, all subgroups of $\mathbb{Z}$ are cyclic (why?).

(b) $G = \mathbb{Z}_{12}$ (with operation addition) is a group of order 12, and its subgroups are the cyclic subgroups $\langle 1 \rangle = G$, $\langle 2 \rangle$, $\langle 3 \rangle$, $\langle 4 \rangle$, and $\langle 6 \rangle$, of orders 12, 6, 4, 3, and 2, respectively. Note, for example, that $\langle 5 \rangle = \langle 7 \rangle = G$ (see Exercise 8) and $\langle 3 \rangle = \langle 9 \rangle$.

(c) Consider the two groups $\mathbb{Z}_4$ and $\mathcal{V}$ of order 4 we have encountered. In $\mathbb{Z}_4$, the elements 1, 2, and 3 have orders 4, 2, and 4, respectively. In $\mathcal{V}$, the elements $a$, $b$, and $c$ all have order 2. Presumably, this means the two groups are "different." Also note that $\mathbb{Z}_4$ is a cyclic group, whereas $\mathcal{V}$ cannot be (why?).

(d) The group $\mathcal{T}$ of symmetries of the equilateral triangle has order 6. Note that $R_2 = R_1^2$, and $R_1^3 = \iota$, so the rotation $R_1$ is an element of order 3 (as is its inverse). The flips $F_1, F_2, F_3$ are each elements of order 2.

We come next to the fundamental example of group theory, the set of permutations of a set. By definition, given a set $A$, a **permutation** $\pi$ of $A$ is a function $\pi : A \to A$ that is bijective (one-to-one and onto).

**Lemma 1.3.** Perm$(A) = \{$*permutations of* $A\}$ *is naturally a group with respect to the operation of composition of functions.*

***Proof.*** First, composition of functions is a *bona fide* operation on Perm($A$), since the composition of one-to-one (resp., onto) functions is one-to-one (resp., onto). The identity element is the identity function, $\iota(a) = a$, $a \in A$. Composition of functions is associative: $(f \circ (g \circ h))(a) = f((g \circ h)(a)) = f(g(h(a))) = (f \circ g)(h(a)) = ((f \circ g) \circ h)(a)$. Lastly, if $\pi \colon A \to A$ is bijective, then its inverse function $\pi^{-1} \colon A \to A$ is defined and is again bijective (see Section 2 of Appendix A). □

When $A$ is the finite set $\{1, 2, 3, \ldots, n\}$, we call Perm($A$) the **symmetric group on $n$ letters** and denote it by $S_n$. Note that $|S_n| = n!$. We introduce some notation for permutations of the numbers $1, \ldots, n$. If $\pi \colon \{1, \ldots, n\} \to \{1, \ldots, n\}$ assigns the values $i_1, i_2, \ldots, i_n$ to the numbers $1, 2, \ldots, n$, respectively (i.e., $\pi(1) = i_1$, $\pi(2) = i_2$, etc.), then we write

$$\pi = \begin{pmatrix} 1 & 2 & 3 & \cdots & n-1 & n \\ i_1 & i_2 & i_3 & \cdots & i_{n-1} & i_n \end{pmatrix}.$$

**Example 4.** We can verify that the group $\mathcal{T}$ "is" $S_3$:

$$\iota \longleftrightarrow \begin{pmatrix} 1 & 2 & 3 \\ 1 & 2 & 3 \end{pmatrix} \qquad F_1 \longleftrightarrow \begin{pmatrix} 1 & 2 & 3 \\ 1 & 3 & 2 \end{pmatrix}$$

$$R_1 \longleftrightarrow \begin{pmatrix} 1 & 2 & 3 \\ 2 & 3 & 1 \end{pmatrix} \qquad F_2 \longleftrightarrow \begin{pmatrix} 1 & 2 & 3 \\ 3 & 2 & 1 \end{pmatrix}$$

$$R_2 \longleftrightarrow \begin{pmatrix} 1 & 2 & 3 \\ 3 & 1 & 2 \end{pmatrix} \qquad F_3 \longleftrightarrow \begin{pmatrix} 1 & 2 & 3 \\ 2 & 1 & 3 \end{pmatrix}$$

**Example 5.** We next analyze the group $\mathcal{S}q$ of symmetries of a square. After our experience with $\mathcal{T}$, we might suspect that the symmetries of the square will comprise the group $S_4$, since now we have the *four* vertices to permute. However, when we look at the possible configurations, we find only eight (see Figure 2). Since $|S_4| = 4! = 24$, we are certainly missing a good number of permutations. But there are lots of ways to understand why: 1 and 3 are at opposite ends of the square; we cannot switch 1 and 2 without simultaneously switching 3 and 4, and so forth. More convincingly, there are four possible positions a symmetry can place vertex 1; once that choice is made, vertex 2 can go only in one of *two* possible places (since vertices 1 and 2 must remain adjacent). But now the positions of vertices 3 and 4 are already determined. Thus, there are $4 \cdot 2$ symmetries of the square. It is nevertheless instructive to write down the permutations corresponding to the eight elements of the group. We have the identity, 90°, 180°, and 270° rotations,

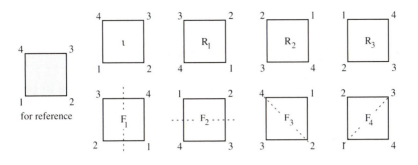

and four "flips." The eight elements are:

$$\iota \longleftrightarrow \begin{pmatrix} 1 & 2 & 3 & 4 \\ 1 & 2 & 3 & 4 \end{pmatrix} \qquad F_1 \longleftrightarrow \begin{pmatrix} 1 & 2 & 3 & 4 \\ 2 & 1 & 4 & 3 \end{pmatrix}$$

$$R_1 \longleftrightarrow \begin{pmatrix} 1 & 2 & 3 & 4 \\ 2 & 3 & 4 & 1 \end{pmatrix} \qquad F_2 \longleftrightarrow \begin{pmatrix} 1 & 2 & 3 & 4 \\ 4 & 3 & 2 & 1 \end{pmatrix}$$

$$R_2 \longleftrightarrow \begin{pmatrix} 1 & 2 & 3 & 4 \\ 3 & 4 & 1 & 2 \end{pmatrix} \qquad F_3 \longleftrightarrow \begin{pmatrix} 1 & 2 & 3 & 4 \\ 3 & 2 & 1 & 4 \end{pmatrix}$$

$$R_3 \longleftrightarrow \begin{pmatrix} 1 & 2 & 3 & 4 \\ 4 & 1 & 2 & 3 \end{pmatrix} \qquad F_4 \longleftrightarrow \begin{pmatrix} 1 & 2 & 3 & 4 \\ 1 & 4 & 3 & 2 \end{pmatrix}$$

It seems clear that both groups $\mathcal{T}$ and $\mathcal{S}q$ are built out of two basic elements: a rotation and a flip. Indeed, a glance at the multiplication table for $\mathcal{T}$ shows that we may express all six elements as various products of $R = R_1$ and $F = F_1$: $\iota = R^0 (= R^3)$, $R_1 = R$, $R_2 = R^2$, $F_1 = F$, $F_2 = FR$, $F_3 = FR^2$. But here we find something quite interesting: we also have the equation $F_2 = R^2F$. The net result $FR = R^2F$ tells us exactly how $F$ and $R$ fail to commute with one another.

What's more, we can express $F_2$ and $F_3$ in a somewhat different manner—one which lends itself to geometric interpretation. Since $FR = R^2F$, we have $F_2 = FR = R^3FR = R(R^2F)R = R(FR)R = RFR^2 = RFR^{-1}$; and, similarly, $F_3 = R^{-1}FR$. It is amusing to check these out formally in terms of permutations:

$$RFR^{-1} \longleftrightarrow \begin{pmatrix} 1 & 2 & 3 \\ 2 & 3 & 1 \end{pmatrix} \begin{pmatrix} 1 & 2 & 3 \\ 1 & 3 & 2 \end{pmatrix} \begin{pmatrix} 1 & 2 & 3 \\ 3 & 1 & 2 \end{pmatrix} = \begin{pmatrix} 1 & 2 & 3 \\ 3 & 2 & 1 \end{pmatrix},$$

which is indeed the symmetry $F_2$. Here is the interpretation we wish to offer: $R^{-1}$ rotates 2 into position 1, about which $F$ is the flip, and then $R$ rotates 1 back to 2. (Thus, we have performed a "change of coordinates," carrying 2 into 1, performing some geometric action, and then carrying 1 back to 2, returning to our original frame of reference.) The net effect is to transport the geometric action originally centered at position 1 to position 2. The reader should follow this reasoning through to see why $F_3 = R^{-1}FR$.

There is a corresponding presentation of the group $Sq$ in terms of $\rho = R_1 =$ rotation through $90°$ and $\psi = F_1$. Of course, $\iota = \rho^0 = \rho^4$, $R_1 = \rho$, $R_2 = \rho^2$, and $R^3 = \rho^3$; in addition, $F_2 = \rho^2\psi$, $F_3 = \rho\psi$, and $F_4 = \rho^3\psi$. The fundamental relation between $\psi$ and $\rho$ is $\psi\rho\psi^{-1} = \rho^3$ (or $\psi\rho = \rho^3\psi$). How does this jibe with the analysis we performed of the group $T$?

## EXERCISES 6.1

0.  "What's purple and commutes?"

1.  Which of the following are groups?
    a.  $\{1, 3, 7, 9\} \subset \mathbb{Z}_{10}$, with operation multiplication
    b.  $\{0, 2, 4, 6\} \subset \mathbb{Z}_{10}$, with operation addition
    c.  $\{x \in \mathbb{Q} : 0 < x \leq 1\}$, with operation multiplication
    d.  the set of all positive irrational real numbers, with operation multiplication
    e.  the set of imaginary numbers $ix, x \in \mathbb{R}$, with operation addition
    f.  the set of complex numbers of modulus 1, with operation multiplication
    g.  $\mathbb{Z}$ with operation $a \bullet b = a + b + 1$
    h.  $\mathbb{Z}$ with operation $a \bullet b = a - b$
    i.  $\mathbb{Q} - \{1\}$ with operation $a \bullet b = a + b - ab$

2.  Find all subgroups of the following groups.
    a.  $\mathbb{Z}_{18}$
    b.  $\mathbb{Z}_5$
    c.  $\mathbb{Z}_5^{\times}$
    d.  $\mathbb{Z}_{11}^{\times}$

3.  Complete the proof of Lemma 1.1.

4.  Let $GL(2, \mathbb{Z}_2)$ be the group of invertible $2 \times 2$ matrices with entries in $\mathbb{Z}_2$. List its elements. What is the order of the group? Find all its subgroups.

5.  Prove or give a counterexample. If $a, b, c \in G$ and $ab = bc$, then $a = c$.

6.  Prove that if $G$ is a group and $H \subset G$ is a subgroup, then $H$ is itself a group. (Of course, looking at the appropriate definition, we see it is necessary to check that multiplication is associative and that $H$ has an identity element.)

7.  Suppose $G$ is a group and $H \subset G$ is a nonempty *finite* subset of $G$ that is closed under multiplication. Prove that $H$ is a subgroup of $G$. Give an example to show that the finiteness hypothesis is essential.

8.  Prove that if $\gcd(a, m) = 1$, then the cyclic subgroup $\langle a \rangle$ of $\mathbb{Z}_m$ is the whole group.

9.  Prove that if $H, K \subset G$ are subgroups, then $H \cap K = \{g \in G : g \in H \text{ and } g \in K\}$ is a subgroup.

10. a.  Let $G$ be a group. Prove that $(ab)^2 = a^2 b^2$ for all $a, b \in G$ if and only if $G$ is abelian.
    b.  Prove that if every element (other than the identity element) of a group $G$ has order 2, then $G$ is abelian.

11. Prove that any group of order $\leq 4$ is abelian.

12. Let $a \in G$ be fixed. Prove that $C_a = \{x \in G : ax = xa\}$ is a subgroup of $G$, called the **centralizer** of $a$.

13. a.  Prove that every subgroup of a cyclic group is cyclic.
    b.  Prove that if $k | m$, then $\mathbb{Z}_m$ has a subgroup of order $k$.
    c.  If $a \in G$ has order $n$, prove that the order of $a^k$ is $\frac{n}{\gcd(k,n)}$.

14. Show that the group of symmetries of a general (i.e., non-square) rectangle is the Klein four-group $\mathcal{V}$.

15. Analyze the group $Sq$ carefully, and check all the details of the claims made in the text. Try to express all the elements in terms of $\rho^i$ and $\rho^i \psi \rho^{-i}$. What goes wrong? Also find subgroups of $Sq$ that "are" the four-group $\mathcal{V}$ and the cyclic group of order 4.

16. Discuss the group of symmetries of the regular pentagon and hexagon, and see if you can give the group of symmetries of the regular $n$-gon.

17. a.  A group has four elements $a$, $b$, $c$, and $d$, subject to the rules $ca = a$ and $d^2 = a$. Fill in the entire multiplication table at the left below.

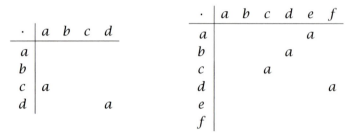

| · | $a$ | $b$ | $c$ | $d$ |
|---|---|---|---|---|
| $a$ |  |  |  |  |
| $b$ |  |  |  |  |
| $c$ | $a$ |  |  |  |
| $d$ |  |  | $a$ |  |

| · | $a$ | $b$ | $c$ | $d$ | $e$ | $f$ |
|---|---|---|---|---|---|---|
| $a$ |  |  |  |  | $a$ |  |
| $b$ |  |  |  | $a$ |  |  |
| $c$ |  |  | $a$ |  |  |  |
| $d$ |  |  |  |  |  | $a$ |
| $e$ |  |  |  |  |  |  |
| $f$ |  |  |  |  |  |  |

   b.  A group has six elements $a$, $b$, $c$, $d$, $e$, and $f$, subject to the rules $ae = a$, $bd = a$, $c^2 = a$, and $df = a$. Fill in the entire multiplication table at the right above.

18. Given the permutations $\sigma, \tau \in S_4$ below, compute the desired elements:

$$\sigma = \begin{pmatrix} 1 & 2 & 3 & 4 \\ 2 & 4 & 3 & 1 \end{pmatrix}, \quad \tau = \begin{pmatrix} 1 & 2 & 3 & 4 \\ 3 & 4 & 1 & 2 \end{pmatrix}$$

   a.  $\sigma^{-1}$
   b.  $\sigma\tau$
   c.  $\tau\sigma$
   d.  $\sigma^2$
   e.  $\sigma^2\tau$
   f.  $\sigma\tau\sigma^{-1}$
   g.  $\tau\sigma\tau^{-1}$

19. Suppose $G$ is an abelian group of order 6 containing an element of order 3. Prove that $G$ must be cyclic.

20. Prove that the isometries of $\mathbb{R}$ form a group; do the same for the isometries of $\mathbb{C}$. (See Section 5 of Chapter 2 for the definitions.)

21. What is the order of the group $\mathbb{Z}_m^\times$? (Hint: see Exercise 1.3.39.)

22. Let $G = GL(2, F)$ be the group of invertible $2 \times 2$ matrices with entries in a field $F$, and assume $F$ has more than 2 elements.
   a.  Let $H = \left\{ \begin{bmatrix} a & b \\ 0 & c \end{bmatrix} : ac \neq 0 \right\} \subset G$. Prove $H$ is a subgroup.
   b.  Find all matrices $A \in G$ so that $AB = BA$ for all $B \in H$.
   c.  Find all matrices $A \in G$ so that $AB = BA$ for all $B \in G$.
   d.  Prove or give a counterexample: For every $A \in G$, there is $P \in G$ so that $PAP^{-1} \in H$.

23.  Let $G$ be a group, and let $a, b \in G$ satisfy $ab = ba$.
  a.  Prove that if the order of $a$ is $m$, the order of $b$ is $n$, and $\gcd(m, n) = 1$, then the order of $ab$ is $mn$.
  b.  Prove or give a counterexample: In general, the order of $ab$ is $\operatorname{lcm}(m, n)$.

24.  a.  Let $G$ be a finite abelian group. Let $m$ be largest of all the orders of elements of $G$. Prove that the order of every element $a \in G$ divides $m$. (Hint: Find a clever way to use Exercise 23a.)
  b.  Let $p$ be prime, and let $\mathbb{F}_q$ be a field with $q = p^n$ elements (see Section 3 of Chapter 5). Deduce from a. that $\mathbb{F}_q^{\times}$ is a cyclic group of order $q - 1$. (Hint: If $m$ is the largest of all the orders of elements of $\mathbb{F}_q^{\times}$, then every element of $\mathbb{F}_q$ is a root of the polynomial $f(x) = x^{m+1} - x$.)

25.  Generalizing Exercise 4, find the order of the group $GL(2, \mathbb{Z}_p)$ of invertible $2 \times 2$ matrices with entries in $\mathbb{Z}_p$ ($p$ prime, of course). What is the order of the subgroup $SL(2, \mathbb{Z}_p)$ of matrices with determinant 1?

## 2. Group Homomorphisms and Isomorphisms

Just as for rings, we need the notion of a mapping between two groups that respects their algebraic structure. We give here the necessary definitions and several examples.

**Definition.** Let $G$ and $G'$ be groups. A function $\phi: G \to G'$ is called a **group homomorphism** if $\phi(ab) = \phi(a)\phi(b)$ for all $a, b \in G$.

**Lemma 2.1.** Let $\phi: G \to G'$ be a homomorphism.

  (i)  Let $e$ and $e'$ be the respective identity elements of $G$ and $G'$; then $\phi(e) = e'$.
  (ii)  $(\phi(a))^{-1} = \phi(a^{-1})$ for all $a \in G$.

**Proof.** $e'\phi(a) = \phi(a) = \phi(ea) = \phi(e)\phi(a) \implies e' = \phi(e)$, by Lemma 1.1(iii). Now $e' = \phi(e) = \phi(aa^{-1}) = \phi(a)\phi(a^{-1}) \implies \phi(a^{-1}) = (\phi(a))^{-1}$, by Lemma 1.1(ii). □

Once again, the kernel of a homomorphism plays a crucial rôle in our work. We define the **kernel** of a group homomorphism

$\phi: G \to G'$ to be

$$\ker \phi = \{a \in G : \phi(a) = e'\}.$$

Here are some easy consequences of the definition.

**Lemma 2.2.** $\ker \phi$ *is a subgroup of* $G$, *and* $\ker \phi = \{e\} \iff \phi$ *is one-to-one.*

**Proof.** To establish that $\ker \phi$ is a subgroup, we must show that if $a, b \in \ker \phi$, then $ab \in \ker \phi$ and $a^{-1} \in \ker \phi$. So suppose $a, b \in \ker \phi$, i.e., $\phi(a) = \phi(b) = e'$; then $\phi(ab) = \phi(a)\phi(b) = e'e' = e'$, whence $ab \in \ker \phi$. Similarly, by Lemma 2.1(ii), $\phi(a^{-1}) = (\phi(a))^{-1} = (e')^{-1} = e'$. Thus, $\ker \phi$ is indeed a subgroup of $G$.

Now suppose $\ker \phi = \{e\}$ and $\phi(a) = \phi(b)$. Then $\phi(ab^{-1}) = \phi(a)\phi(b)^{-1} = e' \implies ab^{-1} \in \ker \phi \implies ab^{-1} = e \implies a = b$. Conversely, suppose $\phi$ is one-to-one; then, since $\phi(e) = e'$ and $\phi$ maps (at most) one element of $G$ to $e' \in G'$, we have $\ker \phi = \{e\}$.  □

We want to say two groups are "the same" if there is a one-to-one correspondence between them respecting their group structures.

**Definition.** A group homomorphism $\phi: G \to G'$ is called an **isomorphism** if it maps one-to-one and onto $G'$. We say the groups $G$ and $G'$ are **isomorphic** (denoted by $G \cong G'$) if there exists an isomorphism between them.

First, we give some necessary conditions for two groups to be isomorphic. These criteria may help us decide when two groups are *not* isomorphic.

**Proposition 2.3.** *Suppose* $\phi: G \to G'$ *is an isomorphism.*

(i) *If* $G$ *is a finite group, then so is* $G'$, *and* $|G| = |G'|$.
(ii) *For each* $a \in G$, *the order of* $a$ *equals the order of* $\phi(a)$.
(iii) *If* $G$ *is abelian, then* $G'$ *is abelian.*

**Proof.** We prove (ii) here and leave the rest for Exercise 3. Recall that the order of an element $a \in G$ is the smallest integer $n \in \mathbb{N}$ so that $a^n = e$. Since $\phi$ is a (group) homomorphism, $\phi(a)^n = \phi(a^n) = \phi(e) = e'$. But we are not done: we must make sure that no smaller positive power of $\phi(a)$ yields $e'$. Suppose $\phi(a)^k = e'$, $0 \le k < n$. Then $\phi(a^k) = e'$, so $a^k \in \ker \phi$. Since $\phi$ is an isomorphism, $\ker \phi = \{e\}$, and so $a^k = e$. Since the order of $a$ is $n$, we know that $k$ must equal 0.  □

As a consequence of Proposition 2.3, two groups $G$ and $G'$ cannot possibly be isomorphic if one is abelian and the other is not. And given two abelian groups $G$ and $G'$ of order 8, once again they cannot possibly be isomorphic if $G$ has an element of order 8 and $G'$ does not.

**Examples 1.** Here now are some examples of group homomorphisms and isomorphisms.

(a) Let

$$\phi: \mathbb{C}^\times \to S = \{w \in \mathbb{C} : |w| = 1\}, \quad \phi(z) = \frac{z}{|z|},$$

be the function that associates to each nonzero complex number $z$ the unique complex number of length one lying on the ray $\overrightarrow{0z}$. The group $S$ has multiplication as its operation; so $\phi$ is a homomorphism, because $\phi(ab) = \frac{ab}{|ab|} = \frac{a}{|a|} \cdot \frac{b}{|b|} = \phi(a)\phi(b)$. Note that $\ker \phi = \{$positive real numbers$\}$.

(b) Let $\mathbb{R}$ be the additive group of real numbers, and let $\mathbb{R}^+$ be the *multiplicative* group of positive real numbers. Let $\phi: \mathbb{R} \to \mathbb{R}^+$ be given by $\phi(a) = \exp(a)$. (Here we write the exponential function as $\exp(x)$ rather than as $e^x$ to avoid confusion with the identity element of a group.) Now, $\phi$ is a homomorphism by the formula $\exp(a + b) = \exp(a)\exp(b)$. Note, moreover, that $\ker \phi = \{a \in \mathbb{R} : \exp(a) = 1\} = \{0\}$, so $\phi$ is one-to-one. And $\phi$ maps onto $\mathbb{R}^+$, with inverse function $\ln x$; therefore, $\phi$ is an isomorphism. (This isomorphism explains the usefulness of log tables and slide rules—or at least it used to, before the advent of the pocket super-computer.)

(c) The determinant function gives a homomorphism

$$\det: GL(n, F) \to F^\times,$$

because of the product rule for determinants (Theorem 2.9 of Appendix B).

(d) We have suggested group homomorphisms: $\mathcal{T} \to S_3$ and $Sq \to S_4$ in Examples 4 and 5 of Section 1. We should check that these correspondences are homomorphisms, and that the former map is in fact an isomorphism (see Exercise 4).

(e) We saw in Proposition 1.2 that the units in a ring form a group. Let's consider the two groups $\mathbb{Z}_5^\times$ and $\mathbb{Z}_8^\times$. There are four elements in each: namely, 1 (mod 5), 2 (mod 5), 3 (mod 5), and 4 (mod 5) in $\mathbb{Z}_5^\times$; and 1 (mod 8), 3 (mod 8),

5 (mod 8), and 7 (mod 8) in $\mathbb{Z}_8^{\times}$. Could these two groups possibly be isomorphic? Proposition 2.3 is not apparently of much help, since both are abelian groups with four elements. However, note that in $\mathbb{Z}_8^{\times}$, we have

$$(3 \text{ (mod 8)})^2 = (5 \text{ (mod 8)})^2 = (7 \text{ (mod 8)})^2 = 1 \text{ (mod 8)},$$

whereas in $\mathbb{Z}_5^{\times}$,

$$(2 \text{ (mod 5)})^2 = (3 \text{ (mod 5)})^2 = 4 \text{ (mod 5)}$$

$$\text{and } (4 \text{ (mod 5)})^2 = 1 \text{ (mod 5)}.$$

Thus, in $\mathbb{Z}_8^{\times}$ there are three elements of order 2, but in $\mathbb{Z}_5^{\times}$ there are only two. So, by part (ii) of Proposition 2.3, the groups cannot be isomorphic.

Becoming more optimistic, we claim that there is an isomorphism $\phi\colon \mathcal{V} \to \mathbb{Z}_8^{\times}$, given by

$$\phi(e) = 1 \text{ (mod 8)}$$

$$\phi(a) = 3 \text{ (mod 8)}$$

$$\phi(b) = 5 \text{ (mod 8)}$$

$$\phi(c) = 7 \text{ (mod 8)}.$$

Note that the rule $ab = c$ in $\mathcal{V}$ translates to $3 \cdot 5 = 7$ in $\mathbb{Z}_8^{\times}$. And, similarly, there is an isomorphism $\psi\colon \mathbb{Z}_4 \to \mathbb{Z}_5^{\times}$, given by

$$\psi(0 \text{ (mod 4)}) = 1 \text{ (mod 5)}$$

$$\psi(1 \text{ (mod 4)}) = 3 \text{ (mod 5)}$$

$$\psi(2 \text{ (mod 4)}) = 4 \text{ (mod 5)}$$

$$\psi(3 \text{ (mod 4)}) = 2 \text{ (mod 5)}.$$

Since $\psi(k \text{ (mod 4)}) = (\psi(1))^k \text{ (mod 5)}$, as the reader should check, $\psi$ is indeed a homomorphism.

(f) In Exercise 6.1.4 the reader was asked to study the group $GL(2, \mathbb{Z}_2)$. There are six elements:

$$\begin{bmatrix} 1 & 0 \\ 0 & 1 \end{bmatrix}, \begin{bmatrix} 1 & 1 \\ 1 & 0 \end{bmatrix}, \begin{bmatrix} 0 & 1 \\ 1 & 1 \end{bmatrix}, \begin{bmatrix} 0 & 1 \\ 1 & 0 \end{bmatrix}, \begin{bmatrix} 1 & 1 \\ 0 & 1 \end{bmatrix}, \begin{bmatrix} 1 & 0 \\ 1 & 1 \end{bmatrix}.$$

This group is not abelian, and we are acquainted with only one non-abelian group of order 6, namely, $S_3$; so we may as well try to prove that $GL(2, \mathbb{Z}_2) \cong S_3$. We start to define

$\phi: GL(2, \mathbb{Z}_2) \to S_3$ as follows:

$$\phi\left(\begin{bmatrix} 1 & 1 \\ 1 & 0 \end{bmatrix}\right) = \begin{pmatrix} 1 & 2 & 3 \\ 2 & 3 & 1 \end{pmatrix},$$

$$\phi\left(\begin{bmatrix} 0 & 1 \\ 1 & 0 \end{bmatrix}\right) = \begin{pmatrix} 1 & 2 & 3 \\ 1 & 3 & 2 \end{pmatrix}.$$

Then we leave it to the reader to check (see Exercise 7) that this definition can be extended compatibly to the remaining elements of the group.

**Example 2.** If $G$ is a cyclic group of order $n$, then we claim $G \cong \mathbb{Z}_n$. (For utter clarity, we return briefly to our old custom of representing elements of $\mathbb{Z}_n$ by barred integers.) By definition, $G = \langle a \rangle$ for some $a \in G$, and $a^n = e$, whereas no lesser power of $a$ is equal to $e$. Define $\phi: \mathbb{Z}_n \to G$ by $\phi(\bar{k}) = a^k$. Note, first, that $\phi$ is well-defined: if $\bar{k} = \bar{\ell} \in \mathbb{Z}_n$, then $k = \ell + jn$ for some $j \in \mathbb{Z}$ and $a^k = a^{\ell+jn} = a^\ell \cdot (a^n)^j = a^\ell \cdot e^j = a^\ell$. Next, $\phi(\overline{k + \ell}) = a^{k+\ell} = a^k \cdot a^\ell = \phi(\bar{k}) \cdot \phi(\bar{\ell})$; so $\phi$ is a homomorphism. And $\phi$ is one-to-one, since if $\phi(\bar{k}) = e$, then $\bar{k} = \bar{0}$, because we know that $a^k \neq e$ for $0 < k < n$. It is immediate that $\phi$ maps onto $G$ (why?).

## EXERCISES 6.2

1. Show that the Klein four-group $\mathcal{V}$ is not isomorphic to $\mathbb{Z}_4$.

2. Prove that $\mathbb{Z}_7^\times \cong \mathbb{Z}_6$. (It is crucial to remember that we multiply in $\mathbb{Z}_7^\times$ and add in $\mathbb{Z}_6$.)

3. Complete the proof of Proposition 2.3.

4. Check that the maps $\mathcal{T} \to S_3$ and $Sq \to S_4$ given in Section 1 are group homomorphisms, and that the former is an isomorphism.

5. To what well-known groups are the following subgroups of $GL(2, \mathbb{C})$ isomorphic?

   a. $\left\{ \begin{bmatrix} 1 & 0 \\ 0 & 1 \end{bmatrix}, \begin{bmatrix} 1 & 0 \\ 0 & -1 \end{bmatrix}, \begin{bmatrix} -1 & 0 \\ 0 & 1 \end{bmatrix}, \begin{bmatrix} -1 & 0 \\ 0 & -1 \end{bmatrix} \right\}$

   b. $\left\{ \begin{bmatrix} 1 & 0 \\ 0 & 1 \end{bmatrix}, \begin{bmatrix} 0 & -1 \\ 1 & 0 \end{bmatrix}, \begin{bmatrix} 0 & 1 \\ -1 & 0 \end{bmatrix}, \begin{bmatrix} -1 & 0 \\ 0 & -1 \end{bmatrix} \right\}$

   c. $\left\{ \begin{bmatrix} 1 & 0 \\ 0 & 1 \end{bmatrix}, \begin{bmatrix} -1 & 0 \\ 0 & -1 \end{bmatrix}, \begin{bmatrix} i & 0 \\ 0 & -i \end{bmatrix}, \begin{bmatrix} -i & 0 \\ 0 & i \end{bmatrix}, \begin{bmatrix} 0 & i \\ i & 0 \end{bmatrix}, \begin{bmatrix} 0 & -i \\ -i & 0 \end{bmatrix}, \right.$

   $\left. \begin{bmatrix} 0 & -1 \\ 1 & 0 \end{bmatrix}, \begin{bmatrix} 0 & 1 \\ -1 & 0 \end{bmatrix} \right\}$

What about
$$\left\{ \begin{bmatrix} 1 & 0 & 0 \\ 0 & 1 & 0 \\ 0 & 0 & 1 \end{bmatrix}, \begin{bmatrix} 1 & 0 & 0 \\ 0 & 0 & 1 \\ 0 & 1 & 0 \end{bmatrix}, \begin{bmatrix} 0 & 1 & 0 \\ 1 & 0 & 0 \\ 0 & 0 & 1 \end{bmatrix}, \begin{bmatrix} 0 & 1 & 0 \\ 0 & 0 & 1 \\ 1 & 0 & 0 \end{bmatrix}, \begin{bmatrix} 0 & 0 & 1 \\ 0 & 1 & 0 \\ 1 & 0 & 0 \end{bmatrix}, \right.$$
$$\left. \begin{bmatrix} 0 & 0 & 1 \\ 1 & 0 & 0 \\ 0 & 1 & 0 \end{bmatrix} \right\} \subset GL(3, \mathbb{R})?$$ (Hint: Let $\mathbf{v} = \begin{bmatrix} 1 \\ 2 \\ 3 \end{bmatrix}$. Compute $g\mathbf{v}$ for each $g \in G$.)

6.  a.  Prove that $\mathbb{Z}_{12}^{\times} \cong \mathcal{V}$.
    b.  Prove that $\mathbb{Z}_{15}^{\times} \cong \mathbb{Z}_{16}^{\times} \cong \mathbb{Z}_{20}^{\times}$. What about $\mathbb{Z}_{24}^{\times}$?

7.  Finish the verification that $GL(2, \mathbb{Z}_2) \cong S_3$. (Hint: Every element of $GL(2, \mathbb{Z}_2)$ can be written as a product of the matrices $\begin{bmatrix} 1 & 1 \\ 1 & 0 \end{bmatrix}$ and $\begin{bmatrix} 0 & 1 \\ 1 & 0 \end{bmatrix}$, and every element of $S_3$ can be written as a product of the permutations $\begin{pmatrix} 1 & 2 & 3 \\ 2 & 3 & 1 \end{pmatrix}$ and $\begin{pmatrix} 1 & 2 & 3 \\ 1 & 3 & 2 \end{pmatrix}$.)

8.  Prove that $G$ is abelian if and only if $\phi \colon G \to G$, $\phi(a) = a^{-1}$, is a homomorphism.

9.  Show that $\phi \colon \mathbb{C}^{\times} \to \mathbb{R}^{\times}$, $\phi(z) = |z|$, is a homomorphism. What is $\ker \phi$?

10. Show that $\phi \colon \mathbb{R} \to \mathbb{C}^{\times}$, $\phi(t) = \cos(2\pi t) + i\sin(2\pi t)$, is a homomorphism. What are its kernel and image?

11. Let $\phi \colon G \to G'$ and $\psi \colon G \to G'$ be two homomorphisms. Let $H = \{a \in G : \phi(a) = \psi(a)\}$. Prove or disprove: $H \subset G$ is a subgroup.

12. Let $a \in G$ be fixed, and define $\phi \colon G \to G$ by $\phi(x) = axa^{-1}$, $x \in G$. Prove that $\phi$ is a homomorphism. Under what circumstances is $\phi$ an isomorphism?

13. Let $G = \left\{ \begin{bmatrix} a & b \\ 0 & 1 \end{bmatrix} : a, b \in \mathbb{R}, a \neq 0 \right\}$.
    a.  Prove that $G$ is a group (under multiplication of matrices).
    b.  Prove that $G$ contains a subgroup isomorphic to $\mathbb{R}$ and a subgroup isomorphic to $\mathbb{R}^{\times}$.
    c.  Let $H = \left\{ \begin{bmatrix} a & b \\ 0 & 1 \end{bmatrix} \in G : a = \pm 1, b \in \mathbb{R} \right\}$. Prove $H$ is a subgroup of $G$.
    d.  Prove $H$ is isomorphic to the group of isometries of $\mathbb{R}$.

14. Decide whether the groups $\mathcal{S}q$ and $\mathcal{Q}$ (see Example 2(b) of Section 1) are isomorphic.

15. The **dihedral group** of order $2n$, denoted $\mathcal{D}_n$, is given by $\{\rho^i \psi^j : 0 \le i < n, \ 0 \le j \le 1\}$ subject to the rules $\rho^n = e$, $\psi^2 = e$, and $\psi \rho \psi^{-1} = \rho^{-1}$.

    a. Check this is really a group. That is, what is $(\rho^i \psi^j)^{-1}$, and what is the product $(\rho^i \psi^j)(\rho^k \psi^\ell)$?

    b. Check that $\mathcal{T} \cong \mathcal{D}_3$ and $\mathcal{S}q \cong \mathcal{D}_4$.

    c. Prove in general that $\mathcal{D}_n$ is the group of symmetries of the regular $n$-gon.

    d. Let $\zeta = e^{\frac{2\pi i}{n}}$. Prove that $\mathcal{D}_n$ is isomorphic to the subgroup of $GL(2, \mathbb{C})$ obtained by taking all products of the two matrices $\begin{bmatrix} 0 & 1 \\ 1 & 0 \end{bmatrix}$ and $\begin{bmatrix} \zeta & 0 \\ 0 & \bar{\zeta} \end{bmatrix}$ and their inverses. (This is called the subgroup generated by the two elements.)

    e. Let $\theta = \frac{2\pi}{n}$. Prove that $\mathcal{D}_n$ is isomorphic to the subgroup of $GL(3, \mathbb{R})$ generated by the two matrices

$$A = \begin{bmatrix} -1 & & \\ & 1 & \\ & & -1 \end{bmatrix} \quad \text{and} \quad B = \begin{bmatrix} \cos\theta & -\sin\theta & \\ \sin\theta & \cos\theta & \\ & & 1 \end{bmatrix}.$$

    (Hint: Compute $A^2$, $B^n$, and $ABA^{-1}$.)

16. Let $\mathbf{v}, \mathbf{w} \in \mathbb{R}^2$ be linearly independent vectors, and let $L = \{m\mathbf{v} + n\mathbf{w} : m, n \in \mathbb{Z}\}$. We call $L$ a **lattice** in the plane. Let $G$ be a finite group of rotations of the plane that map the lattice $L$ to itself. Prove that $|G|$ must be either 1, 2, 3, 4, or 6. (Thus, we can "tile" the plane with equilateral triangles, squares, and regular hexagons, but not, for example, with regular pentagons. Cf. Weyl's book *Symmetry*.)

17. Prove or give a counterexample. Let $G$ and $G'$ be finite groups. If there is a bijection $f : G \to G'$ so that for every $a \in G$ the order of $a$ is equal to the order of $f(a)$, then $G \cong G'$.

## 3. Cosets, Normal Subgroups, and Quotient Groups

The crucial ideas in group theory come from using a subgroup to divide the group into equivalence classes, as we did with an ideal in a ring. Let $H \subset G$ be a subgroup, and let $a \in G$. Then define

$$aH = \{ah : h \in H\};$$

this is called a (left) **coset** of $H$. (When $G$ is an abelian group, we usually write the group operation as addition, and indicate the coset by $a + H = \{a + h : h \in H\}$.)

**Examples 1.**

(a) Let $H = 2\mathbb{Z} = \{$even integers$\} \subset \mathbb{Z}$. Then $0 + H = H = \{$even integers$\}$ and $1 + H = \{$odd integers$\}$. Note that if $a$ is even, then $a + H = 0 + H$; and if $a$ is odd, then $a + H = 1 + H$.

(b) Let $G = \mathcal{T}$, $H = \{\iota, F\}$, where $F = F_1$. Then the cosets of $H$ are the following:

$$
\begin{aligned}
\iota H &= & H &= \{\iota,\ F\} & &= FH \\
RH &= & \{R,\ RF = F_3\} & &= F_3 H \\
R^2 H &= & \{R^2,\ R^2 F = F_2\} & &= F_2 H
\end{aligned}
$$

(c) Consider the vector space $\mathbb{R}^2$ as an abelian group with operation addition. Let $H \subset \mathbb{R}^2$ be a one-dimensional vector subspace (therefore a subgroup). Then its cosets are the "affine subspaces" $\mathbf{a} + H$, $\mathbf{a} \in \mathbb{R}^2$; these are the lines parallel to $H$, as shown in Figure 1.

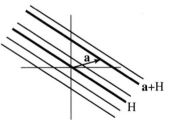

FIGURE 1

Here is the key observation:

**Lemma 3.1.** *Two cosets have a nonempty intersection if and only if they are identical.*

**Proof.** Suppose $c \in aH \cap bH$. Then there are $h, h' \in H$ so that $c = ah = bh'$. Thus, $a = (bh')h^{-1} = b(h'h^{-1}) \in bH$, since $H$ is a subgroup. As a result, $aH \subset bH$ (why?). Similarly, $b = a(hh'^{-1})$, and so $bH \subset aH$. Therefore, $aH = bH$, as required.  $\square$

Given a group $G$ and a subgroup $H$, each element $a \in G$ belongs to some coset—namely, $aH$. By Lemma 3.1, each element belongs to *only one* coset. So $G$ can be expressed as the the union of distinct cosets $aH$; we write this symbolically as

$$G = \bigcup_{\text{disjoint}} aH .$$

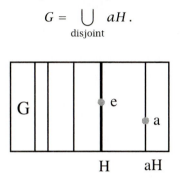

$$\text{H} \qquad \text{aH}$$

FIGURE 2

We define the **index** of $H$ in $G$ to be the number of distinct cosets of $H$ in $G$, denoted $[G : H]$.

### Examples 2.

(a) The subgroup $H = n\mathbb{Z} \subset \mathbb{Z}$ has index $n$, since the cosets are $0 + H, 1 + H, 2 + H, \ldots, (n - 1) + H$.

(b) The subgroup $H \subset \mathcal{T}$ given in Example 1(b) has index 3.

If $G$ is a finite group, we have the following very important counting principle.

**Theorem 3.2** (Lagrange)**.** *If $G$ is a finite group and $H \subset G$ is a subgroup, then*

$$[G : H] = |G| \, / \, |H|.$$

***Proof.*** The union of the $[G : H]$ disjoint cosets is all of $G$. On the other hand, $H$ is in one-to-one correspondence with each coset $aH$ (letting $h$ correspond to $ah$), so every coset has $|H|$ elements. Thus, there are $[G : H]|H|$ elements all together, from which we conclude that $|G| = [G : H]|H|$. $\square$

**Corollary 3.3.** *If $H$ is a subgroup of a finite group $G$, then $|H|$ divides $|G|$.* $\square$

From this, a useful observation follows: if $H$ is a subgroup of a finite group $G$ and $|H| > \frac{1}{2}|G|$, then $H$ must be the entire group $G$.

**Corollary 3.4.** *If G is a finite group and a is an arbitrary element of G, then the order of a divides* $|G|$.

*Proof.* The order of $a$ is the order of the cyclic subgroup $\langle a \rangle \subset G$ it generates. By Corollary 3.3, this number is a divisor of $|G|$. □

**Corollary 3.5.** *If* $|G| = p$ *and p is prime, then G is cyclic.*

*Proof.* Let $a \in G$ be any element other than the identity, and consider the cyclic subgroup $\langle a \rangle$ of $G$ generated by $a$. The order of this subgroup must divide $p$ and is strictly greater than 1 (why?). Therefore, $a$ has order $p$, and so $\langle a \rangle = G$. □

**Corollary 3.6.** *If* $|G| = n$, *and* $a \in G$ *is arbitrary, then* $a^n = e$.

*Proof.* Let the order of the element $a$ be $k$. By Corollary 3.4, $k|n$, so there is an integer $\ell$ with $n = k\ell$. Then $a^n = a^{k\ell} = (a^k)^\ell = e^\ell = e$. □

**Remark.** Be warned that, in general, the converse of Lagrange's Theorem is false. For example, there is a group of order 12 having no subgroup of order 6; we shall get to know this group well in Section 2 of Chapter 7 (see also Section 4 of this chapter).

Just to impress upon you the power of Theorem 3.2, we proceed to classify all groups of order up to seven.

**Theorem 3.7.** *Let G be a group of order* $n \leq 7$.

(1) *If* $n = 2, 3, 5,$ *or 7, then G is cyclic.*
(2) *If* $n = 4$, *then G is isomorphic to either* $\mathbb{Z}_4$ *or* $\mathcal{V}$.
(3) *If* $n = 6$, *then G is isomorphic to either* $\mathbb{Z}_6$ *(if G is abelian) or* $\mathcal{T}$ *(if not).*

*Proof.* Part (1) follows from Corollary 3.5.

So far as (2) is concerned, if $G$ contains an element of order 4, then $G \cong \mathbb{Z}_4$ (see Example 2 of Section 2). If not, by Corollary 3.4, every element other than the identity must have order 2. Suppose $a$ and $b$ are two (distinct) elements of order 2. Then their product $ab$ is an element of the group $G$. It cannot be equal to either $a$ or $b$, by Lemma 1.1(iii), and it cannot be equal to $e$ (since $a^2 = b^2 = e$). We have listed four elements in $G$—$e, a, b,$ and $ab$; since $G$ has order 4, this is a complete list of its elements. Note that $ba = ab$ (by the identical reasoning), so the group is abelian and has the multiplication table of $\mathcal{V}$.

Suppose now that $G$ is an abelian group of order 6. If $G$ contains an element $a$ of order 6, $G$ is cyclic (generated by $a$), and we're done. If every element (other than the identity) of $G$ were to have order 2, we would find that any $a, b \in G$ would generate the subgroup $\{e, a, b, ab\} \subset G$; this contradicts Corollary 3.3. Thus, there must be an element $a \in G$ of order 3 (by the process of elimination, using Corollary 3.4). Let $H = \langle a \rangle$, and choose $b \notin H$. Since $[G : H] = 2$, we have $G = H \cup bH$. It follows that $b^2 \in H$ (why?), and so $b^2 = a$, $b^2 = a^2$ or $b^2 = e$. In either of the first two cases, we may conclude that $G = \langle b \rangle$ (and so $G \cong \mathbb{Z}_6$) by listing the powers of $b$: if $b^2 = a$,

$$\langle b \rangle = \{e,\ b,\ b^2 = a,\ b^3 = ab,\ b^4 = a^2,\ b^5 = a^2 b\};$$

whereas if $b^2 = a^2$,

$$\langle b \rangle = \{e,\ b,\ b^2 = a^2,\ b^3 = a^2 b,\ b^4 = a,\ b^5 = ab\}.$$

(Note that in either case $b^3 \neq e$, since $b \notin H$.) Lastly, if $b^2 = e$, we claim that $G = \langle ba \rangle$ for $\langle ba \rangle = \{e,\ ba,\ (ba)^2 = a^2,\ (ba)^3 = b,\ (ba)^4 = a,\ (ba)^5 = ba^2\}$. This completes the analysis when $G$ is abelian.

Here is the most interesting case: suppose $G$ is a non-abelian group of order 6. First, it cannot be true that every element has order 2 (why?). So, as before, there must be an element $a \in G$ of order 3. Consider the cyclic subgroup $H = \langle a \rangle$, and choose $b \notin H$. As before, either $b^2 = e$, $b^2 = a$, or $b^2 = a^2$. In either of the latter two cases, we would have $a^3 = e \implies (b^2)^3 = e$ (and $b^3 \neq e$); and so $G$ would be cyclic of order 6, therefore abelian. Thus, we must have $b^2 = e$.

Summarizing, so far we have $a, b \in G$ with $a^3 = b^2 = e$. Consider $ab \in G$. It is an element of the coset $bH = \{b, ba, ba^2\}$. Once again, we have the three possibilities: $ab = b$, $ab = ba$, $ab = ba^2$; but only the last is viable (why?). So, $G = \{e, a, a^2, b, ba, ba^2\}$ with $ab = ba^2$ (and likewise $a^2 b = ba$). We define an isomorphism $\phi: G \to \mathcal{T}$ by $\phi(a) = R$, $\phi(b) = F$. We leave the rest of the details to the reader (see Exercise 6.2.7). □

We saw in Chapter 4 that given a ring $R$ and an ideal $I \subset R$, the set of equivalence classes $R/I$ always inherits a ring structure. In the context of groups, given a group $G$ and a subgroup $H \subset G$, we denote by $G/H$ the set of cosets of $H$ in $G$, and we ask when this is again a group. Given two cosets $aH$ and $bH$, we must first define their product. The natural thing to try is

$$(aH)(bH) = (ab)H.$$

Does this work? Suppose $aH = a'H$ and $bH = b'H$; then it had better be the case that $(ab)H = (a'b')H$. If $aH = a'H$, then $a' = ah$ for some $h \in H$; if $bH = b'H$, then $b' = bk$ for some $k \in H$. Computing, we have $a'b' = (ah)(bk) = a(hb)k$, and we are stymied. In order to proceed, we need to be able to "pass $b$ through $h$" somehow; that is, we need a rule that says $hb = b\tilde{h}$ for some $\tilde{h} \in H$. If we have such a rule, then we can continue: $a'b' = (ah)(bk) = a(hb)k = a(b\tilde{h})k = (ab)(\tilde{h}k) \in (ab)H$, since $\tilde{h}k \in H$. In conclusion, $(a'b')H = (ab)H$, as desired. Rephrasing our rule slightly, $hb = b\tilde{h}$ for some $\tilde{h} \in H$ if and only if $b^{-1}hb \in H$. Changing notation, this motivates the following definition.

**Definition.** A subgroup $H \subset G$ is **normal** if $aHa^{-1} \subset H$ for all $a \in G$, i.e., if for all $a \in G$ and $h \in H$, $aha^{-1} \in H$. We call $aha^{-1}$ the **conjugate** of $h$ by $a$; in general, we say $g$ and $g' \in G$ are conjugate provided $g' = aga^{-1}$ for some $a \in G$.

We now summarize these calculations officially.

**Proposition 3.8.** *Let $H \subset G$ be a normal subgroup. Then $G/H$ is naturally a group with identity element $eH$ and multiplication defined by $(aH)(bH) = (ab)H$.*

***Proof.*** We first check that multiplication is well-defined. Suppose, as above, that $aH = a'H$ and $bH = b'H$. Then $a' = ah$ and $b' = bk$ for some $h, k \in H$. Since $H$ is normal, $b^{-1}hb = \tilde{h}$ for some $\tilde{h} \in H$. And so, $a'b' = (ah)(bk) = a(hb)k = a(b\tilde{h})k = (ab)(\tilde{h}k) \in (ab)H$; thus, $(ab)H = (a'b')H$, as required.

Now we check that $eH$ is indeed an identity element for $G/H$: $(aH)(eH) = (ae)H = aH$, and $(eH)(aH) = (ea)H = aH$. We must check associativity: given $aH, bH, cH \in G/H$, we must verify that $((aH)(bH))(cH) = (aH)((bH)(cH))$. But $((aH)(bH))(cH) = ((ab)H)(cH) = ((ab)c)H$, whereas $(aH)((bH)(cH)) = (aH)((bc)H) = (a(bc))H$; by the associative property in $G$, these are indeed equal. Lastly, we claim that $a^{-1}H$ is the desired inverse element of $aH$: $(a^{-1}H)(aH) = (a^{-1}a)H = eH = (aH)(a^{-1}H)$, as required. $\square$

**Remark.** When $[G : H]$ is finite, the group $G/H$ has order $[G : H]$. As a consequence, when $G$ is a finite group, $|G/H| = |G| / |H|$.

**Examples 3.**

(a) Note, first, that if $G$ is abelian, then any subgroup is normal; so $G/H$ makes sense in this case for all subgroups $H$. What's more, $G/H$ will be an abelian group as well (see Exercise 10a.).

(b) Recall that $\mathbb{R}^\times$ is the group of nonzero real numbers with operation multiplication. Since $G = \mathbb{R}^\times$ is abelian, any subgroup is normal. Let $H = \{\text{positive real numbers}\} \subset G$. Since the product of positive numbers and the reciprocal of a positive number are positive, $H$ is a subgroup. Since $G = H \cup (-1)H$, there are two cosets, and the quotient group $G/H$ has two elements, say $\overline{1}$ and $\overline{-1}$. We infer that $\overline{1} \cdot \overline{-1} = \overline{-1}$ and $\overline{-1} \cdot \overline{-1} = \overline{1}$; we have deduced that the product of a positive real number and a negative real number is negative, and that the product of two negative real numbers is positive!

(c) Here is our prototypical normal subgroup. Let $\phi\colon G \to G'$ be a homomorphism, and let $H = \ker \phi$. Suppose $a \in G$ and $h \in H$; we must show that $aha^{-1} \in H$. Compute $\phi(aha^{-1}) = \phi(a)\phi(h)\phi(a^{-1}) = \phi(a)e'\phi(a^{-1}) = \phi(a)\phi(a^{-1}) = \phi(aa^{-1}) = \phi(e) = e'$. Therefore, $aha^{-1} \in \ker \phi$, as promised.

(d) The matrices of determinant 1, written $SL(n, F)$, form a normal subgroup of $GL(n, F)$: if $\det(B) = 1$ and $A \in GL(n, F)$, then $\det(ABA^{-1}) = \det(B) = 1$ (see Corollary 2.11 in Appendix B).

(e) The subgroup $H = \{\iota, R_2\} \subset Sq$ is a normal subgroup. First, $R_2 = \rho^2$ commutes with any rotation (since all rotations are powers of $\rho$). Now the relation $\psi\rho = \rho^3\psi$ implies that

$$\psi\rho^2\psi^{-1} = \psi\rho^2\psi = (\psi\rho)(\rho\psi) = \rho^3(\psi\rho)\psi$$

$$= \rho^3(\rho^3\psi)\psi = \rho^2,$$

$$(\rho\psi)\rho^2(\rho\psi)^{-1} = \rho\psi\rho^2\psi\rho^3 = \rho(\rho^3\psi\rho)\psi\rho^3 = (\psi\rho)\psi\rho^3$$

$$= \rho^3\psi^2\rho^3 = \rho^2;$$

and similarly in the last case. The quotient group $Sq/H$ has order $8/2 = 4$, and so by Theorem 3.7 is isomorphic to either $\mathbb{Z}_4$ or $\mathcal{V}$. But every element of $Sq/H$ has order at most 2, since $Sq/H = \{eH, \rho H, \psi H, (\rho\psi)H\}$ and $(\rho H)^2 = (\psi H)^2 = (\rho\psi H)^2 = H$. Thus, $Sq/H \cong \mathcal{V}$.

**Examples 4.** It might be good to have a few examples of subgroups that are *not* normal.

(a) Let $G = \mathcal{T}$ and $H = \{\iota, F\}$, where $F = F_1$. Consider $RHR^{-1}$: $R\iota R^{-1} = \iota \in H$, of course; but $RFR^{-1} = RFR^2 = R^2F = F_2 \notin H$. That is, the conjugate of $F_1$ by $R$ (which is the flip fixing vertex 1) is $F_2$ (which is the flip fixing vertex 2).

(b) We saw in Exercise 6.1.22 that $H = \left\{\begin{bmatrix} a & b \\ 0 & c \end{bmatrix} : ac \neq 0\right\}$ is a subgroup of $GL(2, \mathbb{R})$. It is, however, not a normal subgroup. To establish this, we need only exhibit a single element $h \in H$ and a single element $a \in G$ so that $aha^{-1} \notin H$. Let $h = \begin{bmatrix} 1 & 1 \\ 0 & 1 \end{bmatrix}$ and $a = \begin{bmatrix} 1 & 0 \\ 1 & 1 \end{bmatrix}$; then $aha^{-1} = \begin{bmatrix} 0 & 1 \\ -1 & 2 \end{bmatrix} \notin H$.

(c) Consider next the smaller subgroup $K = \left\{\begin{bmatrix} a & 0 \\ 0 & c \end{bmatrix} : ac \neq 0\right\} \subset GL(2, \mathbb{R})$. It is easy to check that $K$ is a subgroup, but again is not a normal subgroup. Let $k = \begin{bmatrix} 2 & 0 \\ 0 & 1 \end{bmatrix} \in K$ and $b = \begin{bmatrix} 1 & 1 \\ 0 & 1 \end{bmatrix} \in GL(2, \mathbb{R})$. Then $bkb^{-1} = \begin{bmatrix} 2 & -1 \\ 0 & 1 \end{bmatrix} \notin K$. Indeed, we've shown more. Since $K$ happens to be a subgroup of $H$ (from the preceding example), and since $b \in H$, it follows that $K$ is not a normal subgroup of $H$, either.

It should by now come as no surprise that there is a Fundamental Homomorphism Theorem for Groups, analogous to Theorem 2.2 of Chapter 4.

**Theorem 3.9** (Fundamental Homomorphism Theorem).    *Let $\phi: G \to G'$ be a homomorphism onto $G'$. Then $G/\ker\phi \cong G'$.*

**Proof.** Let $H = \ker\phi$. As usual, we define a homomorphism $\overline{\phi}: G/H \to G'$ by $\overline{\phi}(aH) = \phi(a)$. We need first to check this is well-defined: if $aH = a'H$, then $a' = ah$ for some $h \in H = \ker\phi$; so $\phi(a') = \phi(ah) = \phi(a)\phi(h) = \phi(a)$, as desired. And $\overline{\phi}$ is a homomorphism:

$$\overline{\phi}((aH)(bH)) = \overline{\phi}((ab)H) \qquad \text{by definition of the group } G/H$$

$$= \phi(ab) \qquad \text{by definition of } \overline{\phi}$$

$$= \phi(a)\phi(b) \qquad \text{because } \phi \text{ is a homomorphism}$$

$$= \overline{\phi}(aH)\overline{\phi}(bH) \qquad \text{again by definition of } \overline{\phi}.$$

The homomorphism $\overline{\phi}$ maps onto $G'$ since $\phi$ does; it remains only to see $\overline{\phi}$ is one-to-one. Suppose $aH \in \ker\overline{\phi}$. Then $\overline{\phi}(aH) = \phi(a) = e' \implies a \in H \implies aH = eH$.   □

**Remark.** *Any* normal subgroup $H$ of $G$ can be realized as the kernel of a homomorphism, since we merely consider the homomorphism $\phi: G \to G/H$ defined by $\phi(a) = aH$. Then $\ker\phi = H$. This explains our earlier remark that $\ker\phi$ is the prototypical normal subgroup.

When the homomorphism fails to map onto $G'$, there is still a useful result. Recall that the **image** of a function $\phi: G \to G'$ is

$$\text{image}(\phi) = \{y \in G' : y = \phi(a) \text{ for some } a \in G\}.$$

It is easy to check that image $(\phi)$ is a subgroup of $G'$.

**Corollary 3.10.** *If $\phi: G \to G'$ is a group homomorphism, then $G/\ker\phi \cong \text{image}(\phi)$.*

***Proof.*** See Exercise 9.   □

**Example 5.** $\mathbb{R}/\mathbb{Z} \cong S = \{z \in \mathbb{C} : |z| = 1\}$. Consider the homomorphism $\phi: \mathbb{R} \to S$ defined by $\phi(t) = \cos(2\pi t) + i\sin(2\pi t) = \exp(2\pi it)$. Since $\phi(t+u) = \exp(2\pi i(t+u)) = \exp(2\pi it)\exp(2\pi iu) = \phi(t)\phi(u)$, it follows that $\phi$ is a homomorphism. Clearly, $\phi$ maps onto $S$. On the other hand, $\ker\phi = \{t \in \mathbb{R} : \cos(2\pi t) = 1 \text{ and } \sin(2\pi t) = 0\} = \mathbb{Z}$, as required.

**Example 6.** Pursuing Examples 3(c) and (d), we consider the determinant homomorphism det: $GL(n,F) \to F^\times$. This homomorphism maps *onto* $F^\times$ (for example, if $a \in F - \{0\}$, let

$$A = \begin{bmatrix} a & 0 & \cdots & 0 \\ 0 & 1 & \cdots & 0 \\ \vdots & & \ddots & \vdots \\ 0 & & \cdots & 1 \end{bmatrix};$$

then $\det A = a$, as desired). On the other hand, the kernel of det is $SL(n,F)$. So, by Theorem 3.9, the quotient group $GL(n,F)/SL(n,F)$ is isomorphic to $F^\times$.

We give an application of this theory to the analysis of the isometries of $\mathbb{C}$ presented in Section 5 of Chapter 2. See also Exercise 29.

**Theorem 3.11.** *The set of translations forms a normal subgroup $H$ of the group $G$ of isometries of $\mathbb{C}$. The quotient group $G/H$ is isomorphic to the subgroup $G_0$ of isometries fixing the origin.*

***Proof.*** Define $\phi \colon G \to G_0$ by

$$\phi(f)(z) = f(z) - f(0) = (f - f(0))(z).$$

(Note that $\phi(f) \in G_0$, since $\phi(f)(0) = 0$.) To check that $\phi$ is a homomorphism, we must show that $\phi(f \circ g) = \phi(f)\phi(g)$. Well, on one hand, $\phi(f \circ g)(z) = (f \circ g)(z) - (f \circ g)(0)$; on the other hand,

$$(\phi(f)\phi(g))(z) = ((f - f(0)) \circ (g - g(0)))(z)$$

(since $f - f(0)$ is a linear function)

$$\begin{aligned} &= (f - f(0))(g(z)) - (f - f(0))(g(0)) \\ &= (f(g(z)) - f(0)) - (f(g(0)) - f(0)) \\ &= f(g(z)) - f(g(0)) = (f \circ g)(z) - (f \circ g)(0), \end{aligned}$$

as required. Next, if $c \in \mathbb{C}$, let $\tau_c(z) = z + c$. Now, for any $\tau = \tau_c \in H$ and any $z \in \mathbb{C}$, $\phi(\tau)(z) = \tau(z) - \tau(0) = (z + c) - c = z$; so $\tau \in \ker \phi$. Conversely, if $f \in \ker \phi$, then $f(z) - f(0) = z$ for all $z \in \mathbb{C}$; so $f(z) = z + f(0)$ and $f = \tau_{f(0)}$, so $f \in H$. Thus, $\ker \phi = H$, from which it follows that $H$ is a normal subgroup of $G$; and since $\phi$ maps onto $G_0$ (why?), we are done.  $\square$

## EXERCISES 6.3

1.  Let $a, h, h' \in G$. Compute $(aha^{-1})(ah'a^{-1})$, $(aha^{-1})^n$, and $(aha^{-1})^{-1}$.

2.  Use Corollary 3.6 to give another proof of Fermat's little theorem, Proposition 3.3 of Chapter 1. (Hint: In our more up-to-date language, the theorem should be restated as follows: given any prime number $p$, $a^p = a$ for all $a \in \mathbb{Z}_p$.)

3.  Let
    $$G = \left\{ \begin{bmatrix} 1 & 0 \\ 0 & 1 \end{bmatrix}, \begin{bmatrix} 0 & 1 \\ 1 & 0 \end{bmatrix}, \begin{bmatrix} 1 & 0 \\ -1 & -1 \end{bmatrix}, \begin{bmatrix} 0 & 1 \\ -1 & -1 \end{bmatrix}, \begin{bmatrix} -1 & -1 \\ 1 & 0 \end{bmatrix}, \begin{bmatrix} -1 & -1 \\ 0 & 1 \end{bmatrix} \right\},$$
    $$H = \left\{ \begin{bmatrix} 1 & 0 \\ 0 & 1 \end{bmatrix}, \begin{bmatrix} 0 & 1 \\ -1 & -1 \end{bmatrix}, \begin{bmatrix} -1 & -1 \\ 1 & 0 \end{bmatrix} \right\},$$
    $$K = \left\{ \begin{bmatrix} 1 & 0 \\ 0 & 1 \end{bmatrix}, \begin{bmatrix} 0 & 1 \\ 1 & 0 \end{bmatrix} \right\}.$$

    Show that $H$ is a normal subgroup of $G$, but $K$ is not.

4.  Let $S = \{z \in \mathbb{C} : |z| = 1\} \subset \mathbb{C}^\times$. Describe the cosets of $S$ in $\mathbb{C}^\times$, and identify the quotient group $\mathbb{C}^\times / S$.

5.  Prove that if $H$ and $K$ are normal subgroups of $G$, then so is $H \cap K$.

6.  Suppose $H$ is a subgroup of $G$ and $K$ is a normal subgroup of $G$. Prove that $H \cap K$ is a normal subgroup of $H$.

7.  Prove that $Z = \{a \in G : ax = xa \text{ for all } x \in G\}$ is a normal subgroup of $G$. (This is called the **center** of $G$.)

8.  Let $H \subset G$ be a subgroup, and let $a \in G$ be given. Prove that $aHa^{-1} \subset G$ is a subgroup (called a **conjugate subgroup** of $H$). Prove, moreover, that it is isomorphic to $H$ (cf. Exercise 6.2.12).

9.  Verify that if $\phi \colon G \rightarrow G'$ is a group homomorphism, then image $(\phi) \subset G'$ is a subgroup. Prove Corollary 3.10.

10. a.  Prove that any quotient of an abelian group is abelian.
    b.  Prove that any quotient of a cyclic group is cyclic.

11. Prove that a group of order $n$ has a proper subgroup if and only if $n$ is composite.

12. Find all the subgroups of the group $\mathcal{Q}$ (see Example 2(b) of Section 1), and decide which are normal. Find the corresponding quotient groups.

13. Suppose $H, K \subset G$ are subgroups of orders 5 and 8, respectively. Prove that $H \cap K = \{e\}$.

14. Prove or give a counterexample. If $H \subset G$ is the only subgroup of $G$ of order $|H|$, then $H$ is a normal subgroup.

15. Show that every element of the quotient group $G = \mathbb{Q}/\mathbb{Z}$ has finite order. Does $G$ have finite order?

16. Suppose $H$ and $K$ are normal subgroups of $G$ with $H \cap K = \{e\}$. Prove that $hk = kh$ for all $h \in H$ and $k \in K$. (Hint: Consider $hk(kh)^{-1} = hkh^{-1}k^{-1}$.)

17. a.  Prove that a group $G$ of even order has an element of order 2. (Hint: If $a \neq e$, $a$ has order 2 if and only if $a = a^{-1}$.)
    b.  Suppose $m$ is odd, $|G| = 2m$, and $G$ is abelian. Prove $G$ has precisely one element of order 2. (Hint: If there were two,

they would provide a Klein four-group.)

c.  Prove that if $G$ has exactly one element of order 2, then it must be in the center of $G$.

18.  Prove that $H \subset G$ is a normal subgroup if and only if every left coset is a right coset, i.e., $aH = Ha$ for all $a \in G$.

19.  Use Exercise 18 to prove that if $[G : H] = 2$, then $H \subset G$ is a normal subgroup.

20.  Here is an alternative proof of the result of Exercise 19. Suppose $[G : H] = 2$. We wish to show that $H$ is normal. Given $a \in G$, we must show that $aha^{-1} \in H$ for all $h \in H$.

a.  Suppose $a \in H$. Deduce the result.

b.  Suppose $a \notin H$. If $aha^{-1} \notin H$, then show that $aha^{-1} \in aH$, and derive a contradiction.

21.  Suppose $H \subset G$ is a normal subgroup of index $k$. Prove that for any $a \in G$, $a^k \in H$. Does this hold without the normality assumption?

22.  Find all groups $G$ with the property that there is a homomorphism $\phi$ mapping $Sq$ onto $G$. (Hint: What are the normal subgroups of $Sq$?)

23.  Suppose $|G| = 12$.

a.  Suppose $\phi: G \rightarrow \mathbb{Z}_5$ is a homomorphism. What can you conclude immediately about $\phi$?

b.  Suppose $\phi: G \rightarrow \mathbb{Z}_{10}$ is a homomorphism. What can you conclude now?

c.  Suppose the only nontrivial normal subgroup of $G$ is of order 4. What do you then conclude in the case of b.?

24.  Let $H$ be a normal subgroup of $G$ of index $k$. Show that if $a \in G$ has order $n$, then the order of $aH$ in $G/H$ divides both $n$ and $k$. What can you conclude?

25.  Let $p$ be an odd prime. Prove that $x^2 + 1$ has a root in $\mathbb{Z}_p$ if and only if $\mathbb{Z}_p^{\times}$ contains an element of order 4. Now infer that $p \equiv 1 \pmod 4$. This gives another proof of the implication $(3) \Rightarrow (1)$ of Theorem 3.7 of Chapter 4.

26.  a.  Recall that $Z = \{a \in G : ax = xa \text{ for all } x \in G\}$ is the center of $G$. Suppose $G/Z$ is cyclic; prove that $G$ is abelian.

b.  If your proof of a. used just the hypothesis that $G/Z$ is abelian, it is flawed. Give an example of a non-abelian group $G$ with the property that $G/Z$ is abelian.

27.  Let $p$ be a prime. Let $G$ be a group of order $p^2$. Taking it for granted that its center $Z$ cannot consist of $\{e\}$ alone (see Exercise 7.1.17a.), use Exercise 26 to prove that $G$ is abelian. (Hint: $G$ is abelian $\iff Z = G$. If $Z \subsetneq G$, consider the quotient group $G/Z$; what is its order?)

28.  Show that the matrices $\begin{bmatrix} 1 & 0 \\ 1 & 1 \end{bmatrix}$ and $\begin{bmatrix} 1 & 1 \\ 0 & 1 \end{bmatrix}$ are conjugate in $GL(2, \mathbb{R})$, but not in $SL(2, \mathbb{R})$. Is the same true in $GL(2, \mathbb{Z}_2)$?

29.  For $c \in \mathbb{C}$, let $\tau_c(z) = z + c$. Let $H = \{\tau_c : c \in \mathbb{C}\}$ be the set of translations of $\mathbb{C}$. Check directly that $H$ is a normal subgroup of the group of isometries of $\mathbb{C}$. (First, check that $H$ is a subgroup. Then, to prove that $H$ is a normal subgroup, conjugate by a general isometry $f$; by Theorem 5.7 of Chapter 2, either $f(z) = \zeta z + y$ or $f(z) = \zeta \bar{z} + y$, where $|\zeta| = 1$ and $y \in \mathbb{C}$.)

30.  a.  Suppose $K \subset H$ is a normal subgroup and $H \subset G$ is a normal subgroup. Is $K$ a normal subgroup of $G$? (Think about $Sq$ or see Exercise 6.4.9 if you get stuck.)
     b.  Suppose $H \subset G$ is an abelian subgroup. Is $H$ a normal subgroup of $G$?

31.  Consider the repeating decimals $\frac{3}{7} = .428571\overline{428571}\ldots$, $\frac{5}{11} = .45\overline{45}\ldots$, etc. These repeating decimals have period 6 and 2, respectively. In general, let $p$ be any prime other than 2 or 5. We wish to derive an analogous result.
     a.  Prove that the remainders obtained when you do the long division $\frac{1}{p}$ form a subgroup of $\mathbb{Z}_p^{\times}$.
     b.  Deduce that the period of the repeating decimal $\frac{1}{p}$ is a divisor of $p - 1$.
     c.  Deduce the corresponding result for the repeating decimal $\frac{k}{p}$, $1 \le k \le p - 1$.

32.  Let $H \subset K \subset G$ be subgroups. Prove that

$$[G : H] = [G : K][K : H].$$

(When $G$ is infinite, the statement includes the observation that $[G : H]$ is finite if and only if $[G : K]$ and $[K : H]$ are both finite.)

33. Let $p$ be an odd prime.
    a. Show that $\phi: \mathbb{Z}_p^\times \to \mathbb{Z}_p^\times$, $\phi(a) = a^2$, is a group homomor-
       phism whose image is a subgroup $H$ of index 2.
    b. Define $\psi: \mathbb{Z}_p^\times \to \{\pm 1\}$ by
       $$\psi(a) = \begin{cases} +1, & \text{if } a \text{ is a square in } \mathbb{Z}_p \\ -1, & \text{if } a \text{ is not a square in } \mathbb{Z}_p \end{cases}.$$
       Prove $\psi$ is a group homomorphism. (Hint: Consider the
       quotient group $\mathbb{Z}_p^\times / H$.)
    c. Conclude that if neither $a$ nor $b$ is a square in $\mathbb{Z}_p$, then their
       product $ab$ is a square in $\mathbb{Z}_p$. (See Exercise 3.3.10.)

34. Prove that if $|G| = 8$ and $G$ is not abelian, then $G \cong \mathcal{Q}$ or $G \cong \mathcal{D}_4$.
    (Sketch of proof: There must be an element $x$ of order 4 (why?).
    Let $H = \langle x \rangle$ and choose $y \notin H$. Show that $yxy^{-1} = x^3$. Now
    what can the order of $y$ be?)

35. Prove that if $G$ is a finite abelian group of order $n$ and a prime
    number $p | n$, then $G$ contains an element of order $p$. (Remark:
    This is a case where the converse of Lagrange's Theorem does
    hold.) (Hint: Proceed by complete induction on $n$, by passing
    to a *quotient* group of $G$.)

## 4. The Symmetric Group $S_n$ and the 15-Puzzle

We now turn to a more detailed study of the symmetric group
$S_n = \mathrm{Perm}\{1, \ldots, n\}$. In Section 1, we introduced the rather cumber-
some notation $\begin{pmatrix} 1 & 2 & 3 & \cdots & n-1 & n \\ i_1 & i_2 & i_3 & \cdots & i_{n-1} & i_n \end{pmatrix}$ for permutations, and we in-
tend first to find a more manageable one. The basic idea is to break
a permutation into its "indecomposable" pieces. When $k \geq 2$, we
call $\pi \in S_n$ a $k$-**cycle** if there are distinct integers $1 \leq i_1, \ldots, i_k \leq n$
so that
$$\pi(i_1) = i_2,$$
$$\pi(i_2) = i_3,$$
$$\vdots$$
$$\pi(i_{k-1}) = i_k,$$
$$\pi(i_k) = i_1, \quad \text{and}$$
$$\pi(i) = i, \quad \text{otherwise}.$$

Casually, we say that $\pi$ permutes the elements $i_1, i_2, \ldots, i_k$ cyclically. As a convincing example, consider the rotation of a regular $k$-gon through angle $\frac{2\pi}{k}$. We then denote such a $k$-cycle by the simpler notation

$$\pi = (i_1 \; i_2 \; i_3 \; \ldots \; i_{k-1} \; i_k).$$

Note, however, that such a representation is not unique; for example, $(i \; j \; \ell) = (j \; \ell \; i) = (\ell \; i \; j)$. We call a 2-cycle a **transposition** (since it transposes the two elements $i_1$ and $i_2$).

**Example 1.** Consider $\pi \in S_6$ given by $\pi = \begin{pmatrix} 1 & 2 & 3 & 4 & 5 & 6 \\ 1 & 5 & 6 & 2 & 3 & 4 \end{pmatrix}$. Then $\pi$ is the 5-cycle $(2 \; 5 \; 3 \; 6 \; 4)$.

**Example 2.** Multiply the two permutations $\sigma = (2 \; 5 \; 3)$ and $\tau = (1 \; 3 \; 4 \; 2) \in S_5$. To compute $\sigma\tau = \sigma \circ \tau$, remembering to read right to left, we must track down the eventual progress of each element. For example, $\tau(1) = 3$ and $\sigma(3) = 2$, so $\sigma\tau(1) = 2$. On the other hand, $\sigma\tau(3) = \sigma(\tau(3)) = \sigma(4) = 4$. The end result is the following:

$$\sigma\tau = \begin{pmatrix} 1 & 2 & 3 & 4 & 5 \\ 2 & 1 & 4 & 5 & 3 \end{pmatrix}.$$

To write this in our new notation, we start with 1 and follow its tracks: $1 \to 2 \to 1$, so we have a 2-cycle $(1 \; 2)$; following 3 next, $3 \to 4 \to 5 \to 3$, and so we have a 3-cycle $(3 \; 4 \; 5)$. So

$$\sigma\tau = \begin{pmatrix} 1 & 2 & 3 & 4 & 5 \\ 2 & 1 & 4 & 5 & 3 \end{pmatrix} = (1 \; 2)(3 \; 4 \; 5).$$

On the other hand, we leave it to the reader to check (see Exercise 1) that if we let $\tau = (1 \; 2 \; 4)$, then $\sigma\tau$ *is* a 5-cycle.

When the elements of two cycles are disjoint (i.e., have no integer in common), the cycles act independently and commute. We now can begin to come to grips with general permutations.

**Proposition 4.1.** *Every permutation $\pi \in S_n$, other than the identity element, can be written uniquely (up to order) as the product of disjoint cycles.*

**Proof.** Proceeding as in Example 2, we compute $\pi(1)$, $\pi^2(1) = \pi(\pi(1))$, $\pi^3(1) = \pi(\pi(\pi(1)))$, $\ldots$. Since $\{1, \ldots, n\}$ is finite, there is some smallest positive integer $k$ so that $\pi^k(1) = 1$. This yields a $k$-cycle $(1 \; \pi(1) \; \pi^2(1) \; \ldots \; \pi^{k-1}(1))$. (Of course, if $k = 1$, then we don't bother to write it down.) We now remove the numbers $1, \pi(1), \ldots, \pi^{k-1}(1)$ from consideration and continue the process. (The fastidious reader may write down a proof by complete induction.) $\quad\square$

**Proposition 4.2.** *Every permutation (other than the identity element) can be written as the product of transpositions.*

**Proof.** By virtue of Proposition 4.1, it suffices to prove that every $k$-cycle can be so written. We just do it with bare hands:

$$(1\ 2\ 3\ \ldots\ k) = (1\ k)(1\ k-1)\ldots(1\ 3)(1\ 2).$$

Obviously, the same construction works for $(i_1\ i_2 \ldots\ i_k)$.  $\square$

Although we sacrifice a good deal in proceeding from the product of disjoint cycles to the product of transpositions (for example, uniqueness and commutativity), we gain one piece of information. For example, $(1\,2\,3) = (1\,3)(1\,2) = (2\,3)(1\,3) = (1\,3)(3\,2)(2\,3)(1\,2)$. When we write a permutation $\sigma$ as a product of transpositions, even though the number of terms in such a product is free to vary, experiments show that the parity is not; that is, given $\sigma$, it cannot be written as the product of both an even number and an odd number of transpositions. We now turn to a proof of this fact.

**Lemma 4.3.** *The identity permutation $\iota$ cannot be expressed as the product of an odd number of transpositions.*

**Proof.** Among all possible representations $\iota = \tau_1\tau_2 \cdots \tau_{k-1}\tau_k$, where each $\tau_i$ is a transposition and $k$ is odd, choose one with the property that $k$ is as small as possible. (Note that $k \geq 3$, since a single transposition is not the identity.) Suppose $\tau_k = (\alpha\ \beta)$ where $\alpha < \beta$; among all such representations of $\iota$, choose the one with the fewest possible appearances of the integer $\alpha$. We will now proceed to arrive at a contradiction.

Since $\iota(\alpha) = \alpha$, there must be another transposition $\tau_j$ that moves $\alpha$; choose the largest such $j$, and suppose $\tau_j = (\alpha\ \gamma)$. Now, if no $\gamma$ appears among $\tau_{j+1}, \ldots, \tau_{k-1}$, then $\tau_j$ commutes with all of these transpositions and we move $\tau_j$ to the $(k-1)^{\text{th}}$ position. If one or more $\gamma$'s appears among $\tau_{j+1}, \ldots, \tau_{k-1}$, then we may use the relation $(\alpha\ \gamma)(\gamma\ \delta) = (\gamma\ \delta)(\alpha\ \delta)$ to move $\tau_j$ to the $(k-1)^{\text{th}}$ position, where it will now have the form $\tau_{k-1} = (\alpha\ \epsilon)$ for some $\epsilon \in \{1, \ldots, n\}$. In any event, we've not changed either $k$ or the (minimal) number of $\alpha$'s in any of these manipulations. Consider now the product $\tau_{k-1}\tau_k = (\alpha\ \epsilon)(\alpha\ \beta)$. If $\epsilon = \beta$, then $\tau_{k-1}\tau_k = (\alpha\ \beta)^2 = \iota$; and we may delete both $\tau_{k-1}$ and $\tau_k$, contradicting the assumption that $k$ was as small as possible. On the other hand, if $\epsilon \neq \beta$, then we use the relation $(\alpha\ \epsilon)(\alpha\ \beta) = (\beta\ \epsilon)(\alpha\ \epsilon)$ to obtain a representation of $\iota$ with fewer $\alpha$'s, contradicting the assumption that we started with the fewest possible. The inevitable contradiction completes the proof that $\iota$ cannot be written as the product of an odd number of transpositions. □

**Proposition 4.4.** *Every permutation can be written as the product of either an even number or an odd number of transpositions, but not both.*

**Proof.** Suppose $\pi = \tau_1 \tau_2 \cdots \tau_k$ and $\pi = \tau_1' \tau_2' \cdots \tau_\ell'$. Then $\iota = (\tau_1 \tau_2 \cdots \tau_k)(\tau_1' \tau_2' \cdots \tau_\ell')^{-1} = \tau_1 \tau_2 \cdots \tau_k \tau_\ell' \cdots \tau_2' \tau_1'$, since every transposition is its own inverse. It follows from Lemma 4.3 that $k + \ell$ must be even, whence $k$ and $\ell$ are either both even or both odd. □

**Definition.** We say a permutation is **even** (resp., **odd**) if it can be written as the product of an even (resp., odd) number of transpositions.

Note that the identity element is an even permutation, and that, moreover, the set of even permutations forms a subgroup of $S_n$ (why?), called the **alternating group** $A_n \subset S_n$.

**Proposition 4.5.** $A_n$ *is a normal subgroup of* $S_n$ *of index* 2.

**Proof.** Consider the homomorphism $\phi \colon S_n \to \{+1, -1\}$ given by $\phi(\pi) = +1$ if $\pi$ is an even permutation and $\phi(\pi) = -1$ if $\pi$ is an odd permutation. Basic rules of arithmetic imply that $\phi$ is a homomorphism, and, by definition, $\ker \phi = A_n$. From the Fundamental Homomorphism Theorem, Theorem 3.9, we now infer that $S_n/A_n \cong \{+1, -1\}$, and so $[S_n : A_n] = 2$. □

One of the beautiful rewards of the cycle notation for working with permutations is that we can easily compute the effect of conjugation in $S_n$.

**Lemma 4.6.** *Let* $\pi \in S_n$ *be a* $k$-*cycle*, $\pi = (i_1\ i_2\ \ldots\ i_k)$. *Then for any* $\sigma \in S_n$, *we have* $\sigma \pi \sigma^{-1} = (\sigma(i_1)\ \sigma(i_2)\ \ldots\ \sigma(i_k))$.

**Proof.** We must merely compute: for $1 \le \ell \le k - 1$, we have $\sigma \pi \sigma^{-1}(\sigma(i_\ell)) = \sigma \pi(i_\ell) = \sigma(i_{\ell+1})$. And $\sigma \pi \sigma^{-1}(\sigma(i_k)) = \sigma \pi(i_k) = \sigma(i_1)$. □

In particular, conjugation preserves the "cycle structure" of a permutation. Using Proposition 4.1 to write a permutation as the product of disjoint cycles, $\pi = \pi_1 \pi_2 \cdots \pi_s$, we have

$$\sigma \pi \sigma^{-1} = \sigma(\pi_1 \pi_2 \cdots \pi_s)\sigma^{-1} = (\sigma \pi_1 \sigma^{-1})(\sigma \pi_2 \sigma^{-1}) \cdots (\sigma \pi_s \sigma^{-1}).$$

Now, the cycle $\sigma \pi_i \sigma^{-1}$ has the same length as $\pi_i$; and it follows from Lemma 4.6 that if the $\pi_i$'s are disjoint cycles, then so are the $\sigma \pi_i \sigma^{-1}$'s. Indeed, if we denote by the cycle structure of a permutation the numbers of cycles of various lengths in its decomposition as a product of disjoint cycles, then two permutations are conjugate if and only if they have identical cycle structures (see Exercise 6).

**Example 3.** Let $\pi = (1\ 5)(2\ 4\ 7\ 3)$ and $\sigma = (3\ 5\ 6\ 2)(1\ 4\ 7) \in S_7$. Then $\sigma \pi \sigma^{-1} = (4\ 6)(3\ 7\ 1\ 5)$.

We have already seen that every permutation is the product of transpositions. It is amusing to ask what permutations can be expressed as the product of 3-cycles. Since a 3-cycle is an even permutation, clearly all such permutations will have to be even. Conversely, we have the following lemma.

**Lemma 4.7.** *Let $n \geq 3$. Every element of $A_n$ can be written as the product of 3-cycles.*

**Proof.** Since we've already proved that every permutation is the product of transpositions, it will suffice to prove that the product of *two* transpositions can be written as the product of 3-cycles. We just check this case by case:

(i) $(\alpha\ \beta)(\gamma\ \delta) = (\alpha\ \beta\ \gamma)(\beta\ \gamma\ \delta)$, when $\alpha$, $\beta$, $\gamma$, and $\delta$ are distinct;

(ii) $(\alpha\ \beta)(\alpha\ \gamma) = (\alpha\ \gamma\ \beta)$, when $\alpha$, $\beta$, and $\gamma$ are distinct; and

(iii) $(\alpha\ \beta)(\alpha\ \beta) = \iota$.

Therefore, any even permutation is the product of 3-cycles.  □

We now use the theory of the symmetric and alternating groups now to settle the age-old problem of the 15-puzzle. (We are gratefully borrowing the presentation of this material from the book *Introduction to Modern Algebra* by McCoy and Janusz.) The 15-puzzle is a $4 \times 4$ frame holding squares numbered from 1 to 15, with one space left empty. The squares may be moved horizontally or vertically into the empty space, but all moves are confined to the plane. We shall call the process of moving one square into the empty space (thereby swapping the positions of the square and the empty space) a simple move. The puzzle is usually posed as follows: given an initial position, say Position I below, can we arrive—after a finite number of simple moves—at the "standard position"?

Standard Position

| 1  | 2  | 3  | 4  |
|----|----|----|----|
| 5  | 6  | 7  | 8  |
| 9  | 10 | 11 | 12 |
| 13 | 14 | 15 |    |

Position I

| 12 | 5  | 10 | 2  |
|----|----|----|----|
| 1  | 15 | 6  | 3  |
| 4  | 9  | 13 | 8  |
| 7  |    | 14 | 11 |

If we number the squares by the locations of the numbers in the standard position, letting 16 represent the empty space, then we see that new arrangements of the squares correspond uniquely to permutations of the numbers 1 through 16, i.e., to elements of $S_{16}$. For example, starting at the standard position, the result of the permutation (15 16) is to swap the empty space and the square marked 15. Starting at Position I, the effect of the permutation (9 10 14) is to permute cyclically the empty space, square 4, and square 9. We now focus on those positions having the empty space still in square 16.

**Theorem 4.8.** *Let $H \subset S_{16}$ correspond to those arrangements that can be obtained from the starting position by a sequence of simple moves, ending with the empty space in square 16. Then $H$ is a subgroup of $S_{16}$ and is isomorphic to $A_{15}$.*

**Proof.** The proof proceeds in two steps: first, to show $H$ is a subgroup of $A_{15}$; second, to show $H$ contains $A_{15}$.

Step 1. The simple moves are in one-to-one correspondence with transpositions $(\alpha\ \beta)$, where the empty space is in square $\alpha$, say, before the move occurs. If we perform a sequence of simple moves, starting and ending with the empty space in square 16, then such a permutation is of the form

$$(*) \qquad \pi = (16\ i_k)(i_k\ i_{k-1}) \cdots (i_2\ i_1)(i_1\ 16),$$

where each pair $i_j, i_{j-1}$ of integers corresponds to touching squares. From this description it is clear that $H$ is a subgroup of $S_{15}$. But we claim that $\pi$ is the product of an even number of transpositions and therefore is an element of $A_{15}$. Each transposition appearing in $(*)$ corresponds to one move of the empty space, either left, right, up, or down. From the fact that the empty space must return to its original position, we infer that the net number of moves is even. Letting $S_{15} \subset S_{16}$ denote those permutations fixing 16, we now see that $\pi \in S_{15} \cap A_{16} = A_{15}$, as required.

Step 2. We will prove $H \supset A_{15}$ by applying Lemma 4.7: we show that every 3-cycle belongs to $H$. First, note that four particular elements $\alpha$, $\beta$, $\gamma$, and $\delta$ belong to $H$:

$$\alpha = (16\ 15)(15\ 14)(14\ 13)(13\ 9)(9\ 5)(5\ 1)(1\ 2)(2\ 3)(3\ 4)(4\ 8)\cdot$$
$$\cdot(8\ 12)(12\ 16) = (1\ 2\ 3\ 4\ 8\ 12\ 15\ 14\ 13\ 9\ 5)$$
$$\beta = (16\ 15)(15\ 14)(14\ 10)(10\ 6)(6\ 7)(7\ 8)(8\ 12)(12\ 16)$$
$$= (6\ 7\ 8\ 12\ 15\ 14\ 10)$$
$$\gamma = (16\ 15)(15\ 11)(11\ 12)(12\ 16) = (11\ 12\ 15)$$
$$\delta = (16\ 15)(15\ 14)(14\ 10)(10\ 6)(6\ 2)(2\ 3)(3\ 4)(4\ 8)(8\ 12)(12\ 16)$$
$$= (4\ 8\ 12\ 15\ 14\ 10\ 6\ 2\ 3).$$

These arrangements look like this:

$\alpha$

| 5 | 1 | 2 | 3 |
|----|----|----|----|
| 9 | 6 | 7 | 4 |
| 13 | 10 | 11 | 8 |
| 14 | 15 | 12 | |

$\beta$

| 1 | 2 | 3 | 4 |
|----|----|----|----|
| 5 | 10 | 6 | 7 |
| 9 | 14 | 11 | 8 |
| 13 | 15 | 12 | |

$\gamma$

| 1 | 2 | 3 | 4 |
|----|----|----|----|
| 5 | 6 | 7 | 8 |
| 9 | 10 | 15 | 11 |
| 13 | 14 | 12 | |

$\delta$

| 1 | 6 | 2 | 3 |
|----|----|----|----|
| 5 | 10 | 7 | 4 |
| 9 | 14 | 11 | 8 |
| 13 | 15 | 12 | |

The entire argument is based on the fact that the 3-cycle $\gamma$ is an element of $H$. Whenever $\sigma \in H$, by applying Lemma 4.6, we find that $\sigma\gamma\sigma^{-1} = (\sigma(11)\ \sigma(12)\ \sigma(15)) \in H$ as well. Using $\sigma = \beta^5 = (12\ 7\ 10\ 15\ 8\ 6\ 14)$, we infer that $(11\ 7\ 8) \in H$. Since $\alpha$ and $\delta$ both fix 11 and 7, it follows that for any $j$ and $\sigma = \alpha^j$ or $\delta^j$, $\sigma(11\ 7\ 8)\sigma^{-1} = (11\ 7\ \sigma(8))$. Here is the neat observation: for any $x$ between 1 and 15, other than 7 and 11, we can write $x$ as $\alpha^j(8)$ or $\delta^j(8)$ for appropriate values of $j$. Thus, every 3-cycle of the form $(11\ 7\ x)$ belongs to $H$.

Now we conjugate one such element by another: $(11\ 7\ x)(11\ 7\ y)$

$(11\ 7\ x)^{-1} = (7\ x\ y) \in H$ as well. Playing this game one more time, we obtain $(x\ y\ z) = (11\ 7\ z)(7\ x\ y)(11\ 7\ z)^{-1} \in H$; and so, an arbitrary 3-cycle belongs to $H$, as desired.  □

From the proof we've just given, we can deduce the following, more general result.

**Proposition 4.9.** *If $n \geq 4$, $H$ is a normal subgroup of $A_n$ and $H$ contains one 3-cycle, then $H = A_n$.*

**Proof.** See Exercise 16a.  □

**Example 4.** In order to apply Theorem 4.8 to an arbitrary arrangement of the 15-puzzle, we must first by hand move the empty space to square 16. Let's consider, in particular, Position I above. With two (obvious) transpositions, we arrive at the following:

Position I'

| 12 | 5  | 10 | 2 |
|----|----|----|---|
| 1  | 15 | 6  | 3 |
| 4  | 9  | 13 | 8 |
| 7  | 14 | 11 |   |

Now we must decide whether the permutation

$$\pi = \begin{pmatrix} 1 & 2 & 3 & 4 & 5 & 6 & 7 & 8 & 9 & 10 & 11 & 12 & 13 & 14 & 15 \\ 5 & 4 & 8 & 9 & 2 & 7 & 13 & 12 & 10 & 3 & 15 & 1 & 11 & 14 & 6 \end{pmatrix}$$

is an element of $A_{15}$. Well, $\pi = (1\ 5\ 2\ 4\ 9\ 10\ 3\ 8\ 12)(6\ 7\ 13\ 11\ 15)$ is the product of a 9-cycle and a 5-cycle, and is therefore an even permutation. We invite the reader to put the square back into standard position!

With only a bit more work, we can prove one of the fundamental results in elementary group theory, one that will re-emerge in a different context in Section 2 of Chapter 7.

**Theorem 4.10.** *Let $n \geq 5$. The alternating group $A_n$ has no proper normal subgroup.*

**Sketch of proof.** Let $H \subset A_n$ be a normal subgroup containing more than the identity element. If we can find a single 3-cycle in $H$, then it will follow from Proposition 4.9 that $H = A_n$. Let $\pi \in H$, $\pi \neq \iota$, and write $\pi = \pi_1 \pi_2 \cdots \pi_s$ as a product of disjoint cycles.

   Case 1. Let $k \geq 4$, and suppose that some factor, say $\pi_1$, of $\pi$ is a $k$-cycle. We may as well assume that $\pi_1 = (1\ 2\ \ldots\ k)$. Since $H$ is normal, $(1\ 2\ 3)\pi(1\ 2\ 3)^{-1} \in H$, and $(1\ 2\ 3)$ commutes with all the factors of $\pi$ except $\pi_1$ (since we represented $\pi$ as a product of disjoint cycles). Thus, $\sigma = (1\ 2\ 3)\pi(1\ 2\ 3)^{-1} = (2\ 3\ 1\ 4\ \ldots\ k)\pi_2 \cdots \pi_s$. Then $\sigma\pi^{-1} \in H$, but $\sigma\pi^{-1} = (2\ 3\ 1\ 4\ \ldots\ k)(1\ 2\ 3\ \ldots\ k)^{-1} = (2\ 3\ 1\ 4\ \ldots\ k)(k\ \ldots\ 4\ 3\ 2\ 1) = (1\ 2\ 4)$.

   Case 2. Suppose $\pi$ has at least two 3-cycles as factors, say $\pi_1 = (1\ 2\ 3)$ and $\pi_2 = (4\ 5\ 6)$. Then $\sigma = (3\ 4\ 5)\pi(3\ 4\ 5)^{-1} = (1\ 2\ 4)(3\ 6\ 5) \cdot \pi_3 \cdots \pi_s \in H$; but then $\sigma\pi^{-1} = (1\ 2\ 4)(3\ 6\ 5)(4\ 5\ 6)^{-1}(1\ 2\ 3)^{-1} = (1\ 6\ 3\ 4\ 5) \in H$. Since $H$ contains a 5-cycle, we are done, by Case 1.

   Case 3. Suppose $\pi$ has precisely one 3-cycle as a factor, and all the other factors are transpositions. If the 3-cycle is $\pi_1 = (1\ 2\ 3)$, say, then $\pi^2 = (1\ 2\ 3)^2 = (1\ 3\ 2)$ is a 3-cycle.

   Case 4. Suppose $\pi$ is the product of disjoint transpositions. Say $\pi_1 = (1\ 2)$ and $\pi_2 = (3\ 4)$. Then, as before, let $\sigma = (1\ 2\ 4)\pi(1\ 2\ 4)^{-1}$, and compute $\sigma\pi^{-1} = (1\ 4)(2\ 3) \in H$. Since $n \geq 5$, we may now use the permutation $\tau = (2\ 3\ 5)$: $\tau(1\ 4)(2\ 3)\tau^{-1} = (1\ 4)(3\ 5) \in H$; and so $(1\ 4)(3\ 5)(1\ 4)(2\ 3) = (2\ 5\ 3) \in H$. This last case completes the proof. $\square$

## EXERCISES 6.4

1. Let $\sigma = (2\ 5\ 3)$, $\tau = (1\ 2\ 4) \in S_5$. Compute the following:
   a.  $\sigma\tau$
   b.  $\tau\sigma$
   c.  $\sigma\tau\sigma^{-1}$
   d.  $\tau\sigma\tau^{-1}$

2. A perfect shuffle of a deck of $2n$ cards is the permutation

$$\begin{pmatrix} 1 & 2 & 3 & \ldots & n & n+1 & n+2 & \ldots & 2n \\ 2 & 4 & 6 & \ldots & 2n & 1 & 3 & \ldots & 2n-1 \end{pmatrix}.$$

   a.  What is the fewest perfect shuffles required with a deck of 6 cards to return the cards to their original position?
   b.  What is the fewest perfect shuffles required with a deck of 10 cards to return the cards to their original position?
   c.  What is the fewest perfect shuffles required with a deck of 20 cards to return the cards to their original position?

    d.   What is the fewest perfect shuffles required with a deck of 50 cards to return the cards to their original position?

    e.   What is the fewest perfect shuffles required with a deck of 52 cards to return the cards to their original position?

3.   a.   Prove that a transposition is its own inverse.

    b.   Prove that a $k$-cycle is its own inverse if and only if $k = 2$.

    c.   When we express a permutation $\sigma \in S_n$ as a product of disjoint cycles, what is a necessary and sufficient condition for $\sigma$ to be its own inverse?

4.   a.   Prove that a $k$-cycle in $S_n$ is an element of order $k$.

    b.   Prove that when we represent a permutation as a product of disjoint cycles, the order of the permutation is the least common multiple of the lengths of these cycles.

5.   Prove that $A_n$ contains an $n$-cycle if and only if $n$ is odd.

6.   Prove that two permutations are conjugate if and only if they have the same cycle structure.

7.   Prove that every element of $A_n$ can be written as the product of an appropriate number of the 3-cycles (1 2 3), (1 2 4), ..., (1 2 $n$).

8.   Prove or disprove:

    a.   (1 2 3) and (3 4 5) are conjugate in $A_5$.

    b.   (1 2 3) and (1 3 2) are conjugate in $A_5$.

    c.   (1 2 3 4 5) and (1 2 4 3 5) are conjugate in $A_5$.

9.   Show that $A_4$ has a normal subgroup $H$ of order 4, and that $H \cong \mathcal{V}$. Show, moreover, that $H$ has a subgroup $K$ of order 2, which is necessarily normal in $H$ (why?). Is $K$ normal in $A_4$?

10.  a.   Prove that every element of $S_n$ can be written as the product of an appropriate number of the transpositions (1 2), (2 3), (3 4), ..., ($n-1$ $n$).

    b.   Prove that if a subgroup $H \subset S_n$ contains the transposition (1 2) and the $n$-cycle (1 2 3 ... $n$), then $H = S_n$.

    c.   Prove that when $p$ is prime, if a subgroup $H \subset S_p$ contains a transposition and a $p$-cycle, then $H = S_p$.

    d.   Prove or give a counterexample: If a subgroup $H \subset S_n$ contains a transposition and an $n$-cycle, then $H = S_n$.

11. Decide whether each of the following positions can be obtained from the standard position by a sequence of simple moves:

Position II

| 15 | 14 | 13 | 12 |
|----|----|----|----|
| 11 | 10 | 9  | 8  |
| 7  | 6  | 5  | 4  |
| 3  | 2  | 1  |    |

Position III

| 12 | 14 | 5  | 7  |
|----|----|----|----|
| 10 |    | 3  | 9  |
| 8  | 2  | 1  | 11 |
| 6  | 4  | 15 | 13 |

Position IV

| 8  | 10 | 7  | 9  |
|----|----|----|----|
| 6  | 12 | 5  | 11 |
| 4  | 14 | 3  | 13 |
| 2  |    | 1  | 15 |

Position V

| 1  | 2  | 3  | 4  |
|----|----|----|----|
| 8  | 7  | 6  | 5  |
| 9  | 10 | 11 | 12 |
|    | 15 | 14 | 13 |

12. Show that $A_4$ has no subgroup of order 6. (Hint: First show that any such subgroup would have to contain a 3-cycle.)

13. Examine all parts of the proof of Theorem 4.10 carefully to see what goes wrong when $n = 4$. Can you find the normal subgroups of $A_4$ using this proof?

14. Prove that for $n \geq 5$, the only proper normal subgroup of $S_n$ is $A_n$. (Hint: see Exercise 6.3.6.)

15. If $\phi: S_n \to S_n$ is a group homomorphism, prove that $\phi(A_n) \subset A_n$. (Hint: Use Lemma 4.7.)

16. a. Prove Proposition 4.9. (Hints: Show first that when $n \geq 5$, then $(1\ 2\ 3) \in H \implies (1\ 2\ z) \in H$ for all $z \neq 1, 2 \implies (1\ y\ z) \in H$ for all $y, z \neq 1$ and distinct $\implies (x\ y\ z) \in H$ for all $x, y, z$ distinct. When $n = 4$, show that $(1\ 2\ 3) \in H \implies (1\ 3\ 2) \in H \implies (1\ 2\ 4) \in H$, and now deduce that every 3-cycle is in $H$.)

    b. Prove that if $H \subset S_n$ is a normal subgroup and $H$ contains one 3-cycle, then $H = A_n$ or $H = S_n$.

17. Let $p$ be prime. Count the distinct subgroups of order $p$ in $S_p$. (Hint: First count the $p$-cycles, and then figure out how many

times each subgroup is represented.)

18.    Here we pursue (with thanks to Will Kazez) the investigation of the perfect shuffle defined in Exercise 2. Let

$$\sigma = \begin{pmatrix} 1 & 2 & 3 & \ldots & n & n+1 & n+2 & \ldots & 2n \\ 2 & 4 & 6 & \ldots & 2n & 1 & 3 & \ldots & 2n-1 \end{pmatrix} \in S_{2n}.$$

   a.    Show that $\sigma(k) \equiv 2k \pmod{2n+1}$ for all $k = 1, 2, \ldots, 2n$.
   b.    Let $v$ be the smallest positive integer so that $2^v = 1$ in $\mathbb{Z}_{2n+1}$, i.e., the order of $2 \in \mathbb{Z}_{2n+1}^\times$. Show that $v$ is the length of the cycle $(1\ 2\ 4\ \ldots)$ in the disjoint cycle decomposition of $\sigma$ given by Proposition 4.1. (Hint: What is $\sigma^j(1)$?)
   c.    Prove that for any $k = 1, 2, \ldots, 2n$, the length of the cycle $(k\ \sigma(k)\ \sigma^2(k)\ \ldots)$ is a divisor of $v$.
   d.    Conclude that the order of $\sigma \in S_{2n}$ is $v$. Compare this with your answers to Exercise 2 (parts c. and d. especially).

19.    Here is an alternative definition of even and odd permutations not relying on Lemma 4.3 and Proposition 4.4. To each $\sigma \in S_n$ we associate a matrix $P_\sigma \in GL(n, \mathbb{R})$ given by

$$(P_\sigma)_{ij} = \begin{cases} 1, & i = \sigma(j) \\ 0, & \text{otherwise} \end{cases}.$$

   a.    Check that $\pi: S_n \to GL(n, \mathbb{R})$, $\pi(\sigma) = P_\sigma$, is a group homomorphism (called the **permutation representation**).
   b.    Prove that if $\tau$ is a transposition, then $\det P_\tau = -1$ (see the discussion following Corollary 2.1 of Appendix B).
   c.    Define $\text{sign}(\sigma) = \det P_\sigma$. Check that $\sigma$ is an even permutation if $\text{sign}(\sigma) = +1$ and an odd permutation if $\text{sign}(\sigma) = -1$. Now reinterpret the proof of Proposition 4.5.

20.    Develop a theory of the 8-puzzle, and decide what arrangements can be reached by a sequence of simple moves from the obvious starting position.

21.    Count the elements of $S_n$ having no fixed point. (As an amusing application, given $n$ letters, when each is placed at random in one of $n$ envelopes, determine the probability that none is placed in its matching envelope. What is particularly interesting is the limiting value of this probability as $n \to \infty$.)

# CHAPTER 7

# Group Actions and Symmetry

In this chapter the theme is symmetry, of both geometric and algebraic objects. When a group acts on a set, it partitions the set into orbits (analogous to the partition of a group itself into the cosets of a subgroup). We obtain a variety of results just by counting: How many orbits are there and how many elements are in the different orbits? We determine in §2 the symmetry groups of the Platonic solids (which have fascinated mathematicians, philosophers, and even astronomers for centuries), and in §3 give a classic combinatorial application credited to Burnside and Pólya. Further geometric applications appear in §4, first generalizing the classification of isometries of the Euclidean plane (presented in Chapter 2, §5) to three-dimensional space, and then briefly using the counting arguments to find all possible finite subgroups of the group of rotations of three-space. In §5 we give the standard applications to finite groups, starting with $p$-groups and ending with the Sylow Theorems, which in principle allow one to determine the number of nonisomorphic groups of a given order.

Lastly, in §6 we return to the problem solved by Abel and Galois in the early nineteenth century: gaining an understanding of splitting fields of polynomials by studying their symmetry groups. In particular, we conclude the chapter with one of the crowning achievements of mathematics. We saw in Chapter 2, §4, the classic formulas for roots of quadratic and cubic polynomials; although we have not presented it, there is also a formula for quartic (degree 4) polynomials. However, the fact that there is no explicit formula for solving a quintic (degree 5) polynomial is, as we shall see, intimately related to the group of isometries of the icosahedron (which the student will analyze in detail in the exercises in §2).

## 1. Group Actions on a Set

We have seen that groups arise in our studies of symmetries of objects. We wish now to pursue this theme, in both geometric and algebraic directions. The **action** of a group $G$ on a set $S$ is given by a homomorphism

$$\phi: G \to \text{Perm}(S),$$

where we normally write $\phi(g)(s) = g \cdot s$ for $g \in G$ and $s \in S$. Note that since $\phi$ is a homomorphism, we have $\phi(e) = \iota$, and so $e \cdot s = s$ for all $s \in S$; also, $g \cdot (g' \cdot s) = (gg') \cdot s$ for all $g, g' \in G$ and $s \in S$.

**Examples 1.**

(a) $G$ acts on $G$ itself by left multiplication: $g \cdot x = gx$, for $g, x \in G$. This is an action because of the associative law in $G$: $(gg') \cdot x = (gg')x = g(g'x) = g \cdot (g' \cdot x)$.

(b) $G$ acts on $G$ itself by conjugation: $g \cdot x = gxg^{-1}$, $g, x \in G$. Here $\phi: G \to \text{Perm}(G)$ is given by $\phi(g) = \psi_g$, where $\psi_g(x) = gxg^{-1}$. Note that $\psi_{gg'}(x) = (gg')x(gg')^{-1} = g(g'xg'^{-1})g^{-1} = \psi_g(\psi_{g'}(x)) = (\psi_g \circ \psi_{g'})(x)$, as required.

(c) Given a subgroup $H \subset G$, $G$ acts on $S = G/H$, the set of cosets of $H$, by left multiplication. Here $\phi: G \to \text{Perm}(G/H)$ is defined by $\phi(g)(aH) = (ga)H$.

(d) The group $G$ of isometries of $\mathbb{C}$ acts on $\mathbb{C}$; the isometries comprise a (geometrically meaningful) subgroup of $\text{Perm}(\mathbb{C})$.

(e) Given a polynomial $f(x) \in F[x]$, let $K$ be the splitting field of $f(x)$, and let $S = \{\alpha_1, \ldots, \alpha_n\} \subset K$ be the set of roots of $f(x)$. Let $G$ (the so-called Galois group of the polynomial $f(x)$) be the group of all isomorphisms of $K$ with itself leaving $F$ fixed; these are called $F$-automorphisms of $K$. Then $G$ acts on $S$ (see Exercise 4.2.10), and, in this manner, can be viewed as a subgroup of $S_n$. Here are a few explicit examples:

   (i) Let $f(x) = x^2 - 2 \in \mathbb{Q}[x]$. Then $K = \mathbb{Q}[\sqrt{2}]$ and $G = \{\iota, \psi\}$, where $\psi(a + b\sqrt{2}) = a - b\sqrt{2}$. Note that $\psi$ interchanges the two roots of $f(x)$ in $K$.

   (ii) Let $f(x) = x^3 - 2 \in \mathbb{Q}[x]$. Then $K = \mathbb{Q}[\sqrt[3]{2}, \omega]$, where $\omega = -\frac{1}{2} + \frac{\sqrt{3}}{2}i$ is a primitive cube root of unity. Two types of $\mathbb{Q}$-automorphisms are suggested by the geometry: reflection in the real axis ($\psi$), and rotation of the roots through $120°$ ($\phi$). These are determined by the respective rules:

$$\psi(\sqrt[3]{2}) = \sqrt[3]{2} \qquad \phi(\sqrt[3]{2}) = \sqrt[3]{2}\omega$$
$$\psi(\omega) = \overline{\omega} \qquad \phi(\omega) = \omega.$$

Then note that these $\mathbb{Q}$-automorphisms do act on the roots of $f(x)$:

| root | $\psi$(root) | $\phi$(root) |
|------|--------------|--------------|
| $\alpha_1$ | $\alpha_1$ | $\alpha_2$ |
| $\alpha_2$ | $\alpha_3$ | $\alpha_3$ |
| $\alpha_3$ | $\alpha_2$ | $\alpha_1$ |

FIGURE 1

It follows that the group $G$ acts on the set $S$ of roots of $f(x)$, and the reader has no doubt guessed by now that $G \cong \mathcal{T}$.

(f) Recall from Section 1 of Chapter 6 that the group of symmetries of the square is the group $Sq$ of order 8. Earlier, we considered its action on the set of four vertices of the square. We can equally well consider its action on the set of (four) edges and on the set of (two) diagonals. We leave this analysis to the reader (see Exercise 1a.).

(g) The group of proper symmetries of the tetrahedron (see Figure 2), which we shall analyze in great detail in Section 2, acts on the set of (four) vertices, the set of (six) edges, and the set of (four) faces. The reader is now invited to find the *twelve* symmetries and analyze the resulting actions (see Exercise 1b.).

We now introduce some vocabulary.

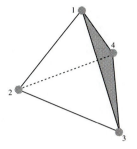

FIGURE 2

**Definition.** Let $G$ act on a set $S$, and let $s \in S$. The **orbit** of $s$ is

$$\mathcal{O}_s = \{s' \in S : s' = g \cdot s \text{ for some } g \in G\},$$

and the **stabilizer** of $s$ is

$$G_s = \{g \in G : g \cdot s = s\}.$$

We may think of dividing $S$ into its various orbits under the $G$-action. As was the case with cosets, distinct orbits cannot intersect. Indeed, much of our earlier work in Chapter 6 follows from our more general setup here (see Exercise 9).

FIGURE 3

**Lemma 1.1.** *If $\mathcal{O}_s \cap \mathcal{O}_{s'} \neq \emptyset$, then $\mathcal{O}_s = \mathcal{O}_{s'}$. Thus, $S$ is a union of distinct orbits; i.e., $S = \bigcup\limits_{\text{disjoint}} \mathcal{O}_s$.*

**Proof.** Suppose $x = g \cdot s = g' \cdot s'$. Then $s' = (g'^{-1}) \cdot (g \cdot s) = (g'^{-1}g) \cdot s$, and so $s' \in \mathcal{O}_s$. Now it follows that $\mathcal{O}_{s'} \subset \mathcal{O}_s$; by symmetry, $\mathcal{O}_s \subset \mathcal{O}_{s'}$, and so the two orbits are equal. Since for each $s \in S$, $s \in \mathcal{O}_s$, it follows that $S$ is the union of all (distinct) orbits.  □

In fact, we see once again the notion of an equivalence relation: We call two elements $s$ and $s'$ of $S$ equivalent (denoted $s \sim s'$) if they belong to the same orbit (i.e., if $\mathcal{O}_s = \mathcal{O}_{s'}$). Then it is easy to see that $\sim$ is an *equivalence relation* on S, and thus the disjoint orbits give a *partition* of $S$ (see Section 3 of Appendix A).

**Lemma 1.2.** *The stabilizer $G_s$ is a subgroup of $G$.*

**Proof.** The identity element belongs to $G_s$. Now suppose $g, g' \in G_s$. Then $(gg') \cdot s = g \cdot (g' \cdot s) = g \cdot s = s$, so $gg' \in G_s$. Similarly, $g^{-1} \cdot s = g^{-1} \cdot (g \cdot s) = (g^{-1}g) \cdot s = e \cdot s = s$, so $g^{-1} \in G_s$.  □

**Examples 2.**

(a) Let $G_0$ be the group of isometries of the plane fixing the origin. Then $G_0$ acts on the plane, and the orbits consist of concentric circles centered at the origin *and* the origin itself.

(b) Let $G$ be the group of all isometries of the plane, and let $S$ be the set of all triangles in the plane. Then, given a particular triangle $s \in S$, $\mathcal{O}_s$ consists of all the triangles congruent to $s$. The stabilizer $G_s$ of this triangle is the subgroup of $G$ consisting of elements that map the triangle $s$ to itself: this subgroup is isomorphic to $\mathcal{T}$, $\mathbb{Z}_2$, or $\{e\}$, depending on whether the triangle is, respectively, equilateral, isosceles, or scalene.

(c) Let $G = GL(2, \mathbb{R})$ be the group of invertible $2 \times 2$ real matrices. Then $G$ acts on itself by conjugation (see Example 1(b)). The orbit of $A \in G$ is

$$\mathcal{O}_A = \{B \in GL(2, \mathbb{R}) : B = PAP^{-1} \text{ for some } P \in GL(2, \mathbb{R})\},$$

which is the set of all matrices similar to $A$. The stabilizer $G_A$ of $A$ consists of those matrices $P \in GL(2, \mathbb{R})$ so that $PAP^{-1} = A$, i.e., the matrices $P$ that commute with $A$.

By the way, we say $G$ acts **transitively** on $S$ if there is just one orbit: i.e., given $s, s' \in S$, there is $g \in G$ so that $s' = g \cdot s$. For example, the group of symmetries of a regular $n$-gon acts transitively on the set of vertices of the $n$-gon.

We now come to the fundamental result.

**Proposition 1.3.** *Let $G$ act on $S$, and let $s \in S$. There is a one-to-one correspondence*

$$G/G_s \to \mathcal{O}_s .$$

*In particular, $\#(\mathcal{O}_s) = [G : G_s]$, when these numbers are finite.*

*Proof.* We define a function $f: G/G_s \to \mathcal{O}_s$ by $f(gG_s) = g \cdot s$. We must check that $f$ is well-defined, one-to-one, and onto. (Note that it is only a function between sets; there need not be a group structure on either $G/G_s$ or $\mathcal{O}_s$.) First, suppose $gG_s = g'G_s$; this means $g' = gh$ for some $h \in G_s$. Now $g' \cdot s = (gh) \cdot s = g \cdot (h \cdot s) = g \cdot s$, since $h \cdot s = s$. Thus $f$ is indeed well-defined. Also, $f$ maps onto $\mathcal{O}_s$ by the definition of orbit. Now we must check that $f$ is one-to-one. Suppose $f(gG_s) = f(g'G_s)$, i.e., $g \cdot s = g' \cdot s$. Then $s = g^{-1} \cdot (g \cdot s) = g^{-1} \cdot (g' \cdot s) = (g^{-1}g') \cdot s$, so $g^{-1}g' \in G_s$, by the definition of $G_s$. Therefore, $g' = gh$ for some $h \in G_s$, and so $gG_s = g'G_s$, as required.   □

**Corollary 1.4.** *If a finite group $G$ acts on a set $S$, then for any $s \in S$, $|G| = \#(\mathcal{O}_s)|G_s|$. In particular, $\#(\mathcal{O}_s)$ divides $|G|$.*

*Proof.* This is immediate from Theorem 3.2 of Chapter 6.   □

**Example 3.** The group $\mathcal{T}$ acts transitively on the set $S$ of vertices of an equilateral triangle. The stabilizer $G_s$ of a vertex $s$ is a subgroup of order 2; $\#(\mathcal{O}_s) = \#(S) = 3$, and so we recover the fact that $|\mathcal{T}| = 6$.

**Example 4.** The group *Tetra* of (proper) symmetries of a regular tetrahedron acts transitively on the set of 4 vertices; the stabilizer of a vertex is a subgroup of order 3 (consisting of the rotations of the opposite face about the axis passing through the given vertex and the centroid of the face). As a result, $|\textit{Tetra}| = 4 \cdot 3 = 12$. *Tetra* also acts transitively on the set of 6 edges, and we leave it to the reader to find the stabilizer subgroup of an edge (whose order, of course, must be $\frac{12}{6} = 2$).

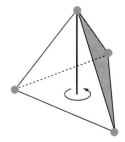

FIGURE 4

**Example 5.** Let $G = S_n = \text{Perm}\{1,\ldots,n\}$ be the symmetric group, and let $S$ be the set of all $k$-element subsets of $\{1,\ldots,n\}$. Then $G$ acts on $S$ in the obvious way: if $s = \{i_1,\ldots,i_k\} \in S$ and $\sigma \in G$, then $\sigma(s) = \{\sigma(i_1),\ldots,\sigma(i_k)\} \in S$. It is immediate from the definition that $G$ acts transitively on $S$. Now, let $s = \{1,\ldots,k\}$. Then $G_s \cong \{\sigma \in S_n : \sigma(\{1,\ldots,k\}) = \{1,\ldots,k\}\}$. Such permutations $\sigma$ therefore have the property that $\sigma(\{k+1,\ldots,n\}) = \{k+1,\ldots,n\}$ as well, and an easy counting argument yields $|G_s| = k!(n-k)!$. From Corollary 1.4, we conclude that

$$\#(S) = \#(\mathcal{O}_s) = \frac{|G|}{|G_s|} = \frac{n!}{k!(n-k)!}.$$

(Cf. the discussion of the binomial coefficient $\binom{n}{k}$ on p. 6.)

A very important application of Corollary 1.4 arises when $G$ acts on itself by conjugation (see Example 1(b)). Recall that $a, a' \in G$ are **conjugate** if $a' = gag^{-1}$ for some $g \in G$, so that the orbit $\mathcal{O}_a$ consists of all elements $a' \in G$ so that $a$ and $a'$ are conjugate. The stabilizer subgroup $G_a$ consists of those elements of $G$ that commute with $a$ (since $gag^{-1} = a \iff ga = ag$).

**Corollary 1.5.** *Let $G$ be a finite group, and let $a \in G$. Then*

$$\#(\text{elements conjugate to } a) = \frac{|G|}{\#(\text{elements commuting with } a)}. \quad \square$$

Combining Proposition 1.3 and Lemma 1.1, we obtain the following powerful counting formula.

**Proposition 1.6.** *Suppose $S$ is a finite set, and $G$ acts on $S$. Then*

$$\#(S) = \sum_{\text{distinct orbits}} \#(\mathcal{O}_s)$$

$$= \sum_{\text{distinct orbits}} \frac{|G|}{|G_s|} \quad \text{if } G \text{ is finite.}$$

*Proof.* The former equality comes from decomposing $S$ as a union of the disjoint orbits $\mathcal{O}_s$ by Lemma 1.1. The latter equality follows from Corollary 1.4.  $\square$

**Example 6.** Suppose $G$ is a group of order 8 acting on a set $S$ with 15 elements. Then this action must have a fixed point; i.e., there must be an element $s \in S$ with the property that $g \cdot s = s$ for all $g \in G$. Rephrasing this, there must be an element $s \in S$ whose orbit $\mathcal{O}_s$ consists of $s$ alone—or, equivalently, whose stabilizer subgroup $G_s$ is the whole group $G$.

By Corollary 1.4, $\#(\mathcal{O}_s)\,|\,|G|$, so if $\#(\mathcal{O}_s) \neq 1$, it is even. If this is true for all $s \in S$, then, by Proposition 1.6, $\#(S)$ must be even. Since 15 is not even, we conclude that the action must indeed have a fixed point.

It is natural to ask how the stabilizer subgroups of two elements $s, s' \in S$ are related if $\mathcal{O}_s = \mathcal{O}_{s'}$. The following result will be crucial in the next section for our applications to symmetry groups of polyhedra.

**Proposition 1.7.** *Let $G$ act on a set $S$, and let $s \in S$; and suppose $s' = a \cdot s$ for some $a \in G$. Then $G_{s'} = aG_s a^{-1}$.*

**Proof.** If $g \in G_{s'}$, then $(a^{-1}ga) \cdot s = (a^{-1}g) \cdot s' = a^{-1} \cdot (g \cdot s')$ $= a^{-1} \cdot s' = s$, so $a^{-1}ga \in G_s$. Therefore, $a^{-1}G_{s'}a \subset G_s$, or $G_{s'} \subset aG_s a^{-1}$. Conversely, if $g \in G_s$, then $(aga^{-1}) \cdot s' = (ag) \cdot s = a \cdot (g \cdot s) = a \cdot s = s'$; so $aga^{-1} \in G_{s'}$, whence $aG_s a^{-1} \subset G_{s'}$. Thus, $G_{s'} = aG_s a^{-1}$, as claimed. $\square$

**Example 7.** Consider the group $G$ of isometries of the plane. Denote by $G_0$ the subgroup of isometries fixing the origin, and by $G_P$ the subgroup of isometries fixing the point $P \in \mathbb{R}^2$. Let $\tau \in G$ be the translation carrying 0 to $P$. Then $G_P = \tau G_0 \tau^{-1}$. (Cf. the remarks on pp. 75 and 77.)

### EXERCISES 7.1

1.  a.  Give the action of $Sq$ on the sets of edges and diagonals of a square.
    b.  Give the action of the group of symmetries $\mathcal{T}\!etra$ on the sets of vertices, edges, and faces of the tetrahedron. (It will help you to make a model. Be alert: You should find elements of order 2 that simultaneously interchange pairs of vertices.)

2.  Using Corollary 1.4, determine the order of the group of symmetries of the cube. Check your answer in as many ways as possible.

3.  Let $G = \mathcal{D}_5$ be the group of symmetries of a regular pentagon. What is the action of $G$ on the set $S$ of the five diagonals of the pentagon?

4.  a.  Suppose a group $G$ has order 35 and acts on a set $S$ consisting of four elements. What can you say about the action? (Hint: Must there be a fixed point?)

    b.  What happens if $|G| = 28$? $|G| = 30$?

5.  Let $p$ be prime, and suppose $|G| = p^m$. If $G$ acts on a finite set $S$, and $p \nmid \#(S)$, prove that the action must have a fixed point.

6.  If a group $G$ of order 35 acts on a set $S$ with 16 elements, show that the action must have a fixed point. Indeed, list all the possible numbers of fixed points of the action.

7.  Give a new proof that $\mathcal{T} \cong S_3$, as follows. Let $G = \mathcal{T}$, $H = \{e, F\}$, and consider the action of $G$ on $G/H$ by left multiplication given in Example 1(c). Show that this action gives an isomorphism $G \to \text{Perm}(G/H) \cong S_3$.

8.  Let $G$ act by left multiplication on the set of cosets $G/H$. Let $a \in G$.

    a.  What is the orbit of the coset $aH$?

    b.  What is the stabilizer of the coset $aH$?

    c.  Interpret Proposition 1.7 in this context.

9.  Let $H$ be a subgroup of $G$. Define an action of $H$ on $G$ by $h \cdot g = gh^{-1}$.

    a.  Prove that this is a *bona fide* action. (Why do we use $h^{-1}$?)

    b.  Prove that the orbits are the cosets of $H$ in $G$.

    c.  Deduce Lagrange's Theorem, Theorem 3.2 of Chapter 6.

10. Let the group $G = Sq$ of symmetries of a square act on itself by conjugation. How many orbits are there? What are the orbits?

11. Let the dihedral group $D_n$ act on itself by conjugation. For each case

    a.  $n = 7$,

    b.  $n = 8$,

    list the orbits, verify Proposition 1.6, and give all the normal subgroups of the group.

12. Is the converse of Proposition 1.7 valid? Give a proof or counterexample.

13. Let $G$ act on itself by conjugation. Describe the orbits and verify Proposition 1.6 for $G = A_4$, $S_4$, $A_5$, and $S_5$.

14. Let a group $G$ act on itself by conjugation. Recall that the center of $G$ is defined to be $Z = \{a \in G : ag = ga \text{ for all } g \in G\}$.
    a. Prove that $a \in Z \iff \mathcal{O}_a = \{a\} \iff G_a = G$.
    b. Prove that $Z \subset G_a$ for any $a \in G$ and that if $a \notin Z$, then $Z \subsetneq G_a \subsetneq G$.
    c. Suppose that $[G : Z]$ is finite. Prove that if $a \notin Z$, then $1 < \#(\mathcal{O}_a) < [G : Z] \ (= |G|/|Z| \text{ if } G \text{ is finite})$. Prove, moreover, that $[G : Z]$ must be an integral multiple of $\#(\mathcal{O}_a)$. (See Exercise 6.3.32.)

15. Suppose $|G| = 10$. Prove that if its center $Z$ is nontrivial, then $G$ is abelian. (Hint: see Exercise 14.)

16. Let $G$ be a group of odd order, and suppose $H \subset G$ is a normal subgroup of order 3. Prove that $H \subset Z$. How can you generalize this result? (Hint: Let $G$ act on $H$ by conjugation.)

17. a. Let $p$ be prime, and suppose $G$ is a group of order $p^m$. Use the fact that $|Z| \geq 1$ (why?) and Proposition 1.6 to prove that $|Z| > 1$. (See Exercise 14.)
    b. Use Exercise 14c. to prove that a group $G$ of order $p^2$ must be abelian. (See also Exercise 6.3.27.)

18. Let $G$ be a group, and let $S = \{\text{subgroups of } G\}$. Let $G$ act on $S$ by conjugation.
    a. If $H \in S$, the stabilizer subgroup $G_H$ is called the **normalizer** of $H$. Prove that $H$ is a normal subgroup of $G_H$.
    b. Prove that $H$ is normal $\iff \mathcal{O}_H = \{H\}$ and $G_H = G$.
    c. Prove or give a counterexample: The normalizer of any subgroup $H \subset G$ is a normal subgroup of $G$.

19. Let $G$ be a finite group of order $n$. Prove that $G$ is isomorphic to a subgroup of $S_n$. Proceed as follows:
    a. For each $a \in G$, let $L_a : G \to G$ be defined by $L_a(g) = ag$. Prove that $L_a$ is a permutation of $G$.
    b. Define $\phi : G \to \text{Perm}(G)$ by $\phi(a) = L_a$. Prove $\phi$ is a one-to-one homomorphism.
    c. Deduce the result.

## 2. The Symmetry Groups of the Regular Polyhedra

Although we shall launch a thorough study of the group of isometries of $\mathbb{R}^3$ in Section 4, we begin here by analyzing the symmetries of the five regular polyhedra (Platonic solids) pictured below: the tetrahedron, cube, octahedron, dodecahedron, and icosahedron. For simplicity, we consider only the *proper* symmetries of these objects (i.e., symmetries arising from rotations of $\mathbb{R}^3$). In Section 4 we'll find all possible finite subgroups of the group of rotations of $\mathbb{R}^3$, and we'll infer that there can be no other regular polyhedra.

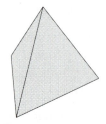

TETRAHEDRON

CUBE

OCTAHEDRON

DODECAHEDRON

ICOSAHEDRON

FIGURE 1

Our discussion will be based on the notion of conjugate elements of a group. Since the group of symmetries of a regular polyhedron acts transitively on the sets of vertices, edges, and faces, it follows from Proposition 1.7 that the stabilizer subgroup of any given vertex is conjugate to that of another vertex (and the same for edges and faces). This observation makes it far easier to find (and count) the various elements of one of these symmetry groups.

We will also be interested in all the normal subgroups of our various symmetry groups. Recall that a subgroup $H \subset G$ is normal if and only if, given any $h \in H$, all its conjugates $ghg^{-1}$, $g \in G$, belong to $H$. In particular, if $G$ acts on a set $S$, and the stabilizer subgroup $G_s$ of some element $s \in S$ is contained in a normal subgroup $H$, then $H$ must contain the stabilizer subgroups $G_{s'}$ of all the elements $s' \in \mathcal{O}_s$ (see Exercise 2).

As a warm-up exercise, we analyze the group $\mathcal{T}$ of symmetries of an equilateral triangle. This group contains three elements of order 2 (the reflections fixing each of the three vertices), and these are all conjugate; two elements of order 3 (the two nontrivial rotations); and the identity element. Note that the two rotations $R$ and $R^2$ are conjugate, since $R^2 = FRF^{-1}$ (see p. 177). We summarize this discussion in a table:

| order of element | geometric description | #(conjugates) |
| :---: | :---: | :---: |
| 1 | identity | 1 |
| 2 | reflections | 3 |
| 3 | 120° rotations | 2 |

How do we use these data to find normal subgroups of $\mathcal{T}$? Let $N$ be a putative normal subgroup of $\mathcal{T}$. It certainly must contain the identity element. If it contains an element of order 2, it must contain all its conjugates, and this would mean $N$ has at least $1 + 3 = 4$ elements. Since $4 > \frac{1}{2}(6)$, we conclude that if $N$ is a normal subgroup containing an element of order 2, then $N = \mathcal{T}$. On the other hand, if $N$ contains an element of order 3, and hence its conjugates, this will account for $1 + 2 = 3$ elements. Indeed, the subgroup $\langle R \rangle$ is a normal subgroup of $\mathcal{T}$, and the only nontrivial one.

Let us now analyze the symmetry group $\mathcal{T}etra$. We can easily find eight elements of order 3: the stabilizer of a vertex is a group of order 3, and there are four such conjugate subgroups. Removing the identity element from each yields $4 \times 2 = 8$ elements of order

3. Now, the $120°$ rotations about the four vertices are all conjugate: let $\rho_i$ and $\rho_j$ be $120°$ rotations about vertices $i$ and $j$, respectively; if $\sigma$ is any symmetry of the tetrahedron carrying vertex $i$ to vertex $j$, then $\rho_j = \sigma \rho_i \sigma^{-1}$. Likewise, the $240°$ rotations are all conjugate.

On the other hand, unlike the case of the symmetries of an equilateral triangle, a $120°$ rotation is *not* conjugate to any $240°$ rotation. Suppose the $120°$ rotation about vertex 4 were conjugate to some $240°$ rotation; then it would be conjugate to the $240°$ rotation about vertex 4. So the vertex permutations (1 2 3) and (1 3 2) would have to be conjugate in the group *Tetra*. And in order for this to happen, there would have to be a symmetry of the tetrahedron fixing one pair of vertices (e.g., 1 and 4) and interchanging the remaining pair (2 and 3); there is no such proper symmetry. Should the reader be somewhat dissatisfied by this *ad hoc* argument, a much more succinct and elegant argument follows in a moment.

Next, we claim there are three elements of order 2, for consider Figure 2. We rotate the tetrahedron $180°$ about the axis joining

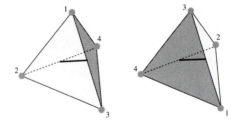

FIGURE 2

the midpoints of two opposite edges (e.g., edges $\overline{13}$ and $\overline{24}$); this element of *Tetra* swaps the pairs of vertices, but stabilizes each of the two chosen edges. There are three pairs of opposite edges, and so this accounts for three *conjugate* elements of order 2. Are there any other elements of *Tetra*? Well, from Corollary 1.4 we infer that $|Tetra| = 12$, and we've accounted for $1 + 8 + 3 = 12$ elements, so that's it. Summarizing, we have the table:

| order of element | geometric description | #(conjugates) |
|:---:|:---:|:---:|
| 1 | identity | 1 |
| 2 | flips | 3 |
| 3 | $120°$ rotations | 4 |
| 3 | $240°$ rotations | 4 |

We can now see quickly that the 120° and 240° rotations cannot be conjugate. By Corollary 1.5, the number of conjugates of a given element must be a divisor of the order of the group. In our case, if all 8 rotations were conjugate, we would have 8|12, which is non-sensical.

Once again, we search for normal subgroups. The fact that a proper subgroup contains at most half the elements of the group leaves us only one choice: the subgroup $N$ containing the identity and the three elements of order 2 is a subgroup of order 4 and is normal (why?). (To which standard group of order 4 is $N$ isomorphic?)

We next consider the group *Cube* of symmetries of a cube. To count $|Cube|$, it is easiest to let the group act (transitively) on the set of six faces. The stabilizer subgroup of a face has order 4, and so $|Cube| = 24$. Since three faces meet at a vertex, a subgroup of order 3 stabilizes each vertex (and its diagonal opposite). Note that

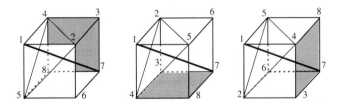

FIGURE 3

this checks with Corollary 1.4: There are eight vertices, and the stabilizer subgroup of a vertex has order 3. A typical such symmetry is pictured in Figure 3. On the other hand, there are two types of elements of order 2: those that stabilize a face (and its opposite), and those that stabilize an edge (and its diagonal opposite). These are pictured on the left and on the right, respectively, in Figure 4. Of the former there are three elements (for there are $\frac{6}{2} = 3$ pairs of opposite faces), whereas of the latter there are six elements (since there are $\frac{12}{2} = 6$ pairs of opposite edges). There are $4 \times 2$ elements of order 3, and $3 \times 2$ elements of order 4. We have therefore accounted for $1 + 3 + 6 + 8 + 6 = 24$ elements, and there are no more. Summarizing, we have

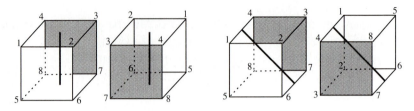

FIGURE 4

| order of element | geometric description | #(conjugates) |
|:---:|:---:|:---:|
| 1 | identity | 1 |
| 2 | 180° rotations preserving edges | 6 |
| 2 | 180° rotations preserving faces | 3 |
| 3 | ±120° rotations | 8 |
| 4 | ±90° rotations | 6 |

Now, what are the proper normal subgroups of *Cube*? Just trial-and-error arithmetic shows what combinations of numbers in the right column of the table will produce a divisor of 24. Of course, we are obliged to include the identity element in any trial; and if we include an element of order 4, we must include its square, which is the second type of element of order 2 listed in the table. The only possibilities are these:

(i) $1 + 3 = 4$: the subgroup consisting of the 180° face rotations; and

(ii) $1 + 3 + 8 = 12$: the subgroup consisting of the 180° rotations and the 120° rotations.

The reader should check that these are in fact subgroups (i.e., are closed under multiplication). This will be particularly easy to do after we learn the following result (and complete Exercise 6), for we can actually identify the group *Cube* quite concretely. Better yet, we can use a model!

**Proposition 2.1.** *Cube* $\cong S_4$.

*Proof.* The group *Cube* acts on the set $S$ of the four major diagonals of the cube, pictured in Figure 5. (We will number them $1, 2, 3, 4$, by their upper-level endpoints.) The action of *Cube* on $S$ gives a homomorphism $\phi \colon Cube \to \mathrm{Perm}(S) \cong S_4$. It remains only to show that $\phi$ is an isomorphism: i.e., it has trivial kernel and maps onto $S_4$. If an element of *Cube* does not exchange any of the diagonals, it is

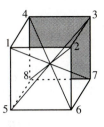

FIGURE 5

obviously the identity element (for example, consider the action of each of the elements shown in Figures 3 and 4). Proving $\phi$ maps onto $S_4$ seems a difficult task, but we need only count. Both *Cube* and $S_4$ have twenty-four elements, and a one-to-one map between sets with the same number of elements must be onto (why?).   □

The analysis of the group *Icosa* of symmetries of a regular icosahedron is left to the exercises.

## EXERCISES 7.2

1.  Let *Tetra* act on the set $S$ of pairs of opposite edges of the tetrahedron. What is the stabilizer of one such pair? Use Corollary 1.4 to recompute $|$*Tetra*$|$.

2.  Suppose $G$ acts on a set $S$; let $G_s$ be the stabilizer of $s \in S$. Suppose $H \subset G$ is a normal subgroup and $G_s \subset H$. Prove that $G_{s'} \subset H$ for all $s' \in \mathcal{O}_s$.

3.  Give an example of a group of order 12 having no subgroup of order 6. (See the remark following Corollary 3.6 of Chapter 6.)

4.  a.  Show that a regular tetrahedron may be inscribed in a cube, and that the subgroup of *Cube* that preserves the tetrahedron is isomorphic to *Tetra*. (Query: How many such inscribed tetrahedra are there?)
    b.  Show that *Tetra* $\cong A_4$.
    c.  Interpret Exercise 6.4.9 in terms of symmetries of the tetrahedron.

5.  Explain why the group of symmetries of the octahedron is *Cube*, and why the group of symmetries of the dodecahedron is *Icosa*.

(Hint: List the number of vertices, edges, and faces for each of these pairs. What do your lists suggest?)

6.  Write down the isomorphism $\phi\colon \mathit{Cube} \to S_4$ explicitly. Apply these results to your study of Exercise 4.

7.  Let $\mathcal{D}_6$ be the group of symmetries of the regular hexagon (see Exercise 6.2.15).
    a.  Determine the orders of the elements of $\mathcal{D}_6$, and count the elements of each order. Decide which ones are conjugate (make a table summarizing your results, as in the text).
    b.  What are the normal subgroups of $\mathcal{D}_6$?
    c.  If we inscribe an equilateral triangle in the hexagon, and prescribe a symmetry of the triangle, what further information is required to specify a symmetry of the hexagon?

8.  A regular icosahedron has twenty faces, each an equilateral triangle.
    a.  Conclude that it has 30 edges and 12 vertices.
    b.  Use Corollary 1.4 to compute $|\mathit{Icosa}|$ several ways.
    c.  Determine the orders of the elements of $\mathit{Icosa}$, and count the elements of each order. Decide which ones are conjugate.
    d.  Find the normal subgroups of $\mathit{Icosa}$. (It will help to make a table summarizing your results of c., as we did in the text.)
    e.  Fix a vertex of the icosahedron; it belongs to five faces. The centroids of these five faces form a pentagon, which has five diagonals. Show that each of these diagonals is one edge of a cube that is inscribed in the icosahedron. Thus, we obtain a set $S$ of five inscribed cubes. The action of $\mathit{Icosa}$ on $S$ therefore determines a homomorphism $\phi\colon \mathit{Icosa} \to S_5$. Prove that $\ker \phi = \{e\}$ and that the image of $\phi$ is a subgroup of $S_5$ of index 2. Alternatively, make use of the diagram of an "unfolded" icosahedron in Figure 6.
    f.  Prove that $\mathit{Icosa} \cong A_5$. (See Section 4 of Chapter 6 for the relevant definition.) (Hint: see Exercises 6.3.5 and 6.3.19, and use your answer to d.)

9.  Let $(1, 1, 1)$, $(-1, -1, 1)$, $(-1, 1, -1)$, and $(1, -1, -1)$ be the vertices of a tetrahedron. Find the matrices representing the symmetries of the tetrahedron. (Hint: Each symmetry is a rotation

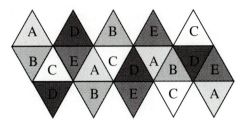

FIGURE 6

of $\mathbb{R}^3$, hence a linear map $\mathbb{R}^3 \to \mathbb{R}^3$, hence can be represented—with respect to the standard basis—by a $3 \times 3$ matrix. It should suffice to so represent one 120° rotation about a vertex and one 180° rotation fixing a pair of opposite edges. Why? For example, the symmetry pictured in Figure 2 is $\begin{bmatrix} -1 & & \\ & 1 & \\ & & -1 \end{bmatrix}$.)

10. Let the eight points $(\pm 1, \pm 1, \pm 1)$ be the vertices of a cube. Find the matrices representing the symmetries of the cube. (Hint: It should suffice to so represent one 120° rotation about a vertex, one 90° rotation fixing a face, and one 180° rotation fixing a pair of opposite edges. Why?)

11. What shape is obtained when a cube is revolved about the (long) diagonal joining a pair of opposite vertices?

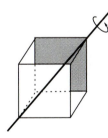

FIGURE 7

12. Let $G = GL(3, \mathbb{Z}_2)$.
    a. Show that $|G| = 168$. (Hint: The columns of $g \in G$ must be linearly independent.)

b. Let $G$ act on itself by conjugation. Find the orders of the stabilizer subgroups of each of the following elements:

$$\begin{bmatrix} 1 & 0 & 0 \\ 0 & 1 & 0 \\ 0 & 0 & 1 \end{bmatrix}, \begin{bmatrix} 1 & 1 & 0 \\ 0 & 1 & 0 \\ 0 & 0 & 1 \end{bmatrix}, \begin{bmatrix} 1 & 1 & 0 \\ 0 & 1 & 1 \\ 0 & 0 & 1 \end{bmatrix},$$

$$\begin{bmatrix} 0 & 1 & 0 \\ 0 & 0 & 1 \\ 1 & 0 & 0 \end{bmatrix}, \begin{bmatrix} 0 & 1 & 0 \\ 0 & 0 & 1 \\ 1 & 1 & 0 \end{bmatrix}, \begin{bmatrix} 0 & 1 & 0 \\ 0 & 0 & 1 \\ 1 & 0 & 1 \end{bmatrix}.$$

c. Show that the last two elements listed above are not conjugate. (Hint: Consider their characteristic polynomials.)

d. Show that every element of $G$ is conjugate to one of the matrices listed above. (Hint: Compute the number of elements in each orbit.)

e. Prove that $G$ has no proper normal subgroup.

### 3. Burnside's Theorem and Enumeration

In this section we give a different way of counting the orbits of the action of a finite group $G$ on a finite set $S$. The result is due to Frobenius and Cauchy, but is credited to Burnside, who proved it in 1911 and applied it to group theoretic questions; somewhat later, the famous mathematician Pólya used it to solve combinatorial problems. That will be our main application here.

**Theorem 3.1** (Burnside). *Let $G$ be a finite group acting on a finite set $S$. For each $g \in G$, let $\mathrm{Fix}(g) = \{s \in S : g \cdot s = s\}$ be the fixed-point set of $g$. If $N$ is the number of distinct orbits in $S$, then*

$$N = \frac{1}{|G|} \sum_{g \in G} \#(\mathrm{Fix}(g)).$$

**Proof.** By Corollary 1.4, $|G_s| = |G|/\#(\mathcal{O}_s)$, and so

$$\sum_{s \in S} |G_s| = \sum_{s \in S} \frac{|G|}{\#(\mathcal{O}_s)} = \sum_{\text{distinct orbits } \mathcal{O}} \left( \sum_{s \in \mathcal{O}} \frac{|G|}{\#(\mathcal{O}_s)} \right)$$

$$= \sum_{\text{distinct orbits } \mathcal{O}} |G| = N |G|.$$

Now the key observation is this: $s \in \mathrm{Fix}(g) \iff g \in G_s$. Thus,

(*)                          $\displaystyle \sum_{s \in S} |G_s| = \sum_{g \in G} \#(\mathrm{Fix}(g)),$

and the result follows.  $\square$

**Remark.** The best way to understand (∗) is this: we create a table, listing the elements of $G$ down the left and the elements of $S$ across the top. In the position corresponding to $(g, s)$, we place a check mark provided $g \in G_s$ or, equivalently, $s \in \text{Fix}(g)$. Now either sum is just the total number of check marks. The left-hand side of (∗) is obtained by first summing the number of check marks in each column, whereas the right-hand side of (∗) is obtained by first summing the check marks in each row. For example, we have the following table:

|   | $V$ | $W$ | $X$ | $Y$ | $Z$ | $\ldots$ | total |
|---|---|---|---|---|---|---|---|
| $e$ | ✓ | ✓ | ✓ | ✓ | ✓ |  | 5 |
| $a$ |  | ✓ |  | ✓ | ✓ |  | 3 |
| $b$ |  | ✓ | ✓ |  | ✓ |  | 3 |
| $c$ | ✓ | ✓ |  |  |  |  | 2 |
| $\vdots$ |  |  |  |  |  |  |  |
| total | 2 | 4 | 2 | 2 | 3 |  | 13 |

**Example 1.** Suppose we are going to paint the faces of a cube with six different (but specified) colors. If we are allowed to move the cube around before placing it on the table, how many different designs can we obtain?

This is not too difficult to figure out directly: Put the brown face on the table. Either the green face is opposite the brown (in which case there will be six different designs—why?), or else the green face is adjacent to the brown (in which case there will be twenty-four different designs—why?). Altogether, there are thirty different designs.

This is, however, a particularly simple application of Theorem 3.1. Assign a number to each face, and let $S$ be the set of all the possible ways of assigning the six colors to the six numbered faces. Obviously $\#(S) = 6!$. The group $Cube$ acts on $S$, and since all six faces are painted different colors, only the identity element has any fixed points. Indeed, it is always the case that $\text{Fix}(e) = S$. Here $N$, the number of distinct orbits, is the number of different designs, and we have

$$N = \frac{1}{|Cube|} \sum_{g \in Cube} \#(\text{Fix}(g)) = \frac{6!}{24} = 30.$$

**Example 2.** Suppose four keys—two red and two yellow—are on a circular key ring. How many (distinguishable) configurations are there?

Common sense tells us that the answer is two: either the red keys are adjacent to each other (hence, the yellow as well), or the colors of the keys alternate:

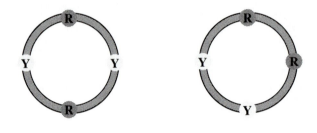

FIGURE 1

Let $S$ be the set of all possible configurations of the keys. It is easiest to visualize the keys as lying at the vertices of a square. To count $S$, if we choose the positions of the two red keys, then the configuration is completely determined; there are $\binom{4}{2} = 6$ ways to select two objects (the positions of the red keys) from a set of four (the total number of positions). The group of symmetries of our arrangement of keys is $Sq$; two configurations are indistinguishable if they lie in the same orbit of the group action. We thus wish to know the number of distinct orbits when $Sq$ acts on $S$.

Recall that $Sq$ consists of the identity element; rotations through 90° ($\rho$), 180° ($\rho^2$), and 270° ($\rho^3$); and four reflections. Fix($\rho$) and Fix($\rho^3$) are empty (since to have a fixed point in either case, all the keys would have to be the same color). Of course, Fix($e$) = $S$. On the other hand, #(Fix($\rho^2$)) = 2, since we can arrange the red keys opposite one another and the yellow keys opposite one another in two distinct ways (RYRY and YRYR). For any of the four reflections, the fixed-point set consists of the same two configurations: a reflection of the square interchanges at least one pair of vertices of the square, and that pair must be either red or yellow. So, totting up, Burnside's Theorem tells us that

$$N = \frac{1}{8} \sum_{g \in Sq} \#(\text{Fix}(g)) = \frac{1}{8} \left( \underset{\substack{\uparrow \\ \text{identity}}}{6} + \underset{\substack{\uparrow \\ \text{180° rotation}}}{2} + \underset{\substack{\uparrow \\ \text{reflections}}}{4 \times 2} \right) = 2.$$

There are two distinguishable arrangements of the keys, as common sense told us in the first place.

**Example 3.** Suppose now we have five keys—two red, two yellow, and one green—on a circular key ring. We picture them at the vertices of a regular pentagon, and the symmetry group is now $\mathcal{D}_5$. We must first count the set $S$ of configurations of the keys. We have $\binom{5}{2}$ possible locations for the red keys; having chosen them, there are three remaining vertices. Once we select one for the green key, the location of the yellow keys is determined. There are obviously three choices, then, for the green key; and so $\#(S) = \binom{5}{2} \cdot 3 = 10 \cdot 3 = 30$.

Once again, Fix($e$) = $S$. Since there is only one green key, it is clear that no nontrivial rotation can have a fixed point. On the other hand, any reflection $\psi$ has $\#(\text{Fix}(\psi)) = 2$: fix the green key, and then interchange the pair of red keys and the pair of yellow keys, as pictured in Figure 2. There are five reflections in $\mathcal{D}_5$, and

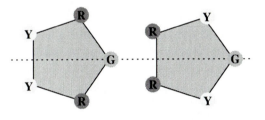

FIGURE 2

so the number of distinct key arrangements is

$$N = \frac{1}{10}\left( \underset{\substack{\uparrow \\ \text{identity}}}{30} + \underset{\substack{\uparrow \\ \text{reflections}}}{5 \times 2} \right) = 4.$$

Try it. Draw the four configurations, and check it out.

**Example 4.** Suppose we are given four colors of paint, and we paint each edge of an equilateral triangle one of these colors (with repeated colors allowed). How many distinguishable designs will there be?

Let $S$ consist of the $4^3$ possible paint jobs. We wish to know how many orbits there are in $S$ under the action of the triangle group $\mathcal{T}$. As usual, $\#(\text{Fix}(e)) = \#(S) = 64$. For either rotation element in $\mathcal{T}$, the fixed-point set has four elements (since we must paint all three edges the same color, and there are four ways to do this). For any flip in $\mathcal{T}$, we can paint the two edges that are interchanged the same color and the remaining edge any color we want; there are sixteen ways to do this. Thus, we find

$$N = \frac{1}{6} \left( \underset{\substack{\uparrow \\ \text{identity}}}{64} + \underset{\substack{\uparrow \\ \text{rotations}}}{2 \times 4} + \underset{\substack{\uparrow \\ \text{reflections}}}{3 \times 16} \right) = 20$$

possible designs.

**Example 5.** Suppose we are going to paint two faces of a die red, two faces white, and two faces blue. How many different dice can we manufacture?

Let $S$ be the set of possible colorings of a die. We have $\binom{6}{2}$ choices for the red faces; having chosen them, we must paint two of the remaining four faces white; there are $\binom{4}{2}$ possibilities. The location of the blue faces is now determined. Therefore, $\#(S) = \binom{6}{2}\binom{4}{2} = 90$. We consider the orbits of $S$ under the action of *Cube*.

As always, $\#(\text{Fix}(e)) = \#(S) = 90$. Now, every element $g$ of order two interchanges pairs of faces of the cube (see Section 2): the $180°$ face rotations interchange two pairs and leave two faces fixed (let's agree to call these a pair as well); the $180°$ "edge rotations" interchange all the faces in pairs. If we paint the die with the faces in each pair the same color, then this coloring will be a fixed point of $g$; there are $3! = 6$ such colorings for each element of order 2. Now the elements of *Cube* of order 3 and 4 have no fixed points (as the orbit of a face consists of 3 and 4 faces, respectively, all of which would have to be the same color). Recalling that there are nine elements of order 2 in *Cube*, we therefore have $N = \frac{1}{24}(90 + 9 \times 6) = 6$ possible dice. How mystic!

**Example 6.** Now suppose we paint three faces of a die red and three faces white. How many different dice can we now manufacture?

As is now customary, let $S$ be the set of possible colorings of our die. Clearly, $\#(S) = \binom{6}{3} = 20$. As before, we consider the orbits of $S$ under *Cube*. Of course, $\#(\text{Fix}(e)) = 20$, but now things get a bit more interesting. First, each element $g \in$ *Cube* of order 3 has $\#(\text{Fix}(g)) = 2$: each such element stabilizes a pair of diagonally opposite vertices (see Section 2) and cyclically permutes the three faces meeting at each of the vertices. We paint one such triad red and the other white; there are therefore two colorings fixed by each such $g$.

As before, the elements of order 4 have no fixed points. Nor do the 180° edge rotations, since they interchange all the faces in pairs.

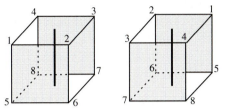

FIGURE 3

However, the 180° rotations are of greater interest; to fix ideas, consider the rotation fixing the top and bottom faces, as pictured in Figure 3. We can paint the sides in pairs (using up two red and two white faces), and then paint the top either red or white (and the bottom whatever's left over). For each such element $g$ of order 2, then, #(Fix($g$)) = 2 × 2 = 4. Summing up, there are

$$N = \frac{1}{24}\left(\underset{\substack{\uparrow \\ \text{identity}}}{20} + \underset{\substack{\uparrow \\ \text{120° rotations}}}{8 \times 2} + \underset{\substack{\uparrow \\ \text{180° face rotations}}}{3 \times 4}\right) = 2$$

possible dice to manufacture. The reader should figure out why this was obvious from the outset.

### EXERCISES 7.3

1. Use Burnside's Theorem to count the number of distinguishable ways of placing four red, two yellow, and two green keys (identical except for their colors) on a circular key ring.

2. How many dodecahedral dice are there? That is, in how many distinguishable ways can the twelve faces of a dodecahedron be numbered 1 through 12? (The group of symmetries of the dodecahedron is *Icosa*.)

3. Find the number of different colorings of a cube with two white, one black, and three red faces.

4.  How many different chemical compounds can be made by attaching $H$, $CH_3$, $C_2H_5$ or $Cl$ radicals to the four bonds of a carbon atom? (The radicals lie at the vertices of a regular tetrahedron.)

5.  A toy pyramid in the shape of a regular tetrahedron is built out of six pegs. Use Burnside's Theorem to count the number of distinct designs if there are:
    a.  two each of red, white, and blue pegs;
    b.  three each of red and white pegs.

6.  The skeleton of a cube is made out of twelve pegs. How many distinguishable such cubes can be made from
    a.  seven blue and five white pegs?
    b.  six blue, two white, and four red pegs?

7.  Count the number of distinguishable ways to paint the faces of a regular dodecahedron with five green faces, five magenta faces, and two orange faces.

8.  How many ways can the vertices of a regular tetrahedron be colored using at most three colors? (Hint: see Example 4.)

9.  How many different patchwork quilts, four patches long and three patches wide, can be made from five red and seven blue squares, assuming the quilts
    a.  cannot be turned over?
    b.  can be turned over?

10. Find the number of different types of circular necklaces that can be made from
    a.  seven black and five white beads;
    b.  six red and four green beads;
    c.  five black, six white, and three red beads;
    d.  ten beads of at most two colors (Answer: 78).

11. Two persons from each of two feuding families and five neutral parties are to be seated around a circular table. In how many inequivalent ways (under the action of $\mathcal{D}_9$) can they be seated if no two members of opposing factions sit next to each other? (You should treat the people of each of the three groups as "indistinguishable.")

12. In how many distinguishable ways can you paint the faces of a cube using green and magenta paint? (You are allowed any number $k$ ($0 \le k \le 6$) of green faces, with the obvious restriction that you must have $6 - k$ magenta faces.)

13. How many different ways can the faces of a regular octahedron be colored using at most five colors?

14. Prove that if $g$ and $g'$ are conjugate, then $\#(\mathrm{Fix}(g)) = \#(\mathrm{Fix}(g'))$.

15. Suppose a finite group $G$ acts on a finite set $S$. Suppose, moreover, that for each $s \in S$, $G_s = \{e\}$. (Such an action is called *free*.) Prove that $N|G| = \#(S)$. (See Lagrange's Theorem, Theorem 3.2 of Chapter 6.)

## 4. Isometries of $\mathbb{R}^3$ and Classification of the Regular Polyhedra

The purpose of this section is twofold: to study the group of isometries of $\mathbb{R}^n$ (but $\mathbb{R}^3$ in particular) by linear algebra; and to apply the counting techniques of Section 1 to find all the finite subgroups of the group of rotations of $\mathbb{R}^3$. One consequence of this work will be a proof that the five Platonic solids we studied in Section 2 are the only regular polyhedra. This may be an appropriate time to recommend the book *Symmetry* by Hermann Weyl, which makes fascinating—and nontechnical—reading.

We recall that the usual dot product of vectors in $\mathbb{R}^n$ is used to define length and angle in $\mathbb{R}^n$. If $\mathbf{x} = (x_1, \ldots, x_n)$, $\mathbf{y} = (y_1, \ldots, y_n) \in \mathbb{R}^n$, then we define $\mathbf{x} \cdot \mathbf{y} = \sum_{i=1}^{n} x_i y_i$, and set $|\mathbf{x}| = \sqrt{\mathbf{x} \cdot \mathbf{x}} = \left( \sum_{i=1}^{n} x_i^2 \right)^{1/2}$. We define the angle between vectors $\mathbf{x}$ and $\mathbf{y}$ to be $\arccos\left( \frac{\mathbf{x} \cdot \mathbf{y}}{|\mathbf{x}||\mathbf{y}|} \right)$ (cf. Exercise 4).

**Definition.** An isometry of $\mathbb{R}^n$ is a distance-preserving function $f$ from $\mathbb{R}^n$ to $\mathbb{R}^n$, i.e., a function $f \colon \mathbb{R}^n \to \mathbb{R}^n$ satisfying

$$|f(\mathbf{x}) - f(\mathbf{y})| = |\mathbf{x} - \mathbf{y}|, \quad \text{for all } \mathbf{x}, \mathbf{y} \in \mathbb{R}^n.$$

First, we should observe that the set of all isometries of $\mathbb{R}^n$ forms a group (under composition of functions), denoted $\mathrm{Isom}(\mathbb{R}^n)$; see Exercise 1a.. Recall that an $n \times n$ matrix $A$ is called **orthogonal** if $AA^T = A^T A = \mathrm{Id}$, where $A^T$ is the transpose of the matrix $A$ (i.e., the $ij^{\mathrm{th}}$ entry of $A^T$ is the $ji^{\mathrm{th}}$ entry of $A$). Denote by $O(n)$ the set of all

$n \times n$ orthogonal matrices; and by $SO(n)$, those of determinant one. It is easy to check that these are subgroups of $GL(n, \mathbb{R})$, called, respectively, the **orthogonal group** and the **special orthogonal group** (see Exercise 1b.).

**Example 1.** The rotation matrix $\begin{bmatrix} \cos\theta & -\sin\theta \\ \sin\theta & \cos\theta \end{bmatrix}$ is an orthogonal $2 \times 2$ matrix.

The fundamental observation is this:

**Lemma 4.1.** *Let $A$ be an orthogonal $n \times n$ matrix. The linear function $T: \mathbb{R}^n \to \mathbb{R}^n$ given by $T(\mathbf{x}) = A\mathbf{x}$ is an isometry.*

**Proof.** Since $|T(\mathbf{x})|^2 = A\mathbf{x} \cdot A\mathbf{x} = (A^T A\mathbf{x}) \cdot \mathbf{x} = \mathbf{x} \cdot \mathbf{x} = |\mathbf{x}|^2$ (see Exercise 2), we have $|T(\mathbf{x})| = |\mathbf{x}|$. But since $T$ is linear, $T(\mathbf{x}) - T(\mathbf{y}) = T(\mathbf{x} - \mathbf{y})$; and so $|T(\mathbf{x}) - T(\mathbf{y})| = |T(\mathbf{x} - \mathbf{y})| = |\mathbf{x} - \mathbf{y}|$, as required. $\square$

We call such a linear function an orthogonal linear function. More surprisingly, any isometry that fixes the origin must be an orthogonal linear function.

**Proposition 4.2.** *If $f \in \mathrm{Isom}(\mathbb{R}^n)$ is an isometry with $f(\mathbf{0}) = \mathbf{0}$, then there is an orthogonal matrix $A$ so that $f(\mathbf{x}) = A\mathbf{x}$.*

**Proof.** <u>Step 1.</u> Given that $f$ is an isometry fixing $\mathbf{0}$, we have $|f(\mathbf{x})| = |f(\mathbf{x}) - f(\mathbf{0})| = |\mathbf{x} - \mathbf{0}| = |\mathbf{x}|$, so that $f$ preserves lengths of vectors. Using this fact, we prove that $f(\mathbf{x}) \cdot f(\mathbf{y}) = \mathbf{x} \cdot \mathbf{y}$ for all $\mathbf{x}, \mathbf{y} \in \mathbb{R}^n$. We have $|f(\mathbf{x}) - f(\mathbf{y})|^2 = |\mathbf{x} - \mathbf{y}|^2 = |\mathbf{x}|^2 - 2\mathbf{x} \cdot \mathbf{y} + |\mathbf{y}|^2$; on the other hand, $|f(\mathbf{x}) - f(\mathbf{y})|^2 = |f(\mathbf{x})|^2 - 2f(\mathbf{x}) \cdot f(\mathbf{y}) + |f(\mathbf{y})|^2 = |\mathbf{x}|^2 - 2f(\mathbf{x}) \cdot f(\mathbf{y}) + |\mathbf{y}|^2$. We conclude that $f(\mathbf{x}) \cdot f(\mathbf{y}) = \mathbf{x} \cdot \mathbf{y}$, as desired.

<u>Step 2.</u> We next prove that $f$ must be a linear function. Let $\mathbf{e}_1, \ldots, \mathbf{e}_n$ be the standard orthonormal basis for $\mathbb{R}^n$, and let $f(\mathbf{e}_j) = \mathbf{v}_j$, $j = 1, \ldots, n$. It follows from Step 1 that $\mathbf{v}_1, \ldots, \mathbf{v}_n$ is again an orthonormal basis. Given an arbitrary vector $\mathbf{x} \in \mathbb{R}^n$, write $\mathbf{x} = \sum_{i=1}^n x_i \mathbf{e}_i$ and $f(\mathbf{x}) = \sum_{j=1}^n \alpha_j \mathbf{v}_j$. Then

$$x_i = \mathbf{x} \cdot \mathbf{e}_i = f(\mathbf{x}) \cdot \mathbf{v}_i = \left( \sum_{j=1}^n \alpha_j \mathbf{v}_j \right) \cdot \mathbf{v}_i = \alpha_i,$$

and so

$$f(\mathbf{x}) = f\left( \sum_{i=1}^n x_i \mathbf{e}_i \right) = \sum_{i=1}^n x_i \mathbf{v}_i = \sum_{i=1}^n x_i f(\mathbf{e}_i).$$

This formula makes it clear that $f$ is linear.

Step 3. We now know that $f$ is a linear function that carries the standard basis $\mathbf{e}_1, \ldots, \mathbf{e}_n$ for $\mathbb{R}^n$ to another orthonormal basis $\mathbf{v}_1, \ldots, \mathbf{v}_n$. Let $A$ be the matrix representing $f$ with respect to the standard basis. Then the $j^{\text{th}}$ column of $A$ is $A\mathbf{e}_j = f(\mathbf{e}_j) = \mathbf{v}_j$ (see Section 1 of Appendix B). Since the columns form an orthonormal basis for $\mathbb{R}^n$, we have $A^T A = \text{Id}$ (see Exercise 3).   □

**Corollary 4.3.** *The subgroup* $\text{Isom}_0(\mathbb{R}^n) \subset \text{Isom}(\mathbb{R}^n)$ *of isometries of* $\mathbb{R}^n$ *fixing* $\mathbf{0}$ *is isomorphic to* $O(n)$.   □

We now deduce, in a manner reminiscent of our work in Section 5 of Chapter 2, that any isometry of $\mathbb{R}^n$ can be written as the product of a translation and an element of $O(n)$. We first update the definition: a **translation** of $\mathbb{R}^n$ is a function of the form $\tau = \tau_{\mathbf{c}} \colon \mathbb{R}^n \to \mathbb{R}^n$, $\tau_{\mathbf{c}}(\mathbf{x}) = \mathbf{x} + \mathbf{c}$, for some vector $\mathbf{c} \in \mathbb{R}^n$. Note that all translations are isometries, and that the set of translations forms a subgroup *Trans* $\subset \text{Isom}(\mathbb{R}^n)$ (see Exercise 1a.).

**Proposition 4.4.** *Any isometry* $f \in \text{Isom}(\mathbb{R}^n)$ *can be written uniquely in the form* $f = \tau \circ \Upsilon$ *for some translation* $\tau$ *and some* $\Upsilon \in O(n)$.

***Proof.*** Let $f$ be an isometry of $\mathbb{R}^n$, and let $\mathbf{c} = f(\mathbf{0})$. Let $\tau = \tau_{\mathbf{c}}$. Then $\tau^{-1} f(\mathbf{0}) = \mathbf{0}$, so the isometry $\tau^{-1} f$ fixes $\mathbf{0}$ and hence is an element $\Upsilon \in O(n)$. Therefore, $f = \tau\Upsilon$, as required.

Uniqueness is evident, since $\tau$ is uniquely determined and $\Upsilon = \tau^{-1} \circ f$. Here is a more abstract proof, however: Suppose we had $f = \tau\Upsilon = \tau'\Upsilon'$, where $\tau, \tau' \in$ *Trans* and $\Upsilon, \Upsilon' \in O(n)$. Then $\tau'^{-1}\tau = \Upsilon'\Upsilon^{-1}$. But *Trans* $\cap \text{Isom}_0(\mathbb{R}^n)$ consists only of the identity, and so both $\tau'^{-1}\tau$ and $\Upsilon'\Upsilon^{-1}$ are the identity. This means that $\tau = \tau'$ and $\Upsilon = \Upsilon'$.   □

Generalizing the result of Theorem 3.11 of Chapter 6, we have the following theorem.

**Theorem 4.5.** *Trans* $\subset \text{Isom}(\mathbb{R}^n)$ *is a normal subgroup, and*

$$\text{Isom}(\mathbb{R}^n)/\textit{Trans} \cong \text{Isom}_0(\mathbb{R}^n) \cong O(n).$$

***Proof.*** Let $f \in \text{Isom}(\mathbb{R}^n)$ be an arbitrary isometry, and let $\tau = \tau_{\mathbf{c}} \in$ *Trans* be a translation. We must check that $f\tau f^{-1}$ is again a translation. Using Proposition 4.4, we write $f = \tau'\Upsilon$, where $\tau' \in$ *Trans* and $\Upsilon \in O(n)$. Then $f\tau f^{-1} = (\tau'\Upsilon)\tau(\tau'\Upsilon)^{-1} = \tau'(\Upsilon\tau\Upsilon^{-1})\tau'^{-1}$.

Now, since $\Upsilon$ is a linear map, $\Upsilon \tau \Upsilon^{-1}(\mathbf{x}) = \Upsilon(\Upsilon^{-1}(\mathbf{x}) + \mathbf{c}) = \mathbf{x} + \Upsilon(\mathbf{c}) = \tilde{\tau}(\mathbf{x})$, where $\tilde{\tau} = \tau_{\Upsilon(\mathbf{c})} \in \mathit{Trans}$ as well. Thus, $f\tau f^{-1} = \tau'\tilde{\tau}\tau'^{-1} \in \mathit{Trans}$, as required.

Now define a homomorphism $\phi \colon \mathrm{Isom}(\mathbb{R}^n) \to O(n)$ by applying Proposition 4.4: Given $f \in \mathrm{Isom}(\mathbb{R}^n)$, write $f = \tau\Upsilon$, and define $\phi(f) = \Upsilon$. First, $\phi$ is a homomorphism, because if $g = \tau'\Upsilon'$, then $fg = (\tau\Upsilon)(\tau'\Upsilon') = \tau(\Upsilon\tau')\Upsilon'$. By normality of $\mathit{Trans}$, $\tilde{\tau} = \Upsilon\tau'\Upsilon^{-1} \in \mathit{Trans}$, and so $fg = \tau(\Upsilon\tau')\Upsilon' = \tau(\tilde{\tau}\Upsilon)\Upsilon' = (\tau\tilde{\tau})(\Upsilon\Upsilon')$, from which we conclude that $\phi(fg) = \Upsilon\Upsilon' = \phi(f)\phi(g)$, as required. Now $\ker\phi = \mathit{Trans}$, and so we need only to cite the Fundamental Homomorphism Theorem, Theorem 3.9 of Chapter 6. $\square$

What further can we say about the isometries fixing the origin in $\mathbb{R}^3$, i.e., the group $O(3)$? If $A \in O(3)$, then $AA^T = \mathrm{Id}$; and so $\det(AA^T) = (\det A)^2 = 1$ (using Theorem 2.8 of Appendix B), whence $\det A = \pm 1$. We concentrate first on the situation when $\det A = 1$. Recall that $SO(3) = \{A \in O(3) : \det A = +1\}$. This is also called the **rotation group** of $\mathbb{R}^3$, because of the following proposition.

**Proposition 4.6.** *Suppose $A \in SO(3)$. Then there is a nonzero vector $\mathbf{v} \in \mathbb{R}^3$ so that $A\mathbf{v} = \mathbf{v}$ (called an* eigenvector *of $A$); and the linear map $f(\mathbf{x}) = A\mathbf{x}$ is rotation through some angle $\theta$ about the axis determined by $\mathbf{v}$.*

***Proof.*** There is such a vector $\mathbf{v}$ if and only if the matrix $A - \mathrm{Id}$ has nontrivial kernel, i.e., if and only if $\det(A - \mathrm{Id}) = 0$. But since $A \in SO(3)$,

$$\det(A - \mathrm{Id}) = \det(A - AA^T) = \det A(\mathrm{Id} - A^T) = \det A \det(\mathrm{Id} - A^T)$$
$$= \det(\mathrm{Id} - A^T) = \det(\mathrm{Id} - A) = (-1)^3 \det(A - \mathrm{Id}),$$

and hence $\det(A - \mathrm{Id}) = 0$. (If you find this proof deceitful, a more standard proof is suggested in Exercise 14.)

Choose a unit eigenvector $\mathbf{v}$. Now choose an orthonormal basis $\mathbf{v}_1, \mathbf{v}_2, \mathbf{v}_3$ for $\mathbb{R}^3$ with $\mathbf{v}_1 = \mathbf{v}$. Since $f(\mathbf{v}_1) = \mathbf{v}_1$, $f$ must map the vectors $\mathbf{v}_2, \mathbf{v}_3$ to mutually orthogonal unit vectors likewise orthogonal to $\mathbf{v}_1$. This means that $f$ must give an orthogonal linear map from the plane spanned by $\mathbf{v}_2$ and $\mathbf{v}_3$ to itself, i.e., either a rotation of this plane or the product of a rotation and a reflection (see Exercise 6). Since $\det A = +1$, the latter is ruled out. $\square$

To extend this result to all of $O(3)$, we can check that if $\det A = -1$, then there is a nonzero vector $\mathbf{v}$ with $A\mathbf{v} = -\mathbf{v}$ (i.e., $\mathbf{v}$ is an eigenvector with corresponding eigenvalue $-1$) and vectors orthogonal to $\mathbf{v}$ are rotated through some angle (see Exercise 17).

We can now obtain a complete classification of the isometries of $\mathbb{R}^3$, analogous to that obtained of the isometries of $\mathbb{R}^2$ in Chapter 2. Obvious examples of isometries of $\mathbb{R}^3$ are:

**translation:** Given a vector $\mathbf{c} \in \mathbb{R}^3$, let $\tau_{\mathbf{c}}\colon \mathbb{R}^3 \to \mathbb{R}^3$ be given by $\tau_{\mathbf{c}}(\mathbf{x}) = \mathbf{x} + \mathbf{c}$.

**rotation:** We have seen that if $A \in SO(3)$, then the linear map $T(\mathbf{x}) = A\mathbf{x}$ corresponds to a rotation about some axis $\ell \subset \mathbb{R}^3$ through some angle $\theta$.

**reflection:** The map $R(x_1, x_2, x_3) = (x_1, x_2, -x_3)$ is a reflection in the $x_1 x_2$-plane. In the terminology of Section 5 of Chapter 2, this is an improper isometry (since orientation is reversed).

Note that by reflection we always mean reflection in a *plane*. Now, when we consider the composition of these three typical mappings, we are led to define three more types of isometries:

**glide reflection:** Reflect in a plane and then translate by a vector lying in that plane. For example, consider $T(x_1, x_2, x_3) = (x_1 + 1, x_2, -x_3)$.

**rotatory reflection:** Reflect in a plane and then rotate about an axis perpendicular to that plane. For example, consider $T(x_1, x_2, x_3) = (-x_2, x_1 + 1, -x_3)$.

**screw:** Rotate about an axis $\ell$ and then translate by a vector parallel to that axis. For example, consider $T(x_1, x_2, x_3) = (-x_2, x_1, x_3 + 1)$.

It now remains to show that every isometry of $\mathbb{R}^3$ is of one of these six types. The idea is to use Proposition 4.4 to analyze all the possible forms of an isometry. As the first step, we draw the following conclusions from Proposition 4.6.

**Corollary 4.7.** *If $A \in O(3)$, then the linear map $f(\mathbf{x}) = A\mathbf{x}$ is one of the following: the identity, a rotation, a reflection, or the product of a rotation and a reflection.*

**Proof.** If $\det A = +1$, then $f$ is a rotation (if the angle $\theta = 0$, then we get the identity). Now let $\rho(x_1, x_2, x_3) = (x_1, x_2, -x_3)$ be reflection in the $x_1 x_2$-plane; then $\det \rho = -1$. By the product rule for determinants, if $\det A = -1$, then $\rho \circ f$ is a rotation. If we write $\rho \circ f = R$, then $f = \rho^{-1} \circ R = \rho \circ R$ (since $\rho^2 = \mathrm{Id}$); and so $f$ is the product of a rotation and a reflection. $\square$

If $\det A = -1$, is $f(\mathbf{x}) = A\mathbf{x}$ either a reflection or a rotatory reflection? The answer comes from the following lemma.

**Lemma 4.8.** *Let $\ell$ and $H$ be a line and a plane, respectively, passing through the origin. Let $\rho_H$ be reflection in the plane $H$, and let $R_\ell$ be a rotation about the axis $\ell$. Let $f = \rho_H \circ R_\ell$. Then, if $\ell \subset H$, $f$ is a reflection (in a plane); otherwise, $f$ is a rotatory reflection.*

***Proof.*** Suppose first that $\ell \subset H$, and let $\mathbf{v} \in \ell$ be a nonzero vector. Note that $f(\mathbf{v}) = \mathbf{v}$; consider next the action of $f$ on the plane $\Pi$ orthogonal to $\ell$. We see that $R_\ell|_\Pi$ is a rotation of $\Pi$ and $\rho_H|_\Pi$ is a reflection (in the line $\Pi \cap H$). It follows from our work in Chapter 2 that the composition $(\rho_H \circ R_\ell)|_\Pi$ is a reflection of $\Pi$ in a line $\ell' \subset \Pi$. Thus, $f$ is a reflection of $\mathbb{R}^3$ in the plane spanned by $\ell$ and $\ell'$.

Now consider the general case $f = \rho_H \circ R_\ell$. We begin by writing $R_\ell = \rho_{H_1} \circ \rho_{H_2}$ (see Exercise 15a.), with $H_1$ chosen orthogonal to $H$. Now choose $H_3$ orthogonal to both $H$ and $H_1$. By Exercise 15c., $\rho_H \circ \rho_{H_1} \circ \rho_{H_3} = -\mathrm{Id}$, and so

$$f = \rho_H \circ R_\ell = \rho_H \circ (\rho_{H_1} \circ \rho_{H_2}) = (\rho_H \circ \rho_{H_1}) \circ \rho_{H_2}$$
$$= (\rho_H \circ \rho_{H_1}) \circ (\rho_{H_3} \circ \rho_{H_3}) \circ \rho_{H_2}$$
$$= (\rho_H \circ \rho_{H_1} \circ \rho_{H_3}) \circ (\rho_{H_3} \circ \rho_{H_2}) = -\rho_{H_3} \circ \rho_{H_2} = -R_{\ell'},$$

where $\ell' = H_3 \cap H_2$. Now we need only observe that $-R_{\ell'}$ is a rotatory reflection (see Exercise 18).

(Note that in the latter argument we never used the condition that $\ell$ not be contained in $H$. We leave it to the reader to check that if $\ell \subset H$, then the planes $H_2$ and $H_3$ are orthogonal, so that $\rho_{H_3} \circ \rho_{H_2}$ is a rotation through angle $\pi$, from which it follows that $-\rho_{H_3} \circ \rho_{H_2}$ is indeed a reflection.) $\square$

When we introduce translations, we have the analogous question: Is the composition of a translation and a reflection always another reflection or a glide reflection? (Cf. Proposition 5.6 of Chapter 2.) And is the composition of a translation and a rotation always another rotation or a screw?

**Lemma 4.9.** *Let $\tau_{\mathbf{c}}$ be a translation, and let $\rho_H$ be reflection in a plane H (not necessarily passing through the origin). If $\mathbf{c}$ is orthogonal to H, then $\tau_{\mathbf{c}} \circ \rho_H$ is reflection (in a plane parallel to H); otherwise, $\tau_{\mathbf{c}} \circ \rho_H$ is a glide reflection (in a plane parallel to H).*

**Proof.** See Exercise 16.  □

**Lemma 4.10.** *Let $\tau_{\mathbf{c}}$ be a translation, and let $R_\ell$ be rotation about an axis $\ell$ (not necessarily passing through the origin). If $\mathbf{c}$ is orthogonal to $\ell$, then $\tau_{\mathbf{c}} \circ R_\ell$ is a rotation about an axis $\ell'$ parallel to $\ell$; otherwise, $\tau_{\mathbf{c}} \circ R_\ell$ is a screw.*

**Proof.** If $\mathbf{c}$ is orthogonal to $\ell$, then let $\Pi$ be the plane through the origin and orthogonal to $\ell$. Since $\tau_{\mathbf{c}}|_\Pi$ is a translation by the vector $\mathbf{c} \in \Pi$ and $R_\ell|_\Pi$ is a rotation of $\Pi$, we are reduced to the two-dimensional case: the composition $\tau_{\mathbf{c}} \circ R_\ell$ has a fixed point $P$ in $\Pi$, and so is a rotation about $P$. It follows immediately that $\tau_{\mathbf{c}} \circ R_\ell$ is a rotation about the line $\ell'$ parallel to $\ell$ and passing through $P$.

If $\mathbf{c}$ is not orthogonal to $\ell$, write $\mathbf{c} = \mathbf{c}^\| + \mathbf{c}^\perp$, where $\mathbf{c}^\|$ is parallel to $\ell$ and $\mathbf{c}^\perp$ is orthogonal to $\ell$. Since $\tau_{\mathbf{c}} = \tau_{\mathbf{c}^\|} \circ \tau_{\mathbf{c}^\perp}$, we have $\tau_{\mathbf{c}} \circ R_\ell = \tau_{\mathbf{c}^\|} \circ (\tau_{\mathbf{c}^\perp} \circ R_\ell)$, and $\tau_{\mathbf{c}^\perp} \circ R_\ell = R_{\ell'}$, by the first argument, is a rotation about an axis $\ell'$ parallel to $\ell$. But then $\tau_{\mathbf{c}} \circ R_\ell = \tau_{\mathbf{c}^\|} \circ R_{\ell'}$ is a screw, by definition, since $\mathbf{c}^\|$ is parallel to $\ell'$.  □

Finally, we have the following theorem.

**Theorem 4.11.** *Every isometry of $\mathbb{R}^3$ is one of the following:*
(0) *the identity map,·*
(1) *a translation,*
(2) *a rotation,*
(3) *a reflection,*
(4) *a glide reflection,*
(5) *a rotatory reflection, or*
(6) *a screw.*

*Of these, motions (0), (1), (2), and (6) are proper (or orientation-preserving); (3), (4), and (5) are improper (or orientation-reversing).*

**Proof.** By Proposition 4.4, every isometry of $\mathbb{R}^3$ can be written in the form $\tau \circ \Upsilon$, where $\tau$ is a translation and $\Upsilon \in O(3)$. Since $\Upsilon$ is either a rotation, a reflection, or a rotatory reflection (by Lemma 4.8), when $\tau = \mathrm{Id}$, we have nothing to check. Now consider what happens when we follow any of the possible $\Upsilon$'s by a translation $\tau$. When $\Upsilon = \mathrm{Id}$, we obtain a translation. When $\Upsilon$ is a rotation, by

Lemma 4.10 we obtain either a screw or another rotation. When $\Upsilon$ is a reflection, by Lemma 4.9, $\tau \circ \Upsilon$ is either a glide reflection or another reflection. And when $\Upsilon$ is a rotatory reflection, $\tau \circ \Upsilon$ is again a rotatory reflection.

To establish the last statement, we take $\tau = \tau_{\mathbf{c}}$ and $\Upsilon = \rho_H \circ R_\ell$, where $\ell$ is orthogonal to $H$. Since $\tau \circ \Upsilon = (\tau_{\mathbf{c}} \circ \rho_H) \circ R_\ell$, there are two cases to examine. When $\tau_{\mathbf{c}} \circ \rho_H$ is a reflection $\rho_{H'}$ (in a plane $H'$ parallel to $H$), then $\tau \circ \Upsilon = \rho_{H'} \circ R_\ell$ is still obviously a rotatory reflection. When $\tau_{\mathbf{c}} \circ \rho_H$ is a glide reflection, then $\tau_{\mathbf{c}} \circ \rho_H = \tau_{\mathbf{c}'} \circ \rho_{H'}$, where $H'$ is parallel to $H$ and $\mathbf{c}'$ is parallel to $H'$. Moreover, this translation and reflection commute. Thus, $\tau \circ \Upsilon = \rho_{H'} \circ (\tau_{\mathbf{c}'} \circ R_\ell) = \rho_{H'} \circ R_{\ell'}$ (by Lemma 4.10), where $\ell'$ is parallel to $\ell$. Since $\ell'$ is still orthogonal to $H'$, this is once again a rotatory reflection, as required.    □

In summary, we have the following table of the isometries of $\mathbb{R}^3$:

| fixed-point set | proper isometry | improper isometry |
|---|---|---|
| none | translation *or* screw | glide reflection |
| point | ——— | rotatory reflection |
| line | rotation | ——— |
| plane | ——— | reflection |
| all | identity | ——— |

Next, we wish to find all *finite* subgroups of the rotation group $SO(3)$. The result is the following theorem.

**Theorem 4.12.** *If $G \subset SO(3)$ is a finite subgroup, then $G$ is one of the following:*

$C_n$: *the cyclic group of order n (the rotational symmetries of a regular n-gon), $n \geq 1$;*

$\mathcal{D}_n$: *the dihedral group of order 2n (the full symmetry group of a regular n-gon), $n \geq 2$;*

$\mathit{Tetra}$: *the group of symmetries of the regular tetrahedron;*

$\mathit{Cube}$: *the group of symmetries of the cube; or*

$\mathit{Icosa}$: *the group of symmetries of the icosahedron.*

**Proof.** Let $|G| = N$. Every element $g \neq e$ of $G$ is, by Proposition 4.6, a nontrivial rotation about some axis, and thus fixes precisely two points $\mathbf{v}$ and $-\mathbf{v}$ on the unit sphere; we call these the **poles** of $g$. If we let $\mathbf{P}$ denote the set of all the poles, then clearly $G$ acts on $\mathbf{P}$: if $\mathbf{v}$ is a pole of $g$, then for any $h \in G$, $h\mathbf{v}$ is a pole of $hgh^{-1}$, since $hgh^{-1}(h\mathbf{v}) = h\mathbf{v}$. We now apply our counting principles to the action of $G$ on $\mathbf{P}$.

Write $\mathbf{P}$ as the union of disjoint orbits, and let $\mathbf{v}_1,\ldots,\mathbf{v}_r$ be representative poles, one from each orbit. Let $n_j = \#(\mathcal{O}_{\mathbf{v}_j})$, $q_j = |G_{\mathbf{v}_j}|$. Then our first relation is

$$(*) \qquad\qquad |G| = N = n_j q_j.$$

Each element of $G$ besides the identity has two poles; so, counting repetitions, there are $2(N-1)$ poles. What elements have pole $\mathbf{v}$? The answer is obviously the stabilizer subgroup of $\mathbf{v}$, less the identity element; so there are $q_j - 1 \geq 1$ elements of $G$ with pole $\mathbf{v}_j$. On the other hand, there are $n_j$ poles in the same orbit as the pole $\mathbf{v}_j$ (and their respective stabilizer subgroups are conjugate, by Proposition 1.7, and hence have the same number of elements). Thus, using $(*)$), we have

$$2(N-1) = \sum_{j=1}^{r} n_j(q_j - 1) = \sum_{j=1}^{r} (N - n_j);$$

dividing by $N$,

$$(**) \qquad\qquad 2 - \frac{2}{N} = \sum_{j=1}^{r} \left(1 - \frac{1}{q_j}\right).$$

This formula, although at first forbidding, is fabulous. The left-hand side is less than 2 (and at least 1 provided $G$ has at least two elements), whereas the right-hand side is a sum of terms less than 1 and at least $\frac{1}{2}$. Thus, there can be either two or three orbits; i.e., $r = 2$ or $3$. We now consider various cases.

  Two orbits: Then $2 - \frac{2}{N} = (1 - \frac{1}{q_1}) + (1 - \frac{1}{q_2}) \iff \frac{2}{N} = \frac{1}{q_1} + \frac{1}{q_2} \iff 2 = n_1 + n_2$. Thus, $n_1 = n_2 = 1$: There are two poles, each fixed by all elements of the group $G$. Obviously, $G$ is the cyclic group of order $N$ of rotations about the line through the two poles (see Exercise 7).

  Three orbits: Formula $(**)$ reduces in this case to

$$\frac{2}{N} = \frac{1}{q_1} + \frac{1}{q_2} + \frac{1}{q_3} - 1.$$

We may order the $q_j$'s so that $q_1 \leq q_2 \leq q_3$. Since each pole is stabilized by some element other than the identity, we must have $q_j \geq 2$, $j = 1, 2, 3$. But because $q_1, q_2, q_3 \geq 3 \implies \frac{2}{N} \leq 0$, we must have $q_1 = 2$.

  Case 1. Suppose $q_1 = q_2 = 2$. Then $q_3 = q$ can be arbitrary; $N = 2q$ and $n_3 = 2$, so there are two poles $P$ and $P'$ forming one orbit $\mathcal{O}$. Thus, every element $g \in G$ either fixes $P$ and $P'$ or switches them. It is easy to check that $G \cong \mathcal{D}_q$ is the group of symmetries of a regular $q$-gon lying in the plane that is the perpendicular bisector of the line segment $\overline{PP'}$.

<u>Case 2</u>. Suppose $q_1 = 2, q_2 \geq 3$. Note that since $\frac{1}{2} + \frac{1}{4} + \frac{1}{4} - 1 = 0$ and $\frac{1}{2} + \frac{1}{3} + \frac{1}{6} - 1 = 0$, we are left with only three possibilities: $q_2$ must equal 3, and $q_3$ may equal 3, 4, or 5.

(i) $q_1 = 2, q_2 = 3, q_3 = 3$: $N = 12$: $n_1 = 6$, $n_2 = 4$, $n_3 = 4$
(ii) $q_1 = 2, q_2 = 3, q_3 = 4$: $N = 24$: $n_1 = 12$, $n_2 = 8$, $n_3 = 6$
(iii) $q_1 = 2, q_2 = 3, q_3 = 5$: $N = 60$: $n_1 = 30$, $n_2 = 20$, $n_3 = 12$

In case (i), there are four poles $P_1$, $P_2$, $P_3$, $P_4$ in the second orbit, say. The stabilizer subgroup of $P_1$ permutes $P_2$, $P_3$, and $P_4$, and so these points are equidistant from $P_1$. Permuting this reasoning, in fact, all four points are equidistant, and hence they form the vertices of a tetrahedron. Thus we see that $G \subset \textit{Tetra}$, but since $|G| = |\textit{Tetra}| = 12$, it follows that $G = \textit{Tetra}$.

In case (ii), there are six poles $P_1$, $P_2$, ..., $P_6$ in the third orbit. Note, first, that the stabilizer subgroup of $P_1$ must be a subgroup of $SO(2)$ and therefore (see Exercise 7) must be cyclic. Since $q_3 = 4$, a generator of this subgroup must fix $P_1$ and rotate $P_2$, $P_3$, $P_4$, and $P_5$ (say) cyclically; it must therefore fix $P_6$ as well. Also, $P_2$, ..., $P_5$ must be the vertices of a square and equidistant from each of $P_1$ and $P_6$. Reasoning as in the previous case, we see that the six poles must lie at the vertices of a regular octahedron. The eight poles in the second orbit are, correspondingly, the vertices of a cube.

In case (iii), let $P_1$, ..., $P_{12}$ be the twelve poles in the third orbit. The stabilizer subgroup of $P_1$ is of order 5 and must cyclically permute the five nearest neighbors of $P_1$, providing a regular pentagon. Indeed, considering the action of this cyclic group of order 5 on the set of twelve poles, Proposition 1.6 tells us that it must, in fact, fix two poles and cyclically permute the vertices of two parallel pentagons. Pursuing this line of reasoning, we find that the twelve poles must be the vertices of an icosahedron and $G = \textit{Icosa}$. (Cf. *Groups and Symmetry* by Armstrong for more details.)   □

If we take an appropriately group-theoretic definition of a regular polyhedron, it will follow from Theorem 4.12 that the five Platonic solids are the only possible regular polyhedra. It is easy to concoct definitions of a regular polygon: most people would say that all sides of a regular polygon are of equal length and all angles of equal measure. By analogy, all the faces of a regular polyhedron should be congruent regular polygons; but this is not sufficient (consider the hexahedron pictured in Figure 1b. on p. 252). We must require that the polyhedron appear "the same" near each of its vertices (for example, at each vertex, a certain number of faces always come together, and the dihedral angle between two adjacent faces is always the same). In particular, an amazing consequence of a theorem of Cauchy is this: If $\mathcal{P}$ is a polyhedron so that (i) all faces are regular polygons having the same number of edges, and (ii) each vertex belongs to the same number of faces, then $\mathcal{P}$ is a regular polyhedron.

Rather than try to formalize this, we will make the following definition based on group actions. A **polyhedron** in $\mathbb{R}^3$ consists of vertices, edges, and faces. Each edge joins two vertices and belongs to exactly two faces; in addition, any two edges intersect in *at most* one vertex and two faces intersect in *at most* one edge. Given a polyhedron $\mathcal{P} \subset \mathbb{R}^3$, denote by $\text{Isom}(\mathcal{P})$ the subgroup of $\text{Isom}(\mathbb{R}^3)$ preserving $\mathcal{P}$, and by $\text{Isom}^+(\mathcal{P})$ the corresponding subgroup of $\text{Isom}^+(\mathbb{R}^3)$, the group of proper isometries of $\mathbb{R}^3$. Let

$$X = \{(v, e) : v \text{ is a vertex of } \mathcal{P}, e \text{ is an edge of } \mathcal{P}, \text{ and } v \in e\}.$$

We say $\mathcal{P}$ is a **regular polyhedron** if $\text{Isom}^+(\mathcal{P})$ acts transitively on $X$.

**Remark.** The best motivation is to consider the case of a (planar) polygon. It is not sufficient just to require that the symmetry group of the polygon act transitively on the set of vertices (consider a general rectangle) or on the set of edges (consider a general rhombus). However, imposing the requirement that the isometry group of the polygon act transitively on $X$ forces the polygon to be regular. Since the isometry group acts transitively on the set of edges, all the edges have the same length. And since the isometry group acts transitively on $X$, it acts transitively on the angles of the polygon, and so they must all be congruent.

Now note that any isometry preserving $\mathcal{P}$ must fix the center of mass of $\mathcal{P}$ (see Exercise 8.1.3), and we take this to be the origin.

**Lemma 4.13.** *The only proper isometry of $\mathcal{P}$ fixing an element $(v, e) \in X$ is the identity element.*

**Proof.** If an isometry $f$ fixes an edge $e$ of $\mathcal{P}$ and one vertex of that edge, it must fix the other vertex of the edge as well. Since $f$

also fixes the center of mass $O$ of $\mathcal{P}$, it fixes the plane spanned by $O$ and $e$. By Theorem 4.11, the only proper isometry fixing a plane is the identity.  □

From our definition, we are able to deduce the following proposition.

**Proposition 4.14.** *Let $V$, $E$, and $F$ denote the number of vertices, edges, and faces, respectively, of the regular polyhedron $\mathcal{P}$. Suppose each face has $k$ edges and $\ell$ edges meet at every vertex. Then*

(†)                     $|\text{Isom}^+(\mathcal{P})| = \ell V = kF = 2E$.

*$\text{Isom}^+(\mathcal{P})$ acts transitively on the set of faces of $\mathcal{P}$. The stabilizer subgroup of a vertex $v$ of $\mathcal{P}$ has order $\ell$, the stabilizer subgroup of an edge $e$ has order $2$, and the stabilizer subgroup of a face has order $k$.*

*Proof.* By definition, $\text{Isom}^+(\mathcal{P})$ acts transitively on $X$. Since, by Lemma 4.13, only the identity element stabilizes an element of $X$, it follows from Proposition 1.3 that $\#(X) = |\text{Isom}^+(\mathcal{P})|$. But $\#(X) = \#(\text{vertices}) \cdot \#(\text{edges containing a given vertex}) = V \cdot \ell$. In addition, since each edge contains two vertices, $\ell V = 2E$; likewise, each edge is contained in two faces, so $2E = kF$.

Also, $\text{Isom}^+(\mathcal{P})$ acts on the set

$Y = \{(e, f) : e \text{ is an edge of } \mathcal{P}, \ f \text{ is a face of } \mathcal{P}, \text{ and } e \subset f\}$.

Since the only proper isometry of $\mathcal{P}$ that fixes both an edge and a face containing it is the identity, and since $\#(Y) = |\text{Isom}^+(\mathcal{P})|$, it follows that the action is transitive. In particular, $\text{Isom}^+(\mathcal{P})$ acts transitively on the set of faces.

Now the orders of the stabilizer subgroups can be calculated, using Proposition 1.3, immediately from (†).  □

We observe first that $k$ and $\ell$ are at least 3. In order for a face to be a polygon, it must have at least three edges, so $k \geq 3$. Since each edge belongs to precisely two faces, and since those two faces overlap in only that edge, $\ell \geq 3$ as well. Since $\text{Isom}^+(\mathcal{P})$ acts transitively on the set of vertices of $\mathcal{P}$ (and since a vertex is stabilized by a group of order $\ell > 1$), all the vertices form one orbit in the set **P** of poles (see the proof of Theorem 4.12). Since $\text{Isom}^+(\mathcal{P})$ acts transitively on the set of faces of $\mathcal{P}$ (and since a face is stabilized by a group of order $k > 1$), the centers of mass of all the faces form another orbit in **P**. And, lastly, the midpoints of all the edges form a third orbit. So, considering the end of the proof of Theorem 4.12, we have only the following possibilities:

| $|\text{Isom}^+(\mathcal{P})|$ | $V$ | $\ell$ | $F$ | $k$ | polyhedron |
|---|---|---|---|---|---|
| 12 | 4 | 3 | 4 | 3 | tetrahedron |
| 24 | 6 | 4 | 8 | 3 | octahedron |
| 24 | 8 | 3 | 6 | 4 | cube |
| 60 | 12 | 5 | 20 | 3 | icosahedron |
| 60 | 20 | 3 | 12 | 5 | dodecahedron |

In conclusion, the five Platonic solids are the only regular polyhedra.

Schläfli considered these questions in higher dimensions as well in 1850; cf. the books by Berger (especially §1 and §12) and Hilbert and Cohn-Vossen (especially §14 and §23) in the Supplementary Reading.

## EXERCISES 7.4

1.  a.  Prove that $\text{Isom}(\mathbb{R}^n)$ is a group and that *Trans* is a subgroup of $\text{Isom}(\mathbb{R}^n)$.
    b.  Prove that $O(n)$ and $SO(n)$ are subgroups of $GL(n, \mathbb{R})$.

2.  Prove that for any $n \times n$ matrix $A$ and $\mathbf{x}, \mathbf{y} \in \mathbb{R}^n$, $\mathbf{x} \cdot A\mathbf{y} = A^T\mathbf{x} \cdot \mathbf{y}$.

3.  Check that the columns of the $n \times n$ matrix $A$ form an orthonormal basis for $\mathbb{R}^n \iff A^T A = \text{Id} \iff AA^T = \text{Id}$.

4.  Given $\mathbf{x}, \mathbf{y} \in \mathbb{R}^n$, prove that $|\mathbf{x} \cdot \mathbf{y}| \le |\mathbf{x}||\mathbf{y}|$. (Hint: Consider $Q(t) = |\mathbf{x} + t\mathbf{y}|^2$, and observe that this quadratic has at most one real root.)

5.  Let $H \subset \mathbb{R}^3$ be a plane passing through the origin. Let $\mathbf{n}$ be a unit vector orthogonal to $H$. Show that reflection in $H$ is given by the formula $\rho_H(\mathbf{x}) = \mathbf{x} - 2(\mathbf{n} \cdot \mathbf{x})\mathbf{n}$.

6.  Here we dispense with the isometries of $\mathbb{R}^2 = \mathbb{C}$.
    a.  Prove that an orthogonal $2 \times 2$ matrix $A$ can be written either as $\begin{bmatrix} \cos\theta & -\sin\theta \\ \sin\theta & \cos\theta \end{bmatrix}$ or as $\begin{bmatrix} \cos\theta & \sin\theta \\ \sin\theta & -\cos\theta \end{bmatrix}$ for some $\theta \in \mathbb{R}$.
    b.  Show that the second matrix above is the product of a rotation and a reflection. Show, moreover, that if $A \in SO(2)$, then $A$ is necessarily a rotation.

c.  Give a new proof of Theorem 5.5 of Chapter 2.

7.  Prove that any finite subgroup of $SO(2)$ must be cyclic. (Hint: By Exercise 6, each element $g_j$ is given by rotation through some angle $\theta_j$. There must be a number $\theta_j$ of smallest absolute value.) What are all the finite subgroups of $O(2)$?

8.  Prove that any two reflections of $\mathbb{R}^3$ are conjugate in $\mathrm{Isom}(\mathbb{R}^3)$.

9.  Let $O(n)$ act on itself by conjugation. What are the orbits?

10. Determine the type of the isometry $T(\mathbf{x}) = A\mathbf{x}$ explicitly for $A =:$

a.
$$\begin{bmatrix} \frac{2}{3} & \frac{1}{3} & -\frac{2}{3} \\ -\frac{1}{3} & -\frac{2}{3} & -\frac{2}{3} \\ \frac{2}{3} & -\frac{2}{3} & \frac{1}{3} \end{bmatrix}$$
b.
$$\begin{bmatrix} 1 & 0 & 0 \\ 0 & 0 & -1 \\ 0 & -1 & 0 \end{bmatrix}$$

11. What are the poles of the group $Tetra \subset SO(3)$? $Cube$?

12. What is the geometric interpretation of the poles in the first orbits in case 2 in the proof of Theorem 4.12?

13. Carry through the reasoning at the end of the proof of Theorem 4.12 carefully. In particular, show that the eight poles in the second orbit in case 2(ii) must be the vertices of a cube.

14. Prove that if $A \in SO(3)$, then 1 is an eigenvalue of $A$, as follows.
    a.  Let $p(t) = \det(A - t\mathrm{Id})$ be the characteristic polynomial of $A$. This is a real cubic polynomial, and hence must have a real root (why?).
    b.  Prove that if $A \in O(3)$ and $A\mathbf{v} = \lambda\mathbf{v}$ for some $\mathbf{v} \neq 0$, then $|\lambda| = 1$. (Hint: Compute $A\mathbf{v} \cdot A\mathbf{v}$.)
    c.  Prove that if $\det A = 1$, then $+1$ must be a root of $p(t)$. (Hint: The product of the roots of $p(t)$ is $\det A$.)

15. a.  Prove that any rotation $R_\ell$ of $\mathbb{R}^3$ is the product of two reflections $\rho_{H_1}$ and $\rho_{H_2}$ (see Exercise 2.5.6). (Hint: For starters, choose planes $H_1$ and $H_2$ intersecting in $\ell$.) Show, moreover, that the angle between the two planes will be one-half the rotation angle.
    b.  Prove that two reflections $\rho_H$ and $\rho_{H'}$ commute if and only if $H$ and $H'$ are orthogonal planes.

    c.   Check that if $H_1$, $H_2$, and $H_3$ are mutually orthogonal planes through the origin in $\mathbb{R}^3$, then $\rho_{H_1} \circ \rho_{H_2} \circ \rho_{H_3} = -\mathrm{Id}$.

16.    Prove Lemma 4.9. (Hint: Decompose $\mathbf{c} = \mathbf{c}^{\|} + \mathbf{c}^{\perp}$, where $\mathbf{c}^{\|}$ is parallel to $H$ and $\mathbf{c}^{\perp}$ is orthogonal to $H$. If necessary, see the proof of Proposition 5.6 of Chapter 2.)

17.    Prove directly that if $A \in O(3)$ and $\det A = -1$, then there is an orthonormal basis $\mathbf{v}_1, \mathbf{v}_2, \mathbf{v}_3$ for $\mathbb{R}^3$ with respect to which the matrix for $A$ is $\begin{bmatrix} \cos\theta & -\sin\theta & \\ \sin\theta & \cos\theta & \\ & & -1 \end{bmatrix}$. (Hint: Start by proving that $-1$ is an eigenvalue of $A$.)

18.    Let $R_\ell$ be a rotation about axis $\ell$ through angle $\theta$. Show that $-R_\ell$ is a rotatory reflection. (Hint: Assuming $\ell$ passes through the origin, show that with respect to an appropriate basis, the matrix for $R_\ell$ is $\begin{bmatrix} \cos\theta & -\sin\theta & \\ \sin\theta & \cos\theta & \\ & & 1 \end{bmatrix}$. Now cf. Exercise 6.)

19.    Let $A$ be a symmetric $n \times n$ matrix.
    a.   Show that if $A\mathbf{x} \cdot \mathbf{y} = 0$ for all $\mathbf{x}$ and $\mathbf{y} \in \mathbb{R}^n$, then $A = \mathbf{0}$.
    b.   Show that if $A\mathbf{x} \cdot \mathbf{x} = 0$ for all $\mathbf{x} \in \mathbb{R}^n$, then $A = \mathbf{0}$.
        (Hint: Consider $A(\mathbf{x} + \mathbf{y}) \cdot (\mathbf{x} + \mathbf{y})$.)
    c.   Which of parts a. and b. remains true for a general $n \times n$ matrix? Give proofs or counterexamples.

20.    Give a self-contained proof of the converse of Lemma 4.1: if $A$ is an $n \times n$ matrix such that $|A\mathbf{x}| = |\mathbf{x}|$ for all $\mathbf{x} \in \mathbb{R}^n$, then $A$ is orthogonal. (Hint: Show that $A^T A \mathbf{x} \cdot \mathbf{x} = \mathbf{x} \cdot \mathbf{x}$ for all $\mathbf{x} \in \mathbb{R}^n$.)

21.    Give solid figures in $\mathbb{R}^3$ whose isometry groups are:
    a.   $SO(2)$
    b.   $O(2)$
    c.   $\left\{ \begin{bmatrix} \pm 1 & \\ \hline & A \end{bmatrix} : A \in SO(2) \right\}$

22.    Prove that a polyhedron $\mathcal{P}$ is regular if and only if its *full* group of isometries (proper and improper) acts transitively on the set

$$Z = \{(v, e, f) : v \text{ is a vertex of } \mathcal{P}, \ e \text{ is an edge of } \mathcal{P},$$
$$f \text{ is a face of } \mathcal{P}, \text{ and } v \in e \subset f\}.$$

23. Determine the isometry groups of the following figures:
    a. a rectangular parallelepiped (box) with precisely two square faces
    b. a hexahedron (obtained by gluing two regular tetrahedra together)
    c. a truncated cube
    d. a truncated octahedron
    e. a great dodecahedron

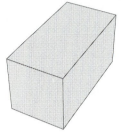

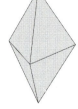

a.                          b.                          c.

d.                          e.

FIGURE 1

## 5. Direct Products, $p$-Groups, and Sylow Theorems

In this section, we will study some results in group theory that are more technical, but exhibit the power of analyzing group actions. The goal is to understand when a finite group will have a subgroup of a given order. We know a necessary condition from Corollary 3.3 of Chapter 6: the order must be a divisor of $|G|$. But we have also seen that this condition is not sufficient, since $|A_4| = 12$ and $A_4$ has no subgroup of order 6 (see Exercises 6.4.12 and 7.2.3).

We begin by introducing a simple way of constructing more complicated groups, the notion of direct product of groups; we've already
seen this construction for rings in Section 2 of Chapter 4.

**Definition.** Let $G$ and $G'$ be groups. Define their **direct product**
$G \times G'$ to be the set of ordered pairs $(g, g')$, $g \in G$, $g' \in G'$, together
with the operation $(g_1, g_1') \cdot (g_2, g_2') = (g_1 g_2, g_1' g_2')$. Note that the
products are computed in each component separately.

**Lemma 5.1.** *Using this definition, $G \times G'$ is a group.*

**Proof.** Associativity follows from associativity in the respective
factors $G$ and $G'$. If $e \in G$ and $e' \in G'$ are the respective identity
elements, then it is easy to check that $e_{G \times G'} = (e, e')$ is the identity
element in $G \times G'$. Inverses are given component-by-component:
$(g, g')^{-1} = (g^{-1}, g'^{-1})$, since $(g, g') \cdot (g^{-1}, g'^{-1}) = (g g^{-1}, g' g'^{-1}) =
(e, e') = e_{G \times G'}$. $\square$

**Examples 1.**

(a) Consider the group $\mathbb{Z}_2 \times \mathbb{Z}_3$. Since the operation in both
groups is written as addition, we shall do the same in the
product. The elements of the group are: $(0, 0)$, $(1, 0)$, $(0, 1)$,
$(0, 2)$, $(1, 1)$, and $(1, 2)$. Since $\mathbb{Z}_2 \times \mathbb{Z}_3$ is an abelian group
of order 6, it follows from Theorem 3.7 of Chapter 6 that
$\mathbb{Z}_2 \times \mathbb{Z}_3 \cong \mathbb{Z}_6$. Of course, we can check this directly, since

$$
\begin{aligned}
(1, 1) &= (1, 1) \\
2(1, 1) &= (0, 2) \\
3(1, 1) &= (1, 0) \\
4(1, 1) &= (0, 1) \\
5(1, 1) &= (1, 2) \\
6(1, 1) &= (0, 0) .
\end{aligned}
$$

(b) In contrast, consider the group $\mathbb{Z}_3 \times \mathbb{Z}_3$. This group has order 9, but cannot be isomorphic to $\mathbb{Z}_9$, since every element
$(a, b) \in \mathbb{Z}_3 \times \mathbb{Z}_3$ (other than the identity) has order 3.

(c) The stabilizer subgroup $G_s$ in Example 5 of Section 1 is naturally the direct product of the permutation group of $\{1, \ldots, k\}$
and the permutation group of $\{k + 1, \ldots, n\}$. Since the latter
set has $n - k$ elements, we see that $G_s \cong S_k \times S_{n-k}$.

Motivated by the first two examples, we prove the following general proposition.

**Proposition 5.2.** *The direct product $\mathbb{Z}_m \times \mathbb{Z}_n$ is isomorphic to the cyclic group $\mathbb{Z}_{mn}$ if and only if $\gcd(m, n) = 1$.*

**Proof.** Consider the multiples of the identity element $(1, 1) \in \mathbb{Z}_m \times \mathbb{Z}_n$. If $r(1, 1) = (0, 0)$, then it must be that $m | r$ and $n | r$, and so $r$ is a common multiple of $m$ and $n$. If $m$ and $n$ are relatively prime, then the smallest possible such $r$ is $mn$ (see Exercise 1.2.13). Therefore, the cyclic subgroup $\langle (1, 1) \rangle$ must be the whole group $\mathbb{Z}_m \times \mathbb{Z}_n$, and the group is cyclic.

On the other hand, if $\gcd(m, n) = d > 1$, then given any $(a, b) \in \mathbb{Z}_m \times \mathbb{Z}_n$, $\frac{mn}{d}(a, b) = (\frac{n}{d}ma, \frac{m}{d}nb) = (0, 0)$. Thus, every element of $\mathbb{Z}_m \times \mathbb{Z}_n$ has order at most $\frac{mn}{d}$, and the group cannot be cyclic. $\quad\square$

Observe that the direct product $G \times G'$ contains obvious copies of $G$ and $G'$ as normal subgroups: $G \times \{e'\}$ and $\{e\} \times G' \subset G \times G'$ (see Exercise 2). This leads us to ask when a given group $G$ might be isomorphic to the direct product of two of its normal subgroups.

**Proposition 5.3.** *Suppose $H, K \subset G$ are normal subgroups so that*

     (i) *$H \cap K = \{e\}$, and*
     (ii) *every element $g \in G$ can be written in the form $g = hk$ for some $h \in H$ and $k \in K$.*

*Then $G \cong H \times K$.*

**Proof.** Define a function $\phi: H \times K \to G$ by $\phi(h, k) = hk$. We must first prove $\phi$ is a homomorphism. We have $\phi((h_1, k_1)(h_2, k_2)) = \phi(h_1 h_2, k_1 k_2) = (h_1 h_2)(k_1 k_2)$. But why should this be equal to $\phi(h_1, k_1)\phi(h_2, k_2) = (h_1 k_1)(h_2 k_2)$? The key is that when $H$ and $K$ are normal subgroups intersecting only in $e$, then elements of $H$ must commute with elements of $K$. This was the content of Exercise 6.3.16, but we give the proof here: given $h \in H$ and $k \in K$, consider $hkh^{-1}k^{-1}$. On one hand, $hkh^{-1}k^{-1} = (hkh^{-1})k^{-1} \in K$ by normality of $K$; on the other, $hkh^{-1}k^{-1} = h(kh^{-1}k^{-1}) \in H$ by normality of $H$. Therefore, $H \cap K = \{e\} \implies hkh^{-1}k^{-1} = e \implies hk = kh$. Thus, returning to our proof, $h_1(h_2 k_1)k_2 = h_1(k_1 h_2)k_2 = (h_1 k_1)(h_2 k_2)$, and so $\phi((h_1, k_1)(h_2, k_2)) = \phi(h_1, k_1)\phi(h_2, k_2)$, as required.

The homomorphism maps onto $G$ by hypothesis (ii). And the homomorphism is one-to-one because, by (i), $\phi(h, k) = e \implies hk = e \implies k = h^{-1} \in H \cap K \implies h = k = e$. $\quad\square$

**Example 2.** Consider the group $\mathbb{Z}_{16}^{\times}$ of units in $\mathbb{Z}_{16}$ (with operation multiplication). By inspection, $\mathbb{Z}_{16}^{\times} = \{1, 3, 5, 7, 9, 11, 13, 15\}$ has order 8. Since the group is abelian, any subgroup is normal. Let $H = \{1, 3, 9, 11\}$ and $K = \{1, 15\}$. Then $H$ and $K$ are both cyclic; also, writing $15 = -1$, it is easy to see that every element of $\mathbb{Z}_{16}^{\times}$ can be written as the product of an element of $H$ and an element of $K$. Thus, the hypotheses of Proposition 5.3 are satisfied, and therefore $\mathbb{Z}_{16}^{\times} \cong H \times K \cong \mathbb{Z}_4 \times \mathbb{Z}_2$.

In the rest of this section we will focus on the action of a finite group $G$ on itself, first of all by conjugation. The orbit $\mathcal{O}_x$ of an element $x$ consists of all the conjugates of $x$, and hence is called its **conjugacy class**. Recall from Corollary 1.5 that $\#(\mathcal{O}_x) = |G|/|G_x|$, where the stabilizer subgroup $G_x = \{g : gxg^{-1} = x\} = \{g : gx = xg\}$ consists of all elements commuting with $x$. The **center** $Z$ of the group $G$ consists of the elements that commute with *all* elements of the group: i.e., $Z = \{x : gx = xg \text{ for all } g \in G\}$. Thus, $x \in Z \iff G_x = G \iff \mathcal{O}_x = \{x\}$. (Note that $Z \subsetneq G_x$, unless $G$ is abelian.)

We can rewrite the beautiful formula in Proposition 1.6 as follows, the so-called **class equation**:

$$(*) \qquad |G| = |Z| + \sum_{\substack{\text{distinct orbits} \\ \#(\mathcal{O}_x) > 1}} \#(\mathcal{O}_x).$$

We now apply this formula to the study of groups $G$ whose order is a power of the prime number $p$, called $p$-**groups** for short.

**Proposition 5.4.** *Let $p$ be prime. The center $Z$ of a nontrivial $p$-group contains more than one element.*

**Proof.** Let $|G| = p^s$, $s \geq 1$. If $Z = G$, we are done. If not, there is $x \in G$ so that $\#(\mathcal{O}_x) > 1$. Then, as we said earlier, $\#(\mathcal{O}_x) = |G|/|G_x|$, and hence must be a positive power of $p$. Considering formula $(*)$, $p$ divides the left-hand side and the giant sum on the right-hand side. It follows that $p$ divides $|Z|$. Since $e \in Z \implies |Z| \geq 1$, we now infer that $|Z|$ is a positive power of $p$. $\square$

**Proposition 5.5.** *Let $p$ be prime and $|G| = p^2$. Then $G$ is abelian.*

**Proof.** We must show that the center $Z$ is the whole group $G$. From Proposition 5.4 we infer that $|Z|$ must be either $p$ or $p^2$. Suppose $|Z| = p$; choose $x \notin Z$, and note that $G_x$, the set of elements of $G$ that commute with $x$, contains $x$ *and* all the elements of the center. Therefore, $|G_x| > |Z| = p$; but since $G_x$ is a subgroup of $G$, $|G_x|$ divides $|G| = p^2$. The only possibility is that $G_x = G$; but this means that $x \in Z$, contradicting our assumption. It follows, then, that $|Z| = p^2$, i.e., $Z = G$; and so, $G$ is abelian. $\quad\square$

**Remark.** Groups of order $p^3$ may well be non-abelian, however. The groups $\mathcal{Q}$ and $\mathcal{D}_4$ of order 8 come to mind.

We conclude this section with a discussion of the Sylow Theorems, which are a powerful tool for understanding finite groups. In particular, they allow us to come to terms with subgroups of prime power order. If $|G| = n$ and $p^\alpha$ is the largest power of $p$ dividing $n$, we call a subgroup of order $p^\alpha$ a **Sylow $p$-subgroup** of $G$.

**Theorem 5.6** (Sylow I). *Let $G$ be a finite group of order $n$, and suppose $p|n$. Then there is a Sylow $p$-subgroup $H$ of $G$.*

**Corollary 5.7** (Cauchy's Theorem). *Let $G$ be a finite group of order $n$, and suppose $p|n$, $p$ prime. Then there is an element $a \in G$ of order $p$.*

***Proof of Corollary 5.7.*** Let $H$ be a Sylow $p$-subgroup, and let $h \in H$ be an element other than the identity. Then the order of $h$ is a power of $p$, say $p^s$, $s \geq 1$. Let $a = h^{p^{s-1}}$; then $a$ is an element of order $p$. $\quad\square$

***Proof of Theorem 5.6.*** Set $n = p^\alpha m$, where $p \nmid m$. We consider the action of $G$ by left multiplication on the set $S = \{$subsets of $G$ with $p^\alpha$ elements$\}$: if $\xi = \{a_1, a_2, \ldots, a_{p^\alpha}\}$ is an element of $S$, then $g \cdot \xi = \{ga_1, ga_2, \ldots, ga_{p^\alpha}\}$. We will unearth the desired subgroup by finding an element $\xi \in S$ whose stabilizer subgroup has order $p^\alpha$. Now, $\#(S) = \binom{n}{p^\alpha}$, and it is straightforward to check that, since $p^\alpha$ is the largest power of $p$ dividing $n$, $p \nmid \#(S)$ (see Exercise 12). We apply Proposition 1.6: $\#(S) = \sum_{\text{distinct orbits } \mathcal{O}} \#(\mathcal{O})$. Since $p \nmid \#(S)$, it follows that there is some orbit $\mathcal{O}_\xi$ so that $p \nmid \#(\mathcal{O}_\xi)$. But then we infer from Corollary 1.4 that $|G_\xi| = n/\#(\mathcal{O}_\xi)$ is divisible by $p^\alpha$.

On the other hand, if a subgroup $H \subset G$ stabilizes $\xi \subset G$, this means that $\xi$ is the union of orbits of the action of $H$ on $G$ by left multiplication. Every such orbit is of the form $Ha$, $a \in G$, and contains $|H|$ elements. It follows, then, that $\#(\xi)$ is divisible by $|H|$. (The reader should compare this argument with the proof of Lagrange's Theorem, Theorem 3.2 of Chapter 6.) In our case, we set $H = G_\xi$; and so $|G_\xi| \mid \#(\xi)$, i.e., $|G_\xi| \mid p^\alpha$. But we have already proved $p^\alpha \mid |G_\xi|$. Thus, $|G_\xi| = p^\alpha$, as desired. $\quad\square$

Now come the remaining theorems. We give some easy applications before the proofs.

**Theorem 5.8** (Sylow II and III). *All Sylow p-subgroups are conjugate, and the number s of Sylow p-subgroups divides n and satisfies $s \equiv 1 \pmod{p}$.*

**Corollary 5.9.** *Let H be a Sylow p-subgroup of the group G. Then H is a normal subgroup if and only if it is the only Sylow p-subgroup.*

**Proof.** Suppose $H$ is the unique Sylow $p$-subgroup. Since, for any $a \in G$, the subgroup $aHa^{-1}$ is again a Sylow $p$-subgroup, we must have $aHa^{-1} = H$ for all $a \in G$. This means $H$ is normal. Conversely, if $H$ is normal, $aHa^{-1} = H$ for all $a \in G$; but since by Theorem 5.8 all the Sylow $p$-subgroups are conjugate, $H$ can be the only one.   □

**Examples 3.** The group $\mathcal{T}$ (of order 6) has one Sylow 3-subgroup (which is necessarily normal), and three Sylow 2-subgroups. The group $\mathcal{T}etra$ (of order 12) has one Sylow 2-subgroup (see Section 2) and four Sylow 3-subgroups. On the other hand, the group $\mathcal{D}_6$ (the group of symmetries of the regular hexagon, which also has order 12) has three Sylow 2-subgroups and one Sylow 3-subgroup. (The Sylow 2-subgroups are all Klein 4-groups and are the symmetry groups of the three rectangles formed by pairs of opposite edges.)

**Example 4.** Every group of order 15 is cyclic.

**Proof.** Let $G$ be a group of order 15. By Theorem 5.8 the number of Sylow 3-subgroups divides 15 and is congruent to 1 (mod 3), and therefore must equal 1. The same is true for $p = 5$. Thus we have, by Corollary 5.9, *normal* subgroups $H$ and $K$ of orders 3 and 5, respectively. Now we claim that $G \cong H \times K$; to apply Proposition 5.3, we must check that $H \cap K = \{e\}$ and that every element of $G$ can be written as the product of elements of $H$ and $K$. Since $H \cap K$ is a subgroup of both $H$ and $K$, its order must divide both 3 and 5; thus $H \cap K = \{e\}$. On the other hand, by Exercise 13, $HK = \{hk : h \in H, k \in K\}$ is a subgroup of $G$ strictly containing $K$; therefore, it must equal $G$. To finish the proof, we merely observe that $\mathbb{Z}_3 \times \mathbb{Z}_5 \cong \mathbb{Z}_{15}$ by Proposition 5.2, and so $G \cong \mathbb{Z}_{15}$, as required.   □

**Example 5.** A group of order 14 is isomorphic either to $\mathbb{Z}_{14}$ or to $\mathcal{D}_7$ (see Exercise 6.2.15).

***Proof.*** Let $G$ be a group of order 14. By Theorem 5.8, the Sylow 7-subgroup of $G$ is normal, and there are either one or seven Sylow 2-subgroups.

Case 1. If there is one Sylow 2-subgroup, then it too is normal; and, reasoning as in Example 4, $G \cong \mathbb{Z}_2 \times \mathbb{Z}_7 \cong \mathbb{Z}_{14}$.

Case 2. If there are seven Sylow 2-subgroups $H_1, H_2, \ldots, H_7$, then let $\psi$ be the generator of $H_1$. Let $K$ be the Sylow 7-subgroup, generated by $\rho$. Then the conjugate elements $\rho^i \psi \rho^{-i}$, $i = 0, 1, 2, \ldots, 6$, generate the subgroups $H_1, \ldots, H_7$. Since $K$ is normal, $\psi \rho \psi^{-1} = \rho^j$ for some $j = 1, 2, \ldots, 6$. As a result, $\rho = \psi^2 \rho \psi^{-2} = \psi(\psi \rho \psi^{-1})\psi^{-1} = \psi(\rho^j)\psi^{-1} = (\psi \rho \psi^{-1})^j = (\rho^j)^j = \rho^{j^2}$. Thus, $j^2 \equiv 1 \pmod 7$. The only solutions are $j = \pm 1$; since $G$ is not abelian, $j \neq 1$. This leaves $\psi \rho \psi^{-1} = \rho^{-1}$, and, by definition, $G \cong \mathcal{D}_7$. $\square$

***Proof of Theorem 5.8.*** We must first establish that if $H$ and $K$ are Sylow $p$-subgroups of $G$, then $K = aHa^{-1}$ for some $a \in G$. The key idea is to let $K$ act on $G/H$ by left multiplication. By Proposition 1.6, $\#(G/H) = \sum_{\text{distinct orbits } \mathcal{O}} \#(\mathcal{O})$ is a sum of factors of $|K|$ (all of which are powers of $p$). Since $\#(G/H) = |G|/|H|$ is not divisible by $p$, there must be an element $aH \in G/H$ with $\#(\mathcal{O}_{aH}) = 1$. That is, there is a coset $aH$ whose $K$-orbit is just itself; but this means that $k(aH) \in aH$ for all $k \in K$, and so $K \subset aHa^{-1}$ (see Proposition 1.7). Since $|K| = |H| = |aHa^{-1}|$, it follows that $K = aHa^{-1}$, as desired.

Next, we must prove that the number $s$ of Sylow subgroups divides $n$ and satisfies $s \equiv 1 \pmod p$. What could be more natural, then, than to consider the set of Sylow subgroups? Let $S = \{H = H_1, H_2, \ldots, H_s\}$ be this set; by the first part of the theorem, we know that this is also the set of conjugate subgroups $aHa^{-1}$ of $H$. From the latter observation we infer that $G$ acts transitively (by conjugation) on $S$, and so $s \mid n$ (why?). It will be important in a moment to know the stabilizer subgroup of the element $H_i \in S$: it is the subgroup $N_i = \{g \in G : gH_ig^{-1} = H_i\}$, called the **normalizer** of $H_i$; and it is immediate from the definition that $H_i$ is a normal subgroup of $N_i$.

To prove that $s \equiv 1 \pmod p$, we need a different group action: now let $H$ (rather than all of $G$) act on $S$ by conjugation. As usual, $s = \#(S) = \sum_{\text{distinct orbits } \mathcal{O}} \#(\mathcal{O})$, and every summand is a power of $p$. As before, we ask when $\#(\mathcal{O}_{H_i}) = 1$. Well, $\#(\mathcal{O}_{H_i}) = 1$ if and only if the stabilizer subgroup of $H_i \in S$ is $H$, and this happens (by our earlier remarks) if and only if $H \subset N_i$. Here comes the clincher: if $H \subset N_i$, then $H$ and $H_i$ are both Sylow subgroups of $N_i$; and so, by the first part of the theorem, $H$ is conjugate to $H_i$ in $N_i$. But $H_i$ is a normal subgroup of $N_i$, and so $H = H_i$. Thus, $\#(\mathcal{O}_{H_i}) = 1 \iff H_i = H$; there is only one summand equal to 1, all the others being higher powers of $p$. This proves that $s \equiv 1 \pmod{p}$, and we are done.  $\square$

## EXERCISES 7.5

1. Prove that $\mathcal{V} \cong \mathbb{Z}_2 \times \mathbb{Z}_2$.

2. Prove that $G \times \{e'\}$ and $\{e\} \times G'$ are normal subgroups of the direct product $G \times G'$.

3. Suppose $g \in G$ has order $m$ and $g' \in G'$ has order $n$. What is the order of the element $(g, g') \in G \times G'$?

4. Decide in each of the following cases whether $G \cong H \times K$:
   a.  $G = \mathbb{C}^{\times}$, $H = \{z \in \mathbb{C} : |z| = 1\}$, $K = \mathbb{R}^{+}$
   b.  $G = \mathcal{D}_n$, $H = \langle \rho \rangle$, $K = \langle \psi \rangle$ (see Exercise 6.2.15)
   c.  $G = \left\{ \begin{bmatrix} a & b \\ 0 & c \end{bmatrix} : a, b, c \in \mathbb{R},\ ac \neq 0 \right\}$,
   
   $H = \left\{ \begin{bmatrix} a & 0 \\ 0 & c \end{bmatrix} : ac \neq 0 \right\}$, $K = \left\{ \begin{bmatrix} 1 & b \\ 0 & 1 \end{bmatrix} : b \in \mathbb{R} \right\}$

5. Is the group $\mathcal{T}$ isomorphic to the direct product of two nontrivial groups?

6. a.  Prove that $G \times G'$ is an abelian group if and only if $G$ and $G'$ are abelian groups.
   b.  Prove that $\Delta = \{(g,g) : g \in G\} \subset G \times G$ is a normal subgroup if and only if $G$ is abelian.

7. Identify the following groups, and decide which are isomorphic.
   a.  $\mathbb{Z}_8^{\times}$
   b.  $\mathbb{Z}_{12}^{\times}$
   c.  $\mathbb{Z}_{15}^{\times}$
   d.  $\mathbb{Z}_{16}^{\times}$
   e.  $\mathbb{Z}_{21}^{\times}$

8. For those who have studied Section 4:
   a.  Is $O(n) \cong SO(n) \times \{\pm \operatorname{Id}\}$?

b. Do Proposition 4.4 and Theorem 4.5 imply that $\text{Isom}(\mathbb{R}^n) \cong \textit{Trans} \times O(n)$?

9. Let $p$ be a prime. Prove that a group of order $p^2$ is either cyclic or isomorphic to the direct product of two groups of order $p$.

10. Suppose $|G| = 81$ and $|Z| \geq 27$. Prove $G$ is abelian.

11. Here is another proof of Cauchy's Theorem (Corollary 5.7). Let

$$S = \{(a_1, a_2, \ldots, a_p) : a_i \in G \text{ and } a_1 a_2 \cdots a_p = e\}.$$

a. Show that $\#(S) = |G|^{p-1}$.
b. Let $\mathbb{Z}_p$ be the cyclic group of order $p$ with generator $\sigma$, and define an action of $\mathbb{Z}_p$ on $S$ by

$$\sigma(a_1, \ldots, a_p) = (a_p, a_1, a_2, \ldots, a_{p-1}).$$

Note that $(e, e, \ldots, e)$ is a fixed point of the action. Apply Proposition 1.6 to prove that there must be another fixed point $(a, a, \ldots, a)$, $a \neq e$.
c. Deduce Cauchy's Theorem.

12. Prove that if $n = p^\alpha m$ and $p \nmid m$, then $p \nmid \binom{n}{p^\alpha}$.

13. Let $H$ and $K$ be subgroups of $G$ with $H$ a *normal* subgroup. Prove that $HK = \{hk : h \in H, k \in K\}$ is a subgroup of $G$.

14. Prove that if $p$ and $q$ are primes, $p < q$, and $p \nmid q - 1$, then any group of order $pq$ is cyclic.

15. Let $p$ be a prime. What are the Sylow $p$-subgroups of $S_p$? How many are there?

16. Prove that every group of order 45 has a normal subgroup of order 9.

17. Prove that a group of order
   a. 42
   b. 56
   must have a proper normal subgroup.

18. Suppose $p$ is prime, $\alpha \geq 1$, and $1 \leq m < p$. If $G$ is a group of order $n = p^\alpha m$, then prove $G$ has a proper normal subgroup.

19. Show that $S_4$ has three Sylow 2-subgroups; these are probably most easily found if we think of $S_4$ as *Cube*. Then $S_4$ acts by

conjugation on the set of these subgroups. This gives a homo-morphism $S_4 \to S_3$. What are its kernel and image?

20.  Prove that when $n$ is odd, $\mathcal{D}_{2n} \cong \mathcal{D}_n \times \mathbb{Z}_2$. (Hint: Let $\mathbb{Z}_2 = \{e, \rho^n\}$, where $\mathcal{D}_{2n}$ is presented as in Exercise 6.2.15.) Is the same true when $n$ is even?

21.  Prove that $\binom{2n}{n} = \sum\limits_{k=0}^{n} \binom{n}{k}^2$. (Hint: see Example 5 of Section 1. Let $G = S_n \times S_n$ act on $\{1, \ldots, 2n\}$ as follows:

$$(\sigma, \tau)(i) = \begin{cases} \sigma(i), & \text{if } 1 \le i \le n \\ n + \tau(i - n), & \text{if } n + 1 \le i \le 2n \end{cases}.$$

Now let $S$ be the set of $n$-element subsets of $\{1, \ldots, 2n\}$; note that every such subset is of the form $\{i_1, \ldots, i_k, j_{k+1}, \ldots, j_n\}$, where $0 \le k \le n$, $1 \le i_1, \ldots, i_k \le n$, and $n + 1 \le j_{k+1}, \ldots, j_n \le 2n$. Analyze the action of $G$ on $S$.)

22.  Prove (by induction on $n$) that a group of order $p^n$ contains a normal subgroup of order $p^{n-1}$.

23.  Suppose $|G| = 595 = 5 \cdot 7 \cdot 17$ and $H$ is a subgroup of $G$ of order 5.
     a.  Prove that $H$ is a normal subgroup.
     b.  Prove that $H$ must be contained in the center of $G$.

24.  Find the Sylow $p$-subgroups of $GL(2, \mathbb{Z}_p)$.

25.  Suppose $H$ is a *normal* Sylow $p$-subgroup of $G$. Prove that if $a \in G$ is an element of $p$-power order, then $a \in H$. (Hint: Consider $aH \in G/H$.)

26.  Classify the groups of order
     a.  21
     b.  22
     c.  18
     d.  12

27.  Recall that if $H \subset G$ is a subgroup, then the normalizer of $H$ is defined by $N(H) = \{g \in G : gHg^{-1} = H\}$.
     a.  Prove that $H$ has $[G : N(H)]$ conjugate subgroups. (Hint: Let $G$ act by conjugation on the set $S$ of conjugate subgroups of $H$.)

b.  Prove that if $p$ is the smallest prime dividing $|G|$, then any subgroup $H$ of *index* $p$ is normal. (Hint: see the end of the proof of Theorem 5.8.)

28.  Generalizing Theorem 5.8, prove that if $K$ is a subgroup of $G$ and $|K|$ is a power of $p$, then $K$ is contained in a Sylow $p$-subgroup of $G$.

29.  Prove that a finite abelian group is the direct product of its Sylow $p$-subgroups (for all the different primes $p$ dividing its order). (With a bit more work, one can prove the Fundamental Theorem of Abelian Groups: Any finite abelian group is a direct product of cyclic groups of prime-power order.) Proceed as follows:

   a.  Suppose $|G| = p^\alpha m$, where $p \nmid m$; let $H = \{a \in G : a^{p^\alpha} = e\}$ and $K = \{a \in G : a^m = e\}$. Then $H$ is the Sylow $p$-subgroup of $G$ (why?). Prove that $H \cap K = \{e\}$.

   b.  If $a \in G$, then the order of $a$ is of the form $p^s j$, where $s \le \alpha$ and $j|m$. Show that there are integers $k$ and $\ell$ so that $kp^s + \ell j = 1$, and observe that $a = (a^{p^s})^k \cdot (a^j)^\ell$. Conclude that $a$ is the product of an element of $H$ and an element of $K$.

   c.  Use Proposition 5.3 to prove that $G \cong H \times K$.

   d.  Finish the proof by complete induction on $|G|$.

30.  a.  Let $G = \{u^j v^k : 0 \le j \le 7, 0 \le k \le 1\}$ subject to the rules $u^8 = v^2 = e$, $vu = u^5 v$. Check that $G$ is a group of order 16, and determine the number of elements of order 2, 4, and 8.

   b.  Let $G' = \mathbb{Z}_2 \times \mathbb{Z}_8$. Determine the number of elements of order 2, 4, and 8 in $G'$.

   c.  Are $G$ and $G'$ isomorphic? (See Exercise 6.2.17.)

## 6. Some Remarks on Galois Theory

We now wish to return to the discussion of the roots of polynomials undertaken in Chapters 3 and 4. The deep idea of Évariste Galois was to study certain permutations of the roots of the polynomial, now called its **Galois group**, as we introduced in Example 1(e) of Section 1. In fact, he was the first to formalize the notion of a group. We hope the reader will appreciate the beautiful inter-weaving of group theory, polynomials, and field extensions. Our goal is twofold: to understand why we cannot find the real roots of an irreducible real cubic using only real numbers, and to understand why the general polynomial of degree $\geq 5$ cannot be "solved by radicals."

Let $K$ be a field extension of $F$ (which we always assume to be of finite degree). We say a (ring) isomorphism $\phi: K \to K$ is an $F$-**automorphism of** $K$ if $\phi(a) = a$ for all $a \in F$. We denote by $G(K/F)$ the set of all $F$-automorphisms of $K$; the first observation is that this is a group, called the **Galois group of** $K$ **over** $F$.

**Lemma 6.1.** $G(K/F) = \{F\text{-}automorphisms\ of\ K\}$ is a group.

**Proof.** The identity map $\iota: K \to K$ certainly is an element of $G(K/F)$. Let $\phi, \psi \in G(K/F)$. Then $\phi \circ \psi$, which is the composition of two isomorphisms, is an isomorphism. Then $(\phi \circ \psi)(a) = \phi(\psi(a)) = \phi(a) = a$ for all $a \in F$, so $\phi \circ \psi \in G(K/F)$. Now $\phi^{-1}$ is likewise an $F$-automorphism of $K$, so we are done. $\square$

**Example 1.** Let $K = \mathbb{Q}[\sqrt{2}]$ and $F = \mathbb{Q}$. Then, given any $a, b \in \mathbb{Q}$ and $\phi \in G(\mathbb{Q}[\sqrt{2}]/\mathbb{Q})$, we have $\phi(a + b\sqrt{2}) = \phi(a) + \phi(b\sqrt{2}) = a + b\phi(\sqrt{2})$; so $\phi$ is completely determined by the one number $\phi(\sqrt{2})$. Now let $\phi(\sqrt{2}) = \alpha + \beta\sqrt{2}$, for some $\alpha, \beta \in \mathbb{Q}$. We have $2 = \phi(2) = \phi(\sqrt{2} \cdot \sqrt{2}) = (\phi(\sqrt{2}))^2 = (\alpha + \beta\sqrt{2})^2 = (\alpha^2 + 2\beta^2) + 2\alpha\beta\sqrt{2}$. Therefore, $\alpha\beta = 0$ and $\alpha^2 + 2\beta^2 = 2$; since $\sqrt{2}$ is irrational, we must have $\alpha = 0$, $\beta = \pm 1$. In conclusion, there are precisely two $\mathbb{Q}$-automorphisms of $\mathbb{Q}[\sqrt{2}]$: the identity map $\iota(a + b\sqrt{2}) = a + b\sqrt{2}$ and the "conjugation" map $\phi(a + b\sqrt{2}) = a - b\sqrt{2}$. In particular, $|G(\mathbb{Q}[\sqrt{2}]/\mathbb{Q})| = 2$.

**Example 2.** Let $K = \mathbb{Q}[\sqrt[3]{2}]$ and $F = \mathbb{Q}$. For convenience, let $\xi = \sqrt[3]{2}$. Since $1, \xi, \xi^2$ are a basis for $\mathbb{Q}[\xi]$ over $\mathbb{Q}$, if $\phi \in G(\mathbb{Q}[\xi]/\mathbb{Q})$, then $\phi(a + b\xi + c\xi^2) = a + b\phi(\xi) + c(\phi(\xi))^2$, and so the action of $\phi$ is completely determined by the value of $\phi(\xi)$. Suppose $\phi(\xi) = \alpha + \beta\xi + \gamma\xi^2$ for some $\alpha, \beta, \gamma \in \mathbb{Q}$. Then

$$2 = \phi(2) = \phi(\xi^3) = (\phi(\xi))^3 = (\alpha + \beta\xi + \gamma\xi^2)^3 =$$
$$(\alpha^3 + 2\beta^3 + 4\gamma^3) + 3(\alpha^2\beta + \beta^2\gamma + \gamma^2\alpha)\xi + 3(\alpha\beta^2 + \beta\gamma^2 + \gamma\alpha^2)\xi^2.$$

From the linear independence of $1, \xi, \xi^2$ we infer that

$$\alpha^3 + 2\beta^3 + 4\gamma^3 = 2$$
$$\alpha^2\beta + \beta^2\gamma + \gamma^2\alpha = 0$$
$$\alpha\beta^2 + \beta\gamma^2 + \gamma\alpha^2 = 0.$$

With some thought, the reader can check that the only rational solution of this system of equations is $\alpha = \gamma = 0$, $\beta = 1$. That is, the
only $\mathbb{Q}$-automorphism of $\mathbb{Q}[\sqrt[3]{2}]$ is the identity map.

Let $f(x) = x^n + a_{n-1}x^{n-1} + \cdots + a_1 x + a_0 \in F[x]$ be a monic
polynomial with coefficients in $F$, and let $K$ be a field extension of
$F$. Suppose $f(x)$ has roots $\alpha_1, \ldots, \alpha_m$ lying in $K$ (and perhaps other
roots lying in a further extension). Our first observation is that any
$F$-automorphism of $K$ must permute $\alpha_1, \ldots, \alpha_m$.

**Lemma 6.2.** *Let $K \supset F$, and let $\alpha \in K$ be a root of $f(x) \in
F[x]$. Then for any $\phi \in G(K/F)$, $\phi(\alpha)$ is also a root of $f(x)$ in
$K$. Moreover, if $K$ is the splitting field of $f(x)$ and $f(x)$ has dis-
tinct roots $\alpha_1, \ldots, \alpha_n \in K$, then $G(K/F)$ is realized as a subgroup of
$\mathrm{Perm}\{\alpha_1, \ldots, \alpha_n\}$.*

**Proof.** We claim first that $f(\phi(\alpha)) = 0$. The crucial observation
is that $\phi$ leaves $F$ fixed, and hence leaves fixed every coefficient
of the polynomial $f(x) \in F[x]$. Therefore, $f(\phi(\alpha)) = \phi(f(\alpha)) =
\phi(0) = 0$. Explicitly, if $f(x) = x^n + a_{n-1}x^{n-1} + \cdots + a_1 x + a_0$,

$$f(\phi(\alpha)) = (\phi(\alpha))^n + a_{n-1}(\phi(\alpha))^{n-1} + \cdots + a_1\phi(\alpha) + a_0$$
$$= \phi(\alpha^n + a_{n-1}\alpha^{n-1} + \cdots + a_1\alpha + a_0) = \phi(0) = 0.$$

If $K$ contains all the roots $\alpha_1, \ldots, \alpha_n$ of $f(x)$, then for all $j =
1, \ldots, n$, we know that $\phi(\alpha_j) \in \{\alpha_1, \ldots, \alpha_n\}$. So, there is a map
$G(K/F) \to \mathrm{Perm}\{\alpha_1, \ldots, \alpha_n\}$; this map is in fact a group homomorphism, as the reader should check. Now, if $K$ is the smallest field
containing all the roots—i.e., a splitting field of $f(x)$—then $K =
F[\alpha_1, \ldots, \alpha_n]$; in this event, if $\phi \in G(K/F)$ fixes all the $\alpha_j$'s, then it
must fix all of $K$. This proves that the homomorphism $G(K/F) \to
\mathrm{Perm}\{\alpha_1, \ldots, \alpha_n\}$ is in fact one-to-one, as desired.   $\square$

**Example 3.** Using Lemma 6.2, we get a much simpler solution
to Example 2. Since $\xi = \sqrt[3]{2}$ is a root of the irreducible polynomial
$f(x) = x^3 - 2 \in \mathbb{Q}[x]$, any $\phi \in G(\mathbb{Q}[\xi]/\mathbb{Q})$ must carry $\xi$ to some
root of $f(x)$ in $\mathbb{Q}[\xi]$. Since there is only one, $\phi$ must be the identity.

**Examples 4.** Let $f(x) = x^2 - 2x - 1 \in \mathbb{Q}[x]$. One root of $f(x)$ is $1 + \sqrt{2} \in \mathbb{Q}[\sqrt{2}]$. It follows from Example 1 that $1 - \sqrt{2}$ must be a root as well. Analogously, $G(\mathbb{C}/\mathbb{R})$ is a group of order 2 generated by complex conjugation. It follows from Lemma 6.2 that if $\alpha \in \mathbb{C}$ is any root of a polynomial with real coefficients, then $\overline{\alpha}$ is a root as well. (See Exercise 4.2.10.)

Let $K$ be a field extension of $F$ of degree $n > 1$, and suppose $\alpha \in K$. Since $1, \alpha, \alpha^2, \ldots, \alpha^n$ are $n + 1$ vectors in an $n$-dimensional vector space, they must be linearly dependent; and so $\alpha$ is a root of a polynomial $f(x) \in F[x]$ of degree at most $n$—therefore, a root of an *irreducible* polynomial of degree at most $n$. If we assume, moreover, that $K = F[\alpha]$, then $f(x)$ must have degree exactly $n$. What can we say about $G(K/F)$? Since $1, \alpha, \alpha^2, \ldots, \alpha^{n-1}$ form a basis for $K$ over $F$, if $\phi \in G(K/F)$, the action of $\phi$ on $K$ is completely determined by $\phi(\alpha)$. On the other hand, by Lemma 6.2, $\phi$ must take $\alpha$ to some root of $f(x)$ in $K$, and there are at most $n = [K:F]$ of these. Thus we have $|G(K/F)| \leq [K:F]$. In fact, this argument shows a bit more: even if $F[\alpha] \subsetneq K$, then there are still at most $[F[\alpha]:F]$ choices for the image of $\alpha$ under any $F$-automorphism $\phi$ of $K$.

In order to see that $|G(K/F)| \leq [K:F]$ in general, we need the following result about "extending" isomorphisms of fields. An isomorphism $\phi \colon F \to F'$ of fields is, by definition, a ring isomorphism between them. Note, further, that such an isomorphism gives rise to a ring isomorphism $F[x] \to F'[x]$, which, by abuse of notation, we denote by $\phi$ as well. (A note of warning: The notation $F'$, $f'$, etc., in what follows *in no way* connotes derivatives.)

**Lemma 6.3.** *Let $F$ and $F'$ be fields, and let $\phi \colon F \to F'$ be an isomorphism of fields. Suppose $f(x) \in F[x]$ is an irreducible polynomial with a root $\alpha$ in some field extension $K \supset F$. Correspondingly, let $f'(x) = \phi(f(x)) \in F'[x]$, and let $\alpha'$ be a root of $f'(x)$ in some field extension $K' \supset F'$. Then there is a* unique *isomorphism*

$$\tilde{\phi} \colon F[\alpha] \to F'[\alpha']$$

*extending $\phi$ (by which we mean that $\tilde{\phi}(a) = \phi(a)$ for all $a \in F$).*

**Proof.** An arbitrary element of $F[\alpha]$ is of the form $p(\alpha)$, where $p(x) = a_n x^n + a_{n-1} x^{n-1} + \cdots + a_1 x + a_0 \in F[x]$. We define $\tilde{\phi}$ by

$$\tilde{\phi}(p(\alpha)) = \tilde{\phi}(a_n \alpha^n + a_{n-1} \alpha^{n-1} + \cdots + a_1 \alpha + a_0)$$
$$= \phi(a_n) \alpha'^n + \phi(a_{n-1}) \alpha'^{n-1} + \cdots + \phi(a_1) \alpha' + \phi(a_0).$$

(A shorthand notation for this would be: $\tilde{\phi}(p(\alpha)) = p'(\alpha')$, where $p'(x) = \phi(p(x))$.) Note first that $\tilde{\phi}$ is well-defined: If $p(\alpha) = q(\alpha)$, then $p(x) \equiv q(x) \pmod{\langle f(x) \rangle}$; and so $\phi(p(x)) \equiv \phi(q(x))$ $\pmod{\langle f'(x) \rangle}$, whence $\tilde{\phi}(p(\alpha)) = \tilde{\phi}(q(\alpha))$. By its very definition, $\tilde{\phi}$ is a homomorphism $F[\alpha] \to F'[\alpha']$ extending $\phi$, and we leave it to the reader to check that $\tilde{\phi}$ maps onto $F'[\alpha']$. If $\tilde{\phi}(p(\alpha)) = 0$, then $p'(\alpha') = 0$, whence $p'(x) \in \langle f'(x) \rangle$, and so $p(x) \in \langle f(x) \rangle$, implying that $p(\alpha) = 0$; thus, $\ker \tilde{\phi} = \langle 0 \rangle$. (Cf. Theorem 2.4 of Chapter 4 for a more conceptual view of this entire proof.)  □

We now use Lemma 6.3 inductively to obtain the following proposition.

**Proposition 6.4.** *Let $F \subset K$ and $F' \subset K'$ be fields, and let $\phi: F \to F'$ be a given isomorphism of fields. Then there are at most $[K : F]$ field isomorphisms $\tilde{\phi}: K \to K'$ extending $\phi$.*

**Proof.** Suppose $\alpha \in K$ is a root of an irreducible polynomial $f(x) \in F[x]$ of degree $> 1$; recall that $\deg(f(x)) = [F[\alpha] : F]$. Let $f'(x) = \phi(f(x)) \in F'[x]$, and let $\alpha'$ be a root of $f'(x)$ in $K'$. Then there will be a unique field isomorphism $\tilde{\phi}: F[\alpha] \to F'[\alpha']$ extending $\phi$ and carrying $\alpha$ to $\alpha'$. In the event that $F[\alpha] = K$, since $f'(x)$ can have at most $\deg(f'(x)) = [K : F]$ roots in $K'$, there are indeed at most $[K : F]$ possible field isomorphisms extending $\phi$.

If $F[\alpha] \subsetneq K$, we proceed by complete induction on $[K : F]$. That is, if $[K : F] = n$ we assume the Proposition to be valid for all field extensions of degree less than $n$. Let $L = F[\alpha]$, $L' = F'[\alpha']$, and let $\psi = \tilde{\phi}: L \to L'$ be the isomorphism already obtained. By induction hypothesis, since $[K : L] < n$, there are at most $[K : L]$ field isomorphisms $\tilde{\psi}: K \to K'$ extending $\psi$. Note, in particular, that each of these is an extension of the original field isomorphism $\phi$.

Taking stock, there are at most $[L : F]$ possible extensions $\psi$ of our original isomorphism $\phi$; and for each one of them, there are at most $[K : L]$ possible extensions $\tilde{\psi}$. Therefore, there are at most $[K : L][L : F]$ possible extensions of $\phi$. By Proposition 1.5 of

Chapter 5, $[K : F] = [K : L][L : F]$; and so there are at most $[K : F]$ isomorphisms extending $\phi$, as desired.   □

As a crucial consequence of this, we obtain the following proposition.

**Proposition 6.5.** *Let $K$ be a field extension of $F$. Then $|G(K/F)| \leq [K : F]$.*

**Proof.** Apply Proposition 6.4 with $F = F'$, $K = K'$, and $\phi = \iota$. Since the isomorphisms $K \to K$ extending $\iota$ are the $F$-automorphisms of $K$, we have $|G(K/F)| \leq [K : F]$.   □

We now come to the main definition.

**Definition.** We say $K$ is a **Galois** extension of $F$ if $|G(K/F)| = [K : F]$.

### Examples 5.

(a) $\mathbb{Q}[\sqrt{2}]$ is a Galois extension of $\mathbb{Q}$.
(b) $\mathbb{Q}[\sqrt[3]{2}]$ is not a Galois extension of $\mathbb{Q}$, since, as we saw in Example 2, $|G(\mathbb{Q}[\sqrt[3]{2}]/\mathbb{Q})| = 1$; but $[\mathbb{Q}[\sqrt[3]{2}] : \mathbb{Q}] = 3$.
(c) Let $K = \mathbb{Q}[\sqrt[3]{2}, i\sqrt{3}]$ be the splitting field of $f(x) = x^3 - 2$. Then $[K : \mathbb{Q}] = 6$, and, as we saw in Example 1(e) of Section 1, $G(K/\mathbb{Q}) \cong S_3$ has order 6. To be more precise, we exhibited six $\mathbb{Q}$-automorphisms of $K$ at the time, but now we know from Proposition 6.5 that there can be no more.

These examples suggest that $K$ will be a Galois extension of $F$ whenever $K$ is a splitting field of some polynomial $f(x) \in F[x]$. To establish this, we must produce (at least) $[K : F]$ $F$-automorphisms of $K$ under these circumstances.

**Proposition 6.6.** *Let $F$ and $F'$ be fields containing $\mathbb{Q}$, and let $\phi : F \to F'$ be an isomorphism. Suppose that $K$ is a splitting field of a polynomial $f(x) \in F[x]$, and $K'$ is a splitting field of the corresponding polynomial $\phi(f(x)) \in F'[x]$. Then there are $[K : F]$ isomorphisms $\tilde{\phi} : K \to K'$ extending $\phi$.*

**Proof.** We proceed by complete induction on $n = [K : F]$. Suppose first that $\alpha \in K$ is a root of an irreducible factor $g(x)$ of $f(x)$. Assume $\deg(g(x)) = \ell > 1$. Let $\alpha' = \alpha'_1, \ldots, \alpha'_\ell$ be the roots in $K'$ of the corresponding $g'(x)$. For each $j = 1, \ldots, \ell$, we obtain by Lemma 6.3 an isomorphism $\tilde{\phi} : F[\alpha] \to F'[\alpha'_j]$ extending $\phi$ and sending

$\alpha$ to $\alpha'_j$. Since the $\ell$ roots are distinct (see Exercise 2), we obtain $\ell = [F[\alpha] : F]$ extensions in this manner.

Now, view $K$ (resp., $K'$) as an extension of $L = F[\alpha]$ (resp., $L' = F'[\alpha_j]$); $K$ is likewise a splitting field of $f(x) \in L[x]$. Given an isomorphism $\psi = \tilde{\phi} \colon L \to L'$, since $[K : L] < n$, by the induction hypothesis, there are $[K : L]$ extensions of $\psi$ to an isomorphism $\tilde{\psi} \colon K \to K'$. Since this is true for each of the $[L : F]$ choices of $\psi$, there are a total (again by Proposition 1.5 of Chapter 5) of $[K : F]$ choices of isomorphisms extending $\phi$. $\square$

**Remark.** Although we took $F$ and $F'$ to be field extensions of $\mathbb{Q}$, we see from the proof of Proposition 6.6 that the appropriate hypothesis is that irreducible polynomials have distinct roots. This holds not only for fields containing $\mathbb{Q}$ (as was stipulated), but also for finite fields.

As a corollary, we can now deduce that splitting fields of polynomials are unique up to isomorphism, a fact which has been tacit in our work throughout. Thus, we are justified in referring to "the" splitting field of a polynomial.

**Corollary 6.7.** *Let $f(x) \in F[x]$ and let $K$ and $K'$ be two splitting fields of $f(x)$. Then $K \cong K'$.*

**Proof.** We apply Proposition 6.6 with $F = F'$ and $\phi = \iota$. Then we obtain $[K : F]$ isomorphisms $K \to K'$ extending $\iota$. In particular, there is at least one isomorphism $\tilde{\iota} \colon K \to K'$ fixing $F$. $\square$

We now come to the result we've been seeking. The proof requires no effort at this point.

**Theorem 6.8.** *Let $F$ be any field containing $\mathbb{Q}$. If $K$ is the splitting field of a polynomial $f(x) \in F[x]$, then $K$ is a Galois extension of $F$. We call $G(K/F)$ the* **Galois group** *of $f(x)$.*

**Proof.** If $K$ is the splitting field of a polynomial in $F[x]$, then we apply Proposition 6.6 with $F = F'$, $K = K'$, and $\phi = \iota$ to obtain $|G(K/F)| = [K : F]$. $\square$

Now we address the converse. Suppose $K$ is a Galois extension of $F$. Is $K$ necessarily the splitting field of some polynomial in $F[x]$? We begin with the following lemma.

**Lemma 6.9.** *Suppose $K$ is a Galois extension of an arbitrary field $F$, and suppose $g(x) \in F[x]$ is an irreducible polynomial with some root $\alpha$ in $K$. Then $g(x)$ splits in $K$.*

*Proof.* Let $\sigma_j$, $j = 1, \ldots, n$, be the elements of $G(K/F)$, and set $\alpha_j = \sigma_j(\alpha)$, $j = 1, \ldots, n$. Define $h(x) = (x - \alpha_1) \cdots (x - \alpha_n) \in K[x]$. Since the coefficients of $h(x)$ are fixed by any element of $G(K/F)$ (why?), it follows that $h(x) \in F[x]$ (see Exercise 3). Since $g(x)$ and $h(x)$ have a common root, namely $\alpha$, in $K$; and since $g(x)$ is irreducible in $F[x]$, by Exercise 3.2.15, $g(x) | h(x)$. Therefore, $g(x) | (x - \alpha_1) \cdots (x - \alpha_n)$ in $K[x]$. It is now obvious that $g(x)$ splits in $K[x]$. □

**Example 6.** To illustrate Lemma 6.9, let $K = \mathbb{Q}[\sqrt{5}, i]$. As the reader can check, $K$ is a Galois extension of $\mathbb{Q}$. The elements of $G(K/\mathbb{Q})$, according to Lemma 6.2, must send $\sqrt{5}$ to either $\sqrt{5}$ or $-\sqrt{5}$, and $i$ to either $i$ or $-i$. If $\alpha = \sqrt{5} + 2i \in K$, then the proof of Lemma 6.9 shows us how to construct an irreducible polynomial in $\mathbb{Q}[x]$ having $\alpha$ as a root: every element of the orbit of $\alpha$ under the Galois group must also be a root, namely $\alpha_1 = \alpha$, $\alpha_2 = -\sqrt{5} + 2i$, $\alpha_3 = \sqrt{5} - 2i$, and $\alpha_4 = -\sqrt{5} - 2i$. A quick computation shows that $\alpha$ is thus a root of the irreducible polynomial $f(x) = x^4 - 2x^2 + 81$.

On the other hand, if $K = \mathbb{Q}[\sqrt[3]{5}, i]$, then $K$ is not a Galois extension of $\mathbb{Q}$, and the orbit of $\alpha = \sqrt[3]{5} + i$ consists only of $\alpha_1 = \alpha$ and $\alpha_2 = \sqrt[3]{5} - i$. The polynomial $(x - \alpha_1)(x - \alpha_2) = x^2 - 2\sqrt[3]{5}\,x + (\sqrt[3]{5})^2 + 1$, of course, does not belong to $\mathbb{Q}[x]$.

What will be most important to us is the following converse of Theorem 6.8.

**Theorem 6.10.** *Let $F$ be an arbitrary field, and let $K$ be a Galois extension of $F$. Then $K$ is the splitting field of some polynomial $f(x) \in F[x]$.*

*Proof.* Let $\alpha_1, \ldots, \alpha_k$ be a basis for $K$ over $F$; and for $j = 1, \ldots, k$, let $g_j(x) \in F[x]$ be an irreducible polynomial with root $\alpha_j$ (cf. the proof of Lemma 6.9 or Exercise 5.1.19). Then $K$ is the splitting field of $f(x) = g_1(x)g_2(x) \cdots g_k(x) \in F[x]$: by Lemma 6.9, $f(x)$ does in fact split in $K$, and it does so in no smaller field (since any splitting field must contain all the $\alpha_j$'s). □

We can now state the central theorem in the subject.

**Theorem 6.11** (Fundamental Theorem of Galois Theory). *Let $K$ be a Galois extension of $\mathbb{Q}$, and let $G = G(K/\mathbb{Q})$. Then there is a one-to-one correspondence between intermediate fields $\mathbb{Q} \subset E \subset K$ and* subgroups $H \subset G$ *given as follows:*

$$E \longrightarrow \left\{ \begin{array}{c} \text{the elements of } G \\ \text{fixing } E \end{array} \right\}$$

$$K^H = \left\{ \begin{array}{c} \text{the elements of } K \\ \text{left fixed by } H \end{array} \right\} \longleftarrow H.$$

*Under this correspondence, we have:*

   (i)  *K is a Galois extension of E, and $G(K/E) = H$;*
   (ii)  $[K : E] = |H|$ *and* $[E : \mathbb{Q}] = [G : H]$;
   (iii)  *E is a Galois extension of $\mathbb{Q}$ if and only if H is a normal subgroup of G, in which case $G(E/\mathbb{Q}) \cong G/H$.*

*We summarize this schematically with the following diagram:*

$$
\begin{array}{cc}
\underline{\textit{fields}} & \underline{\textit{groups}} \\
K & \{e\} \\
\cup & \cap \\
E & H \\
\cup & \cap \\
\mathbb{Q} & G
\end{array}
\quad
\begin{array}{l}
\Big] G(K/E) = H \\[2em]
\Big] G(E/\mathbb{Q}) \cong G/H \ \textit{when } H \textit{ is normal}
\end{array}
$$

   **Sketch of proof.** The first point to realize is this: by Theorem 6.10, if $K$ is a Galois extension of $\mathbb{Q}$, then $K$ is the splitting field of some polynomial $f(x) \in \mathbb{Q}[x]$. Since $f(x) \in E[x]$ as well, $K$ must therefore be a Galois extension of $E$. By definition, the corresponding subgroup $H$ is the subgroup of $E$-automorphisms of $K$, i.e., $H = G(K/E)$. But the equality $E = K^H$ (= {elements of $K$ fixed by $H$}) follows now from Exercise 3. Conversely, given a subgroup $H \subset G$, let $E = K^H$ be the "fixed field" of $H$. Then by Exercise 17, $H = G(K/E)$. This demonstrates that there is a one-to-one correspondence between subgroups and intermediate fields, and (i) is established.
   Part (ii) of the theorem is immediate: since $H = G(K/E)$, and since $K$ is a Galois extension of $E$, we have $|H| = |G(K/E)| = [K : E]$. By Lagrange's Theorem (Theorem 3.2 of Chapter 6) and Proposition 1.5 of Chapter 5, $[E : \mathbb{Q}] = [K : \mathbb{Q}]/[K : E] = |G|/|H| = [G : H]$.
   Suppose $H$ is a subgroup of $G$ with fixed field $K^H = E$. If $\sigma \in G$, then the subgroup $\sigma H \sigma^{-1}$ has fixed field $\sigma(E)$: if $h \in H$ and $e \in E$, then $(\sigma h \sigma^{-1})(\sigma(e)) = \sigma(h(e)) = \sigma(e)$. If $H$ is a normal subgroup of $G$, then $\sigma H \sigma^{-1} = H$ for all $\sigma \in G$; and, since there is a one-to-one correspondence between subgroups and intermediate fields, $\sigma(E) = E$. (Note this does not mean that $\sigma$ fixes each element of $E$; it

means merely that it maps the entire set $E$ back to itself.) Therefore, every $\mathbb{Q}$-automorphism $\sigma$ of $K$ restricts to give a $\mathbb{Q}$-automorphism of $E$; the restriction map is a homomorphism $\pi: G \to G(E/\mathbb{Q})$ whose kernel consists of those elements of $G$ that act by the identity on $E$, which is the subgroup $H$. Thus, by Corollary 3.10 of Chapter 6, we have image $(\pi) \cong G/H$, as desired. Since $|G(E/\mathbb{Q})| \geq |\text{image}(\pi)| = |G/H| = [E : \mathbb{Q}]$, it follows from Proposition 6.5 that $|G(E/\mathbb{Q})| = |\text{image}(\pi)| = [E : \mathbb{Q}]$; so, $G(E/\mathbb{Q}) \cong G/H$ and $E$ is a Galois extension of $\mathbb{Q}$.

Conversely, suppose $E$ is a Galois extension of $\mathbb{Q}$. Then by Theorem 6.10, $E$ is the splitting field of some polynomial $f(x) \in \mathbb{Q}[x]$. Therefore, any $\mathbb{Q}$-automorphism $\sigma$ of $K$ maps the roots of $f(x)$ to each other, and hence preserves $E$. From the preceding discussion, it follows that since $\sigma(E) = E$, $\sigma H \sigma^{-1} = H$, and $H$ is a normal subgroup. This completes the proof of (iii), and, in turn, that of the theorem. □

**Example 7.** Consider the Galois group of $f(x) = x^3 - 2 \in \mathbb{Q}[x]$. We saw in Chapter 3 that the splitting field of $f(x)$ is $K = \mathbb{Q}[\sqrt[3]{2}, i\sqrt{3}] = \mathbb{Q}[\sqrt[3]{2}, \omega]$, where $\omega = \frac{-1+i\sqrt{3}}{2}$ is a primitive cube root of unity. We saw earlier that $G(K/\mathbb{Q}) \cong \mathcal{T}$. We now wish to use Theorem 6.11 to find all the intermediate fields $\mathbb{Q} \subset E \subset K$, given that we know well all the subgroups of $\mathcal{T} \cong S_3$.

We begin by enumerating the subgroups $H$ of $G = \mathcal{T}$, and then we'll determine the corresponding intermediate fields $E = K^H$. Using the notation from Example 1(e) of Section 1, we have

$$G = \mathcal{T} = \{\iota, \phi, \phi^2, \psi, \psi\phi, \psi\phi^2\}, \quad \text{where } \phi^3 = \psi^2 = \iota.$$

We have, aside from the trivial subgroups $\{\iota\}$ and $G$, the normal subgroup $\langle \phi \rangle$ of order 3, and the subgroups of order 2 generated by each of the "flips" $\psi$, $\psi\phi$, and $\psi\phi^2$. The action of $G$ on $K$ by $\mathbb{Q}$-automorphisms is given by:

$$\psi(\sqrt[3]{2}) = \sqrt[3]{2} \qquad \phi(\sqrt[3]{2}) = \sqrt[3]{2}\,\omega$$
$$\psi(\omega) = \overline{\omega} \qquad \phi(\omega) = \omega.$$

- It is clear that $\psi$ fixes the field $\mathbb{Q}[\sqrt[3]{2}]$; to check with (ii) of Theorem 6.11, $\langle \psi \rangle$ is a subgroup of order 2. This is a subgroup of *index* $3 = [\mathbb{Q}[\sqrt[3]{2}] : \mathbb{Q}]$, as required. Note, moreover, that $\mathbb{Q}[\sqrt[3]{2}]$ is *not* a Galois extension of $\mathbb{Q}$, mirroring in accord with (iii) the fact that $\langle \psi \rangle$ is not a normal subgroup of $G$.

- It is equally clear that $\phi$ fixes the field $\mathbb{Q}[\omega] = \mathbb{Q}[i\sqrt{3}]$, which is an extension of degree 2 over $\mathbb{Q}$. Since $\langle\phi\rangle$ is a normal subgroup, $\mathbb{Q}[i\sqrt{3}]$ is supposed to be a Galois extension of $\mathbb{Q}$. It is, indeed, as it's the splitting field of $f(x) = x^2 + 3 \in \mathbb{Q}[x]$.

- What is the fixed field of the subgroup $\langle\psi\phi\rangle$? Well, we begin by computing the action of this element on $\sqrt[3]{2}$ and $\omega$:

$$\psi\phi(\sqrt[3]{2}) = \psi(\sqrt[3]{2}\omega) = \psi(\sqrt[3]{2})\psi(\omega) = \sqrt[3]{2}\,\overline{\omega}$$

$$\psi\phi(\omega) = \psi(\omega) = \overline{\omega}.$$

  So we see that $\psi\phi(\sqrt[3]{2}\,\omega) = (\psi\phi(\sqrt[3]{2}))(\psi\phi(\omega)) = \sqrt[3]{2}\,\omega$. Therefore, the fixed field of $\langle\psi\phi\rangle$ is $\mathbb{Q}[\sqrt[3]{2}\omega]$ (another field extension of degree 3 over $\mathbb{Q}$).

- Lastly, we leave it to the reader to check that the fixed field of $\langle\psi\phi^2\rangle$ is $\mathbb{Q}[\sqrt[3]{2}\,\overline{\omega}]$ (see Exercise 5).

- **Remark**. The reader may wonder why elements of $G(K/\mathbb{Q})$ must either leave $\omega$ fixed or send it to $-\omega$. Well, $\omega$ and $-\omega$ are the roots of the irreducible polynomial $g(x) = x^2 + x + 1$; and by Lemma 6.2, $G(K/\mathbb{Q})$ must permute these roots.

**Example 8.** It is essential that $K$ be a Galois extension of $\mathbb{Q}$, for consider $K = \mathbb{Q}[\sqrt[4]{2}]$. We leave it to the reader to determine the Galois group $G$ and thus decide that the extension is not Galois. We have the intermediate field $E = \mathbb{Q}[\sqrt{2}]$, and yet all of $G$ fixes $E$.

**Example 9.** The following sort of problem has arisen earlier (see Exercise 3.2.3, for example). Given the complex numbers $\sqrt{3 + 4\sqrt{-1}}$ or $\sqrt{23 + 4\sqrt{15}}$, or more generally, $\sqrt{a + \sqrt{b}}$, $a, b \in \mathbb{Q}$, we would like to rewrite them in the form $\sqrt{c} \pm \sqrt{d}$, $c, d \in \mathbb{Q}$. When is this possible? How do we actually do it?

Let $\alpha = \sqrt{a + \sqrt{b}} \in \mathbb{C}$ (and let's assume that $\sqrt{b} \notin \mathbb{Q}$). Then $\alpha$ is a root of the (irreducible) quartic polynomial $f(x) = x^4 - 2ax^2 + (a^2 - b)$. Indeed, letting $\beta = \sqrt{a - \sqrt{b}}$, we have the factorization $f(x) = (x - \alpha)(x + \alpha)(x - \beta)(x + \beta) \in \mathbb{C}[x]$, so the splitting field of $f(x)$ is $\mathbb{Q}[\alpha, \beta]$. In general, there are several potential Galois groups of $f(x)$. But if it happens that $\alpha = \sqrt{c} \pm \sqrt{d}$, then we claim that the splitting field must be $K = \mathbb{Q}[\sqrt{c}, \sqrt{d}]$ and the Galois group must be $\mathcal{V}$ (see Exercise 8). At first, it is not obvious that all the roots of $f(x)$ must belong to $K$ just because one, namely $\alpha$, does; but this is a consequence of Lemma 6.9 (why?). Since the orbit of $\alpha$ under $\mathcal{V}$ is $\{\pm\sqrt{c} \pm \sqrt{d}\}$ (check this!), it follows that if $\alpha = \sqrt{c} + \sqrt{d}$ (say), then $\beta = \sqrt{c} - \sqrt{d}$ (assuming $c > d$), and so $\alpha\beta = c - d \in \mathbb{Q}$. On the other hand, $\alpha\beta = \sqrt{a^2 - b}$, and so we might conjecture that

$$\boxed{\alpha \text{ can be expressed as a sum of square roots } \iff \sqrt{a^2 - b} \in \mathbb{Q}.}$$

Let's verify sufficiency. Suppose, then, that $\alpha\beta = \sqrt{a^2 - b} = r \in \mathbb{Q}$. Then from

$$(\alpha + \beta)^2 = \alpha^2 + \beta^2 + 2\alpha\beta = 2a + 2r$$

$$(\alpha - \beta)^2 = \alpha^2 + \beta^2 - 2\alpha\beta = 2a - 2r$$

we infer that $\alpha = \frac{1}{2}\left(\sqrt{2(a + r)} + \sqrt{2(a - r)}\right)$. In the case of the specific examples given above, we find that $\sqrt{3 + 4\sqrt{-1}} = 2 + \sqrt{-1}$ and $\sqrt{23 + 4\sqrt{15}} = \sqrt{3} + \sqrt{20}$.

We next turn to a brief discussion of the Galois group of cubic polynomials in $\mathbb{Q}[x]$. By Lemma 6.2, this group must be a subgroup of $S_3$; and if the polynomial is irreducible, the group must act transitively on the three roots (see Exercise 14) and must therefore be either $A_3$ or $S_3$ (why?). Then the splitting field $K$ has degree either 3 or 6 over $\mathbb{Q}$. (The 6 is certainly believable from Cardano's formula in Section 4 of Chapter 2. We compute the roots in rational terms of a cube root of an expression involving a square root; that is, we first make a field extension of degree 2, and then one of degree 3.)

How can $[K : \mathbb{Q}] = 3$? It must be the case that when we throw in a single one of the roots, $\alpha_1$, say, then the two remaining roots $\alpha_2$ and $\alpha_3$ both lie in $\mathbb{Q}[\alpha_1] = K$. To be specific, let's consider $f(x) = x^3 + px + q = (x - \alpha_1)(x - \alpha_2)(x - \alpha_3)$. Then we have

$$0 = \alpha_1 + \alpha_2 + \alpha_3$$

$$p = \alpha_1\alpha_2 + \alpha_1\alpha_3 + \alpha_2\alpha_3$$

$$q = -\alpha_1\alpha_2\alpha_3.$$

We can now eliminate $\alpha_3$ and rewrite $q = \alpha_1\alpha_2(\alpha_1 + \alpha_2)$. Consider for the moment the magical formula

$$\delta = (\alpha_1 - \alpha_2)(\alpha_2 - \alpha_3)(\alpha_3 - \alpha_1) = -2\alpha_1^3 - 3\alpha_1^2\alpha_2 + 3\alpha_1\alpha_2^2 + 2\alpha_2^3.$$

Then using the fact that $\alpha_1^3 + p\alpha_1 + q = \alpha_2^3 + p\alpha_2 + q = 0$, we can solve for $\alpha_2$ (and hence $\alpha_3$ as well) in terms of $\delta$, $\alpha_1$, $p$, and $q$:

$$\alpha_2 = \frac{-\delta + 2p\alpha_1 + 3q}{6\alpha_1^2 + 2p}.$$

(We leave it to the reader to check this in Exercise 6; we leave it to the reader to check as well what happens when the denominator vanishes.) It follows, then, that the Galois group of $f(x)$ has order 3 precisely when the mysterious number $\delta \in \mathbb{Q}[\alpha_1]$. But we can say more. Things become somewhat less mysterious if one computes (see likewise Exercise 6) that

$$\delta^2 = -4p^3 - 27q^2,$$

and so $\delta$ is a root of a quadratic polynomial in $\mathbb{Q}[x]$. Insofar as $[\mathbb{Q}[\alpha_1] : \mathbb{Q}] = 3$, it follows from Corollary 1.7 of Chapter 5 that $\delta \in \mathbb{Q}[\alpha_1]$ if and only if $\delta \in \mathbb{Q}$. Thus, we conclude that

> The Galois group of $f(x) = x^3 + px + q$
> is $A_3$ if and only if $\sqrt{-4p^3 - 27q^2} \in \mathbb{Q}$.

We've shown, moreover, that even if $\delta \notin \mathbb{Q}$, $f(x)$ splits when we adjoin a single root $\alpha_1$ to $\mathbb{Q}[\delta]$.

**Remark.** Here is a hint to how one might think of $\delta$ in the first place. If the Galois group is to be $A_3 \subset S_3$, then only even permutations of the roots will arise. So we try to write down a function of the roots $\alpha_1, \alpha_2, \alpha_3$ that is fixed under even permutations but not under transpositions. The formula for $\delta$ then leaps out at us.

Even better, Theorem 6.11 tells us that if the Galois group is $S_3$, then the fixed field of the subgroup $A_3$ is a field extension $E$ of degree 2 over $\mathbb{Q}$. Thus, we must be on the lookout for a *square root* that is fixed by cyclic permutations and not by transpositions.

One slight mystery remains. The formula for $\delta^2$ looks remarkably like the term $\Delta = q^2/4 + p^3/27$ appearing in Cardano's formula. Indeed, $-4p^3 - 27q^2 = -108\Delta$, and so $\sqrt{-4p^3 - 27q^2} \in \mathbb{Q} \iff \sqrt{-3\Delta} = i\sqrt{3}\sqrt{\Delta} \in \mathbb{Q}$. We cannot help but suspect that the $i\sqrt{3}$ that has popped up must be due to the cube roots of unity omnipresent in Cardano's explicit solution!

We are now in a position to answer the following classical question (the so-called *casus irreducibilis*): Suppose a cubic $f(x) = x^3 + px + q \in \mathbb{Q}[x]$ has *all real roots* (so that, by Proposition 4.4 of Chapter 2, $\Delta \leq 0$); then Cardano's formula for the roots involves complex numbers. But could there be a formula for its roots (in terms of radicals) involving only real numbers?

Suppose the cubic $f(x) \in \mathbb{Q}[x]$ is irreducible and has three real roots, and suppose we are able to express one of its roots by radicals involving only real numbers. Then this root lies in a *real* field $K$

obtained as follows:

$$\mathbb{Q} = K_0 \subset K_1 \subset K_2 \subset \cdots \subset K_i \subset K_{i+1} \subset \cdots \subset K_\ell = K,$$

where $K_1 = \mathbb{Q}[\delta]$, each field $K_i \subset \mathbb{R}$, and for each $i$, $K_{i+1} = K_i[\sqrt[p_i]{\beta_i}]$ for some prime number $p_i$ and some element $\beta_i \in K_i$. (If we wish to adjoin an $n^{\text{th}}$ root and $n$ is not prime, then we factor $n$ as a product of primes and do a sequence of extensions of the type prescribed.) Note that we can begin with $K_1$ since $\delta \in \mathbb{R}$; because $K_1$ cannot contain a root of an irreducible cubic (why?), we must have $t \geq 2$. Now, from our earlier remarks, if $K$ contains one root of $f(x)$, since $K \supset \mathbb{Q}[\delta]$, it contains them all; so $K$ contains the splitting field of $f(x)$.

We may assume that since $K_{\ell-1}$ contains no root of $f(x)$, $f(x)$ is irreducible in $K_{\ell-1}[x]$ and splits in $K_\ell$. It follows from Exercise 20 that $K_\ell$ is an extension of $K_{\ell-1}$ of prime degree $p$. Since it contains a root of the irreducible cubic polynomial $f(x)$, $p$ must be divisible by 3, and hence must equal 3. Therefore, on one hand, $K_\ell = K_{\ell-1}[\sqrt[3]{\beta}]$ for some $\beta \in K_{\ell-1}$. On the other hand, $K_\ell$ is the splitting field of $f(x) \in K_{\ell-1}[x]$, so $K_\ell$ is a Galois extension of $K_{\ell-1}$. But now by Lemma 6.9, if $K_\ell$ contains one cube root of $\beta$ it must contain all the cube roots of $\beta$ and hence must contain a complex number. This contradicts our assumption that $K$ is a real field.

We have therefore proved the following theorem.

**Theorem 6.12** (Casus irreducibilis). *Given an irreducible cubic polynomial $f(x) \in \mathbb{Q}[x]$ having all real roots, there is no formula for these real roots involving only real radicals.* □

We conclude with a discussion of the impossibility of solving the general polynomial of degree $\geq 5$ by radicals. As in our discussion above, we say $f(x) \in \mathbb{Q}[x]$ can be solved by radicals if there is a chain of fields

$$\mathbb{Q} = K_0 \subset K_1 \subset K_2 \subset \cdots \subset K_i \subset K_{i+1} \subset \cdots \subset K_\ell = K,$$

where for each $i$, $K_{i+1} = K_i[\sqrt[n_i]{\beta_i}]$ for some $n_i$ and some element $\beta_i \in K_i$, and where $K$ contains the roots of $f(x)$. The following theorem provides the key (for a proof, cf. the books by Artin, Hadlock, or Stewart).

**Theorem 6.13.** *The polynomial $f(x) \in \mathbb{Q}[x]$ can be solved by radicals if and only if such a chain of fields can be found satisfying the additional requirements that*

(i) *$K$ is a Galois extension of $\mathbb{Q}$,*

(ii) *$K_{i+1}$ is a Galois extension of $K_i$, and $G(K_{i+1}/K_i)$ is an abelian group.*  □

**Corollary 6.14.** *$f(x)$ can be solved by radicals if and only if its Galois group $G$ has the property that there is a chain of subgroups*

$$\{e\} = G_\ell \subset G_{\ell-1} \subset \cdots \subset G_{i+1} \subset G_i \subset \cdots \subset G_0 = G$$

*with the property that $G_i/G_{i+1}$ is abelian for all $i = 0, 1, \ldots, \ell - 1$. (Note, in particular, that $G_{i+1}$ is a normal subgroup of $G_i$ for all $i$.)*

**Proof.** Note first that if the Galois group $G$ has the given property, then $f(x)$ is solvable by radicals by the reverse implication in Theorem 6.13. Now suppose that $f(x)$ can be solved by radicals. Then from Theorem 6.11 and Theorem 6.13 we infer that there is a chain of subgroups

$$\{e\} = G'_\ell \subset G'_{\ell-1} \subset \cdots \subset G'_{i+1} \subset G'_i \subset \cdots \subset G'_0 = G(K/\mathbb{Q})$$

with $G'_{i+1}$ a normal subgroup of $G'_i$ and $G'_i/G'_{i+1}$ abelian for all $i = 0, 1, \ldots, \ell - 1$. Let $E \subset K$ be the splitting field of $f(x)$. Then by Theorem 6.11, $G = G(E/\mathbb{Q}) \cong G(K/\mathbb{Q})/G(K/E)$, so $G$ is a quotient of $G(K/\mathbb{Q})$. Let $\pi \colon G(K/\mathbb{Q}) \to G$ be the quotient homomorphism (modding out by $G(K/E)$), and let $G_i = \pi(G'_i)$. Then $G_{i+1}$ is a normal subgroup of $G_i$ and $G_i/G_{i+1}$ is the image of $G'_i/G'_{i+1}$ (we leave the details for the reader to check in Exercise 18). Since an abelian group must map to an abelian group, we are done.  □

We now obtain our desired conclusion.

**Corollary 6.15.** *If the Galois group of $f(x)$ is $A_n$ or $S_n$, $n \geq 5$, then $f(x)$ cannot be solved by radicals.*

**Proof.** By Theorem 4.10 of Chapter 6, for $n \geq 5$, the group $A_n$ has no nontrivial normal subgroup. Since $A_n$ is not abelian, there can be no such chain of subgroups as indicated in Corollary 6.14 in the case $G = A_n$. In the case $G = S_n$, the only nontrivial normal subgroup is $A_n$ (see Exercise 6.4.14), and we are foiled once again.  □

**Example 10.** We finish with a crowning example. *The polynomial $f(x) = x^5 - 6x + 2 \in \mathbb{Q}[x]$ cannot be solved by radicals.*

By the Eisenstein criterion (Theorem 3.5 of Chapter 3), $f(x)$ is irreducible. As the reader can check, $f(x)$ has exactly three real roots: for example, it has at least three between $-2$ and $2$, but can have no more than three by Descartes' Rule of Signs (Theorem 2.3

13.  Let $K = \mathbb{Q}[\sqrt[8]{2}, i]$. Prove that
   a.  $G(K/\mathbb{Q}[i]) \cong \mathbb{Z}_8$
   b.  $G(K/\mathbb{Q}[\sqrt{2}]) \cong \mathcal{D}_4$
   c.  $G(K/\mathbb{Q}[i\sqrt{2}]) \cong \mathcal{Q}$
   For extra credit, give $G(K/\mathbb{Q})$.

14.  Suppose $f(x) \in \mathbb{Q}[x]$ is an irreducible polynomial of degree $n$. Prove that the Galois group of $f(x)$ must act transitively on the set of $n$ roots. Give an example to show this statement is false for reducible polynomials. (Hint: Cf. the proof of Proposition 6.6.)

15.  Exhibit another quintic polynomial that cannot be solved by radicals.

16.  Suppose $K$ is a Galois extension of $\mathbb{Q}$ and $G(K/\mathbb{Q})$ is abelian. Suppose that $f(x) \in \mathbb{Q}[x]$ is irreducible and has a root $\alpha \in K$. Prove that $f(x)$ splits in $\mathbb{Q}[\alpha]$. Give an example. Also give an example to show the result may be false when $f(x)$ is not irreducible.

17.  Suppose $H$ is a (finite) group of automorphisms of a field $K$ and $E = K^H$ is the subfield fixed under $H$. Supply the details in the following proof that $[K : E] \leq |H|$. (It follows then that, since $H \subset G(K/K^H)$, $|H| \leq |G(K/K^H)| \leq [K : K^H] \leq |H|$, and so $H = G(K/K^H)$.) We suppose that $n = [K : E] > |H| = m$ and derive a contradiction.

   Let $\alpha_1, \ldots, \alpha_n$ be a basis for $K$ over $E$; let $\phi_1, \ldots, \phi_m$ be the elements of $H$; we assume throughout that $\phi_1 = \iota$. Consider the system of equations

   $$\phi_1(\alpha_1)x_1 + \phi_1(\alpha_2)x_2 + \cdots + \phi_1(\alpha_n)x_n = 0$$
   $$\phi_2(\alpha_1)x_1 + \phi_2(\alpha_2)x_2 + \cdots + \phi_2(\alpha_n)x_n = 0$$

(A)
   $$\vdots$$

   $$\phi_m(\alpha_1)x_1 + \phi_m(\alpha_2)x_2 + \cdots + \phi_m(\alpha_n)x_n = 0$$

   Since $n > m$, this system has a nontrivial solution $(x_1, \ldots, x_n)$ $\in K^n$. Since $\phi_1 = \iota$, the $x_i$'s cannot all lie in $E$, or else the first equation in (A) would imply linear dependence of $\alpha_1, \ldots, \alpha_n$ over $E$. Now, among all nontrivial solutions $(x_1, \ldots, x_n)$ to (A), we choose one with the minimal number of nonzero entries. By renumbering, we may assume that $x_1, \ldots, x_r \neq 0$, and by dividing through by $x_r$, we may assume that $x_r = 1$. Now we can rewrite (A):

$$\phi_1(\alpha_1)x_1 + \cdots + \phi_1(\alpha_{r-1})x_{r-1} + \phi_1(\alpha_r) = 0$$
$$\phi_2(\alpha_1)x_1 + \cdots + \phi_2(\alpha_{r-1})x_{r-1} + \phi_2(\alpha_r) = 0$$

(B)
$$\vdots$$

$$\phi_m(\alpha_1)x_1 + \cdots + \phi_m(\alpha_{r-1})x_{r-1} + \phi_m(\alpha_r) = 0$$

We remarked earlier that at least one $x_i$ cannot lie in $E$, so assume now that $x_1 \notin E$; then there is some automorphism $\phi \in H$ with $\phi(x_1) \neq x_1$. Apply $\phi$ to the equations in (B); then we obtain the equations

(C)  $\phi\phi_i(\alpha_1)\phi(x_1) + \cdots + \phi\phi_i(\alpha_{r-1})\phi(x_{r-1}) + \phi\phi_i(\alpha_r) = 0,$

$$i = 1, \ldots, m.$$

Now we use the fact that $H$ is a group to note that the elements $\phi$, $\phi\phi_2$, ..., $\phi\phi_m$ are just a relisting of the elements $\phi_1 = \iota$, $\phi_2$, ..., $\phi_m$ of $H$; and so the equations (C) can be rewritten more simply in the form

(D)  $\phi_i(\alpha_1)\phi(x_1) + \cdots + \phi_i(\alpha_{r-1})\phi(x_{r-1}) + \phi_i(\alpha_r) = 0,$

$$i = 1, \ldots, m.$$

Subtract equations (D) from those in (B), and we obtain the system of equations

$$\phi_i(\alpha_1)[x_1 - \phi(x_1)] + \cdots + \phi_i(\alpha_{r-1})[x_{r-1} - \phi(x_{r-1})] = 0,$$

$$i = 1, \ldots, m.$$

Note that we have eliminated $x_r$ (which was $= 1$), and we now have a solution to (A) with fewer than $r$ nonzero entries. (Note that $x_1 - \phi(x_1) \neq 0$ by our choice of the automorphism $\phi$.) This is a contradiction.

18.  Let $\pi: G \to G'$ be a surjective group homomorphism. Let $K$ be a normal subgroup of $G$, and let $K' = \pi(K)$. Prove that $K'$ is a normal subgroup of $G'$, and, generalizing the Fundamental Homomorphism Theorem (Theorem 3.9 of Chapter 6), prove that there is a surjective homomorphism $\overline{\pi}: G/K \to G'/K'$. Query: When is it an isomorphism?

19. Prove that there is no quartic polynomial whose Galois group is isomorphic to $Q$ (see p. 173 and Exercise 6.3.12).

20. Let $F \subset \mathbb{R}$, $a \in F$, and let $p$ be prime. Prove that the polynomial $f(x) = x^p - a$ either has a root in $F$ or is irreducible in $F[x]$. Conclude that $[F[\sqrt[p]{a}] : F] = 1$ or $p$. (Hint: Let the roots of $f(x)$ be $\alpha, \omega\alpha, \omega^2\alpha, \ldots, \omega^{p-1}\alpha$, where $\omega = e^{2\pi i/p}$. Show that if $g(x)$ is a factor of $f(x)$ in $F[x]$, then the constant term of $g(x)$ is $b = \omega^r \alpha^k$ for some positive integers $r$, $k$. Use the fact that $b \in F$ and that $k$ and $p$ are relatively prime to show that some root $\omega^\ell \alpha$ of $f(x)$ lies in $F$.)

21. Even though $\mathbb{R}$ is not a finite-dimensional vector space over $\mathbb{Q}$, the Galois group $G(\mathbb{R}/\mathbb{Q})$ still makes sense. Calculate it. (Hint: see Exercise 2.2.16.)

22. a. Let $F = \mathbb{Z}_p$ and let $K$ be an extension of $F$ of degree $d$. Prove that $G(K/F) \cong \mathbb{Z}_d$. (Hint: Consider $\sigma : K \to K$, $\sigma(x) = x^p$. Prove that $\sigma$ is an $F$-automorphism of $K$. To see why $\sigma^d = \iota$, use the fact that $K^\times$ is a group.)
    b. It follows from Section 3 of Chapter 5 that Theorem 6.11 applies in this setting. Deduce that $E$ is an intermediate field if and only if $E$ is an extension of $F$ of degree $d'$, where $d' \mid d$.

23. Let $p$ be prime. Prove that the Galois group of $f(x) = x^p - 1$ is cyclic. (Hint: Use Exercise 6.1.24.)

24. a. Suppose $\mathbb{Q} \subset E \subset K$ is a chain of *quadratic* field extensions. Although $K$ need not be Galois over $\mathbb{Q}$, we can consider all fields $L$ containing $K$ that *are* Galois over $\mathbb{Q}$. Denote by $\overline{K}$ the smallest such (called the Galois closure of $K$ over $\mathbb{Q}$). Prove that $[\overline{K} : \mathbb{Q}]$ is a power of 2. (Hint: If $E = \mathbb{Q}[\sqrt{a}]$, $a \in \mathbb{Q}$, and $K = E[\sqrt{b}]$, $b \in E$, then construct $\overline{K}$ explicitly as the splitting field of a polynomial, and check that all its roots can be obtained by adjoining square roots.)
    b. More generally, suppose $\mathbb{Q} = K_0 \subset K_1 \subset \cdots \subset K_\ell = K$ is a chain of quadratic field extensions. Let $\overline{K}$ be the Galois closure of $K$ over $\mathbb{Q}$; prove that $[\overline{K} : \mathbb{Q}]$ is a power of 2.

25. Let $f(x) \in \mathbb{Q}[x]$ be an irreducible quartic polynomial with a real root $\alpha$. Prove that if the Galois group of $f(x)$ is either $A_4$ or

$S_4$, then $\alpha$ cannot be constructed by compass and straightedge. (Hint: Use Exercise 24, Theorem 6.11, and see Exercise 7.2.3.)

**Remark.** One quartic polynomial having Galois group $S_4$ is $f(x) = x^4 - x - 9$. However, to establish this here would take us too far afield.

26. Use the results proved in this section to decide whether every quartic polynomial in $\mathbb{Q}[x]$ can be solved by radicals.

# CHAPTER 8

# Non-Euclidean Geometries

In this, the final chapter, we come full circle and, following the lead of Euclid and Klein, put our algebra to use to study some new ideas from geometry. We make full use of linear algebra and group actions, and come to understand geometries by way of their groups of motions. Euclid's parallel postulate bothered mathematicians for more than two thousand years; only in the eighteenth and nineteenth centuries did mathematicians finally construct models in which all of Euclid's axioms but the parallel postulate are valid. A few names to mention here are Gauss, Lobachevski, and Bolyai. As the field of differential geometry developed, Riemann and Beltrami realized how to construct non-Euclidean geometries in terms of abstract surfaces. This all led naturally to Einstein's work on special and general relativity.

The study of non-Euclidean geometries gives us the flavors of many parts of mathematics—algebraic geometry (the study of solutions of polynomial equations in several variables), differential geometry (the study of higher-dimensional surfaces and calculus on them), Lie groups (the interplay of matrix algebra and geometry). In our work we only touch on these subjects, but try to give a thorough development of projective geometry in one, two and three dimensions, together with a smattering of beautiful classical results. In §3 we treat the spectral theorem, a pivotal theorem of linear algebra important both for understanding conic sections (and quadric surfaces) and for theoretical physics. We study conics in the projective plane (first in §2) and quadric surfaces in three dimensions (in §4), and give an application to one formidable classic problem: how many lines meet four (general) mutually skew lines in space? In §5 we bring notions of distance and angle back into projective geometry and then give a gentle introduction to differential geometry by constructing some models of hyperbolic geometry. At this point, we can only hope the reader is sufficiently enthralled to undertake further study—of algebra, number theory, or geometry.

## 1. Affine Geometry

Euclidean geometry, as Felix Klein would define it, is the study of those properties of geometric figures that are *invariant* under the group of Euclidean motions (isometries of Euclidean space). If there is a Euclidean motion carrying one figure to another, so that we might call them "Euclideanly equivalent," we commonly refer to them as *congruent*. We will now broaden the notion of motion so that, in affine geometry, any triangles are "affinely equivalent," and, in projective geometry, any quadrilaterals are "projectively equivalent." Although Euclidean motions preserve lengths and areas, affine motions, as we shall see, preserve certain ratios of lengths and areas; and projective motions preserve *cross-ratios* (a subtle invariant of a *quadruple* of collinear points).

As we saw in Section 4 of Chapter 7, every isometry of $\mathbb{R}^n$ can be written in the form $f(\mathbf{x}) = A\mathbf{x} + \mathbf{b}$, where $A \in O(n)$ and $\mathbf{b} \in \mathbb{R}^n$. (We can think of it this way: the function $T(\mathbf{x}) = A\mathbf{x}$ is an *isometry* fixing the origin; we then compose with a translation by $\mathbf{b}$.) To begin our study of non-Euclidean geometry, we relax this structure somewhat and consider **affine geometry**, whose group of motions consists of mappings of the form $f(\mathbf{x}) = A\mathbf{x}+\mathbf{b}$, where $A \in GL(n, \mathbb{R})$ and $\mathbf{b} \in \mathbb{R}^n$. So we have replaced the isometry fixing the origin by an arbitrary *invertible* linear map.

**Definition.** An **affine motion** is a function $T: \mathbb{R}^n \to \mathbb{R}^n$ of the form $T(\mathbf{x}) = A\mathbf{x} + \mathbf{b}$, where $A \in GL(n, \mathbb{R})$ and $\mathbf{b} \in \mathbb{R}^n$. An **affine subspace** of $\mathbb{R}^n$ is a set of vectors

$$S = \{\mathbf{x} + \mathbf{c} : \mathbf{x} \text{ belongs to a (linear) subspace } S_o \text{ of } \mathbb{R}^n\}$$

for some $\mathbf{c} \in \mathbb{R}^n$. That is, an affine subspace $S$ is the translate of a subspace $S_o$ by a vector $\mathbf{c} \in \mathbb{R}^n$. We say the **dimension** of $S$ is the dimension $\dim S_o$ of the associated subspace. Two $k$-dimensional affine subspaces are **parallel** if they are translates of the same subspace $S_0$. Points $\mathbf{v}_0, \mathbf{v}_1, \ldots, \mathbf{v}_k \in \mathbb{R}^n$ are **affinely independent** if

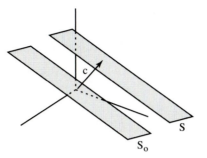

FIGURE 1

$\mathbf{v}_1 - \mathbf{v}_0, \mathbf{v}_2 - \mathbf{v}_0, \ldots, \mathbf{v}_k - \mathbf{v}_0$ are linearly independent; they are **affinely dependent** otherwise.

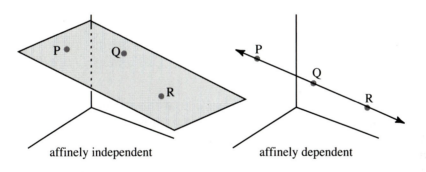

affinely independent                   affinely dependent

FIGURE 2

Affine motions map affine subspaces to affine subspaces and preserve parallelism.

**Lemma 1.1.** *Let $P, Q \in \mathbb{R}^n$ be distinct points. The line passing through $P$ and $Q$ consists of all points of the form*

$$X = \lambda P + \mu Q, \quad \text{where } \lambda + \mu = 1.$$

**Proof.** By definition, the line is obtained by adding to $P$ all real multiples of the direction vector $Q - P$ (see Figure 3). That is, a typical point $X$ of the line is of the form $X = P + t(Q - P) = (1 - t)P + tQ$, as required. $\square$

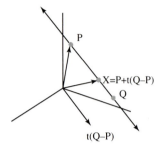

FIGURE 3

**Lemma 1.2.** *Let* $P, Q \in \mathbb{R}^n$ *be distinct points. Then an affine motion* $T \colon \mathbb{R}^n \to \mathbb{R}^n$ *maps the line through* $P$ *and* $Q$ *onto the line through* $T(P)$ *and* $T(Q)$.

**Proof.** The general point $X$ on the line through $P$ and $Q$ is of the form $\lambda P + \mu Q$, where $\lambda + \mu = 1$. Therefore, $T(X) = AX + \mathbf{b} = A(\lambda P + \mu Q) + \mathbf{b} = \lambda A(P) + \mu A(Q) + \mathbf{b} = \lambda(A(P) + \mathbf{b}) + \mu(A(Q) + \mathbf{b}) = \lambda T(P) + \mu T(Q)$, as required. $\square$

**Proposition 1.3.** *Let* $T \colon \mathbb{R}^n \to \mathbb{R}^n$ *be an affine motion, and let* $S$ *and* $S'$ *be parallel affine subspaces. Then* $T(S)$ *and* $T(S')$ *are likewise parallel.*

**Proof.** Since $S$ and $S'$ are parallel affine subspaces, there is a vector $\mathbf{c} \in \mathbb{R}^n$ so that $S' = S + \mathbf{c}$. That is, for each point $P' \in S'$ there is $P \in S$ so that $P' = P + \mathbf{c}$. Write $T(\mathbf{x}) = Ax + \mathbf{b}$. Then $T(P') = A(P') + \mathbf{b} = A(P + \mathbf{c}) + \mathbf{b} = (AP + A\mathbf{c}) + \mathbf{b} = (AP + \mathbf{b}) + A\mathbf{c} = T(P) + A\mathbf{c}$, using the fact that multiplication by $A$ is linear; that is, $T(P')$ is the translate of $T(P)$ by the vector $A\mathbf{c}$. It follows immediately (since the points of $S$ and $S'$ are in one-to-one correspondence) that $T(S') = T(S) + A\mathbf{c}$, and so the affine subspaces are parallel. $\square$

Now we come to the so-called invariants of affine geometry. The proof of Lemma 1.2 establishes the first.

**Corollary 1.4.** *Let* $X \in \overrightarrow{PQ}$, *and let* $T$ *be an affine motion. Then*
$$\frac{|PX|}{|PQ|} = \frac{|T(P)T(X)|}{|T(P)T(Q)|};$$
*i.e., affine motions preserve ratios of lengths along lines.*

**Proof.** Write $X = \lambda P + \mu Q$, with $\lambda + \mu = 1$. Then we have $X - P = (\lambda - 1)P + \mu Q = \mu(Q - P)$, so $\frac{|PX|}{|PQ|} = |\mu|$. Similarly, by the proof of Lemma 1.2, $T(X) - T(P) = \mu(T(Q) - T(P))$, so $\frac{|T(P)T(X)|}{|T(P)T(Q)|} = |\mu|$ as well. $\square$

In particular, we see that the coefficient $\mu$ tells us what fraction of the way from $P$ to $Q$ the point $X$ is located; moreover, $\frac{|PX|}{|XQ|} = \left|\frac{\mu}{\lambda}\right|$. This ratio is preserved by affine motions, even though individual lengths themselves are not. We have ignored the sign of $\mu$ here: when $\mu = 0$, then, of course, $X = P$; when $\mu < 0$, $X$ lies on the opposite side of $P$ from $Q$ (i.e., $X - P$ is a *negative* multiple of $Q - P$).

Next we observe that if $P$, $Q$, and $R$ are affinely independent in $\mathbb{R}^2$, then every point $X \in \mathbb{R}^2$ can be written in the form

$$X = \lambda P + \mu Q + \nu R, \quad \text{where } \lambda + \mu + \nu = 1.$$

This is easy: since $P$, $Q$, and $R$ are affinely independent, the vectors $Q - P$ and $R - P$ are linearly independent and therefore span $\mathbb{R}^2$. Write $X - P = \mu(Q - P) + \nu(R - P)$ for $\mu, \nu \in \mathbb{R}$. Then we have

$$X = P + \mu(Q - P) + \nu(R - P) = (1 - \mu - \nu)P + \mu Q + \nu R =$$
$$\lambda P + \mu Q + \nu R, \quad \text{where } \lambda + \mu + \nu = 1.$$

Note that we can interpret the sign of the coefficients $\lambda$, $\mu$, and $\nu$ geometrically: if, for example, $\lambda = 0$, then $X$ lies on the line $\overleftrightarrow{QR}$; if $\lambda > 0$, then $X$ lies on the same side of the line $\overleftrightarrow{QR}$ as the point $P$, and if $\lambda < 0$, then $X$ lies on the opposite side. But can we interpret the coefficients themselves geometrically?

Recall that area of a triangle in $\mathbb{R}^2$ can be computed by a determinant: if $\mathbf{x} = (x_1, x_2)$, $\mathbf{y} = (y_1, y_2) \in \mathbb{R}^2$ are vectors, then the area of the parallelogram spanned by $\mathbf{x}$ and $\mathbf{y}$ is given by $|x_1 y_2 - x_2 y_2|$ (which is the absolute value of the determinant $\begin{vmatrix} x_1 & x_2 \\ y_1 & y_2 \end{vmatrix}$). If we think of $\mathbb{R}^2$ as a subspace of $\mathbb{R}^3$, the area can be obtained as the length of the *cross product* $\mathbf{x} \times \mathbf{y}$. The upshot is that the area of the triangle with vertices $\mathbf{0}$, $\mathbf{x}$, and $\mathbf{y}$ is given by $\frac{1}{2}|\mathbf{x} \times \mathbf{y}|$.

Returning to our affinely independent points $P$, $Q$, and $R$, we can compute the area of the triangle $PQR$ by any one of the three following formulas:

$$(*) \qquad \text{area } \triangle PQR = \begin{cases} \frac{1}{2}|(Q - P) \times (R - P)| \\ \frac{1}{2}|(R - Q) \times (P - Q)| \\ \frac{1}{2}|(P - R) \times (Q - R)| \end{cases}.$$

Consider now the area of $\triangle XQR$:

$$\text{area} \, \triangle XQR = \tfrac{1}{2}|(X - R) \times (Q - R)|$$
$$= \tfrac{1}{2}| \, (\lambda(P - R) + \mu(Q - R)) \times (Q - R)|$$
$$= |\lambda| \cdot \tfrac{1}{2}|(P - R) \times (Q - R)| = |\lambda| \, \text{area} \, \triangle PQR.$$

(Here we've used the fact that the cross-product of a vector with

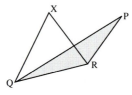

FIGURE 4

itself is zero.) Thus, the absolute value of the coefficient $\lambda$ gives the ratio of the area of $\triangle XQR$ to that of $\triangle PQR$. It follows, as in Corollary 1.4, that this ratio is invariant under affine motions.

In closing, we wish to use affine geometry to derive one of the standard results of Euclidean plane geometry. The crucial point is that *any two triangles are affinely equivalent*; i.e., given two triangles, there is an affine motion carrying one to the other.

**Proposition 1.5.** *Given two triples, P, Q, R and P′, Q′, R′, of affinely independent points in $\mathbb{R}^2$, there is a (unique) affine motion $T: \mathbb{R}^2 \to \mathbb{R}^2$ so that $T(P) = P′$, $T(Q) = Q′$, and $T(R) = R′$.*

***Proof.*** Recall that an affine motion consists of two parts: a translation, and an invertible linear map ($T(\mathbf{x}) = A\mathbf{x} + \mathbf{b}$). Reasoning backwards, if such a $T$ exists, we must have $Q′ - P′ = T(Q) - T(P) = A(Q - P)$ and $R′ - P′ = T(R) - T(P) = A(R - P)$. That is, the matrix $A \in GL(2, \mathbb{R})$ is to map the linearly independent vectors $Q - P$ and $R - P$ to the linearly independent vectors $Q′ - P′$ and $R′ - P′$, respectively; since these are each a basis for $\mathbb{R}^2$, there is a unique $A$ with this property. Now we merely solve for $\mathbf{b}$. We want $T(P) = P′$, so $AP + \mathbf{b} = P′$ will hold if and only if $\mathbf{b} = P′ - AP$. Thus, $A$ and $\mathbf{b}$ exist and are uniquely determined, and this concludes the proof. $\square$

We now combine Corollary 1.4 and Proposition 1.5 to prove the following famous result from Euclidean geometry. Recall that a **median** of a triangle is a line segment joining a vertex and the midpoint of the opposite side.

**Theorem 1.6.** *The medians of a triangle intersect in a point two-thirds the way down each of them.*

**Proof.** For the case of an equilateral triangle, this is easy. Let $\overline{CD}$ and $\overline{BE}$ be two medians, whose intersection point is $M$. Since a median is a perpendicular bisector of the side and the angle bisector of the vertex angle, we have $\triangle MDB \sim \triangle BDC$. Taking the sides of the equilateral triangle to have length 1, it follows that $\frac{MD}{BD} = \frac{BD}{CD}$; and so $MD = \frac{(BD)^2}{CD} = \frac{1}{2\sqrt{3}}$, which is one-third the length of the median $\overline{CD}$. Since $M$ lies two-thirds the way down both medians $\overline{CD}$ and $\overline{BE}$, the same argument shows that the same is true of the third median. Now, given an arbitrary triangle $\triangle PQR$, there is an

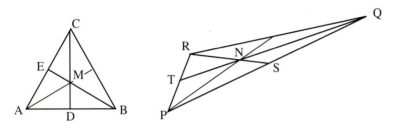

FIGURE 5

affine motion carrying the equilateral triangle $\triangle ABC$ to the triangle $\triangle PQR$. Moreover, by Corollary 1.4, it carries midpoints of sides to midpoints of corresponding sides, hence medians to corresponding medians. In particular, median $\overline{CD}$ is mapped to median $\overline{RS}$, and median $\overline{BE}$ is mapped to median $\overline{QT}$; so, the intersection point $M$ is mapped to the intersection point $N$. Again, by Corollary 1.4, since $\frac{BM}{BE} = \frac{QN}{QT}$, $N$ lies two-thirds the way down median $\overline{QT}$ (and similarly for the other medians). □

The moral of the story is this: any theorem from geometry referring to *ratios* of lengths (or areas)—as opposed to individual lengths or angles—is really a theorem in affine geometry and not Euclidean geometry, and should be proved in that context.

**Remark.** Another way of approaching affine geometry is this: If we "lose" the origin in our vector space $\mathbb{R}^n$, we can no longer add vectors, define linear transformations, linear independence, subspaces, and so forth. But we *can* define affine maps, affine independence, affine span, and so on, by choosing an origin arbitrarily and checking that the corresponding notions are well-defined (i.e., independent of this arbitrary choice).

## EXERCISES 8.1

1. Let $P$, $Q$, and $R$ be affinely independent points in $\mathbb{R}^2$, and let $X \in \mathbb{R}^2$.

   a. Show that if $X$ lies inside $\triangle PQR$, then

   $$\text{area} \triangle PQR = \text{area} \triangle XQR + \text{area} \triangle PXR + \text{area} \triangle PQX.$$

   b. Give an analogous formula when $X$ lies outside $\triangle PQR$.

2. Let $P$, $Q$, and $R$ be affinely independent points in $\mathbb{R}^2$, and let $T$ be an affine motion of $\mathbb{R}^2$, $T(\mathbf{x}) = A\mathbf{x} + \mathbf{b}$. Show that

   $$\text{area} \triangle T(P)T(Q)T(R) = |\det A| \, \text{area} \triangle PQR.$$

3. Suppose point masses $m_1, \ldots, m_k$ are located at points $P_1, \ldots, P_k \in \mathbb{R}^n$. Define the center of mass of this system to be the point $\overline{P} = \left( \sum_{i=1}^{k} m_i P_i \right) \Big/ \left( \sum_{i=1}^{k} m_i \right)$. Suppose $T$ is an affine motion of $\mathbb{R}^n$. If corresponding masses $m_i$ are located at the points $T(P_i)$, $i = 1, \ldots, k$, show that the center of mass of the new system is at $T(\overline{P})$.

4. a. Prove that $\mathbf{v}_0, \mathbf{v}_1, \ldots, \mathbf{v}_k \in \mathbb{R}^n$ are affinely independent if and only if the following criterion holds: if $\lambda_0 + \lambda_1 + \cdots + \lambda_k = 0$ and $\lambda_0 \mathbf{v}_0 + \lambda_1 \mathbf{v}_1 + \cdots + \lambda_k \mathbf{v}_k = 0$, then $\lambda_0 = \lambda_1 = \cdots = \lambda_k = 0$.

   b. Conclude that the notion of affine independence does not depend on the particular ordering of the vectors $\mathbf{v}_0, \mathbf{v}_1, \ldots, \mathbf{v}_k$.

5. Let $\mathbf{v}_0, \mathbf{v}_1, \ldots, \mathbf{v}_k \in \mathbb{R}^n$. We call

   $$\{X \in \mathbb{R}^n : X = \sum_{i=0}^{k} \lambda_i \mathbf{v}_i, \text{ where } \sum_{i=0}^{k} \lambda_i = 1\}$$

   the **affine span** of $\mathbf{v}_0, \mathbf{v}_1, \ldots, \mathbf{v}_k$. Prove the following:

   a. The affine span of $\mathbf{v}_0, \mathbf{v}_1, \ldots, \mathbf{v}_k$ is an affine subspace of $\mathbb{R}^n$.

   b. If $\mathbf{v}_0, \mathbf{v}_1, \ldots, \mathbf{v}_k$ are affinely independent, then their affine span is $k$-dimensional.

c. If $S \subset \mathbb{R}^n$ is a $k$-dimensional affine subspace and $T: \mathbb{R}^n \to \mathbb{R}^n$ is an affine motion, then $T(S)$ is likewise a $k$-dimensional affine subspace of $\mathbb{R}^n$.

6. Use affine motions of the plane to prove the following:
   a. The diagonals of a parallelogram bisect each other.
   b. Let $\triangle ABC$ be arbitrary, and suppose (as pictured in Figure 6) $AD = \frac{1}{2}AB$, $CE = \frac{1}{3}CB$. Prove that $\overline{AE}$ bisects $\overline{CD}$. (Hint: Take $\triangle ABC$ to be a right triangle with $m\angle A = 60°$, $m\angle C = 90°$, and $AC = 1$.)

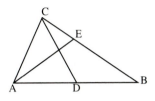

FIGURE 6

(You might also try to give straightforward proofs using vector methods.)

7. Consider the right triangle $\triangle ABC$ pictured in Figure 7, with $AC = AD = 1$, $AB = \alpha \geq 1$. Suppose $\frac{CE}{CB} = \frac{1}{\alpha+1}$. Show that $\overline{AE}$ bisects $\overline{CD}$. (Hint: Introduce cartesian coordinates in the obvious way, and find the intersection of the line $y = x$ with $\overline{CB}$.) What do you conclude about a general triangle? (See Exercise 6b.)

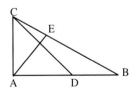

FIGURE 7

8. Let $T$ be an affine motion of $\mathbb{R}^2$, and let $\ell \subset \mathbb{R}^2$ be a line. Prove that as $P$ varies over $\ell$, the points $M = \frac{1}{2}(P + T(P))$ are either all identical or all distinct and collinear. Express the locus of such points $M$ (i.e., the line or the point) in terms of $T$ and $\ell$.

9. Recall that an **altitude** of a triangle $\triangle ABC$ is a line passing through a vertex and perpendicular to the opposite side. Prove that the altitudes are concurrent (i.e., all pass through a common point). (Hint: Let $D$ be the point of intersection of the altitude from $A$ to $\overleftrightarrow{BC}$ and the altitude from $B$ to $\overleftrightarrow{AC}$. Prove that $\overleftrightarrow{CD}$ is orthogonal to $\overleftrightarrow{AB}$. By the way, this is *not* a problem in affine geometry, since angles are involved.)

10. a. Let $P$, $Q$, and $R$ be affinely independent in $\mathbb{R}^2$. Show that the medians of $\triangle PQR$ intersect at the point $\frac{1}{3}(P + Q + R)$. Note that this is the center of mass, if we place equal masses at the vertices $P$, $Q$, and $R$.
    b. Let $P$, $Q$, $R$, and $S$ be affinely independent in $\mathbb{R}^3$. Define a **median** of the tetrahedron $PQRS$ to be a line segment joining a vertex to the center of mass of the opposite face. Prove that the four medians intersect at the center of mass $\frac{1}{4}(P + Q + R + S)$ of the tetrahedron.
    c. Prove that the three line segments joining the midpoints of opposite edges of the tetrahedron $PQRS$ also intersect in the center of mass of the tetrahedron.

11. a. Prove that a parallelepiped in $\mathbb{R}^3$ is affinely equivalent to a cube.
    b. Use your answer to a. to prove that the four diagonals of a parallelepiped intersect in a point that is the midpoint of each diagonal.

12. Prove that in a trapezoid the two diagonals and the line joining the midpoints of the parallel sides go through a common point. (Hint: This is easy for an isosceles trapezoid; then show that any trapezoid is affinely equivalent to an isosceles trapezoid.)

13. Suppose $T$ is an affine motion of $\mathbb{R}^2$ that preserves orthogonality. Must $T$ preserve all angle measures? Give the most specific characterization of $T$ you can.

14. a. Let $P$, $Q$, and $R$ be affinely independent, and suppose $X = \lambda P + \mu Q + \nu R$, $\lambda + \mu + \nu = 1$. Prove that $\overleftrightarrow{PX}$ divides $\overline{QR}$ in a (signed) ratio $\nu : \mu$.
    b. Let $X$ be a point in the interior of $\triangle PQR$. If $\overleftrightarrow{PX}$ divides $\overline{QR}$ in a ratio $x : x'$, $\overleftrightarrow{QX}$ divides $\overline{RP}$ in a ratio $y : y'$, and $\overleftrightarrow{RX}$ divides $\overline{PQ}$ in a ratio $z : z'$, prove that $xyz = x'y'z'$.

15.  Given $\triangle PQR$, $L \in \overrightarrow{PQ}$, $M \in \overrightarrow{QR}$, and $N \in \overrightarrow{RP}$. Write $L = xP + x'Q$, $M = yQ + y'R$, and $N = zR + z'P$ (so $x + x' = y + y' = z + z' = 1$). Prove that $L$, $M$, and $N$ are collinear if and only if $xyz = -x'y'z'$.

16.  Let $\mathbf{x}: (-1, 1) \to \mathbb{R}^2$ be a one-to-one differentiable function with $\mathbf{x}(0) = P$ and $\mathbf{x}'(0) \neq 0$. Let $C$ be the curve $\mathbf{x}((-1, 1))$. Show that an affine motion $T$ of $\mathbb{R}^2$ carries the tangent line to $C$ at $P$ to the tangent line to $T(C)$ at $T(P)$. (Hint: Differentiate $T \circ \mathbf{x}$.)

17.  a.  Prove that any ellipse in $\mathbb{R}^2$ is affinely equivalent to the unit circle centered at the origin.
     b.  Use the result of a. to show that for any diameter $\overline{AB}$ of an ellipse $\mathcal{E}$, the diameter $\overline{CD}$ of $\mathcal{E}$ parallel to the tangent line to $\mathcal{E}$ at $A$ (or $B$) bisects all the chords of $\mathcal{E}$ parallel to $\overline{AB}$. (See Exercise 16 for further discussion of tangent lines.)

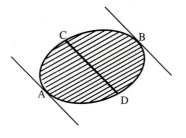

FIGURE 8

18.  Let $P$, $Q$, and $R$ be affinely independent points on the unit circle $x^2 + y^2 = 1$ in $\mathbb{R}^2$.
     a.  Suppose the origin $O$ is in the interior of $\triangle PQR$. Prove that $O$ is also in the interior of the triangle $\triangle(-P)(-Q)(-R)$, but in the exterior of the triangles $\triangle(-P)QR$, $\triangle(-P)(-Q)R$, $\triangle(-P)Q(-R)$, $\triangle P(-Q)R$, $\triangle P(-Q)(-R)$, and $\triangle PQ(-R)$.
     b.  If three points, $P$, $Q$, and $R$, are picked at random on the unit circle, what is the probability that the origin is in the interior of $\triangle PQR$?
     c.  Generalize the result of b. to higher dimensions.

## 2. The Projective Group

Projective geometry grew out of the work of Kepler on conics in the late sixteenth and seventeenth centuries, when an architect named Gérard Desargues published a pamphlet on perspective and then a text on conics. His work had a great deal of impact on Descartes and Pascal, among others. Our obsession with Euclidean geometry is really unfortunate, for in our day-to-day life, we in fact view the world from a non-Euclidean perspective: visualize a pair of train tracks receding to the horizon; in your mind's eye, they appear to meet at the horizon (as it were, "at infinity").

A mathematical model for perspective is indicated in Figure 1. The configurations *abc* and *ABC* are indistinguishable to the eye

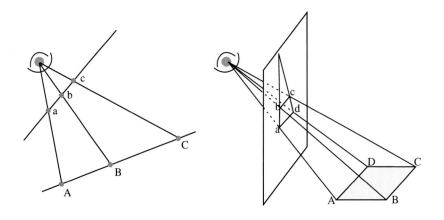

FIGURE 1

in the left diagram in Figure 1, as are *abcd* and *ABCD* to the eye in the right. Two figures are thus *projectively equivalent* if one is obtained from the other by projecting from one line (or plane) to another from some vantage point. Our goal now is to make this all more precise by providing an algebraic formulation, and then to find some "projective invariants."

Consider the statement, "Two lines lying in a plane intersect." In Euclidean geometry, we must stipulate that the lines are not parallel.

We shall see that in projective geometry this condition is unnecessary, since there will be "points at infinity" where parallel lines intersect (as with the train tracks above). What is the algebraic setting for this problem? Consider the inhomogeneous system of linear equations

$$a_{11}x_1 + a_{12}x_2 = b_1$$
$$(*) \qquad a_{21}x_1 + a_{22}x_2 = b_2 \, .$$

If the coefficient matrix $A = \begin{bmatrix} a_{11} & a_{12} \\ a_{21} & a_{22} \end{bmatrix}$ is nonsingular, the equations $(*)$ represent two non-parallel lines in the plane, and the unique solution is the point of intersection. If, however, $\det A = 0$ (e.g., if $a_{11} = ca_{21}$ and $a_{12} = ca_{22}$ for some constant $c$), then the individual equations in $(*)$ represent either the same line (e.g., when $b_1 = cb_2$) or two parallel lines. For example, we have the inconsistent system (in the variables $x_1$ and $x_2$)

$$x_1 = 1$$
$$(\dagger) \qquad x_1 = -1 \, .$$

We recall from elementary linear algebra a technique for solving inhomogeneous systems of linear equations: we form the augmented matrix

$$\left[ \begin{array}{cc|c} a_{11} & a_{12} & b_1 \\ a_{21} & a_{22} & b_2 \end{array} \right] \, ,$$

treat the $b_i$'s as if they were simply the coefficients of another variable, and perform Gaussian elimination. So, let's introduce a new variable, $x_0$, and then the equations $(*)$ can be rewritten as a **homogeneous** linear system, namely:

$$-b_1x_0 + a_{11}x_1 + a_{12}x_2 = 0$$
$$(**) \qquad -b_2x_0 + a_{21}x_1 + a_{22}x_2 = 0 \, .$$

Now there is no longer any need to differentiate between homogeneous and inhomogeneous linear systems. This process, oddly enough, is called the **homogenization** of the equations $(*)$. In the case of our example $(\dagger)$, we obtain

$$-x_0 + x_1 = 0$$
$$(\ddagger) \qquad x_0 + x_1 = 0 \, ,$$

whose solutions are all points of the form $(0,0,t)$, $t \in \mathbb{R}$.

In order to interpret homogenization geometrically, we must stop to define the projective plane. Since it is just as easy to define **projective $n$-space** (for all $n \in \mathbb{N}$), we proceed to do so.

**Definition.** Let $\mathbf{x}, \mathbf{y} \in \mathbb{R}^{n+1} - \{\mathbf{0}\}$; we say $\mathbf{x} \equiv \mathbf{y}$ if $\mathbf{y} = c\mathbf{x}$ for some $c \in \mathbb{R} - \{0\}$. That is, two nonzero vectors are equivalent if and only if they lie on a line through the origin in $\mathbb{R}^{n+1}$. As the reader should check (see Exercise 1), $\equiv$ is an equivalence relation.

Projective $n$-space, $\mathbb{P}^n$, is the set of all $\equiv$ equivalence classes; that is, $\mathbb{P}^n$ is the set of all lines through the origin in $\mathbb{R}^{n+1}$. We will denote the equivalence class of $(x_0, x_1, \ldots, x_n) \in \mathbb{R}^{n+1} - \{\mathbf{0}\}$ by $[x_0, x_1, \ldots, x_n]$, and we refer to $(x_0, x_1, \ldots, x_n)$ as **homogeneous coordinates** of $[x_0, x_1, \ldots, x_n] \in \mathbb{P}^n$.

Let's begin by coming to terms with the projective line $\mathbb{P}^1$. We consider ordered pairs $(x_0, x_1)$, with $x_0$ and $x_1$ not both equal to zero. Whenever $x_0 \neq 0$, we have $(x_0, x_1) \equiv (1, b)$ for some $b \in \mathbb{R}$. If, however, $x_0 = 0$, then $x_1 \neq 0$, and $(0, x_1) \equiv (0, 1)$, so there is one

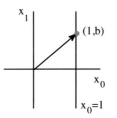

FIGURE 2

additional point "at infinity." Under the identification of $\mathbb{P}^1$ with the lines through the origin in $\mathbb{R}^2$, we have the one-to-one correspondence

$$\{\text{lines through the origin in } \mathbb{R}^2\} \longleftrightarrow \mathbb{R} \cup \{\infty\} \longleftrightarrow \mathbb{P}^1$$

$$\{-bx_0 + ax_1 = 0\} \rightsquigarrow \text{slope } \tfrac{b}{a} \rightsquigarrow [a, b].$$

In particular, the vertical line (with infinite slope) corresponds to the value $b = \infty$, i.e., to the point $[0, 1]$. It is interesting to try to imagine a picture of $\mathbb{P}^1$. The (unit length) direction vectors of lines through the origin, starting with $(1, 0)$ (the $x$-axis), sweep out a semicircle; but, when we come to $(-1, 0)$, we have the same line with which we started. Thus, $\mathbb{P}^1$ can be pictured as a semicircle with its two endpoints identified; this is, in turn, "the same as" a circle.

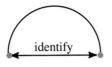

FIGURE 3

We define next the group of motions of $\mathbb{P}^1$. Since points of $\mathbb{P}^1$ are equivalence classes of vectors in $\mathbb{R}^2$, it is natural to consider $GL(2, \mathbb{R})$, the invertible linear maps from $\mathbb{R}^2$ to itself. The matrices $A$ of the form $A = \lambda\,\mathrm{Id}$ ($\lambda \neq 0$) leave every point of $\mathbb{P}^1$ fixed, since $(\lambda\,\mathrm{Id})\begin{bmatrix} x_0 \\ x_1 \end{bmatrix} = \begin{bmatrix} \lambda x_0 \\ \lambda x_1 \end{bmatrix} \equiv \begin{bmatrix} x_0 \\ x_1 \end{bmatrix}$, and conversely (see Exercise 2). Note, moreover, that the set of matrices $\{\lambda\,\mathrm{Id} : \lambda \in \mathbb{R}^\times\}$ forms a normal subgroup of $GL(2, \mathbb{R})$; thus, we define the **projective group**

$$\mathrm{Proj}(1) = GL(2, \mathbb{R})/\{\lambda\,\mathrm{Id} : \lambda \in \mathbb{R}^\times\}.$$

(When we mod out, only the identity element of the quotient group $\mathrm{Proj}(1)$ leaves every point fixed.) We refer to elements of $\mathrm{Proj}(1)$ as **projective transformations**.

**Lemma 2.1.** *Given two trios, $P_1, P_2, P_3$ and $Q_1, Q_2, Q_3$, of points in $\mathbb{P}^1$, there is a unique projective transformation $T \in \mathrm{Proj}(1)$ with $T(P_1) = Q_1$, $T(P_2) = Q_2$, and $T(P_3) = Q_3$.*

***Proof.*** Assume, first, for convenience, that $P_1 = [1, 0]$, $P_2 = [0, 1]$, and $P_3 = [1, 1]$. Let $Q_1 = [\alpha_0, \alpha_1]$, $Q_2 = [\beta_0, \beta_1]$, and $Q_3 = [\gamma_0, \gamma_1]$. Since $Q_1 \neq Q_2$, $(\beta_0, \beta_1)$ is not a scalar multiple of $(\alpha_0, \alpha_1)$, and so the matrix $\begin{bmatrix} \alpha_0 & \beta_0 \\ \alpha_1 & \beta_1 \end{bmatrix}$ is invertible. Therefore, the system of equations

$$\begin{bmatrix} \alpha_0 & \beta_0 \\ \alpha_1 & \beta_1 \end{bmatrix}\begin{bmatrix} \lambda \\ \mu \end{bmatrix} = \begin{bmatrix} y_0 \\ y_1 \end{bmatrix}$$

has a unique nonzero solution $(\lambda, \mu)$. Define $Y \in GL(2, \mathbb{R})$ by

$$Y = \begin{bmatrix} \lambda\alpha_0 & \mu\beta_0 \\ \lambda\alpha_1 & \mu\beta_1 \end{bmatrix};$$

now we calculate:

$$Y\begin{bmatrix} 1 \\ 0 \end{bmatrix} = \lambda\begin{bmatrix} \alpha_0 \\ \alpha_1 \end{bmatrix} \equiv \begin{bmatrix} \alpha_0 \\ \alpha_1 \end{bmatrix} \leftrightarrow Q_1,$$

$$Y\begin{bmatrix} 0 \\ 1 \end{bmatrix} = \mu\begin{bmatrix} \beta_0 \\ \beta_1 \end{bmatrix} \equiv \begin{bmatrix} \beta_0 \\ \beta_1 \end{bmatrix} \leftrightarrow Q_2, \text{ and}$$

$$Y\begin{bmatrix} 1 \\ 1 \end{bmatrix} = \begin{bmatrix} \lambda\alpha_0 + \mu\beta_0 \\ \lambda\alpha_1 + \mu\beta_1 \end{bmatrix} = \begin{bmatrix} y_0 \\ y_1 \end{bmatrix} \leftrightarrow Q_3.$$

Note, moreover, that only the scalar multiples of $Y$ have these properties; and so there is a unique element $T$ of the quotient group Proj(1) carrying $[1,0]$ to $Q_1$, $[0,1]$ to $Q_2$, and $[1,1]$ to $Q_3$. (Here is an important point: if we replace either of the vectors $(\alpha_0, \alpha_1)$ and $(\beta_0, \beta_1)$ by a scalar multiple, the numbers $\lambda$ and $\mu$ adjust accordingly, and the matrix $Y$ remains unchanged. If, however, we replace $(\gamma_0, \gamma_1)$ by a scalar multiple, then $\lambda$ and $\mu$ are both multiplied by the scalar, and the matrix $Y$ is as well.)

The general case follows by composing functions. Given a general triple of distinct points $P_1, P_2, P_3$, let $T'$ be the unique projective transformation carrying $[1,0]$ to $P_1$, $[0,1]$ to $P_2$, and $[1,1]$ to $P_3$. Then $T \circ T'^{-1}$ is the desired transformation carrying $P_i$ to $Q_i$, $i = 1, 2, 3$.  □

As a consequence, we can define an invariant associated to *four* (distinct) points on the projective line. For our present purposes, it will be convenient to think of $\mathbb{P}^1$ as $\mathbb{R} \cup \{\infty\}$. Given points $A$, $B$, $C$, and $D \in \mathbb{P}^1$, there is a unique projective transformation $T \colon \mathbb{P}^1 \to \mathbb{P}^1$ carrying $A$ to $0$, $B$ to $\infty$, and $C$ to $1$. We call the number $T(D) \in \mathbb{R} \cup \{\infty\}$ the **cross-ratio** of $A, B, C, D$, written $|A, B, C, D|$. The importance of the cross-ratio stems from the fact that it is preserved by projective transformations (and is thus a "projective invariant"); cf. Corollary 1.4.

**Lemma 2.2.** *Let $f \in$ Proj(1) be a projective transformation of $\mathbb{P}^1$, and let $A$, $B$, $C$, $D \in \mathbb{P}^1$ be distinct points. Then*

$$|A, B, C, D| = |f(A), f(B), f(C), f(D)| \,.$$

***Proof.*** By definition, there is a unique $T \in$ Proj(1) carrying $A$ to $0$, $B$ to $\infty$, $C$ to $1$, and $D$ to the cross-ratio $|A, B, C, D|$. Now, we merely observe that the projective transformation $T \circ f^{-1}$ carries $f(A)$ to $0$, $f(B)$ to $\infty$, $f(C)$ to $1$, and $f(D)$ to $(T \circ f^{-1})(f(D)) = T(D)$. That is, the cross-ratio $|f(A), f(B), f(C), f(D)|$ is the same as $|A, B, C, D|$, as promised.  □

**Remark.** It is sometimes convenient to have a formula for the cross-ratio $|A, B, C, D|$. We leave it to the reader to check (see Exercise 3):

$$|A, B, C, D| = \frac{D - A}{C - A} \Big/ \frac{D - B}{C - B} \in \mathbb{R} \cup \{\infty\} \,.$$

If we work in that portion of $\mathbb{P}^1$ where $x_0 \neq 0$, then we can write every point in the form $[1, x]$; in these coordinates, projective transformations take the form of so-called **linear fractional** or **Möbius transformations**. If $T \in \mathrm{Proj}(1)$ is represented by $\begin{bmatrix} d & c \\ b & a \end{bmatrix}$, then $T \begin{bmatrix} 1 \\ x \end{bmatrix} = \begin{bmatrix} cx + d \\ ax + b \end{bmatrix} \equiv \begin{bmatrix} 1 \\ \frac{ax+b}{cx+d} \end{bmatrix}$, so that the transformation takes the form of the rational function

$$f(x) = \frac{ax + b}{cx + d}.$$

This explains, by the way, our weird choice of letters in the original matrix representation of $T$.

Next we proceed to the projective plane $\mathbb{P}^2$. Recall that $\mathbb{P}^2$ is the set of equivalence classes $[x_0, x_1, x_2]$ of nonzero ordered triples, where $\mathbf{x} \equiv \mathbf{y} \iff \mathbf{y} = c\mathbf{x}$ for some $c \in \mathbb{R} - \{0\}$. When $x_0 \neq 0$, $(x_0, x_1, x_2) \equiv (1, x, y)$, with $(x, y) \in \mathbb{R}^2$; so we have a copy of $\mathbb{R}^2$. And when $x_0 = 0$, we have the subset $\{[0, x_1, x_2] : (x_1, x_2) \neq (0, 0)\}$, and this is a copy of $\mathbb{P}^1$ "at infinity." That is, we have a copy of $\mathbb{R}^2$ with an additional point at infinity for each "direction" (line through the origin) in that $\mathbb{R}^2$:

$$\mathbb{P}^2 = \mathbb{R}^2 \cup \mathbb{P}^1_\infty.$$

Mathematicians often draw the symbolic picture in Figure 4. It is

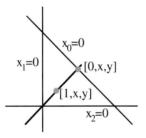

FIGURE 4

interesting to try to give a more precise (geometric?) picture of the projective plane. Each equivalence class $[x_0, x_1, x_2]$ corresponds to a line through the origin in $\mathbb{R}^3$; the line intersects the unit sphere centered at the origin in two points. When $x_0 \neq 0$, exactly one of these points lies in the hemisphere $\{(x_0, x_1, x_2) : x_0^2 + x_1^2 + x_2^2 = 1,\ x_0 \geq 0\}$. However, when $x_0 = 0$, we must identify opposite points on the circle $\{(x_0, x_1, x_2) : x_0 = 0,\ x_1^2 + x_2^2 = 1\}$—this is the projective line at infinity. It is not too hard to see now that the projective plane

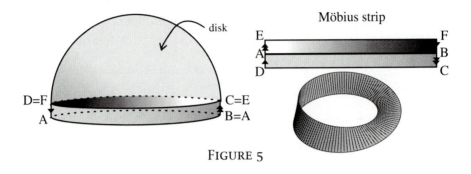

FIGURE 5

looks like a disk with a Möbius strip attached along its boundary. (See Figure 5.)

As in Euclidean and affine geometry, through any two distinct points $P$ and $Q$ of $\mathbb{P}^2$ there is a unique (projective) **line**, given as follows: if $P = [p_0, p_1, p_2]$ and $Q = [q_0, q_1, q_2]$, then

$$\overrightarrow{PQ} = \{[s(p_0, p_1, p_2) + t(q_0, q_1, q_2)] : s, t \in \mathbb{R} \text{ not both zero}\}.$$

That is, $P$ and $Q$ each corresponds to a line through the origin in $\mathbb{R}^3$, and $\overrightarrow{PQ}$ corresponds to the plane they span in $\mathbb{R}^3$ (each point of $\overrightarrow{PQ}$ then corresponding to a line through the origin in that plane). We say three (or more) points of $\mathbb{P}^2$ are **collinear** if they lie on a line.

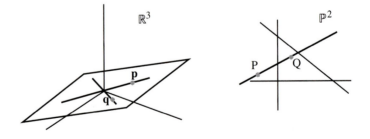

FIGURE 6

**Remark.** Write $\mathbf{p} = (p_0, p_1, p_2)$ and $\mathbf{q} = (q_0, q_1, q_2)$. When $s \neq 0$, $[s\mathbf{p} + t\mathbf{q}] = [\mathbf{p} + \frac{t}{s}\mathbf{q}]$, and as $\frac{t}{s}$ varies over $\mathbb{R}$, we have the usual parametric equations of a line.

Lines originally appeared as the solution sets of linear equations. So we now stop to make it official.

**Lemma 2.3.** *Every line in $\mathbb{P}^2$ is the locus of points $[x_0, x_1, x_2]$ satisfying the equation*

$$(*) \qquad\qquad a_0 x_0 + a_1 x_1 + a_2 x_2 = 0$$

*for some* $\mathbf{a} = (a_0, a_1, a_2) \neq \mathbf{0}$.

**Proof.** If $P$ and $Q$ are distinct points in $\mathbb{P}^2$, then the corresponding vectors $\mathbf{p} = (p_0, p_1, p_2)$ and $\mathbf{q} = (q_0, q_1, q_2)$ are linearly independent, and the nonzero vector $\mathbf{a} = (a_0, a_1, a_2) = \mathbf{p} \times \mathbf{q}$ is orthogonal to them both. Thus, the points $[\mathbf{x}] \in \overrightarrow{PQ}$ are precisely the solutions of the homogeneous linear equation $(*)$. Conversely, if we start with $\mathbf{a} = (a_0, a_1, a_2) \neq \mathbf{0}$, we can choose two linearly independent solutions $\mathbf{p}$ and $\mathbf{q}$ of the equation $(*)$. Letting $P = [\mathbf{p}]$ and $Q = [\mathbf{q}]$, we see then that $\overrightarrow{PQ} = \{[\mathbf{x}] : a_0 x_0 + a_1 x_1 + a_2 x_2 = 0\}$. $\square$

We are now in a position to settle the question with which we began the section.

**Lemma 2.4.** *Any two lines in $\mathbb{P}^2$ intersect.*

**Proof.** If $L = \{[\mathbf{x}] : \mathbf{a} \cdot \mathbf{x} = 0\}$ and $L' = \{[\mathbf{x}] : \mathbf{b} \cdot \mathbf{x} = 0\}$, then $L$ and $L'$ are distinct, provided $\mathbf{a}$ and $\mathbf{b}$ are linearly independent. The set of solutions $\mathbf{x}$ to the homogeneous linear system of equations $\mathbf{a} \cdot \mathbf{x} = \mathbf{b} \cdot \mathbf{x} = 0$ is a one-dimensional subspace of $\mathbb{R}^3$ (spanned by $\mathbf{a} \times \mathbf{b}$), so $[\mathbf{x}]$ is a unique point of $\mathbb{P}^2$. $\square$

In particular, with regard to the equations ($\ddagger$), the two lines intersect, as pictured in Figure 7, at the point $[0, 0, 1]$, i.e., the point vertically at infinity.

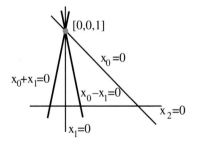

FIGURE 7

**Remark.** Thinking of $\mathbb{P}^2$ as $\mathbb{R}^2 \cup \{\infty\}$, with $[1, x, y] \in \mathbb{R}^2$ as above, any (usual) line in $\mathbb{R}^2$ has an equation of the form $ax + by + c = 0$ for some $a, b, c \in \mathbb{R}$. Writing $x$ and $y$ in terms of homogeneous coordinates $(x_0, x_1, x_2)$ on $\mathbb{P}^2$, we have $x = x_1/x_0$ and $y = x_2/x_0$, so the equation of the line can be written

$$ax + by + c = 0$$
$$a\left(\frac{x_1}{x_0}\right) + b\left(\frac{x_2}{x_0}\right) + c = 0$$
$$ax_1 + bx_2 + cx_0 = 0;$$

that is, every line in $\mathbb{R}^2$ gives rise to a line in the projective plane. In the latter, it acquires an extra point, namely, its point of intersection with the line $\{x_0 = 0\}$ at infinity. In our case, this is the point $[0, -b, a]$. Note that all the parallel lines obtained by varying $c$ intersect in this very point at infinity.

We next define the group of motions of $\mathbb{P}^2$, the projective transformations $\mathrm{Proj}(2) = GL(3, \mathbb{R})/\{\lambda \, \mathrm{Id} : \lambda \in \mathbb{R}^\times\}$. Geometry in the projective plane should be a generalization of affine geometry (see Exercise 23), so we first observe that projective transformations preserve collinearity:

**Lemma 2.5.** Let $T \in \mathrm{Proj}(2)$ be a projective transformation of $\mathbb{P}^2$, and let $P_1, P_2$, and $P_3$ be collinear points. Then $T(P_1), T(P_2)$, and $T(P_3)$ are collinear.

***Proof.*** If we write $P_1 = [x_0, x_1, x_2]$, $P_2 = [y_0, y_1, y_2]$, and $P_3 = [z_0, z_1, z_2]$, let $\mathbf{x} = (x_0, x_1, x_2)$, $\mathbf{y} = (y_0, y_1, y_2)$, and $\mathbf{z} = (z_0, z_1, z_2)$ be choices of corresponding vectors in $\mathbb{R}^3$. Then $P_1$, $P_2$, and $P_3$ are collinear if and only if the vectors $\mathbf{x}, \mathbf{y}, \mathbf{z}$ are linearly dependent. There are scalars $s, t \in \mathbb{R}$ so that $\mathbf{z} = s\mathbf{x} + t\mathbf{y}$. Therefore, if $\Upsilon \colon \mathbb{R}^3 \to \mathbb{R}^3$ is a linear map representing $T \in \mathrm{Proj}(2)$, then $\Upsilon(\mathbf{z}) = s\Upsilon(\mathbf{x}) + t\Upsilon(\mathbf{y})$. Translating everything back, this says precisely that $T(P_3)$ is collinear with $T(P_1)$ and $T(P_2)$. $\square$

We now define the notion of points in general position in the projective plane. We say *three* points in the plane are **in general position** if they are not collinear. We say *four* points in the plane are in general position if no three are collinear. The reader may well want to know what happens if there are five or more points; see Exercise 17.

Now we may generalize Lemma 2.1 in the setting of the projective plane as follows.

**Lemma 2.6.** *Given two quartets, $P_1$, $P_2$, $P_3$, $P_4$ and $Q_1$, $Q_2$, $Q_3$, $Q_4$, of points in general position in $\mathbb{P}^2$, there is a unique projective transformation $T \in Proj(2)$ with $T(P_1) = Q_1$, $T(P_2) = Q_2$, $T(P_3) = Q_3$, and $T(P_4) = Q_4$.*

**Proof.** As in the proof of Lemma 2.1, we first treat the case that $P_1 = [1,0,0]$, $P_2 = [0,1,0]$, $P_3 = [0,0,1]$, and $P_4 = [1,1,1]$. Let $Q_1 = [\alpha_0, \alpha_1, \alpha_2]$, $Q_2 = [\beta_0, \beta_1, \beta_2]$, $Q_3 = [\gamma_0, \gamma_1, \gamma_2]$, and $Q_4 = [\delta_0, \delta_1, \delta_2]$. Since $Q_1$, $Q_2$, and $Q_3$ are not collinear, the matrix $\begin{bmatrix} \alpha_0 & \beta_0 & \gamma_0 \\ \alpha_1 & \beta_1 & \gamma_1 \\ \alpha_2 & \beta_2 & \gamma_2 \end{bmatrix}$ is invertible. Therefore, the system of equations

$$\begin{bmatrix} \alpha_0 & \beta_0 & \gamma_0 \\ \alpha_1 & \beta_1 & \gamma_1 \\ \alpha_2 & \beta_2 & \gamma_2 \end{bmatrix} \begin{bmatrix} \lambda \\ \mu \\ \nu \end{bmatrix} = \begin{bmatrix} \delta_0 \\ \delta_1 \\ \delta_2 \end{bmatrix}$$

has a unique nonzero solution $(\lambda, \mu, \nu)$. Define $Y \in GL(3, \mathbb{R})$ by

$$Y = \begin{bmatrix} \lambda\alpha_0 & \mu\beta_0 & \nu\gamma_0 \\ \lambda\alpha_1 & \mu\beta_1 & \nu\gamma_1 \\ \lambda\alpha_2 & \mu\beta_2 & \nu\gamma_2 \end{bmatrix} ;$$

then we claim $Y$ has the desired properties:

$$Y \begin{bmatrix} 1 \\ 0 \\ 0 \end{bmatrix} = \lambda \begin{bmatrix} \alpha_0 \\ \alpha_1 \\ \alpha_2 \end{bmatrix} \equiv \begin{bmatrix} \alpha_0 \\ \alpha_1 \\ \alpha_2 \end{bmatrix} \leftrightarrow Q_1,$$

$$Y \begin{bmatrix} 0 \\ 1 \\ 0 \end{bmatrix} = \mu \begin{bmatrix} \beta_0 \\ \beta_1 \\ \beta_2 \end{bmatrix} \equiv \begin{bmatrix} \beta_0 \\ \beta_1 \\ \beta_2 \end{bmatrix} \leftrightarrow Q_2,$$

$$Y \begin{bmatrix} 0 \\ 0 \\ 1 \end{bmatrix} = \nu \begin{bmatrix} \gamma_0 \\ \gamma_1 \\ \gamma_2 \end{bmatrix} \equiv \begin{bmatrix} \gamma_0 \\ \gamma_1 \\ \gamma_2 \end{bmatrix} \leftrightarrow Q_3, \text{ and}$$

$$Y \begin{bmatrix} 1 \\ 1 \\ 1 \end{bmatrix} = \begin{bmatrix} \lambda\alpha_0 + \mu\beta_0 + \nu\gamma_0 \\ \lambda\alpha_1 + \mu\beta_1 + \nu\gamma_1 \\ \lambda\alpha_2 + \mu\beta_2 + \nu\gamma_2 \end{bmatrix} = \begin{bmatrix} \delta_0 \\ \delta_1 \\ \delta_2 \end{bmatrix} \leftrightarrow Q_4.$$

Note also that $Y$ is unique, up to scalar multiples, and hence defines a unique element $T$ of the quotient group Proj(2).

Now the general case follows by composing functions. Given a general quartet of distinct points $P_1, P_2, P_3, P_4$ in general position, let $T'$ be the unique projective transformation carrying $[1,0,0]$ to $P_1$, $[0,1,0]$ to $P_2$, $[0,0,1]$ to $P_3$, and $[1,1,1]$ to $P_4$. Then $T \circ T'^{-1}$ is the desired transformation carrying $P_i$ to $Q_i$, $i = 1,2,3,4$. □

At the beginning of the section, we suggested that projective geometry should have something to do with projections. Let's now stop to illustrate this.

**Example.** Let $\ell$ be a line in $\mathbb{P}^2$ and let $P \in \mathbb{P}^2$ be a point not lying on $\ell$. Then there is a **projection** $\pi\colon \mathbb{P}^2 - \{P\} \to \ell$ defined by setting $\pi(Q)$ equal to the point of intersection of $\overleftrightarrow{PQ}$ and $\ell$. The point $P$ is called the **center** of the projection. To give an algebraic formula for $\pi$, we assume that $P = [0, 0, 1]$ and $\ell = \{x_2 = 0\}$ (how may we do so?). If $Q = [x_0, x_1, x_2]$, then $\overleftrightarrow{PQ}$ consists of points of the form $[s(0, 0, 1) + t(x_0, x_1, x_2)] = [tx_0, tx_1, s + tx_2]$; so such a point lies on $\ell$ when $s + tx_2 = 0$. Thus, $\pi[x_0, x_1, x_2] = [tx_0, tx_1, 0] = [x_0, x_1, 0]$.

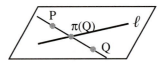

FIGURE 8

We now prove some of the "classical" elementary theorems in projective geometry. See Exercises 14 and 16 for some surprisingly concrete applications of these results.

**Theorem 2.7** (Desargues). *Let $\triangle PQR$ and $\triangle P'Q'R'$ be triangles in $\mathbb{P}^2$. Suppose the three lines $\overleftrightarrow{PP'}$, $\overleftrightarrow{QQ'}$, and $\overleftrightarrow{RR'}$ are concurrent (i.e., intersect in a single point). Then the three points $A = \overleftrightarrow{PQ} \cap \overleftrightarrow{P'Q'}$, $B = \overleftrightarrow{QR} \cap \overleftrightarrow{Q'R'}$, and $C = \overleftrightarrow{PR} \cap \overleftrightarrow{P'R'}$ are collinear.*

**Proof.** By Lemma 2.6, we may assume that $P = [1, 0, 0]$, $Q = [0, 1, 0]$, and $R = [0, 0, 1]$, and that the point of intersection of the lines $\overleftrightarrow{PP'}$, $\overleftrightarrow{QQ'}$, and $\overleftrightarrow{RR'}$ is $X = [1, 1, 1]$. Then, since $P'$ lies on the line $\overleftrightarrow{XP}$, we have $P' = [\alpha, 1, 1]$ for some $\alpha \in \mathbb{R}$ (see Exercise 6); similarly, $Q' = [1, \beta, 1]$, and $R' = [1, 1, \gamma]$ for appropriate $\beta$ and $\gamma$.

Now we must find the coordinates of the points $A$, $B$, and $C$. The line $\overleftrightarrow{PQ}$ has the equation $x_2 = 0$. The line $\overleftrightarrow{P'Q'}$ has the equation $(1 - \beta)x_0 + (1 - \alpha)x_1 + (\alpha\beta - 1)x_2 = 0$. (One way to see this is to find a normal to the plane in $\mathbb{R}^3$ spanned by the vectors $(\alpha, 1, 1)$ and $(1, \beta, 1)$ by computing their cross product.) Thus, the two lines intersect in the point $A = [\alpha - 1, 1 - \beta, 0]$. Similarly (see Exercise 6), $B = [1 - \alpha, 0, \gamma - 1]$ and $C = [0, \beta - 1, 1 - \gamma]$. But it follows now

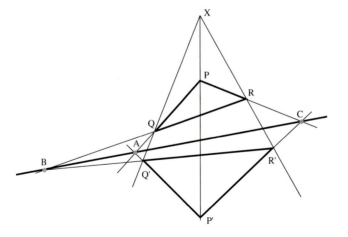

FIGURE 9

that $A$, $B$, and $C$ are collinear, since the sum of their homogeneous coordinate vectors is zero. □

**Theorem 2.8** (Pappus). *Let $P, Q, R$ and $P', Q', R'$ be triples of collinear points. Let $\overleftrightarrow{PQ'} \cap \overleftrightarrow{P'Q} = A$, $\overleftrightarrow{PR'} \cap \overleftrightarrow{P'R} = B$, and $\overleftrightarrow{QR'} \cap \overleftrightarrow{Q'R} = C$. Then $A$, $B$, and $C$ are collinear, as well.*

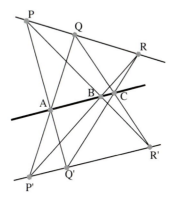

FIGURE 10

**Proof.** By Lemma 2.6, we may assume $P = [1,0,0]$, $P' = [0,1,0]$, $Q = [0,0,1]$, and $Q' = [1,1,1]$. Then $R = [1,0,\alpha]$ and $R' = [\beta,1,\beta]$ for some (nonzero) $\alpha$ and $\beta$. Now we compute the coordinates of $A$, $B$, and $C$ to be the following: $A = [0,1,1]$, $B = [\beta, \alpha, \alpha\beta]$, and

$C = [\beta, 1, \alpha\beta - \alpha + 1]$. It follows that $A$, $B$, and $C$ are collinear, since the following linear combination of their homogeneous coordinate vectors is zero: $(\alpha - 1)\mathbf{A} - \mathbf{B} + \mathbf{C} = 0$. $\quad\square$

At this time we would be remiss if we did not mention one of the fundamental concepts in projective geometry: the notion of duality. Recall that points in $\mathbb{P}^2$ are equivalence classes of nonzero vectors in $\mathbb{R}^3$. A line in $\mathbb{P}^2$ is given by one linear homogeneous equation $\xi_0 x_0 + \xi_1 x_1 + \xi_2 x_2 = 0$, where $\boldsymbol{\xi} = (\xi_0, \xi_1, \xi_2) \neq \mathbf{0}$, and $\boldsymbol{\xi}$ is uniquely defined up to scalar multiples. Thus, there is a one-to-one correspondence between the set of lines in $\mathbb{P}^2$ and another copy of $\mathbb{P}^2$, typically called $\mathbb{P}^{2*}$, the **dual projective plane**:

$$\{\text{lines in } \mathbb{P}^2\} \longrightarrow \mathbb{P}^{2*}$$

$$\text{line } \xi_0 x_0 + \xi_1 x_1 + \xi_2 x_2 = 0 \rightsquigarrow [\xi_0, \xi_1, \xi_2].$$

Alternatively, this is the correspondence between the set of planes (through the origin) in $\mathbb{R}^3$ and the set of lines (through the origin) in $\mathbb{R}^3$, associating to each plane its normal line and *vice versa*.

We can now explore some of the beautiful ramifications of duality. By construction, a point of $\mathbb{P}^{2*}$ corresponds to a line in $\mathbb{P}^2$. We wish to see that a point of $\mathbb{P}^2$ likewise corresponds to a line in $\mathbb{P}^{2*}$. The key observation is this: to each point $P \in \mathbb{P}^2$ we associate the "pencil" of lines passing through $P$, denoted $\lambda_P \subset \mathbb{P}^{2*}$. Then we have the following lemma.

**Lemma 2.9.** *The pencil $\lambda_P \subset \mathbb{P}^{2*}$ is a (projective) line in $\mathbb{P}^{2*}$, and, conversely, every line in $\mathbb{P}^{2*}$ is of this form.*

***Proof.*** Let $P = [x_0, x_1, x_2] \in \mathbb{P}^2$ and $\xi = [\xi_0, \xi_1, \xi_2] \in \mathbb{P}^{2*}$. Then $P$ lies on the line $\xi$ if and only if

$$\boldsymbol{\xi} \cdot \mathbf{p} = 0,$$

where $\boldsymbol{\xi}$ and $\mathbf{p}$ are the corresponding vectors in $\mathbb{R}^3$. But this is a homogeneous linear equation in $\boldsymbol{\xi}$, and thus defines a line in $\mathbb{P}^{2*}$. Conversely, every line in $\mathbb{P}^{2*}$ is given by such an equation. $\quad\square$

The correspondence

$$\mathbb{P}^2 \longrightarrow \{\text{lines in } \mathbb{P}^{2*}\}$$

$$\text{point } P \rightsquigarrow \lambda_P = \text{the pencil of lines through } P$$

is called **projective duality**.

**Proposition 2.10.** *Projective duality enjoys the following properties:*

(i) *Let $\xi, \eta \in \mathbb{P}^{2*}$ correspond to lines $\ell, m \subset \mathbb{P}^2$. The line in $\mathbb{P}^{2*}$ passing through $\xi$ and $\eta$ corresponds to the pencil of lines through the point $P = \ell \cap m$.*

(ii) *Three lines $\ell_1, \ell_2,$ and $\ell_3$ in $\mathbb{P}^2$ are concurrent (i.e., have a single point of intersection) if and only if the corresponding points $\xi_1, \xi_2,$ and $\xi_3$ in $\mathbb{P}^{2*}$ are collinear.*

(iii) *Dually, three points $P, Q,$ and $R$ in $\mathbb{P}^2$ are collinear if and only if the lines $\lambda_P, \lambda_Q,$ and $\lambda_R$ in $\mathbb{P}^{2*}$ are concurrent.*

**Proof.** (i) The line through $\xi$ and $\eta$ is the set of all points of $\mathbb{P}^{2*}$ of the form $[s(\xi_0, \xi_1, \xi_2) + t(\eta_0, \eta_1, \eta_2)]$, where $s$ and $t$ are not both 0. If this is the pencil of lines through $P \in \mathbb{P}^2$, then it follows that $\xi \cdot \mathbf{p} = 0$ and $\eta \cdot \mathbf{p} = 0$; i.e., $P \in \ell$ and $P \in m$. Since the lines $\ell$ and $m$ intersect in precisely one point, we have $P = \ell \cap m$, as claimed.

(ii) This is immediate from (i). Let $P = \ell_1 \cap \ell_2$; then the line $\lambda$ through $\xi_1$ and $\xi_2$ in $\mathbb{P}^{2*}$ corresponds to the pencil of lines through $P$. If $\ell_3$ passes through $P$, then $\xi_3$ lies on $\lambda$, and conversely.

(iii) Apply the same proof as in (ii). Let $\xi = \lambda_P \cap \lambda_Q \in \mathbb{P}^{2*}$. The line $\ell$ through $P$ and $Q$ in $\mathbb{P}^2$ corresponds to the pencil of lines through $\xi$. If $\lambda_R$ passes through $\xi$, then $R$ lies on $\ell$, and conversely. $\square$

The import of projective duality, then, is that any theorem about points and lines in $\mathbb{P}^2$ has a dual statement—about lines and points in $\mathbb{P}^{2*}$, which is, after all, just another projective plane. Interestingly enough, the dual statements of Desargues' and Pappus' Theorems are their respective converses. We check the former here (and leave the latter to the reader; see Exercise 12).

**Theorem 2.11** (Desargues' dual). *Let $\triangle PQR$ and $\triangle P'Q'R'$ be triangles in $\mathbb{P}^2$. Suppose the three points $A = \overleftrightarrow{PQ} \cap \overleftrightarrow{P'Q'}, B = \overleftrightarrow{QR} \cap \overleftrightarrow{Q'R'},$ and $C = \overleftrightarrow{PR} \cap \overleftrightarrow{P'R'}$ are collinear. Then the three lines $\overleftrightarrow{PP'}, \overleftrightarrow{QQ'},$ and $\overleftrightarrow{RR'}$ are concurrent (i.e., intersect in a single point).*

**Proof.** Let $\xi = \overleftrightarrow{PQ}, \eta = \overleftrightarrow{QR},$ and $\rho = \overleftrightarrow{PR} \in \mathbb{P}^{2*}$, and let the primed objects correspond in the obvious manner. The point $A = \overleftrightarrow{PQ} \cap \overleftrightarrow{P'Q'}$ corresponds under duality to the line $\overleftrightarrow{\xi\xi'}$, and similarly for $B$ and $C$. Now by (iii) of Proposition 2.10, if $A, B,$ and $C$ are collinear, then $\overleftrightarrow{\xi\xi'}, \overleftrightarrow{\eta\eta'},$ and $\overleftrightarrow{\rho\rho'}$ are concurrent. (See Figure 11.)

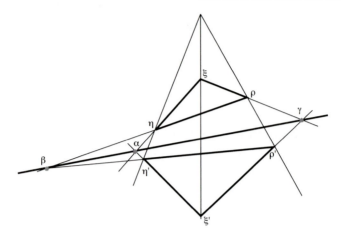

FIGURE 11

The line $\overleftrightarrow{PP'}$ is an element of both pencils $\lambda_P$ and $\lambda_{P'}$. But the pencil of lines through $P$ corresponds under projective duality to the line $\overleftrightarrow{\xi\rho}$, and the pencil of lines through $P'$ corresponds to the line $\overleftrightarrow{\xi'\rho'}$. Thus, the point $\alpha = \overleftrightarrow{\xi\rho} \cap \overleftrightarrow{\xi'\rho'}$ corresponds to the line $\overleftrightarrow{PP'}$. The points $\beta$ and $\gamma$ are constructed analogously.

Now, by Theorem 2.7 applied in $\mathbb{P}^{2*}$, if $\overleftrightarrow{\xi\xi'}$, $\overleftrightarrow{\eta\eta'}$, and $\overleftrightarrow{\rho\rho'}$ are concurrent, then $\alpha$, $\beta$, and $\gamma$ are collinear. But this means, by Proposition 2.10(ii), that the lines $\overleftrightarrow{PP'}$, $\overleftrightarrow{QQ'}$, and $\overleftrightarrow{RR'}$ are concurrent.  □

The last topic we consider in this section is a beautiful and classical approach to conic sections—from the viewpoint of projective geometry. We start with two distinct pencils of lines in $\mathbb{P}^2$, say $\lambda_P$ and $\lambda_Q$. The line $\overleftrightarrow{PQ}$ is, of course, common to the two pencils. Since

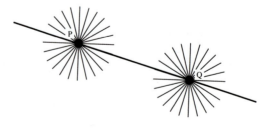

FIGURE 12

by Lemma 2.9 each of these is a line in $\mathbb{P}^{2*}$, there are vectors $\xi$, $\eta$, $\sigma$, and $v \in \mathbb{R}^3$ so that $\lambda_P = \{[s\xi + t\eta] : s, t \in \mathbb{R} \text{ not both zero}\} \subset \mathbb{P}^{2*}$

and $\lambda_Q = \{[s\boldsymbol{\sigma} + t\boldsymbol{v}] : s, t \in \mathbb{R} \text{ not both zero}\} \subset \mathbb{P}^{2*}$. Since these two lines in $\mathbb{P}^{2*}$ intersect (in the point corresponding to the line $\overleftrightarrow{PQ}$), we may take $\boldsymbol{\sigma} = \boldsymbol{\eta}$. Now for fixed $[s, t] \in \mathbb{P}^1$, the equation

(A) $$(s\boldsymbol{\xi} + t\boldsymbol{\eta}) \cdot \mathbf{x} = 0$$

defines a line $L_{[s,t]}$ in $\mathbb{P}^2$ passing through $P$, and the equation

(B) $$(s\boldsymbol{\eta} + t\boldsymbol{v}) \cdot \mathbf{x} = 0$$

likewise defines a line $M_{[s,t]}$ in $\mathbb{P}^2$ passing through $Q$. Thus, a non-trivial simultaneous solution $X_{[s,t]}$ must be the point of intersection $L_{[s,t]} \cap M_{[s,t]}$. We wish to find the locus of these points $X_{[s,t]}$ as $[s, t]$ varies through $\mathbb{P}^1$. (See Figure 13.)

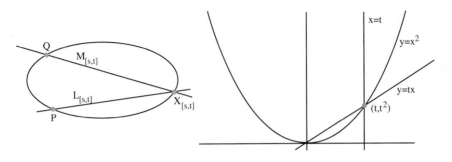

FIGURE 13

We change our perspective and interpret the equations (A) and (B) as a system of homogeneous equations in $(s, t)$:

(C)
$$s(\boldsymbol{\xi} \cdot \mathbf{x}) + t(\boldsymbol{\eta} \cdot \mathbf{x}) = 0$$
$$s(\boldsymbol{\eta} \cdot \mathbf{x}) + t(\boldsymbol{v} \cdot \mathbf{x}) = 0,$$

which will have a nontrivial solution if and only if

(D) $$\begin{vmatrix} (\boldsymbol{\xi} \cdot \mathbf{x}) & (\boldsymbol{\eta} \cdot \mathbf{x}) \\ (\boldsymbol{\eta} \cdot \mathbf{x}) & (\boldsymbol{v} \cdot \mathbf{x}) \end{vmatrix} = 0.$$

Since the pencils are distinct, the vectors $\boldsymbol{\xi}$, $\boldsymbol{\eta}$, and $\boldsymbol{v}$ must be linearly independent; and so we may choose a basis so that $\boldsymbol{\xi} = (1, 0, 0)$, $\boldsymbol{\eta} = (0, 1, 0)$, and $\boldsymbol{v} = (0, 0, 1)$. Thus, the equation (D) becomes

(E) $$\begin{vmatrix} x_0 & x_1 \\ x_1 & x_2 \end{vmatrix} = 0.$$

It may help to set $x_0 = 1$, $x_1 = x$, and $x_2 = y$; then we obtain $y = x^2$, which is the familiar equation of a parabola.

Let's examine this example with hindsight. Note that the line $\eta \cdot \mathbf{x} = 0$ is the line $x_1 = 0$, which contains both the points $P = [1, 0, 0]$ and $Q = [0, 0, 1]$; $P$ and $Q$ satisfy equation (E) as well. If we continue to use coordinates $[1, x, y]$ in the portion of the projective plane given by $x_0 = 1$, the pencil $\lambda_P$ consists of the lines $y = tx$, and the pencil $\lambda_Q$ consists of the lines $x = t$, $[1, t] \in \mathbb{P}^1$. Thus, the point of intersection is, as one might expect, $X_t = [1, t, t^2]$. See Figure 13.

Conversely, suppose we start with any **conic**

$$\mathcal{C} = \{ax_1^2 + bx_1x_2 + cx_2^2 + dx_0x_1 + ex_0x_2 + fx_0^2 = 0\} \subset \mathbb{P}^2,$$

where $(a, b, c, d, e, f) \neq \mathbf{0}$. That is, a conic in the projective plane is the set of solutions to a (nonzero) homogeneous quadratic equation in $\mathbf{x} = (x_0, x_1, x_2)$. (In the affine plane, when $x_0 \neq 0$ and we set $[x_0, x_1, x_2] = [1, x, y]$, this reduces to the customary equation $ax^2 + bxy + cy^2 + dx + ey + f = 0$.) We say the conic $\mathcal{C}$ is **nondegenerate** if it is nonempty and does not consist of either a point or a pair of lines. As Exercise 24 shows, projective transformations map conics to conics and nondegenerate conics to nondegenerate conics.

**Lemma 2.12.** *Let $\mathcal{C}$ be a nondegenerate conic. Then any line $\ell \subset \mathbb{P}^2$ intersects $\mathcal{C}$ in at most two points.*

**Proof.** By a projective transformation we may arrange that $\ell = \{x_2 = 0\}$. Thus, $\mathcal{C} \cap \ell = \{[x_0, x_1, 0] \in \mathbb{P}^2 : ax_1^2 + dx_0x_1 + fx_0^2 = 0\}$. Now, if $[0, 1, 0] \in \mathcal{C} \cap \ell$, this means that $a = 0$ and there is at most one other point of intersection, namely $[d, -f, 0]$. Note that if $d = f = 0$ as well, $\ell \subset \mathcal{C}$, contradicting the hypothesis that $\mathcal{C}$ is a nondegenerate conic. If $[0, 1, 0] \notin \mathcal{C} \cap \ell$, then any point of intersection must be of the form $[1, t, 0]$ for some $t \in \mathbb{R}$. Substituting, we find $at^2 + dt + f = 0$, and this quadratic equation has at most two roots. $\square$

We assume from here on that our conic $\mathcal{C}$ is nondegenerate, and we choose $P$ and $Q \in \mathcal{C}$. By a projective transformation of $\mathbb{P}^2$, we may take $P = [1, 0, 0]$ and $Q = [0, 0, 1]$, and so $c = f = 0$ in the equation of $\mathcal{C}$. Now the pencil $\lambda_P$ of lines through $P$ consists of the lines $x_2 = tx_1$ ($t \in \mathbb{R} \cup \{\infty\}$), and the pencil $\lambda_Q$ of lines through $Q$ consists of the lines $x_1 = t'x_0$ ($t' \in \mathbb{R} \cup \{\infty\}$). By Lemma 2.12, each point of $\mathcal{C}$ (other than $P$ and $Q$) lies on a unique element of $\lambda_P$ and on a unique element of $\lambda_Q$; what is the relation between $t$ and $t'$? By substituting the point $[x_0, t'x_0, tt'x_0]$ into the equation of $\mathcal{C}$, we find that $t' = g(t) = \frac{d+et}{a+bt}$. In other words, the conic is specified by a projective transformation $\lambda_P \rightarrow \lambda_Q$. (Note that $\begin{bmatrix} a & b \\ d & e \end{bmatrix}$ is invertible, because if, for example, $(a, b) = \mu(d, e)$, then the equation of $\mathcal{C}$ becomes $(x_0 + \mu x_1)(dx_1 + ex_2) = 0$, and $\mathcal{C}$ consists of a pair of lines.) It follows that if we replace $t$ by $g^{-1}(t) = \frac{-d+at}{e-bt}$ (which is a projective transformation of $\lambda_P$), then $t' = t$ and then $\mathcal{C}$ will be presented in the form (C), hence in the form (D).

As an immediate consequence, we have the following corollary.

**Corollary 2.13.** *Let $A, B, C, D$ be points on a nondegenerate conic $\mathcal{C}$, and let $P, Q$ be arbitrary points of $\mathcal{C}$. Then*

$$|\overrightarrow{PA}, \overrightarrow{PB}, \overrightarrow{PC}, \overrightarrow{PD}| = |\overrightarrow{QA}, \overrightarrow{QB}, \overrightarrow{QC}, \overrightarrow{QD}|.$$

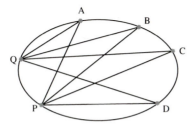

FIGURE 14

**Proof.** There is a unique projective transformation $T: \lambda_P \rightarrow \lambda_Q$ so that $T(\overrightarrow{PA}) = \overrightarrow{QA}$, $T(\overrightarrow{PB}) = \overrightarrow{QB}$, and $T(\overrightarrow{PC}) = \overrightarrow{QC}$. Then for any other point $D \in \mathcal{C}$, $T(\overrightarrow{PD}) = \overrightarrow{QD}$; and this means that the two cross-ratios are equal. $\square$

Here is a pretty consequence of the work we have just done. A nondegenerate conic is uniquely determined by *five* of its points: from Corollary 2.13 it follows that if we fix $A$, $B$, $C$, $P$, and $Q$, then we obtain the projective transformation which identifies $\lambda_P$ and $\lambda_Q$; and we construct the conic as the locus of points of intersection of the corresponding lines. From a slightly different viewpoint (see Exercise 17), through any five points, no three of which are collinear, there passes a unique nondegenerate conic. We conclude this section with a discussion of Pascal's Theorem, which gives a necessary condition for *six* points to lie on a conic. In fact, the condition is sufficient as well (see Exercise 25). Now we can prove our final result. (Cf. Pappus' Theorem, Theorem 2.8.)

**Theorem 2.14** (Pascal). *Let* $P$, $Q$, $R$, $P'$, $Q'$, $R'$ *be six points on a nondegenerate conic. Then the three points* $A = \overleftrightarrow{PQ'} \cap \overleftrightarrow{QP'}$, $B = \overleftrightarrow{PR'} \cap \overleftrightarrow{RP'}$, *and* $C = \overleftrightarrow{QR'} \cap \overleftrightarrow{RQ'}$ *are collinear.*

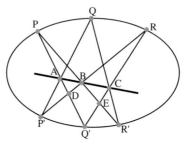

FIGURE 15

***Proof.*** Let $D = \overleftrightarrow{PQ'} \cap \overleftrightarrow{RP'}$ and $E = \overleftrightarrow{PR'} \cap \overleftrightarrow{RQ'}$. Projecting from $P'$ to the line $\overleftrightarrow{PQ'}$, we get (see Exercise 21) $|\overleftrightarrow{P'P}, \overleftrightarrow{P'Q}, \overleftrightarrow{P'R}, \overleftrightarrow{P'Q'}| = |P, A, D, Q'|$. Similarly, projecting from $R'$ to the line $\overleftrightarrow{RQ'}$, we obtain $|\overleftrightarrow{R'P}, \overleftrightarrow{R'Q}, \overleftrightarrow{R'R}, \overleftrightarrow{R'Q'}| = |E, C, R, Q'|$. Since, by Corollary 2.13, $|\overleftrightarrow{P'P}, \overleftrightarrow{P'Q}, \overleftrightarrow{P'R}, \overleftrightarrow{P'Q'}| = |\overleftrightarrow{R'P}, \overleftrightarrow{R'Q}, \overleftrightarrow{R'R}, \overleftrightarrow{R'Q'}|$, we have $|P, A, D, Q'| = |E, C, R, Q'|$; and so $|\overleftrightarrow{BP}, \overleftrightarrow{BA}, \overleftrightarrow{BD}, \overleftrightarrow{BQ'}| = |\overleftrightarrow{BE}, \overleftrightarrow{BC}, \overleftrightarrow{BR}, \overleftrightarrow{BQ'}|$ (why?). Since $\overleftrightarrow{BD} = \overleftrightarrow{BR}$ and $\overleftrightarrow{BP} = \overleftrightarrow{BE}$, it follows that $\overleftrightarrow{BA} = \overleftrightarrow{BC}$. This implies, of course, that $A$, $B$, and $C$ are collinear, as we wished to show. $\square$

**Remark.** Note that when the conic consists of two lines, Theorem 2.14 reduces to Theorem 2.8. (See Exercise 22.)

## EXERCISES 8.2

1. Check that "$\equiv$" (in the definition of $\mathbb{P}^n$ on p. 296) is in fact an equivalence relation.

2. Suppose $A \in GL(n, \mathbb{R})$. Suppose that for every $\mathbf{x} \in \mathbb{R}^n$, there is a constant $c$ so that $A\mathbf{x} = c\mathbf{x}$. Prove that $A = \lambda \, \mathrm{Id}$ for some $\lambda \in \mathbb{R} - \{0\}$. (The hard part here is that $c$ may depend on $\mathbf{x}$.)

3. Prove that
$$|A, B, C, D| = \frac{D - A}{C - A} \Big/ \frac{D - B}{C - B}.$$
Part of your task here may be to interpret things like $\frac{\infty}{\infty}$. (Hint: What is the value of the right-hand side when $D = A$, $B$, or $C$?)

4.  Viewing a line as the solution set of a homogeneous linear equation $\mathbf{a} \cdot \mathbf{x} = 0$, reprove Lemma 2.5: show that a projective transformation $T \in \text{Proj}(2)$ maps the line $\mathbf{a} \cdot \mathbf{x} = 0$ to another line $\tilde{\mathbf{a}} \cdot \mathbf{x} = 0$. (Hint: Represent $T$ by a matrix $Y \in GL(3, \mathbb{R})$, and answer the question $\mathbf{a} \cdot \mathbf{x} = ? \cdot Y\mathbf{x}$ for all $\mathbf{x} \in \mathbb{R}^3$. See Exercise 7.4.2.)

5.  Let $P = [0, 1, 1]$, $Q = [2, -1, 0]$, $R = [1, 1, 1]$, and $S = [1, 0, 2] \in \mathbb{P}^2$.
    a.  Find the point $X = \overleftrightarrow{PQ} \cap \overleftrightarrow{RS}$.
    b.  Show that $X$ lies on the line $\ell = \{4x_0 + 3x_1 - 5x_2 = 0\} \subset \mathbb{P}^2$.
    c.  Give the coordinates of the corresponding points $\overleftrightarrow{PQ}$, $\overleftrightarrow{RS}$, and $\ell$ in the dual projective plane. Use your answer to illustrate part (ii) of Proposition 2.10.

6.  Check the indicated details in the proof of Theorem 2.7.

7.  In this exercise we discover the origins of the mysterious "rationalizing" substitution $u = \tan \theta/2$ (that used to be taught in integral calculus), and we interpret the projective transformations of $\mathbb{P}^1$ in terms of isometries of the circle. We work with the unit circle $x^2 + y^2 = 1$.
    a.  Consider the diagram in Figure 16. Show that $t = \frac{y}{1+x} = \tan \theta/2$. Deduce that $x = \cos \theta = \frac{1-t^2}{1+t^2}$, $y = \sin \theta = \frac{2t}{1+t^2}$

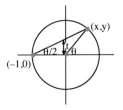

FIGURE 16

(cf. Exercise 1.3.34), and that $d\theta = dy/x = \frac{2}{1+t^2} dt$. Explain now how to evaluate $\int R(\cos \theta, \sin \theta) d\theta$ for any rational function $R(x, y)$.
    b.  Show that in $t$-coordinates, reflection of the circle about the $x$-axis is given by $t \rightsquigarrow -t$, and reflection about the $y$-axis is given by $t \rightsquigarrow 1/t$. What is the form of a general reflection in $t$-coordinates?

    c.  Show that rotation of the circle through angle $\phi \neq \pi$ is given by the Möbius transformation $t \rightsquigarrow \frac{t+A}{1-At}$, where $A = \tan(\phi/2)$. What is the form when $\phi = \pi$?

    d.  In conclusion, what are all the possible isometries of the circle, expressed in $t$-coordinates?

8.  We say two pairs of points $A, B$ and $a, b$ in $\mathbb{P}^1$ are **harmonic** if $|a, b, A, B| = -1$; we say $A$ and $B$ are **harmonic conjugates** if $|0, \infty, A, B| = -1$.

    a.  Show that $A, B$ and $a, b$ are harmonic $\iff$ $A, B$ and $b, a$ are harmonic $\iff$ $a, b$ and $A, B$ are harmonic.

    b.  Suppose $A, B, C \in \mathbb{R}$. Show that $A, B$ and $C, \infty$ are harmonic $\iff$ $C$ is the midpoint of $\overline{AB}$.

    c.  In terms of the geometry of the circle (see Exercise 7), explain what it means for $A$ and $B$ to be harmonic conjugates.

9.  Suppose $\ell$ and $m$ are lines in $\mathbb{P}^2$, and let $P \in \mathbb{P}^2$ be a point lying on neither line. We define a map $f\colon \ell \to m$, called a **perspectivity with center** $P$: for each $Q \in \ell$, let $f(Q)$ be the point of intersection of $\overrightarrow{PQ}$ with $m$ (cf. Figure 1). We may assume (why?) that $\ell = \{x_2 = 0\}$, $m = \{x_1 = 0\}$, $\ell$ and $m$ intersect at the point $[1, 0, 0]$, and take $P = [1, 1, 1]$. Then show that $f$ is given by the linear fractional transformation $f(x) = \frac{x}{x-1}$ (so a perspectivity is a projective transformation $\mathbb{P}^1 \to \mathbb{P}^1$).

10.  As in the treatment on p. 309, write the circle $-x_0^2 + x_1^2 + x_2^2 = 0$ in the form

$$\begin{vmatrix} (\boldsymbol{\xi} \cdot \mathbf{x}) & (\boldsymbol{\eta} \cdot \mathbf{x}) \\ (\boldsymbol{\eta} \cdot \mathbf{x}) & (\boldsymbol{v} \cdot \mathbf{x}) \end{vmatrix} = 0.$$

11.  Let $L, M, N, P, Q$, and $R$ be the consecutive vertices of a hexagon inscribed in a nondegenerate conic $\mathcal{C}$, as pictured in Figure 17. Show that the pairs of opposite sides of the hexagon intersect in three collinear points.

12.  Write out carefully the dual version of Pappus' Theorem (Theorem 2.8), and check that it is equivalent to the following "converse" of the original theorem:

    Let $P, Q, R$ and $P', Q', R'$ be triples of points. Let $\overrightarrow{PQ'} \cap \overrightarrow{P'Q} = A$, $\overrightarrow{PR'} \cap \overrightarrow{P'R} = B$, and $\overrightarrow{QR'} \cap \overrightarrow{Q'R} = C$. Then if $P, Q, R$ are collinear and $A, B, C$ are collinear, it follows that $P', Q'$, and $R'$ are collinear as well.

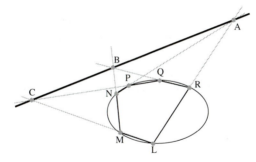

FIGURE 17

13. Prove directly the converse of
    a. Desargues' Theorem, Theorem 2.7;
    b. Pappus' Theorem, Theorem 2.8 (see also Exercise 12 for the appropriate statement).

14. Let $\triangle PQR$ and $\triangle P'Q'R'$ be triangles in $\mathbb{R}^2$. Suppose $\overleftrightarrow{PP'}$, $\overleftrightarrow{QQ'}$, and $\overleftrightarrow{RR'}$ are either parallel or concurrent. Prove that
    a. If $\overleftrightarrow{PQ}$ and $\overleftrightarrow{P'Q'}$ are parallel and $\overleftrightarrow{QR}$ and $\overleftrightarrow{Q'R'}$ are parallel, then so are $\overleftrightarrow{PR}$ and $\overleftrightarrow{P'R'}$.
    b. If $\overleftrightarrow{PQ}$ and $\overleftrightarrow{P'Q'}$ are parallel but $\overleftrightarrow{QR} \cap \overleftrightarrow{Q'R'} = B$, then $\overleftrightarrow{PR}$ and $\overleftrightarrow{P'R'}$ intersect in a point $C$ and $\overleftrightarrow{BC}$ is parallel to $\overleftrightarrow{PQ}$.
    c. If $\overleftrightarrow{PQ} \cap \overleftrightarrow{P'Q'} = A$, $\overleftrightarrow{QR} \cap \overleftrightarrow{Q'R'} = B$, and $\overleftrightarrow{PR} \cap \overleftrightarrow{P'R'} = C$, then $A$, $B$, and $C$ are collinear.

15. Use Theorem 2.11, Desargues' dual, to prove that the medians of a triangle are concurrent.

16. Derive the following very concrete Euclidean results:
    a. Suppose you are given a piece of paper on which someone has drawn a point $A$ and two lines $\ell_1$ and $\ell_2$, neither pass-

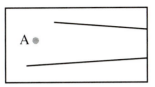

FIGURE 18

ing through $A$ and whose point of intersection lies off the paper. You are asked to draw the line through $A$ which goes through that point of intersection. How do you do it? (Hint: Apply Theorem 2.11 by picking $P, P' \in \ell_1$ and $Q, Q' \in \ell_2$ with $\overrightarrow{PQ} \| \overrightarrow{P'Q'}$.)

b.  You are given a short straightedge and two points $P$ and $Q$ a great distance apart. Use your short straightedge to draw the line joining $P$ and $Q$. (Hint: Use a.)

17.  a.  Show (geometrically) that if $P_1, P_2, P_3$ are collinear, but $P_4$ and $P_5$ do not lie on the line determined by them, then there is a unique conic containing all five points. (Hint: Two distinct lines form a conic. Why? Also see Lemma 2.12.)

b.  By Lemma 2.6, if we are given five distinct points $P_i$, $i = 1, 2, \ldots, 5$, in the plane, no *three* of which are collinear, then there is a projective transformation $T \in \text{Proj}(2)$ carrying $P_1 \rightsquigarrow [1, 0, 0]$, $P_2 \rightsquigarrow [0, 1, 0]$, $P_3 \rightsquigarrow [0, 0, 1]$, and $P_4 \rightsquigarrow [1, 1, 1]$, and which therefore takes $P_5$ to some point $[\alpha, \beta, \gamma]$ distinct from $[1, 0, 0]$, $[0, 1, 0]$, $[0, 0, 1]$, and $[1, 1, 1]$. Show that there is a unique conic passing through any five points, no *three* of which are collinear.* (Hint: Write down the linear equations $a$, $b$, $c$, $d$, $e$, and $f$ must satisfy, and show that the appropriate matrix has rank 5.)

c.  Use the results of a. and b. to conclude that there is a unique conic passing through any five points, no *four* of which are collinear.

d.  Show, however, that in general, six points do not lie on a conic. (Cf. Pascal's Theorem, Theorem 2.14, and Exercise 25.)

Thus, we say *five* points in the plane are in general position if no three are collinear; we say *six* points are in general position if they do not all lie on a conic. What should the notion of general position be, then, for seven or more points in the plane?

18.  Suppose six points in the projective plane are in general position. Show that any five of them must be in general position.

19.  Let $\mathcal{C}$ be a nondegenerate conic in $\mathbb{P}^2$. Show that any other conic $\mathcal{C}'$ intersects $\mathcal{C}$ in at most four points. (Hint: See Exercise 17.)

---

*As a warmup exercise, the reader may wish to prove that through any *three* noncollinear points in $\mathbb{R}^2$ there passes a unique circle $(x - a)^2 + (y - b)^2 = r^2$. Allowing a more general conic buys us two further degrees of freedom.

**Remark.** A function $f \colon \mathbb{R}^n \to \mathbb{R}$ is *homogeneous of degree d* if $f(\lambda \mathbf{x}) = \lambda^d f(\mathbf{x})$ for all $\mathbf{x}$ and $\lambda \in \mathbb{R}$. A curve $C \subset \mathbb{P}^2$ is said to have **degree** $d$ if it is the zero-locus $\{f(x_0, x_1, x_2) = 0\}$ of a homogeneous polynomial $f$ of degree $d$. For example, a line has degree 1 and a conic has degree 2. The result of this exercise is a special case of the famous Bezout's Theorem in algebraic geometry: If $C$ and $D \subset \mathbb{P}^2$ are curves of degree $c$ and $d$, respectively, then, assuming their intersection is a finite set, $C$ and $D$ intersect in at most $cd$ points.

20. Use the methods of Exercise 17 to establish the following. Suppose two conics $\mathcal{C}_1$ and $\mathcal{C}_2$ intersect in the four points $A, B, C, D$. Suppose $\mathcal{C}$ is another conic passing through those four points. Show that there are constants $\lambda$ and $\mu$ so that $\mathcal{C} = \lambda \mathcal{C}_1 + \mu \mathcal{C}_2$ (where by this we mean the corresponding relation among their respective defining quadratic polynomials). Show, moreover, that $\lambda$ and $\mu$ are unique up to multiplication by a common constant.

21. Let $\lambda_P$ be the pencil of lines through $P \in \mathbb{P}^2$. Let $\ell \subset \mathbb{P}^2$ be any line not passing through $P$. Show that the map $f \colon \lambda_P \to \ell$ given by $f(m) = m \cap \ell$ is a projective transformation. (Hint: Let $\alpha$, $\beta \in \lambda_P$, so that the general element of $\lambda_P$ is of the form $[s\alpha + t\beta]$. Let $\alpha \cap \ell = A$ and $\beta \cap \ell = B$, and show that $f[s\alpha + t\beta] = [sA + tB]$.)

22. Deduce Pappus' Theorem (Theorem 2.8) by following the proof of Pascal's Theorem (Theorem 2.14). (Hint: Use Exercise 21 to prove that $|\overrightarrow{P'P}, \overrightarrow{P'Q}, \overrightarrow{P'R}, \overrightarrow{P'R'}| = |\overrightarrow{R'P}, \overrightarrow{R'Q}, \overrightarrow{R'R}, \overrightarrow{R'P'}|$, and then follow the rest of the proof.)

23. We can view affine geometry in the framework of projective geometry. The affine plane is $\mathbb{P}^2$ with the projective line at infinity removed. Show that the group of affine motions is isomorphic to the subgroup of $\mathrm{Proj}(2)$ preserving the line at infinity. Be explicit with matrices.

24. Let $\mathcal{C} = \{ax_1^2 + bx_1x_2 + cx_2^2 + dx_0x_1 + ex_0x_2 + fx_0^2 = 0\} \subset \mathbb{P}^2$ be a conic. Define the symmetric matrix

$$A = \begin{bmatrix} f & d/2 & e/2 \\ d/2 & a & b/2 \\ e/2 & b/2 & c \end{bmatrix}.$$

Then the equation of $\mathcal{C}$ is $\mathbf{x}^T A \mathbf{x} = 0$.

    a.  If $T \in \mathrm{Proj}(2)$ is represented by $Y \in GL(3, \mathbb{R})$ and we write $\mathbf{y} = Y\mathbf{x}$ (so $\mathbf{x} = Y^{-1}\mathbf{y}$), then show that $T(\mathcal{C})$ is given by the equation $\mathbf{y}^T \left((Y^{-1})^T A Y^{-1}\right) \mathbf{y} = 0$. Conclude that $T(\mathcal{C})$ is a conic.

    b.  Prove that $\mathcal{C}$ is a nondegenerate conic if and only if $T(\mathcal{C})$ is a nondegenerate conic.

25.  Here we sketch a completely computational proof of Pascal's Theorem, Theorem 2.14. It has the additional virtue that it will prove the converse as well. That is, if $A = \overleftrightarrow{PQ'} \cap \overleftrightarrow{QP'}$, $B = \overleftrightarrow{PR'} \cap \overleftrightarrow{RP'}$, and $C = \overleftrightarrow{QR'} \cap \overleftrightarrow{RQ'}$ are collinear, the six points $P, Q, R, P', Q'$, and $R'$ necessarily lie on a conic. Here's the setup: Choose coordinates so that $P = [1, 0, 0]$, $Q = [0, 0, 1]$, $R = [0, 1, 0]$, and $Q' = [1, 1, 1]$. Let $P' = [a, b, c]$ and $R' = [\alpha, \beta, \gamma]$.

    a.  Show that $A = [a, b, b]$, $B = [ay, c\beta, cy]$, and $C = [\alpha, \beta, \alpha]$, and that these three points are therefore collinear if and only if

$$\begin{vmatrix} a & ay & \alpha \\ b & c\beta & \beta \\ b & cy & \alpha \end{vmatrix} = 0.$$

    b.  Show that the (degenerate) conics

$$x_1(x_0 - x_2) = 0 \quad \text{and} \quad x_2(x_0 - x_1) = 0$$

contain the four points $P, Q, R$, and $Q'$. Deduce, using Exercise 20, that if $P, Q, R, Q'$ lie on a conic $\mathcal{C}$, then the equation for $\mathcal{C}$ is of the form $\lambda x_1(x_0 - x_2) + \mu x_2(x_0 - x_1) = 0$ for appropriate $\lambda, \mu \neq 0$.

    c.  Show that if $P'$ and $R'$ are to be contained in our conic $\mathcal{C}$, then we must have

$$\begin{vmatrix} b(c-a) & c(a-b) \\ \beta(y-\alpha) & y(\alpha-\beta) \end{vmatrix} = 0.$$

    d.  Conclude that all six points lie on a conic if and only if the points $A, B$, and $C$ are collinear.

26.  Here is a general version of Exercise 17. Let $q_0, \ldots, q_5$ be a basis for the vector space of homogeneous quadratic polynomials in three variables $x_0, x_1, x_2$. (Specifically, $q_0 = x_1^2$, $q_1 = x_1 x_2$, $q_2 = x_2^2$, $q_3 = x_0 x_1$, $q_4 = x_0 x_2$, and $q_5 = x_0^2$.) Write $\mathbf{q} = (q_0, q_1, \ldots, q_5) \in \mathbb{R}^6$.

a.  If $P \in \mathbb{P}^2$, choose $\mathbf{x} \in \mathbb{R}^3$ with $[\mathbf{x}] = P$. Let

$$\mathbf{q}(\mathbf{x}) = (q_0(\mathbf{x}), \dots, q_5(\mathbf{x})) \in \mathbb{R}^6.$$

Check that $[\mathbf{q}(\mathbf{x})] \in \mathbb{P}^5$ is independent of the choice of homogeneous coordinates of $P$.

b.  Let $P_1, \dots, P_5$ be five points in $\mathbb{P}^2$, with corresponding homogeneous coordinate vectors $\mathbf{x}_1, \dots, \mathbf{x}_5 \in \mathbb{R}^6$. Assume that the vectors $\mathbf{q}(\mathbf{x}_1), \dots, \mathbf{q}(\mathbf{x}_5)$ are linearly independent in $\mathbb{R}^6$. Prove that $P_1, \dots, P_5$ lie on a unique conic $\mathcal{C}$ whose equation is

$$\det \begin{bmatrix} \mathbf{q} \\ \mathbf{q}(\mathbf{x}_1) \\ \vdots \\ \mathbf{q}(\mathbf{x}_5) \end{bmatrix} = 0.$$

(Hint: There are scalars $a_0, a_1, \dots, a_5$ so that the equation of $\mathcal{C}$ is $\sum_{i=0}^{5} a_i q_i = 0$.)

<u>Query:</u> Can you use this to prove the results of Exercise 17? Can you generalize this to curves of higher degree?

27.  Reprove Corollary 2.13 as follows: Let the lines $\overrightarrow{PA}$, $\overrightarrow{QA}$, $\overrightarrow{PB}$, and $\overrightarrow{QB} \subset \mathbb{P}^2$ correspond to points $\xi_A$, $\eta_A$, $\xi_B$, and $\eta_B \in \mathbb{P}^{2*}$, respectively, and choose homogeneous coordinates so that the lines $\overrightarrow{PC}$ and $\overrightarrow{QC}$ correspond to $\xi_A + \xi_B$ and $\eta_A + \eta_B$, respectively.

Consider the conics $\mathcal{C}_1 = \overrightarrow{PA} \cup \overrightarrow{QB}$ and $\mathcal{C}_2 = \overrightarrow{PB} \cup \overrightarrow{QA}$. Then $\mathcal{C}_1 \cap \mathcal{C}_2 = \{A, B, P, Q\}$; and so, by Exercise 20, we have constants $\lambda$ and $\mu$ (unique up to a common scalar multiple) so that $\mathcal{C} = \lambda \mathcal{C}_1 + \mu \mathcal{C}_2$. Note that $\mathcal{C}_1$ has the equation $(\xi_A \cdot \mathbf{x})(\eta_B \cdot \mathbf{x}) = 0$ and $\mathcal{C}_2$ has the equation $(\xi_B \cdot \mathbf{x})(\eta_A \cdot \mathbf{x}) = 0$. Therefore, $\mathcal{C}$ has the equation $\lambda(\xi_A \cdot \mathbf{x})(\eta_B \cdot \mathbf{x}) + \mu(\xi_B \cdot \mathbf{x})(\eta_A \cdot \mathbf{x}) = 0$.

a.  Show that $C \in \mathcal{C} \implies \lambda = \mu = 1$, and so

$$\mathcal{C} = \{(\xi_A \cdot \mathbf{x})(\eta_B \cdot \mathbf{x}) = (\xi_B \cdot \mathbf{x})(\eta_A \cdot \mathbf{x})\}.$$

b.  Show that if $\overrightarrow{PD} = \nu \xi_A + \xi_B$ for some $\nu \neq 0, 1$, then

$$|\overrightarrow{PA}, \overrightarrow{PB}, \overrightarrow{PC}, \overrightarrow{PD}| = \frac{1}{\nu} = -\frac{\xi_A \cdot D}{\xi_B \cdot D}.$$

c.  Similarly, show that

$$|\overrightarrow{QA}, \overrightarrow{QB}, \overrightarrow{QC}, \overrightarrow{QD}| = -\frac{\eta_A \cdot D}{\eta_B \cdot D}.$$

d.  Complete the proof.

## 3. The Spectral Theorem and Quadrics

We now turn to one of the most important results from linear algebra, one that is used constantly in physics, advanced mathematics, and—yes—in geometry. We have already witnessed in the preceding section the arrival of projective conics, defined by homogeneous polynomials of degree 2 in three variables. We will now put them in a more general setting, studying symmetric matrices and the homogeneous quadratic polynomials they define. We first introduce some standard terminology.

Recall, first, that an $n \times n$ matrix $A = \left[ a_{ij} \right]$ is **symmetric** if $A = A^T$, i.e., if $a_{ij} = a_{ji}$ for all $i, j = 1, \dots, n$. The importance of the transpose comes from the following formula, which we've already used in Section 4 of Chapter 7 (see Exercise 7.4.2).

**Lemma 3.1.** *For any $m \times n$ real matrix $A$ and vectors $\mathbf{x} \in \mathbb{R}^n$ and $\mathbf{y} \in \mathbb{R}^m$, we have $A\mathbf{x} \cdot \mathbf{y} = \mathbf{x} \cdot A^T\mathbf{y}$. In particular, if $A$ is a symmetric $n \times n$ matrix, then for all $\mathbf{x}, \mathbf{y} \in \mathbb{R}^n$, $A\mathbf{x} \cdot \mathbf{y} = \mathbf{x} \cdot A\mathbf{y}$.*

***Proof.*** By the associativity of matrix multiplication,

$$A\mathbf{x} \cdot \mathbf{y} = (A\mathbf{x})^T\mathbf{y} = (\mathbf{x}^T A^T)\mathbf{y} = \mathbf{x}^T (A^T\mathbf{y}) = \mathbf{x} \cdot A^T\mathbf{y},$$

as required. The latter statement follows immediately upon substituting $A = A^T$. □

**Definition.** Given a symmetric $n \times n$ matrix $A$, we define the (associated) **quadratic form** $Q \colon \mathbb{R}^n \to \mathbb{R}$ by $Q(\mathbf{x}) = A\mathbf{x} \cdot \mathbf{x}$. Note that $Q$ is a homogeneous polynomial of degree 2 in $x_1, \dots, x_n$.

Given a quadratic form, e.g., $Q(x_1, x_2) = x_1^2 + 4x_1x_2 + 3x_2^2$, our problem is this: Can we make a linear change of coordinates in $\mathbb{R}^2$ to "diagonalize" $Q$—that is, so that in terms of new coordinates $x_1', x_2'$, we have $Q(x_1, x_2) = Q'(x_1', x_2') = \lambda_1 x_1'^2 + \lambda_2 x_2'^2$ for some real numbers $\lambda_1, \lambda_2$? Can we make both the $\lambda_i$'s equal to 0 or 1? What will come into play here is the *group* we are willing to use to implement our change of coordinates; the Euclidean group is the most restrictive. Next, we want to describe the conic section $Q(x_1, x_2) = $ constant—is it empty, an ellipse (or, in particular, a circle), a hyperbola, or none of these?

We begin, then, by stating and proving the classical spectral theorem. It will be easier to talk about symmetric linear maps, so, motivated by Lemma 3.1, we make the following definition.

**Definition.** We say a linear map $T\colon \mathbb{R}^n \to \mathbb{R}^n$ is **symmetric** if for all vectors $\mathbf{x}, \mathbf{y} \in \mathbb{R}^n$, $T\mathbf{x} \cdot \mathbf{y} = \mathbf{x} \cdot T\mathbf{y}$.

**Theorem 3.2** (Spectral Theorem). *Let $T\colon \mathbb{R}^n \to \mathbb{R}^n$ be a symmetric linear map. Then there is an orthonormal basis for $\mathbb{R}^n$ consisting of eigenvectors of $T$. That is, there are real numbers $\lambda_1, \ldots, \lambda_n$ and an orthonormal basis $\mathbf{v}_1, \ldots, \mathbf{v}_n$ with the property that $T\mathbf{v}_i = \lambda_i \mathbf{v}_i$ for all $i = 1, \ldots, n$.*

We prepare for the proof with a few lemmas. The first two are rather easy; the last is rather more subtle (but see Exercises 17 and 18 for other proofs).

**Lemma 3.3.** *Let $T\colon \mathbb{R}^n \to \mathbb{R}^n$ be a symmetric linear map, and let $\mathbf{x} \in \mathbb{R}^n$. Then $\mathbf{x} \cdot T^2\mathbf{x} = |T\mathbf{x}|^2$.*

**Proof.** $\mathbf{x} \cdot T^2\mathbf{x} = \mathbf{x} \cdot T(T\mathbf{x}) = (T\mathbf{x}) \cdot (T\mathbf{x}) = |T\mathbf{x}|^2$. $\square$

**Lemma 3.4.** *Let $T\colon \mathbb{R}^n \to \mathbb{R}^n$ be a symmetric linear map, and suppose $V \subset \mathbb{R}^n$ has the property that $T(V) \subset V$. Then for any vector $\mathbf{w}$ orthogonal to $V$, $T(\mathbf{w})$ is likewise orthogonal to $V$.*

**Proof.** Let $\mathbf{v} \in V$ be arbitrary; we wish to show that $T(\mathbf{w}) \cdot \mathbf{v} = 0$. Since $T$ is a symmetric linear map, $T(\mathbf{w}) \cdot \mathbf{v} = \mathbf{w} \cdot T(\mathbf{v})$; now, $T(\mathbf{v}) \in V$ and $\mathbf{w}$ is orthogonal to $V$, so the latter dot product is zero. $\square$

**Proposition 3.5.** *The eigenvalues of a symmetric map $T\colon \mathbb{R}^n \to \mathbb{R}^n$ are all real.*

**Proof.** The proof begins with a trick to turn complex entities into real. Let $\lambda = \alpha + i\beta$ be an eigenvalue of $T$, and consider the *real* linear map $S = (T - \bar{\lambda}\,\mathrm{Id})(T - \lambda\,\mathrm{Id}) = T^2 - 2\alpha T + (\alpha^2 + \beta^2)\mathrm{Id}$. Since $T - \lambda\,\mathrm{Id}$ is not invertible, it follows that $S$ is not invertible (for example, see Section 2 of Appendix B), and so there is a nonzero vector $\mathbf{v} \in \mathbb{R}^n$ so that $S\mathbf{v} = 0$.

Since $S\mathbf{v} = \mathbf{0}$, the dot product $\mathbf{v} \cdot S\mathbf{v} = 0$. So,

$$0 = \mathbf{v} \cdot S\mathbf{v} = \mathbf{v} \cdot (T^2 - 2\alpha T + (\alpha^2 + \beta^2)\mathrm{Id})\mathbf{v}$$

$$= \mathbf{v} \cdot ((T - \alpha\,\mathrm{Id})^2 + \beta^2\,\mathrm{Id})\mathbf{v}$$

$$= |(T - \alpha\,\mathrm{Id})\,\mathbf{v}|^2 + \beta^2|\mathbf{v}|^2 \quad \text{by Lemma 3.3.}$$

(Note that Lemma 3.3 applies, insofar as both $T - \alpha\,\mathrm{Id}$ and $\beta^2\,\mathrm{Id}$ are symmetric maps.) Now, the only way the sum of two nonnegative numbers can be zero is for them both to be zero. That is, since

$\mathbf{v} \neq \mathbf{0}$, $|\mathbf{v}|^2 \neq 0$, and we infer that $\beta = 0$ and $(T - \alpha \mathrm{Id})\mathbf{v} = 0$. So $\lambda = \alpha$ is a real number, and $\mathbf{v}$ is the corresponding (real) eigenvector.  □

***Proof of Theorem 3.2.*** We proceed by induction on $n$. The case $n = 1$ is automatic. Now assume that the result is true for all symmetric linear maps $T' \colon \mathbb{R}^{n-1} \to \mathbb{R}^{n-1}$. Suppose we are given a symmetric linear map $T \colon \mathbb{R}^n \to \mathbb{R}^n$. First, by Proposition 3.5, $T$ has a real eigenvalue $\lambda_1$ and a corresponding eigenvector $\mathbf{v}_1$, which we may assume has unit length. Let $W \subset \mathbb{R}^n$ be the $(n-1)$-dimensional subspace orthogonal to $\mathbf{v}_1$, and let $T' = T|_W$ be the restriction of $T$ to $W$. Applying Lemma 3.4 with $V = \mathrm{Span}(\mathbf{v}_1)$, we infer that $T(W) \subset W$; since $\dim W = n - 1$, it follows from our induction hypothesis that there is an orthonormal basis $\mathbf{v}_2, \dots, \mathbf{v}_n$ for $W$ consisting of eigenvectors of $T'$. Then $\mathbf{v}_1, \mathbf{v}_2, \dots, \mathbf{v}_n$ is the requisite orthonormal basis for $\mathbb{R}^n$, since $T(\mathbf{v}_1) = \lambda_1 \mathbf{v}_1$ and $T(\mathbf{v}_i) = T'(\mathbf{v}_i) = \lambda_i \mathbf{v}_i$ for $i \geq 2$.  □

**Corollary 3.6.** *Let $T \colon \mathbb{R}^n \to \mathbb{R}^n$ be a symmetric linear map. Then there is an orthonormal basis $\mathbf{v}_1, \dots, \mathbf{v}_n$ for $\mathbb{R}^n$ with respect to which the matrix $\Lambda$ for $T$ is diagonal. Moreover, if $A$ is the matrix for $T$ with respect to the standard basis for $\mathbb{R}^n$, there is an orthogonal matrix $P \in O(n)$ so that $P^{-1}AP = \Lambda$ is diagonal.*

**Proof.** Recall (see Section 1 of Appendix B) that the matrix $A$ for $T$ with respect to a given basis $\mathbf{v}_1, \dots, \mathbf{v}_n$ is constructed as follows: the entries of the $j^{\text{th}}$ column vector of $A$ are the coordinates of the vector $T(\mathbf{v}_j)$ with respect to the basis vectors $\mathbf{v}_1, \dots, \mathbf{v}_n$; i.e.,
$$A = \begin{bmatrix} a_{ij} \end{bmatrix}, \text{ where } T(\mathbf{v}_j) = \sum_{i=1}^{n} a_{ij} \mathbf{v}_i.$$
In our case, we take the basis $\mathbf{v}_1, \dots, \mathbf{v}_n$ to be the (orthonormal) basis of eigenvectors provided us by Theorem 3.2. Then $T(\mathbf{v}_j) = \lambda_j \mathbf{v}_j$, and our matrix for $T$ is nothing but the diagonal matrix

$$\Lambda = \begin{bmatrix} \lambda_1 & & & \\ & \lambda_2 & & \\ & & \ddots & \\ & & & \lambda_n \end{bmatrix}.$$

Now, given the matrix $A$ for $T$ with respect to the standard basis for $\mathbb{R}^n$, let $P$ be the matrix whose column vectors are the coordinates of the eigenvectors $\mathbf{v}_1, \dots, \mathbf{v}_n$ with respect to the standard basis (so, in our case, $P$ is an orthogonal matrix; see Exercise 7.4.3). Then the change of basis formula (Theorem 1.1 of Appendix B) tells us that $P^{-1}AP = \Lambda$.  □

We now apply this theory to analyze quadratic forms. Recall that given a symmetric linear map $T: \mathbb{R}^n \to \mathbb{R}^n$, we have the associated quadratic form $Q: \mathbb{R}^n \to \mathbb{R}$ given by $Q(\mathbf{x}) = T(\mathbf{x}) \cdot \mathbf{x}$. From Theorem 3.2 we obtain the following.

**Proposition 3.7.** *Let* $Q: \mathbb{R}^n \to \mathbb{R}$ *be a quadratic form. Then there exist an orthonormal basis* $\mathbf{v}_1, \ldots, \mathbf{v}_n$ *for* $\mathbb{R}^n$ *and real numbers* $\lambda_1, \ldots, \lambda_n$ *so that, writing* $\mathbf{x} = \sum_{i=1}^{n} y_i \mathbf{v}_i$, *we have* $Q(\mathbf{x}) = \tilde{Q}(\mathbf{y}) = \sum_{i=1}^{n} \lambda_i y_i^2$. *That is, in terms of the* $y$*-coordinates, the quadratic form* $Q$ *"has been diagonalized."*

**Proof.** Write $Q(\mathbf{x}) = T(\mathbf{x}) \cdot \mathbf{x}$ for an appropriate symmetric linear map $T$. Then there is an orthonormal basis $\mathbf{v}_1, \ldots, \mathbf{v}_n$ for $\mathbb{R}^n$ consisting of eigenvectors of $T$: i.e., $T\mathbf{v}_i = \lambda_i \mathbf{v}_i$, $i = 1, \ldots, n$. It follows, then, that $Q(\mathbf{v}_i) = T(\mathbf{v}_i) \cdot \mathbf{v}_i = \lambda_i(\mathbf{v}_i \cdot \mathbf{v}_i) = \lambda_i$. Therefore,

$$Q(\mathbf{x}) = Q\left(\sum_{i=1}^{n} y_i \mathbf{v}_i\right) = T\left(\sum_{i=1}^{n} y_i \mathbf{v}_i\right) \cdot \left(\sum_{j=1}^{n} y_j \mathbf{v}_j\right)$$

$$= \left(\sum_{i=1}^{n} \lambda_i y_i \mathbf{v}_i\right) \cdot \left(\sum_{j=1}^{n} y_j \mathbf{v}_j\right) = \sum_{i,j=1}^{n} \lambda_i y_i y_j (\mathbf{v}_i \cdot \mathbf{v}_j)$$

$$= \sum_{i=1}^{n} \lambda_i y_i^2 = \tilde{Q}(\mathbf{y}),$$

as required.  □

**Example 1.** Let $Q(\mathbf{x}) = Q(x_1, x_2) = x_1^2 + 4x_1 x_2 - 2x_2^2$. Then the corresponding symmetric matrix $A$ is

$$A = \begin{bmatrix} 1 & 2 \\ 2 & -2 \end{bmatrix},$$

whose eigenvalues are 2 and $-3$ (see Section 3 of Appendix B). The corresponding eigenvectors are $(2, 1)$ and $(-1, 2)$, which we may normalize to obtain an orthonormal basis $\mathbf{v}_1 = (\frac{2}{\sqrt{5}}, \frac{1}{\sqrt{5}})$, $\mathbf{v}_2 = (-\frac{1}{\sqrt{5}}, \frac{2}{\sqrt{5}})$. In the $y$-coordinates (corresponding to the new basis) we therefore have $Q(\mathbf{x}) = \tilde{Q}(\mathbf{y}) = 2y_1^2 - 3y_2^2$. In particular, if we were asked to identify the conic section $Q(\mathbf{x}) = 10$, we would now know that, in the $y$-coordinates, this conic section has the equation $2y_1^2 - 3y_2^2 = 10$, and this is obviously a hyperbola. (See Figure 1.)

$$x_1^2 + 4x_1x_2 - 2x_2^2 = 10 \qquad\qquad\qquad 2y_1^2 - 3y_2^2 = 10$$

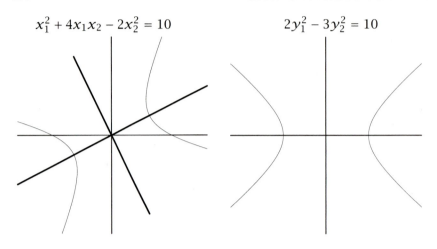

FIGURE 1

**Example 2.** Let $Q(\mathbf{x}) = Q(x_1, x_2, x_3) = 3x_1^2 + 4x_1x_2 + 8x_1x_3 + 4x_2x_3 + 3x_3^2$. The corresponding symmetric matrix $A$ is

$$A = \begin{bmatrix} 3 & 2 & 4 \\ 2 & 0 & 2 \\ 4 & 2 & 3 \end{bmatrix}.$$

The eigenvalues of $A$ are $\lambda_1 = -1$, $\lambda_2 = -1$, and $\lambda_3 = 8$, with associated (normalized) eigenvectors $\mathbf{v}_1 = \frac{1}{\sqrt{2}}(-1, 0, 1)$, $\mathbf{v}_2 = \frac{1}{\sqrt{5}}(-1, 2, 0)$, and $\mathbf{v}_3 = \frac{1}{3}(2, 1, 2)$. In the $y$-coordinates the quadratic form is given by $\tilde{Q}(\mathbf{y}) = -y_1^2 - y_2^2 + 8y_3^2$. Therefore, the quadric surface $Q(\mathbf{x}) = 5$ is a hyperboloid of *two* sheets, and the surface $Q(\mathbf{x}) = -5$ is a hyperboloid of *one* sheet. (See Figure 2.)

Suppose we have the diagonal quadratic form $Q(\mathbf{x}) = \sum\limits_{i=1}^{n} \lambda_i x_i^2$. If all the eigenvalues $\lambda_i$ are positive, then we may make a change of coordinates $y_i = \sqrt{\lambda_i} x_i$ to put the quadratic form in the simpler form $Q(\mathbf{x}) = \sum\limits_{i=1}^{n} y_i^2$. Even if some of the eigenvalues are negative, we can achieve the simpler form $Q(\mathbf{x}) = y_1^2 + \cdots + y_k^2 - y_{k+1}^2 - \cdots - y_n^2$ by making the change of coordinates $y_i = \sqrt{\lambda_i} x_i$. Now these changes of coordinates no longer correspond to an orthonormal change of basis; rather, we have stretched our orthonormal basis vectors, and therefore just have an orthogonal basis. In general, we have the following proposition.

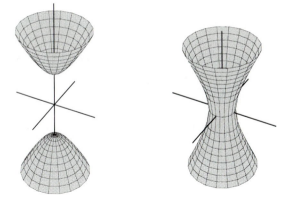

FIGURE 2

**Proposition 3.8.** *Let $Q: \mathbb{R}^n \to \mathbb{R}$ be a quadratic form. Then there is an orthogonal basis $\mathbf{v}'_1, \ldots, \mathbf{v}'_n$ for $\mathbb{R}^n$ so that, writing $\mathbf{x} = \sum_{i=1}^{n} z_i \mathbf{v}'_i$,*

$$Q(\mathbf{x}) = \sum_{i=1}^{n} \epsilon_i z_i^2, \text{ where } \epsilon_i = 0, +1, \text{ or } -1.$$

**Proof.** As in the proof of Proposition 3.7, we begin by writing $Q(\mathbf{x}) = T(\mathbf{x}) \cdot \mathbf{x}$ for an appropriate symmetric linear map $T$. We then obtain an orthonormal basis of eigenvectors $\mathbf{v}_1, \mathbf{v}_2, \ldots, \mathbf{v}_n$ for $T$, as before, with corresponding eigenvalues $\lambda_i$. Suppose $\lambda_1, \ldots, \lambda_m > 0$, $\lambda_{m+1}, \ldots, \lambda_r < 0$, and $\lambda_{r+1}, \ldots, \lambda_n = 0$. Let $\mathbf{v}'_j = \frac{1}{\sqrt{|\lambda_j|}} \mathbf{v}_j$ for $j = 1, \ldots, r$, and $\mathbf{v}'_j = \mathbf{v}_j$ for $j = r+1, \ldots, n$. Then, as before, if $\mathbf{x} = \sum_{i=1}^{n} y_i \mathbf{v}_i$, we have $Q(\mathbf{x}) = \sum_{i=1}^{n} \lambda_i y_i^2$. Now, if we set $\mathbf{x} = \sum_{i=1}^{n} z_i \mathbf{v}'_i$, we have $z_i = \sqrt{\lambda_i} y_i$ for $i = 1, \ldots, r$, and $z_i = y_i$ for $i = r+1, \ldots, n$. Therefore,

$$Q(\mathbf{x}) = \sum_{i=1}^{n} \lambda_i y_i^2 = \sum_{i=1}^{r} \lambda_i y_i^2 = \sum_{i=1}^{r} \frac{\lambda_i}{|\lambda_i|} z_i^2 = \sum_{i=1}^{m} z_i^2 - \sum_{i=m+1}^{r} z_i^2,$$

as required.  □

We shall see the usefulness of this result in our applications to quadric surfaces in projective geometry in the next section.

The following terminology is in common use. We say the quadratic form $Q$ is **nondegenerate** if $r = n$, i.e., if $Q$ has no zero eigenvalues. (Cf. the discussion of nondegenerate conics in the preceding section.) We say $Q$ is **positive definite** if $m = n$ (i.e., if all its eigenvalues are positive), and it is **negative definite** if $m = 0$

and $r = n$ (i.e., if all its eigenvalues are negative). The reader may want to review the maximum/minimum tests for functions of several variables to understand how this concept arises in calculus.

## EXERCISES 8.3

1. Suppose $\mathbf{v}_1, \ldots, \mathbf{v}_k$ are an orthonormal set of vectors in $\mathbb{R}^n$: i.e., $|\mathbf{v}_i| = 1$ for all $i = 1, \ldots, k$, and $\mathbf{v}_i \cdot \mathbf{v}_j = 0$, $i \neq j$. Prove that $\mathbf{v}_1, \ldots, \mathbf{v}_k$ are linearly independent.

2. Let $\mathbf{v}_1 \in \mathbb{R}^n$ be a unit vector. Prove that there are vectors $\mathbf{v}_2$, $\ldots$, $\mathbf{v}_n \in \mathbb{R}^n$ so that $\mathbf{v}_1$, $\mathbf{v}_2$, $\ldots$, $\mathbf{v}_n$ form an orthonormal basis for $\mathbb{R}^n$. (Hints: If $n > 1$, first find a unit vector $\mathbf{v}_2$ orthogonal to $\mathbf{v}_1$. If $n > 2$, continue. Use Exercise 1 to show you have a basis at the end. See Exercise 5.1.9.)

3. Let $T: \mathbb{R}^n \to \mathbb{R}^n$ be a symmetric linear map. Suppose $\mathbf{v}, \mathbf{w} \in \mathbb{R}^n$, $T(\mathbf{v}) = \lambda \mathbf{v}$, and $T(\mathbf{w}) = \mu \mathbf{w}$, where $\lambda \neq \mu$. Prove directly that $\mathbf{v} \cdot \mathbf{w} = 0$.

4. Prove the following converse of Lemma 3.1. If $A$ is an $n \times n$ matrix satisfying $A\mathbf{x} \cdot \mathbf{y} = \mathbf{x} \cdot A\mathbf{y}$ for all $\mathbf{x}, \mathbf{y} \in \mathbb{R}^n$, then $A$ is symmetric. (Hint: see Exercise 7.4.19a.)

5. Diagonalize the following quadratic forms and identify each of the quadrics $Q(\mathbf{x}) = 1$:
   a. $Q(x_1, x_2) = 3x_1^2 - 10x_1x_2 + 3x_2^2$
   b. $Q(x_1, x_2) = 3x_1^2 - 2x_1x_2 + 3x_2^2$
   c. $Q(x_1, x_2) = 6x_1^2 + 4x_1x_2 + 9x_2^2$
   d. $Q(x_1, x_2) = x_1^2 - 4x_1x_2 + 4x_2^2$
   e. $Q(x_1, x_2, x_3) = 3x_1^2 + 2x_1x_2 + 2x_1x_3 + 4x_2x_3$
   f. $Q(x_1, x_2, x_3) = 2x_1^2 + 6x_1x_2 + 5x_2^2 + 2x_2x_3 + 2x_3^2$
   g. $Q(x_1, x_2, x_3) = 4x_1^2 - 10x_1x_2 + 10x_1x_3 - 2x_2^2 - 2x_2x_3 - 2x_3^2$
   h. $Q(x_1, x_2, x_3) = 2x_1^2 + 2x_1x_2 + 2x_1x_3 + 2x_2x_3$

6. Classify the quadric surfaces $Q(\mathbf{x}) = $ constant in $\mathbb{R}^3$.

7. Show that the quadric surface $Q(\mathbf{x}) = 1$ in $\mathbb{R}^3$ is a surface of revolution if and only if $Q$ has a repeated nonzero eigenvalue.

8. Let $Q: \mathbb{R}^2 \to \mathbb{R}$ be a nonzero quadratic form. Classify the plane

conic curves $Q(x_1, x_2) + ax_1 + bx_2 + c = 0$ up to affine equivalence. Can you give a short list (say, consisting of six curves) of the simplest possible equations representing these equivalence classes?

9. Let $Q \colon \mathbb{R}^3 \to \mathbb{R}$ be a nondegenerate quadratic form. Then a (projective) conic $\mathcal{C} \subset \mathbb{P}^2$ is defined by the equation $Q(\mathbf{x}) = 0$. Prove that all (nondegenerate) conics are projectively equivalent. (In particular, be sure you understand why a "parabola," a "hyperbola," and an "ellipse" are "the same." Some pictures may help here, along with a recollection of the term "conic section." For starters, you might check that $-x_0^2 + x_1^2 + x_2^2 = 0$ and $-x_1^2 + x_0 x_2 = 0$ are projectively equivalent.)

10. Suppose $T \colon \mathbb{R}^n \to \mathbb{R}^n$ is a symmetric linear map satisfying $T^2 = T$. Prove that there is an orthonormal basis with respect to which the matrix for $T$ is

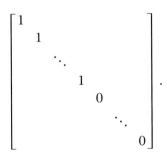

Give a geometric interpretation of such a $T$.

11. Suppose $T \colon \mathbb{R}^n \to \mathbb{R}^n$ is a symmetric linear map satisfying $T^k = \mathrm{Id}$ for some positive integer $k$. Prove that $T^2 = \mathrm{Id}$.

12. Let $\mathcal{C} = \{ax_1^2 + bx_1 x_2 + cx_2^2 + dx_0 x_1 + ex_0 x_2 + fx_0^2 = 0\} \subset \mathbb{P}^2$ be a nondegenerate conic.
    a. Show that there is a projective transformation of $\mathbb{P}^2$ so that this equation becomes $-x_0'^2 + x_1'^2 + x_2'^2 = 0$.
    b. Show that the set of projective transformations of $\mathbb{P}^2$ preserving the conic $\mathcal{C}$ is the subgroup $H = O(3)/\{\pm \mathrm{Id}\} \subset \mathrm{Proj}(2)$.
    c. Show that $H$ acts transitively on $\mathcal{C}$. Therefore, given $P, Q \in \mathcal{C}$, there is a projective transformation $T \in \mathrm{Proj}(2)$ with $T(\mathcal{C}) = \mathcal{C}$ and $T(P) = Q$.

13. Recall the concept of projective duality in $\mathbb{P}^2$. Given a nondegenerate conic $\mathcal{C} \subset \mathbb{P}^2$, we define its **dual** to be $\mathcal{C}^* = \{\ell_P \in \mathbb{P}^{2*} : \ell_P$ is the tangent line to $\mathcal{C}$ at $P\}$.

   a. Show that $\mathcal{C}^* \subset \mathbb{P}^{2*}$ is another nondegenerate conic. (Hint: How do you find the equation of the tangent plane to the level surface $\{f(x, y, z) = 0\} \subset \mathbb{R}^3$?)

   b. Prove (applying projective duality) that $(\mathcal{C}^*)^* = \mathcal{C}$.

   c. Given five lines in $\mathbb{P}^2$ no three of which are concurrent, prove there is a unique conic tangent to these five lines.

   d. Prove or give a counterexample: Given four points $P$, $Q$, $R$, $S$, no three collinear, and a line $\ell$ through $P$ containing none of the others, there is a unique conic $\mathcal{C}$ passing through $P$, $Q$, $R$, $S$ and tangent to $\ell$.

   e. Prove the dual of Pascal's Theorem (Theorem 2.14): If a hexagon $ABCDEF$ is circumscribed about a conic, then the lines $\overleftrightarrow{AD}$, $\overleftrightarrow{BE}$, and $\overleftrightarrow{CF}$ are concurrent.

14. Prove that every invertible matrix $A$ may be written uniquely as a product $A = QR$, where $Q$ is orthogonal and $R$ is symmetric and positive definite. (Hint: Show there is a unique symmetric matrix $R$ with positive eigenvalues satisfying the equation $R^2 = A^T A$.)

15. Here is an alternative proof of Corollary 3.6 (hence, a matrix-oriented proof of Theorem 3.2). Let $A$ be a symmetric $n \times n$ matrix.

   a. Prove that if $P \in O(n)$, then $P^{-1}AP$ is likewise symmetric.

   b. Let $\mathbf{v}_1$ be an eigenvector of $A$, and take it to have length 1. Using Exercise 2, extend to an orthonormal basis $\mathbf{v}_1, \ldots, \mathbf{v}_n$ for $\mathbb{R}^n$, and let $P$ be the orthogonal matrix whose column vectors are $\mathbf{v}_1, \ldots, \mathbf{v}_n$. Show that

$$P^{-1}AP = \left[ \begin{array}{c|ccc} \lambda_1 & 0 & \cdots & 0 \\ \hline 0 & & & \\ \vdots & & A' & \\ 0 & & & \end{array} \right],$$

   where $A'$ is a symmetric $(n-1) \times (n-1)$ matrix.

   c. Proceed by induction.

16. Let $Q_1$ and $Q_2$ be quadratic forms on $\mathbb{R}^n$, and suppose $Q_1$ is positive definite. Show that we can simultaneously diagonalize

$Q_1$ and $Q_2$. (Hint: Use $Q_1$ to define a new dot product on $\mathbb{R}^n$, as follows: If $T: \mathbb{R}^n \to \mathbb{R}^n$ is the symmetric linear map given by $Q_1(\mathbf{x}) = T(\mathbf{x}) \cdot \mathbf{x}$, define $\langle \mathbf{x}, \mathbf{y} \rangle = T\mathbf{x} \cdot \mathbf{y}$. Check that this is in fact a dot product. Now apply Proposition 3.8 working with the dot product $\langle , \rangle$.)

17. Here is a more general variant of Proposition 3.5. Consider the $n$-dimensional *complex* vector space $\mathbb{C}^n$, and define on it a **hermitian inner product** (a generalization of the dot product on real vector spaces) as follows: for $\mathbf{v}, \mathbf{w} \in \mathbb{C}^n$, set $\langle \mathbf{v}, \mathbf{w} \rangle = \sum_{j=1}^{n} v_j \overline{w}_j \ (= \mathbf{v}^T \overline{\mathbf{w}})$. We say a linear map $T: \mathbb{C}^n \to \mathbb{C}^n$ is **hermitian** if $\langle T\mathbf{v}, \mathbf{w} \rangle = \langle \mathbf{v}, T\mathbf{w} \rangle$ for all $\mathbf{v}, \mathbf{w} \in \mathbb{C}^n$.

    a. Prove that the hermitian inner product has the following properties:
   (i) $\langle \mathbf{v} + \mathbf{v}', \mathbf{w} \rangle = \langle \mathbf{v}, \mathbf{w} \rangle + \langle \mathbf{v}', \mathbf{w} \rangle$
   (ii) for all $c \in \mathbb{C}$, $\langle c\mathbf{v}, \mathbf{w} \rangle = c\langle \mathbf{v}, \mathbf{w} \rangle = \langle \mathbf{v}, \overline{c}\mathbf{w} \rangle$
   (iii) $\langle \mathbf{v}, \mathbf{w} \rangle = \overline{\langle \mathbf{w}, \mathbf{v} \rangle}$
   (iv) $\langle \mathbf{v}, \mathbf{v} \rangle = \sum_{j=1}^{n} |v_j|^2 \in \mathbb{R}$ and $\langle \mathbf{v}, \mathbf{v} \rangle = 0 \iff \mathbf{v} = \mathbf{0}$

    b. Prove that the eigenvalues of a hermitian linear map $T$ are real. (Hint: Suppose $T\mathbf{v} = \lambda\mathbf{v}$, $\mathbf{v} \neq 0$. Compute $\langle T\mathbf{v}, \mathbf{v} \rangle$.)

    c. Conclude that the eigenvalues of a symmetric (real) linear map $T$ are real.

    d. Suppose $T: \mathbb{C}^n \to \mathbb{C}^n$ is linear and $\langle T\mathbf{v}, T\mathbf{w} \rangle = \langle \mathbf{v}, \mathbf{w} \rangle$ for all $\mathbf{v}, \mathbf{w} \in \mathbb{C}^n$. What can you prove about the eigenvalues of such a linear map $T$? What can you therefore conclude about (real) isometries $T: \mathbb{R}^n \to \mathbb{R}^n$?

18. In this problem we suggest a proof of the Spectral Theorem, Theorem 3.2, based on the method of Lagrange multipliers in multivariable calculus. There is an "obvious" function whose constrained critical points are the eigenvectors; the corresponding eigenvalues arise as the Lagrange multipliers.

    Let $T: \mathbb{R}^n \to \mathbb{R}^n$ be a symmetric linear map. Define a (differentiable) function $f: \mathbb{R}^n \to \mathbb{R}$ by $f(\mathbf{x}) = T(\mathbf{x}) \cdot \mathbf{x}$. Let $g: \mathbb{R}^n \to \mathbb{R}$ be defined by $g(\mathbf{x}) = |\mathbf{x}|^2$, so that the unit sphere in $\mathbb{R}^n$ is given by $S = \{\mathbf{x} \in \mathbb{R}^n : g(\mathbf{x}) = 1\}$.

    a. Show that if $\mathbf{x} \in S$ is a critical point of $f|_S$, then there is a real number $\lambda$ so that $T(\mathbf{x}) = \lambda\mathbf{x}$. (Hint: Recall from the method of Lagrange multipliers that, at such a constrained critical point, we have $\vec{\nabla f}(\mathbf{x}) = \lambda \vec{\nabla g}(\mathbf{x})$. Now compute that $\vec{\nabla f}(\mathbf{x}) = 2T(\mathbf{x})$ and $\vec{\nabla g}(\mathbf{x}) = 2\mathbf{x}$. Here you will certainly need to use the fact that $T$ is symmetric, and you may find it easier to work with the corresponding symmetric matrix.)

b. As in the proof of Theorem 3.2, proceed by induction.

## 4. Projective Three Space and the Four Skew Line Problem

We now continue the study of projective geometry initiated in Section 2. We first move up to projective three space , $\mathbb{P}^3$, and, after establishing the basics, give a new proof of Desargues' Theorem, Theorem 2.7. We then address what we are calling here the *four skew line problem*:

**Problem.** Given four "general" lines in projective three space, how many lines are there incident to (i.e., intersecting) them all?

Problems like this have motivated a great deal of mathematical work, growing out of a subject called enumerative geometry (in which some key players were Plücker, Cayley, and Schubert). Two other beautiful problems, too advanced for us to deal with here, are these: How many lines lie on a "general" *cubic* surface (i.e., a surface defined by a homogeneous polynomial of degree 3)? (Answer: at most 27.) And, given five "general" nondegenerate conics in $\mathbb{P}^2$, how many nondegenerate conics are tangent to all five? (Answer: at most 3264.)

Recall that $\mathbb{P}^3$ is defined to be the set of equivalence classes $[x_0, x_1, x_2, x_3]$ of points $(x_0, x_1, x_2, x_3) \in \mathbb{R}^4 - \{0\}$ with the equivalence relation $\mathbf{x} \equiv \mathbf{y} \iff \mathbf{y} = c\mathbf{x}$ for some $c \in \mathbb{R} - \{0\}$. As before, we observe that there is a decomposition

$$\mathbb{P}^3 = \mathbb{R}^3 \cup \mathbb{P}^2_\infty,$$

where $\mathbb{R}^3$ is the set of points $[x_0, x_1, x_2, x_3]$ with $x_0 \neq 0$ (so that $[1, x_1, x_2, x_3] \rightsquigarrow (x_1, x_2, x_3) \in \mathbb{R}^3$), and the projective plane "at infinity," $\mathbb{P}^2_\infty$, is given by the points with $x_0 = 0$. We can visualize a "horizon at infinity," with one point for each direction in space, i.e., line through the origin in $\mathbb{R}^3$. (Note that this is not a sphere at infinity, since we identify the points at infinity in the forward and backward directions.)

Exactly as was the case in the projective plane, two points $P$ and $Q$ in $\mathbb{P}^3$ determine a line $\overrightarrow{PQ}$; we call (three or more) points **collinear** if they lie on a line. Now, three noncollinear points $P$, $Q$, and $R$ span a (projective) *plane*, which we denote by $\overline{PQR}$: if $P = [x_0, x_1, x_2, x_3]$, $Q = [y_0, y_1, y_2, y_3]$, and $R = [z_0, z_1, z_2, z_3]$, then

$$\overline{PQR} = \{[s(x_0, x_1, x_2, x_3) + t(y_0, y_1, y_2, y_3) + u(z_0, z_1, z_2, z_3)] :$$
$$s, t, u \in \mathbb{R} \text{ not all zero}\}.$$

We call four or more points in $\mathbb{P}^3$ **coplanar** if they all lie on a plane. It is important to observe that a plane can be described as the solution set to a homogeneous linear equation (cf. Lemma 2.3).

**Lemma 4.1.** *Given a plane* $\Pi \subset \mathbb{P}^3$, *there is* $\mathbf{a} = (a_0, a_1, a_2, a_3) \neq \mathbf{0}$ *so that* $\Pi$ *is the locus of points* $[x_0, x_1, x_2, x_3]$ *satisfying the equation*

$(*)$ $\qquad\qquad\qquad a_0 x_0 + a_1 x_1 + a_2 x_2 + a_3 x_3 = 0.$

**Proof.** This is basic linear algebra. The space of solutions of the linear equation $(*)$, viewed as a linear equation in $\mathbb{R}^4$, is a three-dimensional subspace of $\mathbb{R}^4$. Choose a basis $\mathbf{p} = (p_0, p_1, p_2, p_3)$, $\mathbf{q} = (q_0, q_1, q_2, q_3)$, $\mathbf{r} = (r_0, r_1, r_2, r_3)$ for this subspace. Let $P = [\mathbf{p}]$, $Q = [\mathbf{q}]$, and $R = [\mathbf{r}]$; then all solutions to $(*)$ clearly lie on the plane $\overline{PQR}$.

Conversely, given a plane spanned by three points $P$, $Q$, and $R$, choose $\mathbf{p}$, $\mathbf{q}$, and $\mathbf{r}$ as above. We must find a vector $\mathbf{a} \neq \mathbf{0}$ so that

$$\mathbf{a} \cdot \mathbf{p} = \mathbf{a} \cdot \mathbf{q} = \mathbf{a} \cdot \mathbf{r} = 0.$$

This is a homogeneous system of three linear equations in the four variables $\mathbf{a} = (a_0, a_1, a_2, a_3)$, and so there is a nontrivial solution. Indeed, since the vectors $\mathbf{p}$, $\mathbf{q}$, $\mathbf{r} \in \mathbb{R}^4$ are linearly independent (see Exercise 1), we have a 1-dimensional solution space.  □

The group of motions of $\mathbb{P}^3$ is, as expected, the group of projective transformations $\text{Proj}(3) = GL(4, \mathbb{R})/\{\lambda\, \text{Id} : \lambda \in \mathbb{R}^\times\}$. The results of Section 2 can be easily generalized; we state them here, but leave their proofs as exercises for the reader.

**Lemma 4.2.** *Let* $T \in \text{Proj}(3)$. *Then* $T$ *maps lines to lines and planes to planes.*

**Proof.** See Exercise 2.  □

We say *four* points in $\mathbb{P}^3$ are **in general position** if they are not coplanar (it follows that no three can be collinear); *five* points are in general position if no four are coplanar.

**Lemma 4.3.** *Given two quintets, $P_1, P_2, P_3, P_4, P_5$ and $Q_1, Q_2, Q_3, Q_4, Q_5$, of distinct points in $\mathbb{P}^3$, each set in general position, there is a unique projective transformation $T \in \text{Proj}(3)$ with $T(P_i) = Q_i$, $i = 1, 2, 3, 4, 5$.*

**Proof.** See Exercise 3.   □

Now we have a result analogous to Lemma 2.4.

**Lemma 4.4.** *Let $\ell \subset \mathbb{P}^3$ be a line, and let $\Pi \subset \mathbb{P}^3$ be a plane. Then either $\ell \subset \Pi$ or $\ell$ intersects $\Pi$ in one point.*

**Proof.** By Lemma 4.1, there is a vector $\mathbf{a} \in \mathbb{R}^4 - \{\mathbf{0}\}$ so that $\Pi = \{[\mathbf{x}] : \mathbf{a} \cdot \mathbf{x} = 0\}$. Let $\ell = \overrightarrow{PQ}$, and choose $\mathbf{p}$ and $\mathbf{q} \in \mathbb{R}^4$ with $[\mathbf{p}] = P$ and $[\mathbf{q}] = Q$. Then $\ell = \{[\mathbf{x}] : \mathbf{x} = s\mathbf{p} + t\mathbf{q}, \ s, t \text{ not both } 0\}$.
In order for $[\mathbf{x}]$ to lie in $\ell$ and on $\Pi$, we must have

$$0 = \mathbf{a} \cdot (s\mathbf{p} + t\mathbf{q}) = s(\mathbf{a} \cdot \mathbf{p}) + t(\mathbf{a} \cdot \mathbf{q}).$$

If $\mathbf{a} \cdot \mathbf{p} = \mathbf{a} \cdot \mathbf{q} = 0$, then $\ell \subset \Pi$. If not, up to scalar multiples we must have $(s, t) = (-\mathbf{a} \cdot \mathbf{q}, \mathbf{a} \cdot \mathbf{p})$, and so the unique point of intersection is $[(-\mathbf{a} \cdot \mathbf{q})\mathbf{p} + (\mathbf{a} \cdot \mathbf{p})\mathbf{q}]$.   □

We next generalize the discussion of the Example on p. 304. Let $\Pi \subset \mathbb{P}^3$ be a plane, and let $O \notin \Pi$ be a point. Then the **projection** $\pi \colon \mathbb{P}^3 - \{O\} \to \Pi$ is defined as follows: given $P \in \mathbb{P}^3 - \{O\}$, define $\pi(P)$ to be the point of intersection of the line $\overrightarrow{OP}$ with the plane $\Pi$ (whose existence is guaranteed by Lemma 4.4). See Exercise 5 for an

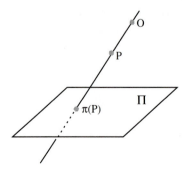

FIGURE 1

algebraic formula for projection. It is a crucial—and not difficult—fact that such projections map lines in $\mathbb{P}^3 - \{O\}$ to lines in $\Pi$ (see Exercise 6).

We now give a three-dimensional generalization of Desargues' Theorem, Theorem 2.7, a special case of which recaptures the two-dimensional version.

**Theorem 4.5** (3-D Desargues). *Let $\triangle PQR$ and $\triangle P'Q'R'$ be triangles in $\mathbb{P}^3$. Suppose the three lines $\overleftrightarrow{PP'}$, $\overleftrightarrow{QQ'}$, $\overleftrightarrow{RR'}$ are concurrent (i.e., intersect in a single point $X$). Then the three points $A = \overleftrightarrow{PQ} \cap \overleftrightarrow{P'Q'}$, $B = \overleftrightarrow{QR} \cap \overleftrightarrow{Q'R'}$, and $C = \overleftrightarrow{PR} \cap \overleftrightarrow{P'R'}$ are collinear.*

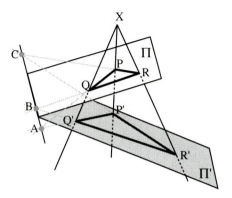

FIGURE 2

**Proof.** Note, first, that the points $A$, $B$, and $C$ really exist. For example, the lines $\overleftrightarrow{PQ}$ and $\overleftrightarrow{P'Q'}$ intersect since they lie in the plane spanned by the (intersecting) lines $\overleftrightarrow{PP'}$ and $\overleftrightarrow{QQ'}$.

Suppose first that the triangles $\triangle PQR$ and $\triangle P'Q'R'$ lie in different planes $\Pi$ and $\Pi'$, respectively. Then these planes intersect in a line $\ell$ (see Exercise 4). Now it remains only to see that the three points $A$, $B$, and $C$ must lie on $\ell$. We give here just the argument that $A \in \ell$. Well, $A \in \overleftrightarrow{PQ} \subset \Pi$ and $A \in \overleftrightarrow{P'Q'} \subset \Pi'$, whence $A \in \Pi \cap \Pi' = \ell$, as required. See Figure 2.

Now comes the tricky part. What if the triangles $\triangle PQR$ and $\triangle P'Q'R'$ lie in a single plane $\Pi$ (which is the classical case of Theorem 2.7)? Pick a point $O \notin \Pi$ and a point $\tilde{R} \in \overleftrightarrow{OR}$, $\tilde{R} \neq R$. Let $\tilde{R}' = \overleftrightarrow{OR'} \cap \overleftrightarrow{X\tilde{R}}$; note that these lines intersect, since they lie in the plane spanned by the lines $\overleftrightarrow{OR}$ and $\overleftrightarrow{X\tilde{R}}$. Then triangles $\triangle PQ\tilde{R}$ and

$\triangle P'Q'\tilde{R}'$ lie in different planes (since $\tilde{R}$ and $\tilde{R}'$ do not lie in the plane $\Pi$), and we may apply the first part of the proof to them. We thus obtain three collinear points $\tilde{A} = A = \overrightarrow{PQ} \cap \overleftrightarrow{P'Q'}$, $\tilde{B} = \overleftrightarrow{QR} \cap \overleftrightarrow{Q'\tilde{R}'}$, and $\tilde{C} = \overrightarrow{PR} \cap \overleftrightarrow{P'\tilde{R}'}$. Let $\tilde{\ell}$ be the line on which $\tilde{A}$, $\tilde{B}$, and $\tilde{C}$ all lie. Here, now, is the key point. We claim that the projection $\pi$ from $O$ to the plane $\Pi$ maps $\tilde{B}$ to $B$ and $\tilde{C}$ to $C$. Since $\pi(\tilde{R}) = R$ and $\pi(\tilde{R}') = R'$, $\pi$ maps the lines $\overleftrightarrow{QR}$ and $\overleftrightarrow{Q'\tilde{R}'}$ to $\overleftrightarrow{QR}$ and $\overleftrightarrow{Q'R'}$, respectively (see Exercise 6), and therefore maps their intersection point $\tilde{B}$ to $B$. The obvious analogous argument shows that $\pi(\tilde{C}) = C$. Since $\tilde{A}$, $\tilde{B}$, and $\tilde{C}$ lie on a line $\tilde{\ell}$, their images must lie on the line $\pi(\tilde{\ell}) = \ell \subset \Pi$, as required (see Exercise 6 once again).  $\square$

Although two lines in the projective plane must intersect, this is no longer true of lines in space. We say that two lines in $\mathbb{P}^3$ are **skew** if they do not intersect. (How does this differ from the notion of skew lines in $\mathbb{R}^3$?)

**Lemma 4.6.** *Suppose $P = [\mathbf{p}]$, $Q = [\mathbf{q}]$, $R = [\mathbf{r}]$, and $S = [\mathbf{s}]$ are distinct points in $\mathbb{P}^3$. Then the lines $\overrightarrow{PQ}$ and $\overrightarrow{RS}$ are skew if and only if the vectors $\mathbf{p}$, $\mathbf{q}$, $\mathbf{r}$, and $\mathbf{s}$ are linearly independent in $\mathbb{R}^4$.*

***Proof.*** It is equivalent to prove that the lines intersect if and only if the vectors $\mathbf{p}$, $\mathbf{q}$, $\mathbf{r}$, and $\mathbf{s}$ are linearly dependent. Suppose the lines intersect in a point $X = [\mathbf{x}]$. Then $\mathbf{x}$ can be written both as a linear combination of $\mathbf{p}$ and $\mathbf{q}$ and as a linear combination of $\mathbf{r}$ and $\mathbf{s}$: $\mathbf{x} = \alpha\mathbf{p} + \beta\mathbf{q} = \gamma\mathbf{r} + \delta\mathbf{s}$, where $(\alpha,\beta) \neq (0,0)$ and $(\gamma,\delta) \neq (0,0)$. Therefore, $\alpha\mathbf{p} + \beta\mathbf{q} - \gamma\mathbf{r} - \delta\mathbf{s} = \mathbf{0}$, and the vectors are linearly dependent.

Suppose, now, that the vectors $\mathbf{p}$, $\mathbf{q}$, $\mathbf{r}$, $\mathbf{s}$ are linearly dependent; then there are scalars $\alpha, \beta, \gamma, \delta$, not all zero, so that $\alpha\mathbf{p}+\beta\mathbf{q}+\gamma\mathbf{r}+\delta\mathbf{s} = \mathbf{0}$. Then let $\mathbf{x} = \alpha\mathbf{p} + \beta\mathbf{q} = -\gamma\mathbf{r} - \delta\mathbf{s}$; $\mathbf{x}$ cannot be the zero vector, since $\mathbf{p}$ and $\mathbf{q}$ must be linearly independent (why?). Thus, $[\mathbf{x}]$ lies on both lines $\overrightarrow{PQ}$ and $\overrightarrow{RS}$, as required.  $\square$

We will say *three lines* in $\mathbb{P}^3$ are **in general position** if they are pairwise skew. No one will be surprised that we can send any pair $\ell_1, \ell_2$ of skew lines to any other pair $\ell_1', \ell_2'$ of skew lines by a projective transformation: pick points $P_1, P_2 \in \ell_1$, $P_1', P_2' \in \ell_1'$, $Q_1, Q_2 \in \ell_2$, $Q_1', Q_2' \in \ell_2'$; and take a projective transformation carrying $P_i \leadsto P_i'$, $Q_i \leadsto Q_i'$, $i = 1, 2$. The following proposition, however, may well come as a shock.

**Proposition 4.7.** *Given two triples $\ell_1, \ell_2, \ell_3$ and $\ell_1', \ell_2', \ell_3'$, of lines in $\mathbb{P}^3$ in general position, there is a projective transformation $T \in Proj(3)$ carrying $\ell_i$ to $\ell_i'$, $i = 1, 2, 3$.*

In order to prove this, we will need one more lemma, which provides a very nice geometric result.

**Lemma 4.8.** *Let $\ell_1, \ell_2, \ell_3$ be lines in general position. Given a point $P_3 \in \ell_3$, there are unique points $P_1 \in \ell_1$ and $P_2 \in \ell_2$ so that $P_1$, $P_2$, and $P_3$ are collinear.*

**Proof.** Consider the plane $\Pi$ spanned by $P_3$ and the line $\ell_2$ (note that $P_3$ cannot lie on $\ell_2$, since $\ell_2$ and $\ell_3$ are skew). By Lemma 4.4, the line $\ell_1$ must either be contained in $\Pi$ or intersect it in a single point. If it were contained in the plane, it would intersect $\ell_2$, contradicting the fact that $\ell_1$ and $\ell_2$ are skew. Thus, there is a unique point $P_1 \in \ell_1$ that is in the plane. Draw the line $\overrightarrow{P_1 P_3} \subset \Pi$. It intersects $\ell_2$ in a unique point $P_2$; obviously, $P_1$, $P_2$, and $P_3$ are collinear. (See Figure 3.)

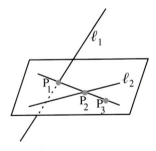

FIGURE 3

Now we argue that $P_1$ and $P_2$ are unique. It is easy to check that given one of them, the other is uniquely determined. So, suppose there are distinct points $P_1, P_1' \in \ell_1$ and $P_2, P_2' \in \ell_2$ satisfying the conclusion of the lemma. Then there are two lines $\overrightarrow{P_1 P_2 P_3}$ and $\overrightarrow{P_1' P_2' P_3}$ intersecting in $P_3$; they therefore lie in a plane, and so $\overleftrightarrow{P_1 P_1'}$ and $\overleftrightarrow{P_2 P_2'}$ must lie in the same plane. But these are the lines $\ell_1$ and $\ell_2$, respectively, and they were skew, by hypothesis.  $\square$

**Proof of Proposition 4.7.** The proof will follow easily once we establish the following interesting result. If $\ell_1$, $\ell_2$, $\ell_3$ are lines in $\mathbb{P}^3$ in general position, then there is a basis $\mathbf{v}_1, \mathbf{w}_1, \mathbf{v}_2, \mathbf{w}_2$ for $\mathbb{R}^4$ so that

    (i) $[\mathbf{v}_1]$ and $[\mathbf{w}_1]$ span $\ell_1$,

    (ii) $[\mathbf{v}_2]$ and $[\mathbf{w}_2]$ span $\ell_2$, and

    (iii) $[\mathbf{v}_1 + \mathbf{w}_1]$ and $[\mathbf{v}_2 + \mathbf{w}_2]$ span $\ell_3$.

We begin by choosing distinct points $P_3$ and $Q_3 \in \ell_3$; by Lemma 4.8, there are corresponding points $P_1, Q_1 \in \ell_1$ and $P_2, Q_2 \in \ell_2$ so that $P_1, P_2, P_3$ are collinear and $Q_1, Q_2, Q_3$ are collinear. We then choose vectors $\mathbf{v}_1, \mathbf{v}_2 \in \mathbb{R}^4$ so that $[\mathbf{v}_1] = P_1$, $[\mathbf{v}_2] = P_2$, and $[\mathbf{v}_1 + \mathbf{v}_2] = P_3$ (see Exercise 7). Similarly, we choose $\mathbf{w}_1, \mathbf{w}_2$ so that $[\mathbf{w}_1] = Q_1$, $[\mathbf{w}_2] = Q_2$, and $[\mathbf{w}_1 + \mathbf{w}_2] = Q_3$. By Lemma 4.6, $\mathbf{v}_1, \mathbf{w}_1, \mathbf{v}_2, \mathbf{w}_2$ are linearly independent; so they form a basis for $\mathbb{R}^4$, as desired.

    Now, given another set $\ell_1', \ell_2', \ell_3'$ of lines in general position, there is a corresponding basis $\mathbf{v}_1', \mathbf{w}_1', \mathbf{v}_2', \mathbf{w}_2'$ for $\mathbb{R}^4$. We choose the linear transformation $Y \in GL(4, \mathbb{R})$ carrying the basis $\mathbf{v}_1, \mathbf{w}_1, \mathbf{v}_2, \mathbf{w}_2$ to the basis $\mathbf{v}_1', \mathbf{w}_1', \mathbf{v}_2', \mathbf{w}_2'$. The corresponding projective transformation $T \in \mathrm{Proj}(3)$ has the desired properties. $\square$

We now come to the solution of the four skew line problem. We begin with the following natural generalization of Lemma 4.4 and Lemma 2.12.

**Lemma 4.9.** *Let $Q \colon \mathbb{R}^4 \to \mathbb{R}$ be a quadratic form, and let $S = \{Q(\mathbf{x}) = 0\} \subset \mathbb{P}^3$ be the corresponding quadric surface. Then any line $\ell \subset \mathbb{P}^3$ either is contained in $S$ or intersects $S$ in at most two points.*

**Proof.** Choose $P = [\mathbf{p}]$ and $R = [\mathbf{r}] \in \ell$. If $\ell$ is contained in $S$, we are done; if not, we may suppose $R \notin S$ and represent $\ell - \{R\}$ by parametric equations $\mathbf{x} = \mathbf{p} + t\mathbf{r}$, $t \in \mathbb{R}$. If such a point $[\mathbf{x}]$ also lies on $S$, then we must have $Q(\mathbf{x}) = 0$.

    Now let $A$ be the symmetric matrix so that $Q(\mathbf{x}) = A\mathbf{x} \cdot \mathbf{x}$. Then

$$Q(\mathbf{x}) = Q(\mathbf{p} + t\mathbf{r}) = Q(\mathbf{p}) + 2t(A\mathbf{p} \cdot \mathbf{r}) + t^2 Q(\mathbf{r}) = 0$$

is a quadratic equation in the variable $t$ (since $Q(\mathbf{r}) \neq 0$), which can have at most two roots. $\square$

**Remark.** If we were working over the complex numbers, we see from the proof that every (complex) line not contained in the (complex) quadric surface $S$ would intersect $S$ in *precisely* two points, counting multiplicities.

**Proposition 4.10.** *Let $\ell_1, \ell_2, \ell_3$ be lines in $\mathbb{P}^3$ in general position. Then there is a unique quadric surface containing them all.*

**_Proof._** Let's start with a specific case. Suppose

$$\ell_1 = \{x_0 = x_1 = 0\}$$
$$\ell_2 = \{x_2 = x_3 = 0\}$$
$$\ell_3 = \{x_0 = x_2, \ x_1 = x_3\}.$$

As in the proof of Lemma 4.8, for each point $P_3 \in \ell_3$, there is a unique line through $P_3$ intersecting both $\ell_1$ and $\ell_2$; we now calculate this line explicitly. Let $P_3 = [\alpha, \beta, \alpha, \beta]$. The plane spanned by $P_3$ and $\ell_2$ consists of all points of the form $[*, *, \alpha, \beta]$ (where $*$ denotes an arbitrary real number), and so it intersects $\ell_1$ in the point $P_1 = [0, 0, \alpha, \beta]$. Now the line $\ell$ through $P_1$ and $P_3$ is given by

$$\overrightarrow{P_1 P_3} = \{[s(0, 0, \alpha, \beta) + t(\alpha, \beta, \alpha, \beta)] : s, t \in \mathbb{R} \text{ not both zero}\}.$$

We can rewrite this explicitly as follows:

(∗)
$$\begin{aligned} x_0 &= t\alpha \\ x_1 &= t\beta \\ x_2 &= (s+t)\alpha \\ x_3 &= (s+t)\beta. \end{aligned}$$

Note that every point on this line satisfies the equation

(∗∗)
$$x_0 x_3 = x_1 x_2.$$

And, conversely, every point on the surface (∗∗) can be expressed in the form (∗) for some values of $s$, $t$, $\alpha$, and $\beta$. (For example, if $x_0 \neq 0$, then we may take $x_0 = 1 = \alpha t$, $x_1 = \beta t$, $x_2 = (s+t)\alpha = 1 + s\alpha$, and solve for $x_3$: $x_3 = x_1 x_2 / x_0 = t\beta(1 + s\alpha) = t\beta + (\alpha t)(s\beta) = (s+t)\beta$, as required. If $x_0 = 0$, then either $x_1 = 0$ or $x_2 = 0$: if $x_0 = x_1 = 0$, then set $t = 0$, and solve $[\alpha, \beta] = [x_2, x_3]$; if $x_0 = x_2 = 0$, then set $\alpha = 0$, $\beta = 1$, and solve $[t, s+t] = [x_1, x_3]$.)

We have proved so far that (∗∗) defines a quadric surface $S$ containing the given three lines, and that every point of $S$ lies on a line such as $\ell$ constructed above. Now we establish uniqueness: suppose there were another quadric surface $\Sigma$ containing the lines $\ell_1, \ell_2, \ell_3$. The general line $\ell$ constructed above intersects all three lines $\ell_1, \ell_2, \ell_3$ and therefore intersects $\Sigma$ in at least three points. It follows from Lemma 4.9 that $\ell \subset \Sigma$, and thus that $S \subset \Sigma$. Since they are both quadric surfaces, we must have $\Sigma = S$ (see Exercises 13 and 14).

But now the general case follows easily from Proposition 4.7. Given an arbitrary set of three lines $m_1, m_2, m_3$ in general position, there is a projective transformation $T \in \operatorname{Proj}(3)$ carrying $\ell_i$ to $m_i$, $i = 1, 2, 3$. If $S$ is the quadric surface above containing the lines $\ell_i$, then $T(S)$ is the quadric surface containing the lines $m_i$ (see Exercise 11), and we are done. $\square$

The quadric surface $S = \{x_0 x_3 = x_1 x_2\}$ has a beautiful geometric property. The portion of the surface away from infinity is the usual "saddle" surface pictured in Figure 4. The key feature

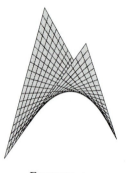

FIGURE 4

to notice is the presence of lines on $S$. In fact, through each point $P = [p_0, p_1, p_2, p_3]$ there pass *two* lines: $\ell_P = \{p_0 x_3 = p_2 x_1, p_1 x_2 = p_3 x_0\}$ and $m_P = \{p_0 x_3 = p_1 x_2, p_2 x_1 = p_3 x_0\}$. Now, every $\ell$-line intersects every $m$-line, but all the $\ell$-lines are mutually skew (as are all the $m$-lines). It follows that, in our construction of the quadric in Proposition 4.10, the three given lines must be either all $\ell$-lines or all $m$-lines. Since the line $\ell_3$ is clearly an $\ell$-line, it follows that all three are (see Exercise 10). Because projective transformations map lines to lines, the same property holds for the image of $S$ under a projective transformation of $\mathbb{P}^3$: it, too, has $\ell$-lines and $m$-lines. For obvious reasons, $S$ is called a **doubly ruled** surface. (In general, we say a surface $S$ is **ruled** if for each point $p \in S$ there is a line passing through $p$ and lying completely in $S$.)

We now can say that *four* lines in $\mathbb{P}^3$ are **in general position** if they are pairwise skew and do not all lie on a quadric surface. That is, by Proposition 4.10, any three of them determine a unique quadric surface $S$; the fourth should not lie on $S$. As a corollary, we obtain at long last the answer to our question.

**Theorem 4.11.** *Given four lines $\ell_1, \ell_2, \ell_3, \ell_4 \subset \mathbb{P}^3$ in general position, there are at most two lines intersecting all four.*

**Proof.** Given $\ell_1, \ell_2, \ell_3$, they are $\ell$-lines on a unique quadric surface $S$. The line $\ell_4$ does not lie on $S$, and, therefore, by Lemma 4.9, intersects it in at most two points $P, Q \in S$. (See Figure 5.) Through $P$ and $Q$ pass $m$-lines $m_P$ and $m_Q$, respectively, which in-

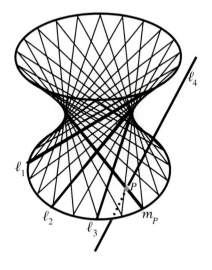

FIGURE 5

tersect our three $\ell$-lines $\ell_1, \ell_2, \ell_3$. Thus, $m_P$ and $m_Q$ are two lines intersecting all four lines $\ell_1, \ldots, \ell_4$. (Could there be another? Well, if a line intersects $\ell_1$, $\ell_2$, and $\ell_3$, it must lie on $S$ (see Exercise 15), and must therefore be an $m$-line; since it must intersect $\ell_4$, it must pass through $P$ or $Q$, and these are the two lines for which we've already accounted.)   □

## EXERCISES 8.4

1.  Let $P, Q, R, S \in \mathbb{P}^3$, and choose vectors $\mathbf{p}, \mathbf{q}, \mathbf{r}, \mathbf{s} \in \mathbb{R}^4$ representing them. Show that $P$, $Q$, and $R$ are collinear if and only if $\mathbf{p}, \mathbf{q}, \mathbf{r}$ are linearly dependent. Show that $P$, $Q$, $R$, and $S$ are coplanar if and only if $\mathbf{p}, \mathbf{q}, \mathbf{r}, \mathbf{s}$ are linearly dependent.

2.  Prove Lemma 4.2.

3.  Prove Lemma 4.3.

4. Let $\Pi, \Pi' \subset \mathbb{P}^3$ be distinct planes. Show that they intersect in a line $\ell$.

5. Let $\Pi = \{x_3 = 0\}$ and let $O = [0,0,0,1]$. Show that the projection $\pi$ from $O$ to $\Pi$ is given in coordinates by $\pi([x_0, x_1, x_2, x_3]) = [x_0, x_1, x_2, 0]$.

6. Let $\pi$ be projection of $\mathbb{P}^3$ from a point $O$ to a plane $\Pi$, $O \notin \Pi$. If $\ell$ is a line in $\mathbb{P}^3$ not passing through $O$, show that $\pi(\ell)$ is a line in $\Pi$.

7. Let $P$, $Q$, and $R \in \mathbb{P}^n$ be collinear. Prove that we may choose homogeneous coordinates $\mathbf{p} = (p_0, \ldots, p_n)$, $\mathbf{q} = (q_0, \ldots, q_n)$, and $\mathbf{r} = (r_0, \ldots, r_n)$ for $P$, $Q$, and $R$, respectively, so that $\mathbf{r} = \mathbf{p} + \mathbf{q}$.

8. Although the saddle surface $z = xy$ and the hyperboloid of one sheet $x^2 + y^2 - z^2 = 1$ are quite different in $\mathbb{R}^3$ (as the pictures in Figure 6 make plain), show that the corresponding projective surfaces $x_0 x_3 - x_1 x_2 = 0$ and $-x_0^2 + x_1^2 + x_2^2 - x_3^2 = 0$ are projectively equivalent. (One approach is to show that the respective quadratic forms are equivalent in the sense of Section 3. But also try to make a plausible geometric argument.)

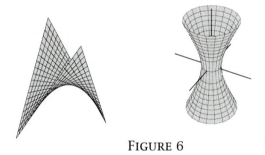

FIGURE 6

9. Let $Q: \mathbb{R}^4 \to \mathbb{R}$ be a nondegenerate quadratic form, and consider the quadric surface $S = \{Q(\mathbf{x}) = 0\} \subset \mathbb{P}^3$. Show that $S$ is either empty, a sphere, or a torus. (Cf. Exercise 8 and Proposition 3.8.)

10. Check that the lines $\ell_1$ and $\ell_2$ defined at the beginning of the proof of Proposition 4.10 are indeed $\ell$-lines on the quadric $S = \{x_0 x_3 = x_1 x_2\}$.

11. Let $Q: \mathbb{R}^4 \to \mathbb{R}$ be a quadratic form, $Q(\mathbf{x}) = A\mathbf{x} \cdot \mathbf{x}$, and let $Y \in GL(4, \mathbb{R})$.

    a.   If $\mathbf{y} = Y(\mathbf{x})$, let $\tilde{A} = (Y^{-1})^T A Y^{-1}$, and $\tilde{Q}(\mathbf{y}) = \tilde{A}\mathbf{y} \cdot \mathbf{y}$. Prove that $\tilde{Q}(Y\mathbf{x}) = Q(\mathbf{x})$.

    b.   Let $T \in \text{Proj}(3)$ be a projective transformation of $\mathbb{P}^3$. Prove that $T$ carries quadric surfaces in $\mathbb{P}^3$ to quadric surfaces.

    c.   Show, moreover, that $T$ carries nondegenerate quadric surfaces to nondegenerate quadric surfaces.

12.   Let $Q: \mathbb{R}^4 \to \mathbb{R}$ be an arbitrary quadratic form in four variables. Show directly that if the quadric surface $\{Q(\mathbf{x}) = 0\} \subset \mathbb{P}^3$ contains $\ell_1$, $\ell_2$, and $\ell_3$ as defined at the beginning of the proof of Proposition 4.10, then $Q(\mathbf{x}) = c(x_0 x_3 - x_1 x_2)$ for some constant $c$. This proves both existence and uniqueness of the quadric surface containing these three lines.

13.   Let $S$ and $\Sigma$ be two quadric surfaces in $\mathbb{P}^3$ with $S \subset \Sigma$. The point of this exercise is to show that $S = \Sigma$. Suppose $Q$ and $Q'$ are quadratic forms defining $S$ and $\Sigma$, respectively. Then $Q(\mathbf{x}) = 0 \implies Q'(\mathbf{x}) = 0$. Choose a point $P = [\mathbf{p}] \notin \Sigma$.

    a.   Let $R = [\mathbf{r}] \in \mathbb{P}^3$ be arbitrary, and consider the line $\ell = \overrightarrow{PR} = \{[\mathbf{p} + t\mathbf{r}]\}$. Let $q_\ell(t)$ (resp., $q'_\ell(t)$) be the restriction of $Q$ (resp., $Q'$) to this line. Deduce that there is a constant $c_\ell$ so that $q_\ell(t) = c_\ell q'_\ell(t)$.

    b.   By considering different lines through $P$, deduce that all the constants $c_\ell$ must be equal, and hence that $Q = cQ'$ for some constant $c$.

    c.   Conclude that $S = \Sigma$.

14.   Here is an alternative proof of the result of Exercise 13, based on the remark on page 336. We can define $\mathbb{P}^3_{\mathbb{C}}$ by using complex numbers (cf. Section 8.5 for the discussion of $\mathbb{P}^1_{\mathbb{C}}$). Given a quadric surface $S \subset \mathbb{P}^3$, it is the zero locus of a quadratic form $Q$, and we can consider the complex quadric surface $S_{\mathbb{C}} \subset \mathbb{P}^3_{\mathbb{C}}$ defined by the *same* quadratic form. Analogously, define $\Sigma_{\mathbb{C}}$. Suppose $P \in S_{\mathbb{C}}$ and $R \in \Sigma_{\mathbb{C}}$ is chosen arbitrarily. Using the aforementioned remark, prove that $R$ must lie in $S_{\mathbb{C}}$, and hence that $S_{\mathbb{C}} = \Sigma_{\mathbb{C}}$. Deduce that $S = \Sigma$.

15.   Let $\ell_1, \ell_2, \ell_3$ be three lines in $\mathbb{P}^3$ in general position. By Proposition 4.10 they lie on a unique quadric surface $S$. Prove that any line intersecting the three lines must also lie on $S$. This shows that (other than a plane) the only doubly ruled surfaces are quadrics.

16.  Give an example of a ruled surface that is not doubly ruled.

17.  What shape is obtained when a cube is revolved about the (long) diagonal joining a pair of opposite vertices?

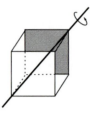

FIGURE 7

18.  Extend the discussion of projective duality in Section 2 to $\mathbb{P}^3$.
     a.  Show that $\mathbb{P}^{3*}$ is the set of planes in $\mathbb{P}^3$, and there is a correspondence $\mathbb{P}^3 \leftrightarrow \{\text{planes in } \mathbb{P}^{3*}\}$.
     b.  Generalize Proposition 2.10.

19.  Let $\ell_1$ and $\ell_2$ be skew lines in $\mathbb{R}^3$. Show that the locus of points equidistant from both lines is a quadric surface. What type quadric surface? Can you determine its rulings geometrically? (We thank Russ Webb for suggesting this problem.)

20.  Let $P = [p_0, p_1, p_2, p_3]$, $Q = [q_0, q_1, q_2, q_3]$. For $0 \le i < j \le 3$, let $a_{ij} = p_i q_j - p_j q_i$. Associate to the line $\overleftrightarrow{PQ}$ the point $\quad = [a_{01}, a_{02}, a_{03}, a_{12}, a_{13}, a_{23}] \in \mathbb{P}^5$.
     a.  Prove that $\quad$ is a well-defined point in $\mathbb{P}^5$ determined only by the line. (There are two things to check here: if we choose different homogeneous coordinates for $P$ and $Q$, the point $\quad \in \mathbb{P}^5$ doesn't change. And if we choose different points $P', Q' \in \overleftrightarrow{PQ}$ and calculate $a'_{ij} = p'_i q'_j - p'_j q'_i$, we have $\quad' = \quad \in \mathbb{P}^5$.)
     b.  Show that $\quad$ satisfies the (homogeneous) quadratic equation
     $$a_{01} a_{23} - a_{02} a_{13} + a_{03} a_{12} = 0.$$
     c.  Given two lines $\ell$ and $m$, let $\quad$ and $\quad$ be the corresponding points in $\mathbb{P}^5$. Show that $\ell$ and $m$ intersect if and only if
     $$a_{01} b_{23} - a_{02} b_{13} + a_{03} b_{12} + a_{12} b_{03} - a_{13} b_{02} + a_{23} b_{01} = 0.$$

21. Using the notation of Exercise 20, show that the two families of rulings on the quadric surface $x_0x_3 - x_1x_2 = 0$ are given by $\{\mathcal{A} : a_{02} = a_{13} = a_{03} + a_{12} = 0\}$ and $\{\mathcal{A} : a_{01} = a_{23} = a_{03} - a_{12} = 0\}$. (It follows from Exercise 20b. that these loci are conics lying in disjoint projective planes in $\mathbb{P}^5$.)

22. Recall from Proposition 4.6 of Chapter 7 that given any $A \in SO(3)$, the linear map $T \colon \mathbb{R}^3 \to \mathbb{R}^3$ defined by $T(\mathbf{x}) = A\mathbf{x}$ is rotation through some angle $\theta$ about some axis. Use this to show that $SO(3)$ is (topologically equivalent to) $\mathbb{P}^3$. (Hint: In analogy with our hemispherical model of $\mathbb{P}^2$, think of $\mathbb{P}^3$ as a three-dimensional ball with the opposite points of its boundary sphere identified. Now think "spherical coordinates.")

## 5. Putting the *metry* back in *Geometry*: Elliptic and Hyperbolic Geometry

In our treatment of projective geometry up to this point, we have made no mention of the two words—length and angle—that we usually associate with the word "geometry." We now rectify this. We need first a model for the projective plane that comes with these notions built in. Recall that each point of $\mathbb{P}^2$ represents a line through the origin in $\mathbb{R}^3$; since each such line intersects the unit sphere $\mathbb{S} = \{(x_0, x_1, x_2) \in \mathbb{R}^3 : x_0^2 + x_1^2 + x_2^2 = 1\}$ in a pair of antipodal points, we can now associate to each point of $\mathbb{P}^2$ a *pair* of *antipodal points* on $\mathbb{S}$ (see the picture on the left in Figure 1). That

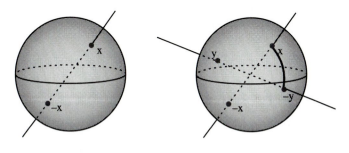

FIGURE 1

is, we think of $\mathbb{P}^2$ as the set of equivalence classes of pairs of antipodal points $\mathbf{x}$ and $-\mathbf{x} \in \mathbb{S}$; we denote by $x$ the equivalence class of $\mathbf{x}$ (or $-\mathbf{x}$). In particular, the equivalence relation gives rise to a mapping $p: \mathbb{S} \to \mathbb{P}^2$, $p(\mathbf{x}) = p(-\mathbf{x}) = x$ (cf. Section 3 of Appendix A).

The spherical distance between two points $\mathbf{x}, \mathbf{y} \in \mathbb{S}$ is given by the angle between the two vectors, so we have $d_{\mathbb{S}}(\mathbf{x}, \mathbf{y}) = \arccos(\mathbf{x} \cdot \mathbf{y})$. Now, given two points $x, y \in \mathbb{P}^2$, we calculate the spherical distances $d_{\mathbb{S}}(\mathbf{x}, \mathbf{y})$, $d_{\mathbb{S}}(-\mathbf{x}, \mathbf{y})$, $d_{\mathbb{S}}(\mathbf{x}, -\mathbf{y})$, and $d_{\mathbb{S}}(-\mathbf{x}, -\mathbf{y})$, and define the distance $d_{\mathbb{P}^2}(x, y)$ to be the smallest of these four numbers (see the picture on the right in Figure 1). Equivalently, but more conveniently, define

$$d_{\mathbb{P}^2}(x, y) = \arccos |\mathbf{x} \cdot \mathbf{y}|.$$

If we are to have any hope of discussing triangles shortly, the distance $d_{\mathbb{P}^2}$ must satisfy the triangle inequality.

**Lemma 5.1.** *Let $x$, $y$, and $z$ be arbitrary points in $\mathbb{P}^2$. Then $d_{\mathbb{P}^2}(x, y) \le d_{\mathbb{P}^2}(x, z) + d_{\mathbb{P}^2}(z, y)$.*

**Proof.** This is a computation based on the addition formula for the cosine function and a pretty formula for the dot product of two cross products (whose proof is left to the reader in Exercise 2):

$(*)$     $\boxed{(\mathbf{x} \times \mathbf{w}) \cdot (\mathbf{z} \times \mathbf{y}) = (\mathbf{x} \cdot \mathbf{z})(\mathbf{w} \cdot \mathbf{y}) - (\mathbf{x} \cdot \mathbf{y})(\mathbf{w} \cdot \mathbf{z}).}$

In particular, taking $\mathbf{w} = \mathbf{z}$, we obtain

$(**)$     $(\mathbf{x} \times \mathbf{z}) \cdot (\mathbf{z} \times \mathbf{y}) = (\mathbf{x} \cdot \mathbf{z})(\mathbf{z} \cdot \mathbf{y}) - (\mathbf{x} \cdot \mathbf{y})|\mathbf{z}|^2$.

Using $(**)$ with $\mathbf{x}, \mathbf{y}, \mathbf{z} \in \mathbb{S}$, Exercise 7.4.4, and the fact that $|\mathbf{v} - \mathbf{w}| \ge |\mathbf{v}| - |\mathbf{w}|$, we obtain

$$|\mathbf{x} \cdot \mathbf{y}| = |(\mathbf{x} \cdot \mathbf{z})(\mathbf{z} \cdot \mathbf{y}) - (\mathbf{x} \times \mathbf{z}) \cdot (\mathbf{z} \times \mathbf{y})| \ge |\mathbf{x} \cdot \mathbf{z}||\mathbf{y} \cdot \mathbf{z}| - |\mathbf{x} \times \mathbf{z}||\mathbf{z} \times \mathbf{y}|.$$

Recall that when $\mathbf{v}$ and $\mathbf{w}$ are unit vectors with angle $\theta$ between them, $|\mathbf{v} \times \mathbf{w}| = \sin \theta$, and so

$$|\mathbf{x} \cdot \mathbf{y}| \ge |\mathbf{x} \cdot \mathbf{z}||\mathbf{y} \cdot \mathbf{z}| - \sin(\arccos |\mathbf{x} \cdot \mathbf{z}|) \sin(\arccos |\mathbf{z} \cdot \mathbf{y}|)$$
$$= \cos(\arccos |\mathbf{x} \cdot \mathbf{z}| + \arccos |\mathbf{z} \cdot \mathbf{y}|).$$

Since arccos is a decreasing function on $[-1, 1]$, it now follows that

$$\arccos |\mathbf{x} \cdot \mathbf{y}| \le \arccos |\mathbf{x} \cdot \mathbf{z}| + \arccos |\mathbf{z} \cdot \mathbf{y}|. \quad \square$$

We refer to $\mathbb{P}^2$, equipped with this distance function, as the **elliptic plane**. It will be convenient to picture the elliptic plane as the upper hemisphere with antipodal points on the equator identified (cf. Figure 5 on p. 300). Before continuing, we should comment that for any $x, y \in \mathbb{P}^2$, $0 \le d_{\mathbb{P}^2}(x, y) \le \pi/2$. Whenever $d_{\mathbb{P}^2}(x, y) < \pi/2$, there is a unique shortest path (the appropriate arc of a great semi-circle) joining $x$ and $y$, and it has length $d_{\mathbb{P}^2}(x, y)$. However, when $d_{\mathbb{P}^2}(x, y) = \pi/2$, there are two shortest paths of equal length, as pictured in Figure 2. In any event, given $x, y \in \mathbb{P}^2$, we can now vi-

FIGURE 2

sualize the line $\overleftrightarrow{xy}$ as the great semicircle spanned by $\mathbf{x}$ and $\mathbf{y}$ (i.e., the intersection of the hemisphere with the plane through the origin spanned by $\mathbf{x}$ and $\mathbf{y}$), with its endpoints on the equator identified.

We next need to measure angles in $\mathbb{P}^2$. Consider the angle $\angle yxz$ formed by the rays $\overrightarrow{xy}$ and $\overrightarrow{xz}$ in the elliptic plane. Since an oriented line in $\mathbb{P}^2$ corresponds to an oriented plane through the origin in $\mathbb{R}^3$, we define the angle in $\mathbb{P}^2$ formed by the two rays to be the angle between the corresponding planes in $\mathbb{R}^3$. This can be computed, for example, by finding the angle between their respective normal vectors. In particular, the angle $\alpha$ between $\overrightarrow{xy}$ and $\overrightarrow{xz}$ is the angle between the vectors $\mathbf{x} \times \mathbf{y}$ and $\mathbf{x} \times \mathbf{z}$. If we choose $y$ and $z$ to be at maximal distance from $x$ (so that $\mathbf{x}$ is orthogonal to both $\mathbf{y}$ and $\mathbf{z}$), then geometry (or formula ($*$)) tells us that $\alpha$ is just the distance between $y$ and $z$.

FIGURE 3

The last statement has an important consequence. An **isometry** $f: \mathbb{P}^2 \to \mathbb{P}^2$ preserves distance between pairs of points; i.e., $d_{\mathbb{P}^2}(x, y) = d_{\mathbb{P}^2}(f(x), f(y))$ for all $x, y \in \mathbb{P}^2$. It now follows that an isometry preserves (unoriented) angles as well. What is the group of isometries of the elliptic plane? By Corollary 4.3 of Chapter 7, the group of isometries of the unit sphere $S$ is $O(3)$. All of these isometries induce isometries of $\mathbb{P}^2$; however, perhaps there could be an isometry of the projective plane that does not arise in this fashion.

**Theorem 5.2.** $\text{Isom}(\mathbb{P}^2) \cong SO(3)$.

***Proof.*** From the comments we've just made and from Exercise 1, it follows that $SO(3) \subset \text{Isom}(\mathbb{P}^2)$. It remains to show the reverse containment. The key idea is this: $SO(3)$ acts transitively not only on $\mathbb{P}^2$, but on the set of directions in $\mathbb{P}^2$ (i.e., points and oriented lines passing through them). So, by composing with an appropriate

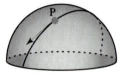

FIGURE 4

element of $SO(3)$, we may assume we are given an isometry $f$ that fixes $P$ and sends every oriented line $\ell$ through $P$ to itself. The claim now is that $f$ must be the identity map. Each point $Q$ with $d_{\mathbb{P}^2}(P, Q) < \pi/2$ must be mapped to itself (since it lies on a unique shortest line segment emanating from $P$); and then it follows by continuity of $f$ that every point at distance $\pi/2$ must be mapped to itself as well.   □

**Example.** Using Proposition 4.6 of Chapter 7 it is straightforward to analyze the isometries of $\mathbb{P}^2$. Every isometry $f$ corresponds to a linear map $A \in SO(3)$, which is rotation about some axis in $\mathbb{R}^3$ through some angle $\theta$. Thus every isometry has a fixed point. There is one especially interesting case: when $\theta = \pi$, we have $A = \begin{bmatrix} -1 & & \\ & -1 & \\ & & 1 \end{bmatrix}$ in an appropriate basis, and the isometry $f$ has a *line* of fixed points as well. See Exercise 21 for an interpretation of this particular isometry from the viewpoint of Section 2.

> **Remark.** There's an interesting (but totally optional) relation between $SO(3) = \text{Isom}(\mathbb{P}^2)$ and the full projective group $\text{Proj}(2)$. Denote by $GL^+(3)$ the set of $3 \times 3$ real matrices with positive determinant. Then $GL^+(3)$ is a *nine-dimensional* group (since there are nine entries to vary,

and det $> 0$ defines an open subset of $\mathbb{R}^9$). On the other hand, $SO(3)$ is a *three*-dimensional group (the first column vector $\mathbf{e}_1$ can vary over a sphere; once it is picked, the second column vector $\mathbf{e}_2$ can vary over the circle of unit vectors orthogonal to it; once $\mathbf{e}_1$ and $\mathbf{e}_2$ are picked, the third column vector $\mathbf{e}_3$ is uniquely determined by the equation $\mathbf{e}_3 = \mathbf{e}_1 \times \mathbf{e}_2$). Now it follows easily from Exercise 8.3.14 that every matrix $A \in GL^+(3)$ can be written uniquely in the form $A = QR$, where $Q \in SO(3)$ and $R$ is symmetric with positive eigenvalues. But the set of symmetric $3 \times 3$ matrices is a *six*-dimensional vector space (why?), and those with all positive eigenvalues comprise an open subset (see Exercise 31).

The upshot is this: to each projective transformation $T \in \mathrm{Proj}(2)$, we associate a matrix $A \in GL^+(3)$ (unique up to multiplication by positive scalars $\lambda$). To $A$ we associate the unique matrix $Q \in SO(3)$ (which does not change as $\lambda$ varies). Thus to each projective transformation is associated a unique isometry. In the language of topology, we have a strong deformation retraction $\mathrm{Proj}(2) \to \mathrm{Isom}(\mathbb{P}^2)$.

Next we turn to a discussion of triangles, the law of cosines, and related matters in the elliptic plane.

**Definition.** A **line** on the sphere $\mathcal{S}$ is a great circle. A **line segment** on the sphere $\mathcal{S}$ is a segment of a line having length $< \pi$. A **spherical triangle** consists of three noncollinear points on the sphere $\mathcal{S}$ and three line segments connecting them in pairs. Lastly,

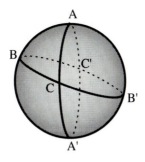

FIGURE 5

a **triangle** in the elliptic plane is the image $p(\triangle)$ for some spherical triangle $\triangle \subset \mathcal{S}$. (Recall that $p \colon \mathcal{S} \to \mathbb{P}^2$ is the obvious two-to-one mapping defined on p. 344.)

It is easy to see that three noncollinear points in the elliptic plane determine *four* triangles. See Figure 5 and Exercise 23.

**Proposition 5.3** (Elliptic Law of Cosines). *Let* $\triangle ABC \subset \mathbb{P}^2$. *Let* $AB = c$, $BC = a$, *and* $CA = b$. *Then*

$$\cos A = \frac{\cos a - \cos b \cos c}{\sin b \sin c}.$$

***Proof.*** As indicated in Figure 6, let vectors **x**, **y**, and **z** $\in \mathcal{S}$, respectively, correspond to the vertices $A$, $B$, and $C$, respectively. Since $A$ is the angle between $\mathbf{x} \times \mathbf{y}$ and $\mathbf{x} \times \mathbf{z}$, we have

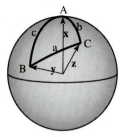

FIGURE 6

$$\cos A = \frac{(\mathbf{x} \times \mathbf{y}) \cdot (\mathbf{x} \times \mathbf{z})}{|\mathbf{x} \times \mathbf{y}||\mathbf{x} \times \mathbf{z}|}.$$

But, by (∗),

$$(\mathbf{x} \times \mathbf{y}) \cdot (\mathbf{x} \times \mathbf{z}) = \mathbf{y} \cdot \mathbf{z} - (\mathbf{x} \cdot \mathbf{z})(\mathbf{y} \cdot \mathbf{x}) = \cos a - \cos b \cos c.$$

Lastly, $|\mathbf{x} \times \mathbf{y}| = \sin c$ and $|\mathbf{x} \times \mathbf{z}| = \sin b$, so we're done.  □

**Corollary 5.4** (Pythagorean Theorem). *Let* $\triangle ABC$ *be a right triangle in the elliptic plane, and assume* $\angle A$ *is a right angle. Then* $\cos a = \cos b \cos c$.  □

The reader has no doubt surmised by now that the sum of the angles of a triangle in the elliptic plane may well exceed $\pi$. Indeed, it is easy to draw a triangle with *three* right angles. But we discover quickly that any two such triangles are congruent. In fact, more is true. As we prove next, the angles of any triangle in elliptic geometry determine the lengths of the sides; thus, there is an angle-angle-angle congruence theorem in elliptic geometry.

**Proposition 5.5.** *Let* $\triangle ABC \subset \mathbb{P}^2$, $a = BC$. *Then*

$$\cos a = \frac{\cos A + \cos B \cos C}{\sin B \sin C}.$$

**Proof.** It is hard to miss the similarity between this formula and the law of cosines. This observation suggests that we should apply the latter to the *dual triangle*, whose vertices are the edges of the original triangle. Because of the sign change, we proceed a bit cautiously: the edges $\overline{AB}$, $\overline{BC}$, and $\overline{CA}$, respectively, become the vertices $C^*$, $A^*$, and $B^*$, respectively. So, in the notation of Proposition 5.3,

$$A^* = \frac{\mathbf{y} \times \mathbf{z}}{|\mathbf{y} \times \mathbf{z}|}, \qquad B^* = \frac{\mathbf{z} \times \mathbf{x}}{|\mathbf{z} \times \mathbf{x}|}, \qquad C^* = \frac{\mathbf{x} \times \mathbf{y}}{|\mathbf{x} \times \mathbf{y}|}.$$

What is $\angle A^*$? By definition we have

$$\cos A^* = \frac{(A^* \times B^*) \cdot (A^* \times C^*)}{|A^* \times B^*||A^* \times C^*|};$$

and so, using the formula in Exercise 3c.,

$$\cos A^* = \frac{(\mathbf{z} \cdot (\mathbf{y} \times \mathbf{x})) \, \mathbf{z} \cdot (\mathbf{y} \cdot (\mathbf{z} \times \mathbf{x})) \, \mathbf{y}}{|(\mathbf{z} \cdot (\mathbf{y} \times \mathbf{x})) \, \mathbf{z}||(\mathbf{y} \cdot (\mathbf{z} \times \mathbf{x})) \, \mathbf{y}|}$$
$$= -\mathbf{y} \cdot \mathbf{z} = -\cos a.$$

Correspondingly, the length of side $\overline{B^*C^*}$, which we name $a^*$, is given by

$$\cos a^* = B^* \cdot C^* = \frac{\mathbf{z} \times \mathbf{x}}{|\mathbf{z} \times \mathbf{x}|} \cdot \frac{\mathbf{x} \times \mathbf{y}}{|\mathbf{x} \times \mathbf{y}|} = \frac{\cos b \cos c - \cos a}{\sin b \sin c} = -\cos A.$$

Thus, the relation between the measurements of the original triangle and those of the dual triangle is very simple: every angle (and, thus, every length) is replaced by its supplement.

Applying the law of cosines to the dual triangle, we now have

$$\cos a = -\cos A^* = -\left( \frac{\cos a^* - \cos b^* \cos c^*}{\sin b^* \sin c^*} \right)$$
$$= \frac{\cos A + \cos B \cos C}{\sin B \sin C}. \qquad \square$$

We now come to one of the remarkable results in non-Euclidean geometry: the formula for the area of a triangle. Since we've just seen that the angles of a triangle determine the triangle uniquely, it will not be a surprise to find that the area of a triangle depends also just on its angles.

**Theorem 5.6.** *Let* $\triangle ABC$ *be a triangle in the elliptic plane. Then*

$$\text{area } \triangle ABC = (A + B + C) - \pi .$$

***Proof.*** First we make a definition: a **lune** is a region on the sphere bounded by two great semicircles with common endpoints. If the angle formed by the lune is $\alpha$, then the area of the region is $\frac{\alpha}{2\pi} \cdot 4\pi = 2\alpha$. Now, as indicated in Figure 7, $\triangle ABC$ is the intersection of three lunes in the elliptic plane, with angles $A$, $B$, and $C$, respectively. On the other hand, the union of these three lunes is

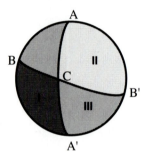

FIGURE 7

the whole elliptic plane, which has area $2\pi$. We therefore have

$$2\pi = \text{area lune } A + \text{area lune } B + \text{area lune } C - 2\,\text{area}\,\triangle ABC,$$

from which it follows easily that area $\triangle ABC = (A + B + C) - \pi$.  $\square$

Now we turn to a different sort of non-Euclidean geometry, *hyperbolic* geometry, in which parallel lines abound. The motions of elliptic geometry consisted of the isometries of the sphere $\mathcal{S}$. There is, however, another natural group action on $\mathcal{S}$, which will lead to hyperbolic geometry.

In Section 2 we studied the projective line $\mathbb{P}^1$ over the real numbers, but the same construction works for any field. In particular, we define the **complex projective line** $\mathbb{P}^1_{\mathbb{C}}$ to be the set of equivalence classes of ordered pairs $(z_0, z_1) \in \mathbb{C}^2 - \{\mathbf{0}\}$, where $(z_0, z_1) \equiv (w_0, w_1) \iff (w_0, w_1) = c(z_0, z_1)$ for some $c \in \mathbb{C} - \{0\}$. As was the case with the real numbers, we have

$$\mathbb{P}^1_{\mathbb{C}} = \mathbb{C} \cup \{\infty\}$$

(with $[z_0, z_1] = [1, z_1/z_0] \leftrightarrow z_1/z_0 \in \mathbb{C}$ when $z_0 \neq 0$ and $[0, 1] \leftrightarrow \infty$). There is a natural identification of $\mathbb{P}^1_{\mathbb{C}} = \mathbb{C} \cup \{\infty\}$ with the sphere $\mathcal{S}$, quite like the identification of $\mathbb{P}^1$ with a circle (cf. Exercise 8.2.7 and Exercise 25). We now have the group of projective transformations of $\mathbb{P}^1_\mathbb{C}$: $\text{Proj}_\mathbb{C}(1) = GL(2, \mathbb{C})/\{\lambda \,\text{Id} : \lambda \in \mathbb{C}^\times\}$, the group of linear fractional (or Möbius) transformations of the complex projective line.

Let $\mathbb{H} = \{z \in \mathbb{C} : \text{Im } z > 0\} \subset \mathbb{P}^1_\mathbb{C}$ be the upper half-plane. What subgroup of $\text{Proj}_\mathbb{C}(1)$ will map $\mathbb{H}$ to itself? The obvious guess (see Lemma 5.7 below) is this: in order to map $\mathbb{H}$ to itself, a projective transformation must map the boundary—namely, the real axis together with the point at infinity—to itself. In order to do this, the transformation must in fact be an element of the *real* projective group $\text{Proj}(1)$ (see Exercise 5). But a bit more must be true: to map the *upper* half-plane to the *upper* half-plane, we must consider only the equivalence classes of linear maps with positive determinant. (Note that if $A \in GL(2, \mathbb{R})$ and $\lambda \in \mathbb{R}^\times$, then $\det(\lambda A) = \lambda^2 \det A$; so the sign of the determinant does not change as we vary over an equivalence class.)

We define the cross-ratio of four points in $\mathbb{P}^1_\mathbb{C}$ exactly as in the real case. Since a line in the complex plane becomes a circle when we include the point at infinity, we will refer to both lines and circles in $\mathbb{C}$ as "generalized circles." We now make two important observations about the geometry of projective transformations.

**Lemma 5.7.** *Elements of* $\text{Proj}_\mathbb{C}(1)$ *map generalized circles to generalized circles.*

***Proof.*** From Lemma 2.2 we know that projective transformations preserve cross-ratio. The key point is this: four points $A$, $B$, $C$, and $D \in \mathbb{P}^1_\mathbb{C}$ lie on a generalized circle if and only if their cross-ratio is *real*. To establish necessity, we argue as follows. If the four points are collinear, then there is a complex number $\zeta$ so that all of $D - A$, $C - A$, $D - B$, and $C - B$ are real multiples of $\zeta$, and so it follows from Exercise 8.2.3 that the cross-ratio is real. If the four points lie on a circle, the diagram in Figure 8 indicates the proof—remember that to divide complex numbers, we must subtract their respective angles. We should be careful to consider other configurations of the points as well. We leave it to the reader to complete the proof, checking details and establishing sufficiency by similar geometric arguments (see Exercise 6). □

**Lemma 5.8.** *Elements of* $\text{Proj}_\mathbb{C}(1)$ *are conformal; i.e., they preserve angles.*

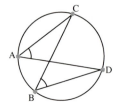

FIGURE 8

***Proof.*** Let $z(t)$ and $w(t)$ be differentiable curves in $\mathbb{C}$, $-\epsilon < t <$ $\epsilon$, with $z(0) = z_o = w(0)$. The angle between the two curves at $z_o$ is, by definition, the angle between the tangent vectors $z'(0)$ and $w'(0)$. Let $f(z) = \frac{az+b}{cz+d}$ be a linear fractional transformation. We need to see that the angle between the tangent vectors $(f \circ z)'(0)$ and $(f \circ w)'(0)$ is the same as that between $z'(0)$ and $w'(0)$. Well, $f'(z) = \frac{ad-bc}{(cz+d)^2}$; and so by the chain rule, $(f \circ z)'(0) = \frac{ad-bc}{(cz_o+d)^2} z'(0)$ and $(f \circ w)'(0) = \frac{ad-bc}{(cz_o+d)^2} w'(0)$. Therefore, $\frac{(f \circ w)'(0)}{(f \circ z)'(0)} = \frac{w'(0)}{z'(0)}$, so the angles between the two sets of vectors are the same. $\quad\square$

For future reference, we include this useful lemma.

**Lemma 5.9.** *There is a projective transformation $T \in \text{Proj}_{\mathbb{C}}(1)$ mapping $\mathbb{H}$ to the interior of the unit disk, $\mathbb{D} = \{z \in \mathbb{C} : |z| < 1\}$.*

***Proof.*** Let $T(z) = \frac{i-z}{i+z}$. Now, $T(0) = 1$, $T(1) = i$, and $T(\infty) = -1$, and so, by Lemma 5.7, the real axis must map to the unit circle $\{|z| = 1\}$. Moreover, since $T(i) = 0$, the upper half-plane must then map to the interior of the circle, and we are done. (See Exercise 7 for a completely computational verification.) $\quad\square$

Thus, we can treat $\mathbb{H}$ and $\mathbb{D}$ interchangeably. We will set up most of our concepts in $\mathbb{H}$, but work with the geometry in $\mathbb{D}$.

Recall (see Exercise 5) that the group of projective transformations which preserve $\mathbb{H}$ is isomorphic to $G = SL(2, \mathbb{R})/\{\pm \text{Id}\}$. The next step is to define a distance function on $\mathbb{H}$ with respect to which this group will act by isometries. If we start with a vertical line segment $y$ from $i$ to $Yi$, its length will be given by the line integral $\int_y ds = \int_1^Y \varphi(y) \, dy$ for some real function $\varphi$. Since $G$ contains transformations of the form $z \rightsquigarrow cz$ for $c > 0$ (see Exercise 8), we must have $\int_1^Y \varphi(y) \, dy = \int_c^{cY} \varphi(y) \, dy$ for all $Y > 0$ and $c > 0$. By differentiating with respect to $Y$, we find that $\varphi(Y) = c\varphi(cY)$; so $\varphi(cy) = \frac{1}{c}\varphi(y)$. If we set $\psi(y) = y\varphi(y)$, then we see that

$\psi(cy) = \psi(y)$ for all $c > 0$, and so $\psi$ must be a constant function. That is, we have $\varphi(y) = C/y$ for some real constant $C$; we choose $C = 1$.

We still need a general formula for the element of arclength $ds$. We impose the further requirement that angles in $\mathbb{H}$ be the usual Euclidean angles. This means that the coordinate axes must be stretched equal amounts at each point $z$; thus, we must have $ds = \lambda(z)\sqrt{dx^2 + dy^2}$ for some function $\lambda(z)$. Since $G$ contains horizontal translations $z \rightsquigarrow z + b$ for any $b \in \mathbb{R}$ (see Exercise 8), it follows that $\lambda(z + b) = \lambda(z)$ for any $b \in \mathbb{R}$, and so $\lambda$ depends just on $Im\ z$. Combining this with our previous calculation, we finally obtain

$$ds = \frac{1}{Im\ z}\sqrt{dx^2 + dy^2} = \frac{|dz|}{Im\ z}.$$

By our construction, $G$ is the group of proper isometries of $\mathbb{H}$, and we aren't going to bother with the improper ones here. By virtue of Lemmas 5.3 and 5.8, $G$ maps a vertical line to lines and circles intersecting the real axis at right angles. This leads us to our definition: the **lines** in $\mathbb{H}$ consist of all vertical half-lines and all semicircles centered on the real axis, as shown in Figure 9. Note that

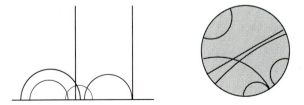

FIGURE 9

given two points in $\mathbb{H}$, there is a unique line segment joining them (see Exercise 9). Given two points $A, B \in \mathbb{H}$, what is the distance between them? We must take the unique line segment $\gamma$ from $A$ to $B$, and compute $\int_\gamma ds$. This is most easily done, however, when $\gamma$ is vertical: if $A = ai$ and $B = bi$, then $d_\mathbb{H}(A, B) = \int_a^b \frac{|dy|}{y} = |\ln \frac{b}{a}|$. (For the general case, see Exercises 10 and 14.)

We mentioned in Lemma 5.9 that there is a projective transformation $T$ carrying $\mathbb{H}$ to the unit disk $\mathbb{D}$. So we can transfer all our work from the half-plane to the disk, and this will be our official model for the **hyperbolic plane**. Angles are Euclidean and lines consist now of arcs of circles meeting the boundary at right angles (as well as diameters). As the reader can check (see Exercise 11), the group of isometries is

$$\text{Isom}(\mathbb{D}) = \left\{ z \rightsquigarrow \zeta \frac{z - a}{1 - \bar{a}z} : |\zeta| = 1, \ |a| < 1 \right\}.$$

(Note that, as in the case of elliptic geometry, the group of isometries involves three real parameters.)

Let's discuss briefly the different types of isometries.

- A *rotation* in the disk model. Rotation about the origin is just $z \rightsquigarrow \zeta z$ (for some $\zeta$ with $|\zeta| = 1$), and the others are conjugate to this. Note each rotation has a unique fixed point inside $\mathbb{D}$. (See the picture at the left in Figure 10.)
- A *horizontal translation* in the half-plane model. In the disk model, points translate along the *horocycles*, circles all tangent at a boundary point. Note that there is a unique fixed point at the boundary. (See the middle picture in Figure 10.)
- A *dilatation* ($z \rightsquigarrow cz$, $c > 0$) in the half-plane model. Such a map has two fixed points at the boundary (0, which is "repelling," and $\infty$, which is "attracting"). In the disk model, points move along arcs of circles (emanating from the repelling fixed point and heading toward the attracting fixed point). (See the picture at the right in Figure 10.)

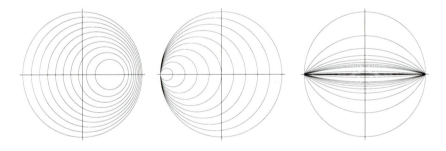

FIGURE 10

Let's consider this from the standpoint of linear algebra. The group of isometries of the half-plane is $G = SL(2, \mathbb{R})/\{\pm\text{Id}\}$. Given a representative $A \in SL(2, \mathbb{R})$, a fixed point $z \in \mathbb{P}^1_\mathbb{C}$ corresponds to an eigenvector $\mathbf{z} \in \mathbb{C}^2$ (why?). If $A$ has a conjugate pair of complex eigenvalues, then, because the entries of $A$ are all real, the eigenvectors also come in a conjugate pair. This means that, passing to $\mathbb{P}^1_\mathbb{C}$, one fixed point is in the upper half-plane and the other is in the lower half-plane; in this case, $A$ has a unique fixed point in $\mathbb{H}$ and $A$ is called *elliptic*. If $A$ has real eigenvalues, it will have real eigenvectors as well, and thus, passing to $\mathbb{P}^1_{\mathbb{C}}$, the fixed points will lie on the real axis. If there is one fixed point on the real axis, $A$ is called *parabolic*; and if there are two, $A$ is called *hyperbolic*.

We summarize our findings in a table (cf. Section 5 of Chapter 2 and Section 4 of Chapter 7):

| fixed point set | invariant set | type isometry | geometric description |
|---|---|---|---|
| 1 point | point | elliptic | rotation |
| 1 point at $\infty$ | none | parabolic | "parallel displacement" |
| 2 points at $\infty$ | line | hyperbolic | "translation" |

For a deeper understanding of these isometries, the reader might consider taking a course in differential geometry.

Now we come to the geometry of triangles in the hyperbolic plane, $\mathbb{D}$. We start with a formula for distance in the disk model. This is a bit of calculus: if the mapping $T\colon \mathbb{H} \to \mathbb{D}$ is given by $w = T(z) = \frac{i-z}{i+z}$, then the inverse mapping is $z = T^{-1}(w) = -i\frac{w-1}{w+1}$. Therefore, $dz = \frac{-2i\,dw}{(w+1)^2}$, $Im\,z = Im\left(-i\frac{w-1}{w+1}\right) = \frac{1-|w|^2}{|w+1|^2}$; and so

$$\frac{|dz|}{Im\,z} = \frac{2\,|dw|}{|w+1|^2} \cdot \frac{|w+1|^2}{1-|w|^2} = \frac{2\,|dw|}{1-|w|^2}.$$

We can now compute the hyperbolic distance from $w = 0$ to $w = R > 0$:

$$d_{\mathbb{D}}(0,R) = \int_0^R \frac{2\,|dw|}{1-|w|^2} = \int_0^R \frac{2\,dr}{1-r^2} = \ln\frac{1+R}{1-R}.$$

Next comes the fun with a bit of "hyperbolic trigonometry." Recall that

$$\cosh t = \tfrac{1}{2}\left(e^t + e^{-t}\right), \ \sinh t = \tfrac{1}{2}\left(e^t - e^{-t}\right), \ \text{and} \ \tanh t = \frac{\sinh t}{\cosh t}.$$

If we denote by $t = \ln\frac{1+R}{1-R}$ the hyperbolic distance between the two points, then we can solve for the Euclidean distance $R$:

$$R = \frac{e^t - 1}{e^t + 1} = \frac{e^{t/2} - e^{-t/2}}{e^{t/2} + e^{-t/2}} = \tanh(\tfrac{t}{2}).$$

**Proposition 5.10.** *Let $\triangle ABC$ be a right triangle with $\angle C$ a right angle. Then*

(a) $\sin A = \dfrac{\sinh a}{\sinh c}$,

(b) $\cos A = \dfrac{\tanh b}{\tanh c}$,

(c) *(Pythagorean Theorem)* $\cosh c = \cosh a \cosh b$, *and*

(d) $\cosh a = \dfrac{\cos A}{\sin B}$.

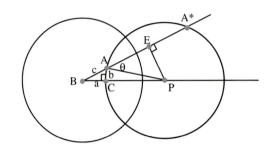

FIGURE 11

***Proof.*** As pictured in Figure 11, we may apply an isometry to move $B$ to the origin and $C$ to the positive real axis. Then $\overline{AC}$ will be an arc of a circle centered at $P$ and intersecting the unit circle at right angles. By the formula above, the Euclidean coordinates of $C$ and $A$ are, respectively, $\tanh \frac{a}{2}$ and $e^{iB} \tanh \frac{c}{2}$.

The first order of business is to compute the coordinates of $P$. Since the two circles intersect at right angles, we may apply the (Euclidean) Pythagorean Theorem (see Figure 12). Letting $k$ denote the (Euclidean) radius of the circle with center $P$, we have $(\tanh \frac{a}{2} + k)^2 = k^2 + 1$, whence $k = \frac{1}{2}(\coth \frac{a}{2} - \tanh \frac{a}{2}) = \frac{1}{\sinh a}$ and $P = \tanh \frac{a}{2} + k = \frac{1}{2}(\tanh \frac{a}{2} + \coth \frac{a}{2}) = \coth a$ (see Exercise 4).

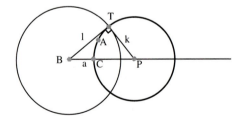

FIGURE 12

Since $\angle A$ and $\theta = \angle EAP$ are complementary, to compute $\sin A$ it suffices to compute $\cos \theta = AE/k$. By Exercise 15, $(BA)(BA^*) = 1$; from $BA = \tanh \frac{c}{2}$, we obtain $BA^* = \coth \frac{c}{2}$ and $AE = \frac{1}{2}AA^* = \frac{1}{2}(\coth \frac{c}{2} - \tanh \frac{c}{2}) = 1/\sinh c$. Thus, $\cos \theta = \sinh a / \sinh c$, as required to establish (a).

Next we find $\cos B = \dfrac{BE}{BP} = \dfrac{\tanh \frac{c}{2} + \frac{1}{\sinh c}}{\coth a}$. Since

$$\tanh \frac{c}{2} + \frac{1}{\sinh c} = \frac{2 \sinh^2 \frac{c}{2} + 1}{\sinh c} = \frac{\cosh c}{\sinh c} = \coth c,$$

we have $\cos B = \tanh a / \tanh c$. Now permuting letters gives (b).

The Pythagorean Theorem is, appropriately, a consequence of the old chestnut $\cos^2 A + \sin^2 A = 1$: Since

$$\frac{\tanh^2 b}{\tanh^2 c} + \frac{\sinh^2 a}{\sinh^2 c} = 1,$$

we have $\cosh^2 c \tanh^2 b + \sinh^2 a = \sinh^2 c = \cosh^2 c - 1$, whence $(\cosh^2 c)(\tanh^2 b - 1) = -(1 + \sinh^2 a) = -\cosh^2 a$, implying that $\cosh^2 c = \cosh^2 a \cosh^2 b$, and that will do it.

Lastly, to prove (d), we use all three previous parts:

$$\frac{\cos A}{\sin B} = \frac{\tanh b}{\tanh c} \cdot \frac{\sinh c}{\sinh b} = \frac{\cosh c}{\cosh b} = \frac{\cosh a \cosh b}{\cosh b} = \cosh a. \quad \square$$

From this we can derive the hyperbolic law of cosines and its "dual," analogous to Propositions 5.3 and 5.5 in elliptic geometry. Indeed, the resemblance to the elliptic results is quite striking. (There is a deep explanation coming from the theory of Lie groups and symmetric spaces. See the comments at the end of this section.)

**Proposition 5.11** (Hyperbolic Laws of Cosines). *Let $\triangle ABC$ be a triangle in the hyperbolic plane. Then*

(a) $\cos A = \dfrac{\cosh b \cosh c - \cosh a}{\sinh b \sinh c}$, *and*

(b) $\cosh a = \dfrac{\cos A + \cos B \cos C}{\sin B \sin C}$.

FIGURE 13

***Proof.*** We drop a perpendicular $\overline{CD}$ from $C$ to $\overleftrightarrow{AB}$ (see Exercise 17) and work with the various formulas for right triangles from Proposition 5.10. Let $AD = x$ and $CD = y$. Then

$$\cosh b = \cosh x \cosh y, \quad \text{and}$$

$$\begin{aligned}
\cosh a &= \cosh(c - x)\cosh y = (\cosh c \cosh x - \sinh c \sinh x)\cosh y \\
&= \cosh b \cosh c - \sinh c \sinh x \cosh y \\
&= \cosh b \cosh c - \sinh c (\cos A \tanh b \cosh x)\cosh y \\
&= \cosh b \cosh c - \cos A \sinh c \tanh b \cosh b \\
&= \cosh b \cosh c - \cos A \sinh b \sinh c \,.
\end{aligned}$$

To prove (b), we drop the perpendicular from $A$ to $\overleftrightarrow{BC}$. Let $D$ be the point of intersection, and let $\angle CAD = \gamma$, $\angle BAD = \beta$, $BD = x$, and $AD = y$, as indicated in Figure 14. We have

FIGURE 14

$$\cosh a = \cosh\left(x + (a - x)\right) = \cosh x \cosh(a - x) + \sinh x \sinh(a - x)$$

using Proposition 5.10(d) and Exercise 18,

$$= \frac{\cos \beta}{\sin B}\frac{\cos \gamma}{\sin C} + \frac{\sin \beta \sinh y}{\sin B}\frac{\sin \gamma \sinh y}{\sin C}$$

since $\sinh^2 y = \cosh^2 y - 1$,

$$= \frac{\cos \beta \cos \gamma - \sin \beta \sin \gamma}{\sin B \sin C} + \frac{\sin \beta \sin \gamma}{\sin B \sin C}\cosh^2 y$$

$$= \frac{\cos A}{\sin B \sin C} + \frac{1}{\sin B \sin C}(\cosh y \sin \beta)(\cosh y \sin \gamma)$$

now, lastly, using Proposition 5.10(d) again,

$$= \frac{\cos A}{\sin B \sin C} + \frac{1}{\sin B \sin C}(\cos B \cos C)$$

$$= \frac{\cos A + \cos B \cos C}{\sin B \sin C}\,,$$

as required. Whew!  □

As a corollary of Proposition 5.11(b), we infer that, as in elliptic geometry, there is an angle-angle-angle congruence theorem: the angles of a hyperbolic triangle uniquely determine the triangle. For our last major result, we compute the area of a hyperbolic triangle. Of course, we must first define area in the hyperbolic plane. Let $r$ denote (Euclidean) distance from the origin. Then a small displacement (Euclidean $dx$) in the $x$ direction has hyperbolic length $\frac{2\,dx}{1-r^2}$, and a small displacement (Euclidean $dy$) in the $y$ direction has hyperbolic length $\frac{2\,dy}{1-r^2}$. So, the element of area $dA$ should be given by

$$dA = \left(\frac{2\,dx}{1-r^2}\right)\left(\frac{2\,dy}{1-r^2}\right) = \frac{4r\,dr\,d\theta}{(1-r^2)^2}.$$

Then the area of a region $R$ in the hyperbolic plane is given simply by

$$\text{area}\,(R) = \iint_R dA.$$

**Theorem 5.12.** *Let* $\triangle ABC \subset \mathbb{D}$. *Then*

$$\text{area}\,\triangle ABC = \pi - (A + B + C).$$

**Proof.** First, it suffices to prove the theorem in the case when $\angle C = \pi/2$ (see Exercise 27). So, we return to our earlier setup with $B$ at the origin of the disk. We denote angles $A$ and $B$ by $\alpha$ and $\beta$, respectively. The idea of the proof is this (see Figure 15): fixing vertices $B$ and $C$, we vary $A$ on the circle centered at $P$ and consider the area $\mathcal{A}$ of $\triangle ABC$ as a function of the angle $\beta$. Now, if $\angle BPA = \psi$, then $\beta + \psi = \frac{\pi}{2} - \alpha$ and $k \sin\alpha = \frac{1}{2}\left(\frac{1}{r} - r\right)$ (see the proof of Proposition 5.10(a)).

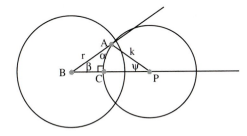

FIGURE 15

The area $\mathcal{A}(\beta)$ of $\triangle ABC$ is given by

$$\mathcal{A}(\beta) = \int_0^\beta \int_0^{r(\theta)} \frac{4r}{(1-r^2)^2}\,dr\,d\theta;$$

so, by the Fundamental Theorem of Calculus,

$$\frac{d\mathcal{A}}{d\beta} = \int_0^{r(\beta)} \frac{4r}{(1-r^2)^2} \, dr = \frac{2r^2}{1-r^2},$$

where we write $r$ for $r(\beta)$. On the other hand, writing the vector equation $\vec{BA} - \vec{BP} = -\vec{AP}$ in terms of complex numbers, we have

$$re^{i\beta} + \text{constant} = -ke^{-i\psi}.$$

Differentiating with respect to $\beta$, we find

$$e^{i\beta} \frac{dr}{d\beta} + rie^{i\beta} = kie^{-i\psi} \frac{d\psi}{d\beta}$$

(multiplying by $-ie^{-i\beta}$)

$$-i\frac{dr}{d\beta} + r = ke^{-i(\psi+\beta)} \frac{d\psi}{d\beta}$$

(taking real parts)

$$r = k\cos(\beta + \psi) \frac{d\psi}{d\beta}.$$

And since $\cos(\beta + \psi) = \sin\alpha$, we have

$$\frac{d\psi}{d\beta} = \frac{r}{k\sin\alpha} = \frac{2r}{\frac{1}{r} - r} = \frac{2r^2}{1-r^2}.$$

The upshot is that $\frac{d\mathcal{A}}{d\beta} = \frac{d\psi}{d\beta}$, and so $\mathcal{A}$ and $\psi$, viewed as functions of $\beta$, differ by a constant. Well, when $\beta = 0$, both $\mathcal{A}$ and $\psi$ are zero, and so we have

$$\mathcal{A} = \psi = \frac{\pi}{2} - (\alpha + \beta) = \pi - \left(\alpha + \beta + \frac{\pi}{2}\right),$$

which completes the proof for a right triangle.  □

**Concluding Remarks.** We have touched upon a lot of beautiful and deep mathematics in this whirlwind tour of two-dimensional geometries. We have discovered that for $\triangle ABC$,

$$A + B + C - \pi \begin{cases} > 0 & \text{in elliptic geometry} \\ = 0 & \text{in Euclidean geometry} \\ < 0 & \text{in hyperbolic geometry} \end{cases}.$$

This is a manifestation of a geometric notion called *curvature*: $\mathbb{P}^2$ is *positively curved*, $\mathbb{R}^2$ is *flat*, and $\mathbb{D}$ is *negatively curved*. What's more, Theorems 5.6 and 5.12 are extremely simple cases of one of the most far-reaching theorems in mathematics, the *Gauss–Bonnet Theorem*, which relates the curvature of a region to the geometry of its boundary curve.

All three spaces—the elliptic plane, the Euclidean plane, and the hyperbolic plane—have isometry groups that depend on three real parameters. The treatment we have chosen here of hyperbolic geometry emphasized the interplay with complex numbers and projective geometry, but there is another approach that "parallels" elliptic geometry. The orthogonal group $O(3)$ consists of all linear maps preserving the standard dot product $\mathbf{x} \cdot \mathbf{y} = x_0 y_0 + x_1 y_1 + x_2 y_2$ in $\mathbb{R}^3$. We can also define the Lorentzian dot product $\mathbf{x} \cdot_L \mathbf{y} = -x_0 y_0 + x_1 y_1 + x_2 y_2$ (important in relativity theory) and study the Lorentz group $O(2,1)$ of all linear maps preserving it. The Lie groups $O(3)$ and $O(2,1)$ are intimately related through their associated Lie algebras. It turns out that $O(2,1)$ has two connected components, but the component of the identity element (denoted $O(2,1)°$) is a group that is isomorphic to $SL(2,\mathbb{R})$. (Why should the minus sign in the dot product turn sin and cos into sinh and cosh? Roughly for the same reason that solutions to the differential equation $y'' + y = 0$ are linear combinations of sin and cos, whereas those of $y'' - y = 0$ are linear combinations of sinh and cosh.)

We want to offer one last comment relating to group actions and coset spaces. The group $SO(3)$ acts transitively on the sphere $\mathbb{S}$; the stabilizer of the north pole is the subgroup of rotations about the $z$-axis. Our fundamental link between orbits and stabilizer subgroups (Proposition 1.3 of Chapter 7) says that there is a one-to-one correspondence between $\mathbb{S}$ and the coset space $SO(3)/SO(2)$. This is the primordial example of a symmetric space. We can study the (differential) geometry of the coset space $\mathbb{S}$ by working with the algebra of the group $SO(3)$. Translating to the Lorentz setting, $\mathbb{H} = SO(2,1)°/SO(2)$ is likewise a symmetric space, called the (noncompact) dual of $\mathbb{S}$. Changing the one sign in the dot product leads to a host of comparisons:

$$\text{Euclidean dot product} \longleftrightarrow \text{Lorentz dot product}$$
$$SO(3) \longleftrightarrow SO(2,1)°$$
$$\mathbb{S} \longleftrightarrow \mathbb{D}$$

positive curvature, compact $\longleftrightarrow$ negative curvature, noncompact

and so on.

We hope that the reader is sufficiently intrigued to go off and study more about algebra and differential geometry, and, of course, one of the links between them, Lie groups, Lie algebras and homogeneous spaces; or, perhaps, sufficiently enraptured with Sections 2 and 4 to go off and learn some algebraic geometry, where commutative ring theory will finally come into its own.

### EXERCISES 8.5

1.  a.  Prove that $\lambda \operatorname{Id} \in O(3) \iff \lambda = \pm 1$.
    b.  Prove that $O(3)/\{\pm \operatorname{Id}\} \cong SO(3)$. (See Exercise 7.4.8a.)

2.  Let $\mathbf{x}$, $\mathbf{y}$, $\mathbf{z}$, and $\mathbf{w}$ be vectors in $\mathbb{R}^3$. Prove that

$$(\mathbf{x} \times \mathbf{w}) \cdot (\mathbf{z} \times \mathbf{y}) = \begin{vmatrix} \mathbf{x} \cdot \mathbf{z} & \mathbf{x} \cdot \mathbf{y} \\ \mathbf{w} \cdot \mathbf{z} & \mathbf{w} \cdot \mathbf{y} \end{vmatrix} = (\mathbf{x} \cdot \mathbf{z})(\mathbf{w} \cdot \mathbf{y}) - (\mathbf{x} \cdot \mathbf{y})(\mathbf{w} \cdot \mathbf{z}).$$

   (Hints: If all else fails, bludgeon it using coordinates. Here's a more elegant solution: both sides are linear separately in each of the four variables ("multilinear"), so it suffices to check the formula as all four vectors vary among an orthonormal basis; but you can use symmetry arguments of various sorts to turn eighty-one verifications into three.)

3.  Let $\mathbf{x}, \mathbf{y}, \mathbf{z} \in \mathbb{R}^3$.
    a.  Prove that $(\mathbf{x} \times \mathbf{y}) \times \mathbf{z} = (\mathbf{x} \cdot \mathbf{z})\mathbf{y} - (\mathbf{y} \cdot \mathbf{z})\mathbf{x}$.
    b.  Prove that $\mathbf{x} \cdot (\mathbf{y} \times \mathbf{z}) = \mathbf{z} \cdot (\mathbf{x} \times \mathbf{y}) = \mathbf{y} \cdot (\mathbf{z} \times \mathbf{x})$, and interpret these equalities geometrically.
    c.  Prove that $(\mathbf{x} \times \mathbf{y}) \times (\mathbf{x} \times \mathbf{z}) = (\mathbf{y} \cdot (\mathbf{z} \times \mathbf{x}))\,\mathbf{x}$.
    d.  Prove the law of sines for an elliptic triangle:

$$\frac{\sin A}{\sin a} = \frac{\sin B}{\sin b} = \frac{\sin C}{\sin c}.$$

4.  Check the following formulas from "hyperbolic trigonometry":
    a.  $\cosh^2 x - \sinh^2 x = 1$
    b.  $\cosh(x + y) = \cosh x \cosh y + \sinh x \sinh y$
    c.  $\sinh(x + y) = \sinh x \cosh y + \cosh x \sinh y$
    d.  $\coth 2x = \frac{1}{2}(\tanh x + \coth x)$

5.  Prove that if $T \in \operatorname{Proj}_{\mathbb{C}}(1)$ maps $\mathbb{H}$ to $\mathbb{H}$, then $T \in \operatorname{Proj}(1)$, and a linear map representing $T$ may be chosen in $SL(2, \mathbb{R})$. Conclude,

finally, that

$$\{T \in \text{Proj}_{\mathbb{C}}(1) : T(\mathbb{H}) = \mathbb{H}\} \cong SL(2, \mathbb{R})/\{\pm\text{Id}\}.$$

6. Complete the proof of Lemma 5.3.

7. Let $T(z) = \frac{i-z}{i+z}$. Show that $\text{Im } z > 0 \implies |T(z)| < 1$. Check, moreover, that $T$ maps $\mathbb{H}$ onto $\mathbb{D}$. (Hint: $|z \pm i|^2 = |z|^2 \pm 2 \text{Im } z + 1$.)

8. Show that every isometry of $\mathbb{H}$ can be written as a composition of translations $z \leadsto z + b$ ($b \in \mathbb{R}$), dilatations $z \leadsto cz$ ($c > 0$), and the "inversion" $z \leadsto -1/z$. Also, check explicitly that each of these in fact belongs to $SL(2, \mathbb{R})/\{\pm\text{Id}\}$.

9. Let $A, B \in \mathbb{H}$. Prove that there is a unique line in $\mathbb{H}$ containing $A$ and $B$. (Remark: The easy proof is to reduce to the case that they lie on a vertical line. However, it's a good exercise to see that in general there's a unique semicircle centered on the real axis containing them.)

10. Consider the points $A = Re^{i\alpha}$ and $B = Re^{i\beta}$ in $\mathbb{H}$. Show that $d_{\mathbb{H}}(A, B) = \left| \ln \frac{(1+\cos\alpha)\sin\beta}{(1+\cos\beta)\sin\alpha} \right|$.

11. Let $G = SL(2, \mathbb{R})/\{\pm\text{Id}\} \subset \text{Proj}_{\mathbb{C}}(1)$ be the subgroup of isometries of $\mathbb{H}$. Let $T \in \text{Proj}_{\mathbb{C}}(1)$ be the map taking $\mathbb{H}$ to $\mathbb{D}$, namely, $T(z) = \frac{i-z}{i+z}$. Check explicitly that the group of isometries of $\mathbb{D}$, the conjugate subgroup $TGT^{-1}$, consists of all linear fractional transformations of the form $z \leadsto \zeta \dfrac{z-a}{1-\bar{a}z}$, where $|\zeta| = 1$ and $|a| < 1$.

12. Determine the type of each of the following isometries of $\mathbb{D}$:
    a. $z \leadsto \frac{(4i-1)z+2(1-i)}{2(i-1)z+(4-i)}$
    b. $z \leadsto \frac{3z-1}{3-z}$
    c. $z \leadsto \frac{(1-2i)z-1}{z-(1+2i)}$

13. Check that the pictures in Figure 10 on p. 354 are in fact correct. In particular, in the parabolic case, check that horizontal lines $\text{Im } z = c$ in $\mathbb{H}$ map under $T(z) = \frac{i-z}{i+z}$ to the horocycles in $\mathbb{D}$, as pictured. In the hyperbolic case, check that the rays emanating from the origin in $\mathbb{H}$ map to arcs of circles from $1$ to $-1$ in $\mathbb{D}$.

14.  Suppose $A, B \in \mathbb{D}$. Let $P$ and $Q$ be the points of intersection of $\overleftrightarrow{AB}$ with the boundary circle. Show that $d_{\mathbb{D}}(A, B) = \ln |P, Q, A, B|$.

15.  Let the distance from a point $P$ outside a circle to the point of tangency be $R$, and let $r$ and $r^*$ be the distances to any pair of points collinear with $P$ on the circle (see Figure 16). Prove that $R^2 = rr^*$. (Hint: Use similar triangles to prove that $\frac{r}{R} = \frac{R}{r^*}$.)

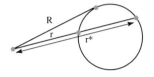

FIGURE 16

16.  Let $P \in \mathbb{D}$ and let $r > 0$. Find the locus of points $\{Q : d_{\mathbb{D}}(P, Q) = r\}$.

17.  Prove that given a point $P \in \mathbb{D}$ and a line $\ell$ not passing through $P$, there is a unique line $m$ passing through $P$ and intersecting $\ell$ orthogonally. In other words, we can drop a (unique) perpendicular from $P$ to $\ell$. (Hint: By using appropriate isometries, arrange for $\ell$ to be the real axis and for $P$ to be on the imaginary axis.)

18.  Prove the law of sines for a hyperbolic triangle:
$$\frac{\sin A}{\sinh a} = \frac{\sin B}{\sinh b} = \frac{\sin C}{\sinh c}.$$

19.  Give a pictorial proof that the sum of the measures of the angles of a hyperbolic triangle is less than $\pi$. Use this to prove the angle-angle-angle congruence theorem: Given two triangles $\triangle ABC$ and $\triangle DEF \subset \mathbb{D}$, if $\angle A \cong \angle D$, $\angle B \cong \angle E$, and $\angle C \cong \angle F$, then $\triangle ABC \cong \triangle DEF$.

20.  Suppose two lines in the hyperbolic plane have a common perpendicular. What can you prove about the two lines?

21.  Consider the interesting isometry $f$ of $\mathbb{P}^2$ given by the matrix
$$A = \begin{bmatrix} -1 & & \\ & -1 & \\ & & 1 \end{bmatrix}.$$
    a.  Check that there are a line $\ell$ and a point $P \notin \ell$ that are fixed by $f$.

b.   Given a point $Q \in \mathbb{P}^2$, let $R$ be the intersection of $\overleftrightarrow{PQ}$ and $\ell$. Prove that the pairs of points $P, R$ and $Q, f(Q)$ are harmonic (see Exercise 8.2.8).

c.   Find the points $Q$ of $\mathbb{P}^2$ that are moved the greatest distance by $f$ (i.e., so that $d_{\mathbb{P}^2}(Q, f(Q))$ is maximized).

22.   Prove that the entries in the following table are correct. Let $C_R$ denote a circle of radius $R$ in each geometry.

| Geometry | Circumference($C_R$) | Area($C_R$) |
| --- | --- | --- |
| Euclidean | $2\pi R$ | $\pi R^2$ |
| Elliptic | $2\pi \sin R$ | $4\pi \sin^2 \frac{R}{2}$ |
| Hyperbolic | $2\pi \sinh R$ | $4\pi \sinh^2 \frac{R}{2}$ |

23.   It might seem that our definition of a triangle in the elliptic plane was overly involved. Why not just take three noncollinear points and *any* three line segments connecting them in pairs? For example, consider Figure 17.

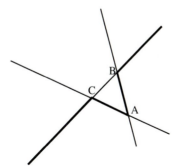

FIGURE 17

24.   Let $T \in SL(2, \mathbb{R})/\{\pm \text{Id}\}$. Prove that $T$ is
a.   parabolic if and only if $|\text{tr}\, T| = 2$,
b.   hyperbolic if and only if $|\text{tr}\, T| > 2$, and
c.   elliptic if and only if $|\text{tr}\, T| < 2$.
(Recall that if $T \colon \mathbb{R}^n \to \mathbb{R}^n$ is linear map, $\text{tr}\, T$ is the sum of the diagonal entries of any matrix representing $T$. Note, moreover, that since $\text{tr}\,(-T) = -\text{tr}\, T$, $|\text{tr}\, T|$ makes sense for an element of the quotient group $SL(2, \mathbb{R})/\{\pm \text{Id}\}$.)

25. Using Figure 18, give the algebraic formulas for the maps $\mathbb{S} \leftrightarrows \mathbb{C} \cup \{\infty\}$. (For those of you who know some topology, we are establishing here a homeomorphism between $\mathbb{S}$ and the one-point compactification of $\mathbb{C}$ (see J. Munkres, *Topology, A First Course*, Prentice Hall, 1975, pp. 183 *ff.*).)

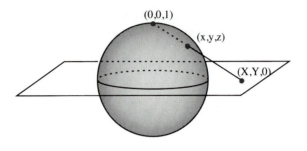

(0,0,1)

(x,y,z)

(X,Y,0)

FIGURE 18

26. Use the hyperbolic law of cosines to prove the triangle inequality in $\mathbb{D}$: for any $A, B, C \in \mathbb{D}$, $d_{\mathbb{D}}(B, C) \leq d_{\mathbb{D}}(B, A) + d_{\mathbb{D}}(A, C)$.

27. a. Prove Theorem 5.12 for general triangles.
    b. Prove that the area of a hyperbolic $n$-gon is $(n - 2)\pi - \sum$ interior angles.

28. Find the area of an asymptotic triangle, a doubly asymptotic triangle, and a trebly asymptotic triangle (one of which is pictured in Figure 19).

FIGURE 19

29. Prove that a horocyclic sector (see Figure 20) has finite area.

30. Give another proof of the hyperbolic Pythagorean Theorem (cf. Proposition 5.10(c)) as follows. We work in the hyperbolic plane. Consider a right triangle formed by a vertical half-line and two half-circles, as pictured in Figure 21.

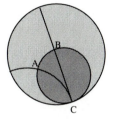

FIGURE 20

a.  If the *Euclidean* radii of the circles are $r$ and $R$, show that
    (see Exercise 10 where appropriate):

$$a = \ln \frac{R \sin \beta}{r}$$

$$b = \ln \frac{1 + \cos \gamma}{\sin \gamma}$$

$$c = \ln \frac{(1 + \cos \alpha) \sin \beta}{(1 + \cos \beta) \sin \alpha}.$$

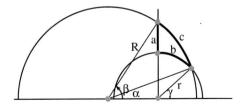

FIGURE 21

b.  Show that $(\frac{r}{R})^2 = 1 + \cos^2 \beta - 2 \cos \alpha \cos \beta$, and deduce that
    $\frac{1}{2} \cosh a \cosh b = \frac{2}{\sin \alpha \sin \beta} (1 - \cos \alpha \cos \beta)$.
c.  Show that $\frac{1}{2} \cosh c = \frac{2}{\sin \alpha \sin \beta} (1 - \cos \alpha \cos \beta)$ as well.

31. a.  Show that $V = \{3 \times 3 \text{ symmetric matrices}\}$ is a 6-dimensional
       vector space.
    b.  Show that the matrices $R$ with all positive eigenvalues form
       an open subset of $V$. (Hint: Let the characteristic polyno-
       mial $p_R(t) = -t^3 + s_1 t^2 - s_2 t + s_3$. Show that all the eigen-
       values are positive if and only if $s_1, s_2, s_3 > 0$.)

32. In this problem we explore the manner in which $SO(3)$, the group of rotations of $\mathcal{S}$, is embedded in $\text{Proj}_\mathbb{C}(1)$. Let

$$G = \left\{ \begin{bmatrix} a & b \\ -\bar{b} & \bar{a} \end{bmatrix} : a, b \in \mathbb{C}, \ |a|^2 + |b|^2 = 1 \right\} \subset SL(2, \mathbb{C}).$$

The group $G$ is often denoted $SU(2)$ (the special unitary group in two dimensions). Let

$$V = \left\{ \begin{bmatrix} ix & z \\ -\bar{z} & -ix \end{bmatrix} : x \in \mathbb{R}, z \in \mathbb{C} \right\}.$$

The real vector space $V$ is three-dimensional (why?), often written $\mathfrak{su}(2)$ (and called the Lie algebra associated to the Lie group $SU(2)$).

a. If $A$ is a matrix with complex entries, denote by $A^*$ the matrix obtained by taking the complex conjugates of the entries of its transpose (i.e., $A^* = (\bar{A})^T$). Show that $A \in G$ if and only if $A^*A = \text{Id}$ and $\det A = 1$.

b. Show that $G$ acts on $V$ by conjugation. If $A \in G$, denote by $\rho(A)$ the associated linear map from $V$ to $V$. Show that the kernel of the map $G \to \text{Perm}(V)$ is $\{\pm\text{Id}\}$.

c. Define an isomorphism of $V$ with $\mathbb{R}^3$ by mapping

$$\mathbf{v} = \begin{bmatrix} ix & z \\ -\bar{z} & -ix \end{bmatrix} \rightsquigarrow X = (x, \text{Re } z, \text{Im } z).$$

Show that the usual dot product $X \cdot Y$ on $\mathbb{R}^3$ corresponds to $\mathbf{v} \cdot \mathbf{w} = \frac{1}{2}\text{tr}(\mathbf{v}^*\mathbf{w})$. Now show that the action of $G$ on $V$ by conjugation is an isometry.

d. Given $A \in G$, write $A = \alpha\,\text{Id} + \beta\,U$, where $\alpha, \beta \in \mathbb{R}$ and $U \in V$ is a unit vector. Show that conjugation by $A$ fixes the line spanned by $U$, so that $\rho(A)$ is a rotation about the axis determined by $U$.

e. Let $A = \begin{bmatrix} e^{i\theta} & \\ & e^{-i\theta} \end{bmatrix}$. Show that $\rho(A)$ is a rotation of $\mathbb{R}^3$ about the first axis through angle $2\theta$. Now show, in general, that if $A = \cos\theta\,\text{Id} + \sin\theta\,U$ (as in part d.), then $\rho(A)$ is a rotation about axis $U$ through angle $2\theta$.

f. Show that $SO(3) \cong G/\{\pm\text{Id}\} \subset \text{Proj}_\mathbb{C}(1)$.

# APPENDIX A

# A Logic Review, Sets and Functions, and Equivalence Relations

### 1. A Few Remarks on Logic

Mathematical reasoning and communication take place by means of sentences of the form

*If* ⟨hypothesis⟩, *then* ⟨conclusion⟩.

Such a sentence is called an *implication* and is often written with the symbol ⟹. To mathematicians, the sentence $P \Rightarrow Q$ (read "$P$ implies $Q$" or "if $P$, then $Q$") means this: If $P$ is true, then $Q$ is also true; but if $P$ is false, $Q$ may be either true or false. This is summarized in the following "truth table":

| $P$ | $Q$ | $P \Rightarrow Q$ |
|:---:|:---:|:---:|
| T | T | T |
| T | F | F |
| F | T | T |
| F | F | T |

Consider the following statements:

(1)  If $x > 0$, then $x^2 \neq 0$.
(2)  If $x \in \mathbb{R}$ and $x^2 < 0$, then $x = \pi$.
(3)  If it is sunny today, then it is not raining.

The first two statements (i.e., implications) are valid; the second, "by default" (since the hypothesis fails for every real number $x$). The last statement seems reasonable ... or does it?

The *converse* of the implication $P \Rightarrow Q$ is the implication $Q \Rightarrow P$, and the two are logically quite distinct:

| $P$ | $Q$ | $Q \Rightarrow P$ |
|:---:|:---:|:---:|
| T | T | T |
| T | F | T |
| F | T | F |
| F | F | T |

(Conv 1)   If $x^2 \neq 0$, then $x > 0$.
(Conv 2)   If $x = \pi$, then $x^2 < 0$.
(Conv 3)   If it is not raining today, then it is sunny.

None of these, of course, is true. (In the last case, it could certainly be overcast without raining.)

Here, however, is a valid statement whose converse is also valid:

(4)   If $x > 0$, then $x^3 > 0$.
(Conv 4)   If $x^3 > 0$, then $x > 0$.

If $P \Rightarrow Q$ and $Q \Rightarrow P$ are both valid, then we say $P$ is *equivalent* to $Q$, or "$P$ if and only if $Q$," written $P \Leftrightarrow Q$. Sometimes, people will say that $P$ is a "necessary and sufficient" condition for $Q$. If we take this apart, $P$ is a *sufficient* condition for $Q$ if $P \Rightarrow Q$ (in other words, in order for $Q$ to hold, it is sufficient for $P$ to hold). Also, $P$ is a *necessary* condition for $Q$ if $Q \Rightarrow P$ (in other words, in order for $Q$ to hold, it is necessary that $P$ hold).

The *contrapositive* of the implication $P \Rightarrow Q$ is $(\text{not } Q) \Rightarrow (\text{not } P)$. Here are the contrapositives of the four example statements:

(Contra 1)   If $x^2 = 0$, then $x \leq 0$.
(Contra 2)   If $x \neq \pi$, then $x^2 \geq 0$.
(Contra 3)   If it is raining today, then it is not sunny.
(Contra 4)   If $x^3 \leq 0$, then $x \leq 0$.

Note that these are all valid, except perhaps for (Contra 3), which may fail in the case that we have a sun shower; but, in this event, we realize that (3) itself has failed, too. In fact, a table of truth values of the original implication and of its contrapositive is quite revealing:

| $P$ | $Q$ | $P \Rightarrow Q$ | not $P$ | not $Q$ | (not $Q$) $\Rightarrow$ (not $P$) |
|---|---|---|---|---|---|
| T | T | **T** | F | F | **T** |
| T | F | **F** | F | T | **F** |
| F | T | **T** | T | F | **T** |
| F | F | **T** | T | T | **T** |

Thus, *an implication and its contrapositive are logically equivalent.* Here is an example of proof by contrapositive. Suppose we know that the sum of two rational numbers is also rational. We are asked now to prove that the sum of a rational number and an irrational number is irrational. Let's state this more precisely in the following form: Given $a \in \mathbb{Q}$, prove that $b \notin \mathbb{Q} \implies a + b \notin \mathbb{Q}$. The contrapositive of the latter implication is $a + b \in \mathbb{Q} \implies b \in \mathbb{Q}$. This is easy: $b = (a + b) + (-a)$; and since the sum of rational numbers is rational, $a + b \in \mathbb{Q}$ and $a \in \mathbb{Q}$ imply $b \in \mathbb{Q}$.

We also want to mention *proof by contradiction*, which is subtly different from proof by contrapositive. To prove $P \Rightarrow Q$, it is logically equivalent to prove that

$$P \text{ and (not } Q) \Rightarrow \langle \text{something known to be false} \rangle .$$

For example, we might prove that $P$ and (not $Q$) $\Rightarrow$ (not $P$), or that $P$ and (not $Q$) $\Rightarrow \langle 0 = 1 \rangle$. A typical example is the proof in the text on p. 50 that $\sqrt{2}$ is irrational. As a hint that we should use proof by contradiction here, we have no specific hypotheses with which to begin our proof. (Be warned, however, that students often introduce unnecessary contradictions at times when a direct proof or proof by contrapositive is clearer. Typically, such a "proof by phony contradiction" has the following structure: to prove $P \Rightarrow Q$, we assume $P$ and (not $Q$); we argue now that $P \Rightarrow Q$, contradicting the assumption (not $Q$). But we *never* use the assumption (not $Q$) in the argument.)

Just to check that proof by contradiction works, we construct a truth table, comparing the validity of $P \Rightarrow Q$ with that of $(P$ and (not $Q$)) $\Rightarrow R$, where $R$ is known to be false.

| $P$ | $Q$ | $P \Rightarrow Q$ | not $Q$ | $P$ and (not $Q$) | $R$ | $P$ and (not $Q$) $\Rightarrow R$ |
|---|---|---|---|---|---|---|
| T | T | **T** | F | F | F | **T** |
| T | F | **F** | T | T | F | **F** |
| F | T | **T** | F | F | F | **T** |
| F | F | **T** | T | F | F | **T** |

The last comments we wish to offer are on the topic of *negation*. There are two issues here: negating an implication, and negation when *quantifiers* ("there exists" and "for all") are involved. First, what does it mean for $P \Rightarrow Q$ to fail? If we translate the original sentence as "Whenever $P$ occurs, so does $Q$," we see that its failure requires that $P$ occur in some instance without $Q$. In mathematical language, the latter is $P$ and (not $Q$). As a check, we make the customary table:

| $P$ | $Q$ | $P \Rightarrow Q$ | not $(P \Rightarrow Q)$ | not $Q$ | $P$ and (not $Q$) |
|-----|-----|-------------------|-------------------------|---------|-------------------|
| T   | T   | T                 | F                       | F       | F                 |
| T   | F   | F                 | T                       | T       | T                 |
| F   | T   | T                 | F                       | F       | F                 |
| F   | F   | T                 | F                       | T       | F                 |

The last issue is negation involving quantifiers. What does it mean for a statement of the form

*For every x,* ⟨something happens⟩

to be false? It means that for *at least one x*, the something does *not* happen. For example, to prove that the statement

For every real number $x$, we have $x^2 + x > 0$

is false, all we have to do is provide a single number $x$, e.g., $x = -1$, for which the inequality $x^2 + x > 0$ fails. Now, what does it mean for a statement of the form

*There exists x such that* ⟨something happens⟩

to be false? It means that no matter what $x$ we try, the something does *not* happen, i.e., *for all x, not* ⟨something happens⟩. For example, to prove that the statement

There exists $x \in \mathbb{R}$ such that $x^2 = -1$

is false, we need only prove that *for all $x \in \mathbb{R}$, $x^2 \neq -1$*. We might establish this by proving that for all $x \in \mathbb{R}$, $x^2 \geq 0$. For emphasis, we close with the final reminder: When your charge is to *disprove* a theorem, you need only provide a counterexample, i.e., an example of the failure of the theorem.

## EXERCISES

1. Negate the following sentences; in each case, indicate whether the original sentence or its negation is a true statement. Be sure to move the "not" through all the quantifiers.
   a. For every integer $n \geq 2$, the number $2^n - 1$ is prime.
   b. There exists a real number $M$ so that for all real numbers $t$, $|\sin t| \leq M$.
   c. For every real number $x > 0$, there exists a real number $y > 0$ so that $xy > 1$.

2. Negate the following sentence: Every student on your college campus knows another student on campus who is at least one year older (than herself or himself).

3. An integer $n$ is called *wonderful* provided that whenever $n$ divides $ab$, then $n$ divides $a$ or $n$ divides $b$. (Here, of course, $a$ and $b$ are integers as well.)
   a. Complete this sentence: $n$ is *not wonderful* if ... .
   b. Decide whether the integer 6 is wonderful. Explain carefully.

4. Suppose $n$ is an odd integer. Prove:
   a. The equation $x^2 + x - n = 0$ has no solution $x \in \mathbb{Z}$.
   b. Prove that for any $m \in \mathbb{Z}$, the equation $x^2 + 2mx + 2n = 0$ has no solution $x \in \mathbb{Z}$.

5. On a tribal island every native either always tells the truth or always lies. A visitor meets three natives, who volunteer the following information:
   >   **A:** "All three of us are liars."
   >   **B:** "Not so; only two of us are liars."
   >   **C:** "Only **A** and **B** are lying."
   Whom can you trust to give you directions to the beach party?

6. In the comic strip *Foxtrot*, July 30, 1994, we find the following exchange:
   **Jason:** "Mom, can Marcus sleep over tonight?"
   **Mom:** "It's OK with me if it's OK with your father."
   **Jason:** "Dad, can Marcus sleep over tonight?"
   **Dad:** "It's OK with me if it's OK with your mother."

We next see Jason poring over a book entitled *Logic.* Help him out. Does Marcus get to sleep over?

7. Prove or give a counterexample:
   a. Every continuous function is differentiable.
   b. The square of every integer is positive.
   c. There exists an integer $n$ so that $n^2 = 9$.
   d. There exists exactly one integer $n$ so that $n^2 = 9$.
   e. For every nonnegative integer $n$, $n^2 \geq n$.
   f. There exists an integer $M$ so that $x^2 + 6x \geq M$ for all real numbers $x$.

## 2. Sets and Functions

A **set** is any collection of objects, and the objects belonging to the set are called its **elements**. If $x$ is an element of the set $X$, we write this as

$$x \in X.$$

If $x$ is *not* an element of $X$, we write

$$x \notin X.$$

The **empty set** (denoted $\emptyset$) is a set with *no* elements. We get used to working with certain sets which we do not describe in any great detail; for example, we have the following common sets of numbers:

| symbol | description |
|:---:|:---:|
| $\mathbb{N}$ | the set of natural numbers |
| $\mathbb{Z}$ | the set of integers |
| $\mathbb{Q}$ | the set of rational numbers |
| $\mathbb{R}$ | the set of real numbers |
| $\mathbb{C}$ | the set of complex numbers |

But most often we describe a set by the attributes we wish its elements to have. For this, the notation we shall employ is:

$$X = \{x \in \langle \text{our universe} \rangle : x \text{ satisfies certain criteria}\}.$$

This is read, "the set of all $x$ in our universe *such that* $x$ satisfies certain criteria." For example, we have:

$$X = \{x \in \mathbb{Z} : x = y^2 \text{ for some } y \in \mathbb{Z}\},$$

$$Y = \{x \in \mathbb{Z} : 2x = 1\},$$

$$Z = \{\text{triangles } \Delta : \text{the sum of the measures of the angles of } \Delta$$
$$\text{is } 200°\}.$$

We recognize that $X$ is the set of perfect squares among all integers, and that $Y$ is a fancy way of writing the empty set; most people's experience with geometry dictates that $Z$ is empty as well (but see Section 5 of Chapter 8).

Given two sets $X$ and $Y$, we say $X$ is a **subset** of $Y$ (written $X \subset Y$) if every element of $X$ is an element of $Y$; i.e.,

$$X \subset Y \quad \text{means} \quad x \in X \implies x \in Y.$$

We say $X$ is a **proper** subset of $Y$ (denoted $X \subsetneq Y$) if $X$ is a subset of $Y$, but $X$ is not equal to $Y$. It occurs often in mathematics that we must prove that two sets $X$ and $Y$ are equal; this is usually accomplished by showing that $X \subset Y$ and $Y \subset X$.

We next review the common ways of constructing new sets from old. Given two sets $X$ and $Y$, their **union** consists of all elements that belong *either* to $X$ *or* to $Y$ (or both):

$$X \cup Y = \{x : x \in X \text{ or } x \in Y\}.$$

Their **intersection** consists of all elements that belong to *both* $X$ *and* $Y$:

$$X \cap Y = \{x : x \in X \text{ and } x \in Y\}.$$

DeMorgan's laws state that for any sets $X$, $Y$, and $Z$,

$$X \cap (Y \cup Z) = (X \cap Y) \cup (X \cap Z);$$

$$X \cup (Y \cap Z) = (X \cup Y) \cap (X \cup Z).$$

The **Cartesian product** of $X$ and $Y$ consists of all ordered pairs $(x, y)$ with $x \in X$ and $y \in Y$:

$$X \times Y = \{(x, y) : x \in X \text{ and } y \in Y\}.$$

A **function** $f: X \to Y$ is a "rule" that assigns to each element $x$ of $X$ an element $y$ of $Y$. More precisely, a function $f: X \to Y$ is a subset $G \subset X \times Y$ with the property that for each $x \in X$ there is a unique element $(x, y) \in G$. (You might recognize $G$ as the *graph* of the function.) We write $y = f(x)$ and call $y$ the image of $x$ under the mapping $f$. (We will occasionally refer to a function by its rule, using the notation $x \rightsquigarrow y$, read "$x$ maps to $y$.") We call $X$ the **domain** of the function, and $Y$ the **range**. The set of all values of the function, i.e., $\{y \in Y : y = f(x) \text{ for some } x \in X\}$, is called the **image** of $f$ and is often denoted $f(X)$. More generally, given any subset $W \subset X$, we can define its image under $f$ to be

$$f(W) = \{y \in Y : y = f(x) \text{ for some } x \in W\}.$$

 Likewise, we can consider the **preimage** of a subset of $Y$: given any subset $Z \subset Y$,

$$f^{-1}(Z) = \{x \in X : f(x) \in Z\}.$$

**Remark.** Be warned that whereas $f(f^{-1}(Z)) \subset Z$ and $W \subset f^{-1}(f(W))$, equality needn't hold in either case.

We say the function $f\colon X \to Y$ is **one-to-one** or **injective** if

$$\boxed{\text{for all } x_1, x_2 \in X, \quad x_1 \neq x_2 \implies f(x_1) \neq f(x_2);}$$

i.e., distinct points of the domain map to distinct points of the range. This is sometimes stated in terms of the contrapositive:

$$\boxed{\text{if } f(x_1) = f(x_2), \text{ then } x_1 = x_2.}$$

We say $f\colon X \to Y$ is **onto** or **surjective** if $f(X) = Y$, i.e., if

$$\boxed{\text{for every } y \in Y, \text{ there is (at least) one } x \in X \text{ so that } f(x) = y.}$$

We say $f\colon X \to Y$ is a **one-to-one correspondence** or **bijective** if $f$ is both one-to-one (injective) and onto (surjective).

If $f\colon X \to Y$ and $g\colon Y \to Z$, then we can define the **composition** of the functions $g$ and $f$ by

$$X \xrightarrow{\;f\;} Y \xrightarrow{\;g\;} Z$$
$$(g \circ f)(x) = g(f(x)).$$

The simplest example of a function is the **identity** function $\iota_X\colon X \to X$ for any set $X$, defined by $\iota_X(x) = x$ for all $x \in X$. Suppose we have functions $f\colon X \to Y$ and $g\colon Y \to X$ so that

$$g \circ f = \iota_X \qquad \text{and} \qquad f \circ g = \iota_Y.$$

Then $g$ is called the **inverse function** of $f$ and *vice versa*. Note that the inverse function is unique if it exists. Indeed, we can prove a somewhat stronger result.

**Lemma.** *Suppose* $f: X \to Y$, $g: Y \to X$ *and* $h: Y \to X$ *are functions satisfying*

$$g \circ f = \iota_X \quad \text{and} \quad f \circ h = \iota_Y.$$

*Then* $g = h$ *and* $g$ *is the inverse function of* $f$.

**Proof.** Composition of functions is associative: for any $y \in Y$,

$$((g \circ f) \circ h)(y) = (g \circ f)(h(y)) = g(f(h(y))) = g((f \circ h)(y))$$
$$= (g \circ (f \circ h))(y).$$

Now using the hypotheses, we have $((g \circ f) \circ h)(y) = (\iota_X \circ h)(y) = h(y)$ and $(g \circ (f \circ h))(y) = (g \circ \iota_Y)(y) = g(y)$. Since $y$ is an arbitrary element of $Y$, $g = h$, as desired.  $\square$

**Examples.**

(a) If $f: \mathbb{Q} \to \mathbb{Q}$ and $g: \mathbb{Q} \to \mathbb{Q}$ are defined by $f(x) = 2x + 3$ and $g(x) = \frac{1}{2}x - \frac{3}{2}$, then $f$ and $g$ are inverse functions.

(b) Let $\mathbb{R}^+$ denote the set of positive real numbers. Let $f: \mathbb{R}^+ \cup \{0\} \to \mathbb{R}^+ \cup \{0\}$ be defined by $f(x) = x^2$. Then the inverse function $g: \mathbb{R}^+ \cup \{0\} \to \mathbb{R}^+ \cup \{0\}$ is given by $g(x) = \sqrt{x}$. Note that if we had taken the domain of $f$ to be all of $\mathbb{R}$, then there would be no inverse function (why?).

(c) Let $f: \mathbb{R} \to \mathbb{R}^+$ be given by $f(x) = e^x$. The inverse function of $f$ is $g: \mathbb{R}^+ \to \mathbb{R}$, $g(x) = \ln x$.

Given a subset $W \subset X$ and a function $f: X \to Y$, we can define a new function $f|_W: W \to Y$ called the **restriction** of $f$ to $W$, just by considering the domain to be the subset $W \subset X$: $f|_W(x) = f(x)$ whenever $x \in W$.

### EXERCISES

1. Let $X$, $Y$ and $Z$ be arbitrary sets. Prove deMorgan's laws:
   a. $X \cap (Y \cup Z) = (X \cap Y) \cup (X \cap Z)$
   b. $X \cup (Y \cap Z) = (X \cup Y) \cap (X \cup Z)$

2. Let $X$ and $Y$ be arbitrary sets. Let $X - Y = \{x : x \in X \text{ and } x \notin Y\}$. Prove that $(X - Y) \cup (Y - X) = (X \cup Y) - (X \cap Y)$.

3. Let $f: X \rightarrow Y$. Let $A, B \subset X$ and $C, D \subset Y$. Prove or give a counterexample (if possible, provide sufficient hypotheses for each statement to be valid):

   a. $f(A) \cup f(B) = f(A \cup B)$

   b. $f(A) \cap f(B) = f(A \cap B)$

   c. $f(A - B) = f(A) - f(B)$

   d. $f^{-1}(C) \cup f^{-1}(D) = f^{-1}(C \cup D)$

   e. $f^{-1}(C) \cap f^{-1}(D) = f^{-1}(C \cap D)$

   f. $f^{-1}(C - D) = f^{-1}(C) - f^{-1}(D)$

   g. $f(f^{-1}(C)) = C$

   h. $f^{-1}(f(A)) = A$

4. Let $f: X \rightarrow Y$ and $g: Y \rightarrow Z$.

   a. Suppose that $f$ and $g$ are injective. Prove that $g \circ f$ is injective.

   b. Suppose that $f$ and $g$ are surjective. Prove that $g \circ f$ is surjective.

   c. Suppose that $f$ and $g$ are bijective. Prove that $g \circ f$ is bijective.

5. Suppose $f: X \rightarrow Y$, $g: Y \rightarrow Z$, and $h = f \circ g$.

   a. Prove that if $h$ is injective, then $g$ is injective.

   b. Prove that if $h$ is surjective, then $f$ is surjective.

   c. Prove that if $h$ is bijective, then $g$ is injective and $f$ is surjective.

   d. Prove or give a counterexample: if $h$ is bijective, then $f$ and $g$ are bijective.

   e. Conclude that if $f$ is to have an inverse function, $f$ must be bijective.

6. Suppose $f: X \rightarrow Y$. Prove that $f$ is bijective if and only if $f$ has an inverse function $g: Y \rightarrow X$.

7. Prove that if $f$ is bijective, then $(f^{-1})^{-1} = f$ and $f^{-1}$ is bijective.

8. Prove that if $g$ is the inverse function of $f$, then

   a. $g^{-1} = f$, and

   b. $g$ is bijective.

9. Suppose $X$ and $Y$ are finite sets, each having $n$ elements.

   a. Prove that if $f: X \rightarrow Y$ is an injection, then $f$ is a bijection.

   b. Prove that if $f: X \rightarrow Y$ is a surjection, then $f$ is a bijection.

10. Prove that if $f: X \to Y$ and $g: Y \to Z$ are bijective, then $(g \circ f)^{-1} = f^{-1} \circ g^{-1}$.

11. Let $X = [1, \infty) \subset \mathbb{R}$, $Y = [2, \infty) \subset \mathbb{R}$. Let $f: X \to Y$ be given by $f(x) = x + 1/x$. Prove that $f$ is bijective. Find its inverse function.

12. Let $X$ and $Y$ be arbitrary sets. Prove that there is a bijection $X \times Y \to Y \times X$.

13. Let $f: X \to Y$, $W \subset X$, and let $f|_W: W \to Y$ denote the restriction of $f$ to $W$. Prove or give a counterexample:
    a. If $f$ is injective, then $f|_W$ is injective.
    b. If $f|_W$ is injective, then $f$ is injective.
    c. If $f$ is surjective, then $f|_W$ is surjective.
    d. If $f|_W$ is surjective, then $f$ is surjective.

## 3. Equivalence Relations

One of the most important foundational concepts for modern algebra is that of an **equivalence relation** on a set. Here we review the definition and give some basic examples. Let $X$ be a set. Recall that the cartesian product $X \times X$ consists of all ordered pairs $(x, y)$ of elements $x, y \in X$, and that a **relation** on $X$ is any subset $R \subset X \times X$.

**Examples 1.**
(a) Let $X$ be any set, and let $R = X \times X$.
(b) Let $X$ be any set, and let $R = \emptyset$.
(c) Let $X$ be any set, and let $R = \{(x, y) \in X \times X : x = y\}$.
(d) Let $X = \mathbb{R}$, and let $R = \{(x, y) \in \mathbb{R} \times \mathbb{R} : x^2 + y^2 = 1\}$.
(e) Let $X = \mathbb{R}$, and let $R = \{(x, y) \in \mathbb{R} \times \mathbb{R} : x^2 = y^2\}$.
(f) Let $X = \{$all human beings$\}$ and define $(x, y) \in R$ if and only if $x$ and $y$ are siblings (i.e., have at least one parent in common).
(g) Let $X = \mathbb{R}$ and define $(x, y) \in R$ if and only if $x < 3$ and $y > 4$.

**Definition.** $R$ is called an **equivalence relation** on $X$ if
    (i) $R$ is **reflexive:** $(x, x) \in R$ for all $x \in X$,
    (ii) $R$ is **symmetric:** $(x, y) \in R \implies (y, x) \in R$, and

(iii) $R$ is **transitive**: $(x, y) \in R$ and $(y, z) \in R \implies (x, z) \in R$.

When $R$ is an equivalence relation, we often write $x \sim y$ to denote that $(x, y) \in R$. As the reader can check, Examples (a), (c), and (e) are equivalence relations. In Example (b), requirement (i) fails, but both (ii) and (iii) hold vacuously. In Example (d), requirements (i) and (iii) fail, and in Example (f) requirement (iii) fails—note that $x$ and $y$ can share a mother while $y$ and $z$ share a father, and $x$ and $z$ need not be siblings. On the other hand, in Example (g), requirements (i) and (ii) fail, but (iii) holds vacuously (inasmuch as there can be no $y$ so that $y > 4$ and $y < 3$).

Here is a standard example (with whose generalizations we shall become extremely familiar). Define a relation on the set $\mathbb{Z}$ of integers by

$$x \sim y \iff x - y \text{ is even.}$$

Note that $\sim$ is reflexive, as $x - x = 0$ is even. The relation $\sim$ is symmetric, because if $x - y$ is even, then $y - x$ is even as well (the negative of an even integer is even). And $\sim$ is transitive, since if $x - y$ is even and $y - z$ is even, then their sum is even, and their sum is $(x - y) + (y - z) = x - z$.

**Remark.** Here is an amusing "proof" that the reflexive property is a consequence of the symmetric and transitive properties. We need to show that $x \sim x$ for any $x \in X$. Suppose $x \sim y$. Then, by symmetry, $y \sim x$; and now by transitivity, $x \sim y$ and $y \sim x$ imply $x \sim x$.

**Example 2.** Here is an important example for our work in Chapter 2. We are going to define an equivalence relation on the set $X = \mathbb{Z} \times (\mathbb{Z} - \{0\})$:

$$(a, b) \sim (c, d) \iff ad = bc.$$

We check the three properties. The reflexive and symmetric properties are immediate: For any $(a, b) \in X$, $(a, b) \sim (a, b)$ since $ab = ab$; if $(a, b) \sim (c, d)$, then $ad = bc$, so $cb = da$ (multiplication of integers being commutative) and $(c, d) \sim (a, b)$. The transitive property is more interesting: suppose $(a, b) \sim (c, d)$ and $(c, d) \sim (e, f)$. Then we have $ad = bc$ and $cf = de$. We must show that $(a, b) \sim (e, f)$, i.e., that $af = be$. Well, $(ad)f = (bc)f = b(cf) = b(de)$. Therefore, $d(af) = d(be)$; and since $d \neq 0$, we must have $af = be$.

Given an equivalence relation $\sim$ on a set $X$, there are natural subsets of $X$ to consider. Namely, for each $x \in X$, let

$$[x] = \{y \in X : x \sim y\}.$$

We call $[x]$ the **equivalence class** of $x$. By the reflexive property, it is always true that $x \sim x$, and so $x \in [x]$.

**Lemma.** *Let $x, y \in X$. If the equivalence classes $[x]$ and $[y]$ have any element $z$ in common, then $[x] = [y]$.*

**Proof.** Suppose $z \in [x] \cap [y]$. This means that $x \sim z$ and $y \sim z$, so $x \sim y$ (why?). Now the transitive property guarantees that if $s \in [y]$, then $s \in [x]$, so that $[y] \subset [x]$. But the identical argument, interchanging $x$ and $y$, establishes that $[x] \subset [y]$, and so $[x] = [y]$. $\square$

Since each $x \in X$ belongs to some equivalence class (namely its own), we can write $X$ as the union of the distinct equivalence classes. In general, a collection $\mathcal{P}$ of subsets $A_j \subset X$ is called a **partition** of $X$ if

(i) $\bigcup_j A_j = X$ (every element of $X$ is in some subset $A_j$), and

(ii) $A_j \cap A_k = \emptyset$ when $j \neq k$ (the subsets are disjoint).

We can turn the process around: given a partition $\mathcal{P}$ of $X$, define an equivalence relation $\sim$ on $X$ as follows:

$$x \sim y \iff \text{there is some subset } A_j \in \mathcal{P} \text{ containing both } x \text{ and } y.$$

Let's check that $\sim$ is truly an equivalence relation. Since, by (i), each $x$ is contained in some $A_j$, $\sim$ is reflexive; symmetry is immediate. Suppose $x \sim y$ and $y \sim z$. There is some $A_j \in \mathcal{P}$ containing both $x$ and $y$. By (ii), $y$ can belong to no other $A_k$, and so $A_j$ must contain $z$ as well; thus, $x \sim z$, establishing transitivity.

Here is one of the most important constructions in mathematics —ubiquitous in algebra and geometry. Let $X$ be a set equipped with an equivalence relation $\sim$. Let $\mathcal{E}$ be the set of distinct equivalence classes. Then there is a natural function

$$\pi : X \to \mathcal{E}$$
$$\pi(x) = [x]$$

which maps each element of $X$ to its equivalence class. Note that $\pi$ is a function because each element belongs to a unique equivalence class. And $\pi$ is also surjective (since equivalence classes are nonempty). Now let $e \in \mathcal{E}$; then $\pi^{-1}(\{e\}) = \{x \in X : \pi(x) = e\} = \{x \in X : [x] = e\}$. That is, $\pi^{-1}(\{e\})$ consists of all elements $x$ whose equivalence class is $e$. So, as we consider various $e \in \mathcal{E}$, the preimages $\pi^{-1}(\{e\})$ run through the equivalence classes (giving us back the partition $\mathcal{P}$ of $X$ determined by the equivalence relation).

**Example 3.** We define $\mathbb{Q}$ to be the set of equivalence classes of the equivalence relation $\sim$ on $\mathbb{Z} \times (\mathbb{Z} - \{0\})$ given in Example 2. It is customary to denote the equivalence class of $(a, b)$ by $\frac{a}{b}$, rather than by the cumbersome $[(a, b)]$. Consider now the following "definition" of a function $f : \mathbb{Q} \to \mathbb{Z}$. We set $f(\frac{a}{b}) = a$ (so that $f(\frac{2}{3}) = 2$, $f(-\frac{5}{8}) = -5$, $f(\frac{10}{15}) = 10$, and so on). You might have noticed something suspicious by now. If we take two different representatives of the same equivalence class—for example, $(2, 3)$ and $(10, 15)$, or $(-5, 8)$ and $(5, -8)$—we get different values for $f(\frac{2}{3})$ and $f(\frac{10}{15})$. In mathematical jargon, we say that the function $f$ is *not* **well-defined**. Remember that a function must associate a *unique* element of its range to each element of its domain. Generally, when we refer to a function as being well-defined, we mean that its values do not depend on any choices that were made along the way.

## EXERCISES

1. Consider the following relations on the set $\mathbb{Z}$:

   (i)   $(a, b) \in R_1$   if $ab \geq 0$

   (ii)  $(a, b) \in R_2$   if $ab > 0$

   (iii) $(a, b) \in R_3$   if $ab > 0$ or $a = b = 0$

   Decide whether each is an equivalence relation. (If not, which requirements fail?)

2. Find the flaw in the so-called proof in the Remark on p. 380.

3. Define a relation on $\mathbb{R}$ as follows: $x \sim y$ if and only if $x - y$ is an integer. Prove that $\sim$ is an equivalence relation and describe the set of equivalence classes.

4. Define a relation on $\mathbb{N}$ as follows: $x \sim y$ if and only if $x$ and $y$ have the same last digit in their base-ten representation. Prove

that ~ is an equivalence relation, and describe the set of equivalence classes.

5.  Define a relation on $\mathbb{R}^2$ as follows: $(x_1, x_2) \sim (y_1, y_2)$ if and only if $x_1^2 + x_2^2 = y_1^2 + y_2^2$. Prove that ~ is an equivalence relation, and describe the set of equivalence classes.

6.  a.  Define a relation on $\mathbb{R}$ as follows: $x$ and $y$ are related if $|x - y| < 1$. Decide whether this is an equivalence relation.
    b.  Define an equivalence relation on $\mathbb{R}$ whose equivalence classes are intervals of length 1.

7.  Which of the following functions $f: \mathbb{Q} \to \mathbb{Q}$ are well-defined?
    a.  $f(\frac{a}{b}) = \frac{a+1}{b+1}$
    b.  $f(\frac{a}{b}) = \frac{a+b}{b}$
    c.  $f(\frac{a}{b}) = \frac{2a^2}{3b^2}$
    d.  $f(\frac{a}{b}) = \frac{b}{a}$
    e.  $f(\frac{a}{b}) = \frac{a^2+ab+b^2}{a^2+b^2}$

8.  Define an equivalence relation on $X = \mathbb{N} \times \mathbb{N}$ as follows:
    $$(a, b) \sim (c, d) \iff a + d = c + b.$$
    a.  Prove that ~ is indeed an equivalence relation.
    b.  Identify the set of equivalence classes.

9.  Define a relation on $\mathbb{N}$ as follows: $x \sim y$ if and only if there are integers $j$ and $k$ so that $x | y^j$ and $y | x^k$.
    a.  Show that ~ is an equivalence relation.
    b.  Determine the equivalence classes $[1]$, $[2]$, $[9]$, $[10]$, and $[20]$.
    c.  Describe explicitly the equivalence classes $[x]$ in general.

10. Given a function $f: S \to T$, consider the following relation on $S$:
    $$x \sim y \iff f(x) = f(y).$$
    a.  Prove that ~ is an equivalence relation.
    b.  Prove that if $f$ maps onto $T$, then there is a one-to-one correspondence between the set of equivalence classes and $T$.

# APPENDIX B

# Miscellaneous Facts from Linear Algebra

### 1. The Matrix of a Linear Map

**N.B.** We assume here a familiarity with the concepts of Section 1 of Chapter 5. As usual, we let $\mathbf{e}_1, \ldots, \mathbf{e}_n$ denote the standard basis for $\mathbb{R}^n$; that is, $\mathbf{e}_i = [0, \ldots, 0, 1, 0, \ldots, 0]^T$, where the 1 appears in the $i^{\text{th}}$ slot.

Let $V$ be a finite-dimensional vector space over a field $F$, and let $T: V \to V$ be a linear map. That is,

$$T(\mathbf{v} + \mathbf{w}) = T(\mathbf{v}) + T(\mathbf{w}) \quad \text{for all vectors } \mathbf{v}, \mathbf{w} \in V, \text{ and}$$
$$T(c\mathbf{v}) = cT(\mathbf{v}) \qquad \text{for all vectors } \mathbf{v} \in V \text{ and scalars } c.$$

Given an ordered basis $\mathcal{B} = \{\mathbf{v}_1, \ldots, \mathbf{v}_n\}$ for $V$, we wish to define the matrix $A$ for $T$ with respect to the basis $\mathcal{B}$. We do this as follows. Define scalars $a_{ij}$, $1 \leq i, j \leq n$, by the equation

$$T(\mathbf{v}_j) = \sum_{i=1}^{n} a_{ij} \mathbf{v}_i.$$

(Since $\mathbf{v}_1, \ldots, \mathbf{v}_n$ form a basis, these exist and are unique.) Let $A$ be the $n \times n$ matrix

$$A = \begin{bmatrix} a_{11} & a_{12} & \cdots & a_{1n} \\ a_{21} & a_{22} & \cdots & a_{2n} \\ \vdots & \vdots & \ddots & \vdots \\ a_{n1} & a_{n2} & \cdots & a_{nn} \end{bmatrix}.$$

Now suppose $\mathbf{v} = \sum\limits_{i=1}^{n} x_i \mathbf{v}_i$ and $T(\mathbf{v}) = \sum\limits_{i=1}^{n} y_i \mathbf{v}_i$. If we set $X = \begin{bmatrix} x_1 \\ x_2 \\ \vdots \\ x_n \end{bmatrix}$

and $Y = \begin{bmatrix} y_1 \\ y_2 \\ \vdots \\ y_n \end{bmatrix}$, then we claim $Y = AX$. Well,

$$\sum_{i=1}^{n} y_i \mathbf{v}_i = T(\mathbf{v}) = T\left(\sum_{j=1}^{n} x_j \mathbf{v}_j\right) = \sum_{j=1}^{n} x_j T(\mathbf{v}_j) = \sum_{j=1}^{n} x_j \left(\sum_{i=1}^{n} a_{ij} \mathbf{v}_i\right)$$

$$= \sum_{i=1}^{n} \left(\sum_{j=1}^{n} a_{ij} x_j\right) \mathbf{v}_i.$$

Thus, $y_i = \sum\limits_{j=1}^{n} a_{ij} x_j$ is the $i^{\text{th}}$ entry of the matrix product $AX$, as desired. Note that the procedure for concocting the matrix $A$ is quite easy to remember:

> *The $j^{th}$ column vector of $A$ consists of the coordinates of $T(\mathbf{v}_j)$ with respect to the basis $\mathcal{B}$.*

When $V = \mathbb{R}^n$ and no basis is explicitly mentioned, it is customary to take the standard basis $\mathbf{e}_1, \ldots, \mathbf{e}_n$.

**Example 1.** Suppose $T \colon \mathbb{R}^3 \to \mathbb{R}^3$ is a linear map satisfying

$$T\left(\begin{bmatrix} 1 \\ 0 \\ 0 \end{bmatrix}\right) = \begin{bmatrix} 2 \\ 3 \\ 1 \end{bmatrix},$$

$$T\left(\begin{bmatrix} 0 \\ 1 \\ 0 \end{bmatrix}\right) = \begin{bmatrix} 1 \\ 5 \\ -2 \end{bmatrix},$$

$$T\left(\begin{bmatrix} 0 \\ 0 \\ 1 \end{bmatrix}\right) = \begin{bmatrix} 1 \\ 1 \\ 0 \end{bmatrix}.$$

What is the matrix $A$ for $T$ with respect to the standard basis $\mathcal{B} = \{\mathbf{e}_1, \mathbf{e}_2, \mathbf{e}_3\}$?

Since we are given the values of $T$ on the standard basis vectors, this is routine: the column vectors of $A$ are the vectors $\begin{bmatrix} 2 \\ 3 \\ 1 \end{bmatrix}, \begin{bmatrix} 1 \\ 5 \\ -2 \end{bmatrix},$

and $\begin{bmatrix} 1 \\ 1 \\ 0 \end{bmatrix}$, respectively, and so

$$A = \begin{bmatrix} 2 & 1 & 1 \\ 3 & 5 & 1 \\ 1 & -2 & 0 \end{bmatrix}.$$

**Example 2.** Suppose $T\colon \mathbb{R}^2 \to \mathbb{R}^2$ is a linear map satisfying

(∗)
$$T\left(\begin{bmatrix} 1 \\ 1 \end{bmatrix}\right) = \begin{bmatrix} 6 \\ 3 \end{bmatrix},$$

$$T\left(\begin{bmatrix} 2 \\ -1 \end{bmatrix}\right) = \begin{bmatrix} 0 \\ -12 \end{bmatrix}.$$

What is the matrix $A$ for $T$ with respect to the standard basis?

At this point, our best bet is to compute $T(\mathbf{e}_1)$ and $T(\mathbf{e}_2)$, where $\mathbf{e}_1, \mathbf{e}_2$ form the standard basis. Letting $\mathbf{v}_1 = \begin{bmatrix} 1 \\ 1 \end{bmatrix}$, $\mathbf{v}_2 = \begin{bmatrix} 2 \\ -1 \end{bmatrix}$, our charge is to use the given information (∗) to accomplish this. By trial and error (or cleverness), we find that

$$\mathbf{e}_1 = \tfrac{1}{3}\mathbf{v}_1 + \tfrac{1}{3}\mathbf{v}_2 \quad \text{and}$$
$$\mathbf{e}_2 = \tfrac{2}{3}\mathbf{v}_1 - \tfrac{1}{3}\mathbf{v}_2,$$

whence

$$T(\mathbf{e}_1) = T(\tfrac{1}{3}\mathbf{v}_1 + \tfrac{1}{3}\mathbf{v}_2) = \tfrac{1}{3}T(\mathbf{v}_1) + \tfrac{1}{3}T(\mathbf{v}_2) = \tfrac{1}{3}\begin{bmatrix} 6 \\ 3 \end{bmatrix} + \tfrac{1}{3}\begin{bmatrix} 0 \\ -12 \end{bmatrix}$$

$$= \begin{bmatrix} 2 \\ -3 \end{bmatrix}, \quad \text{and}$$

$$T(\mathbf{e}_2) = T(\tfrac{2}{3}\mathbf{v}_1 - \tfrac{1}{3}\mathbf{v}_2) = \tfrac{2}{3}T(\mathbf{v}_1) - \tfrac{1}{3}T(\mathbf{v}_2) = \tfrac{2}{3}\begin{bmatrix} 6 \\ 3 \end{bmatrix} - \tfrac{1}{3}\begin{bmatrix} 0 \\ -12 \end{bmatrix}$$

$$= \begin{bmatrix} 4 \\ 6 \end{bmatrix};$$

therefore, the desired matrix for $T$ is

$$A = \begin{bmatrix} 2 & 4 \\ -3 & 6 \end{bmatrix}.$$

Here is a basic problem in linear algebra. Suppose we are given the matrix $A$ of a linear map $T\colon V \to V$ with respect to one basis $\mathcal{B} = \{\mathbf{v}_1, \dots, \mathbf{v}_n\}$ for $V$, and we wish to know the matrix $A'$ with respect to a different basis $\mathcal{B}' = \{\mathbf{v}'_1, \dots, \mathbf{v}'_n\}$. (For example, we might wish to choose a basis $\mathcal{B}'$ to make the matrix $A'$ as simple as possible— this is one possible interpretation of the diagonalization problem.) The answer to the question is the famed "change of basis formula" from elementary linear algebra. It goes as follows. For convenience, let's agree to call the original basis $\mathcal{B}$ the "old" basis, and the basis $\mathcal{B}'$ the "new" basis. Let the columns of the matrix $P$ consist of the coordinates of the *new* basis vectors in terms of the *old*; that is, if the "change of basis matrix" is $P = \left[ p_{ij} \right]$,

$$(\dagger) \qquad\qquad \mathbf{v}_j' = \sum_{i=1}^{n} p_{ij} \mathbf{v}_i.$$

Then we have the

**Theorem 1.1** (Change of Basis Formula).

$$\boxed{A' = P^{-1}AP.}$$

This equation is summarized in Figure 1.

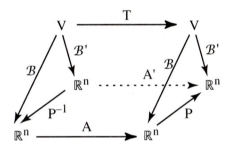

FIGURE 1

***Proof.*** Suppose $X = \begin{bmatrix} x_1 \\ x_2 \\ \vdots \\ x_n \end{bmatrix}$ and $X' = \begin{bmatrix} x_1' \\ x_2' \\ \vdots \\ x_n' \end{bmatrix}$ are the coordinate vectors of $\mathbf{x} \in V$ with respect to the old and new bases, respectively. Then we have the relation

$$X = PX'.$$

To establish this, we note that for $X$ to be the coordinate vector of $\mathbf{x}$ with respect to the basis $\mathcal{B}$ means that $\mathbf{x} = \sum_{i=1}^{n} x_i \mathbf{v}_i$, and similarly for $X'$. So we have (using $(\dagger)$)

$$\mathbf{x} = \sum_{j=1}^{n} x_j' \mathbf{v}_j' = \sum_{j=1}^{n} x_j' \left( \sum_{i=1}^{n} p_{ij} \mathbf{v}_i \right) = \sum_{i=1}^{n} \left( \sum_{j=1}^{n} p_{ij} x_j' \right) \mathbf{v}_i,$$

whence $x_i = \sum\limits_{j=1}^{n} p_{ij}x'_j$, or $X = PX'$.

Now we compute the coordinate vectors of $\mathbf{y} = T(\mathbf{x})$ with respect to the two bases: by definition, we have

$$Y = AX \quad \text{(coming from the basis } \mathcal{B}\text{), and}$$

$$Y' = A'X' \quad \text{(coming from the basis } \mathcal{B}'\text{).}$$

Since $Y = PY'$ and $X = PX'$, we have

$$Y' = P^{-1}Y = P^{-1}(AX) = P^{-1}A(PX') = (P^{-1}AP)X',$$

whence $A' = P^{-1}AP$, as required.  $\square$

**Example 3.** Let $T: \mathbb{R}^2 \to \mathbb{R}^2$ be the linear map defined by reflection in the line $x_1 + 2x_2 = 0$. Give the matrix of $T$ with respect to the standard basis $\mathcal{B} = \{\mathbf{e}_1, \mathbf{e}_2\}$.

Here is an example where the linear map is easiest to understand in terms of a different basis $\mathcal{B}' = \{\mathbf{e}'_1, \mathbf{e}'_2\}$ adapted to the geometry of the situation. Let $\mathbf{e}'_1$ be a vector lying along the line, and let $\mathbf{e}'_2$ be

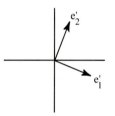

FIGURE 2

a vector orthogonal to the line, as pictured in Figure 2. Then clearly $T(\mathbf{e}'_1) = \mathbf{e}'_1$ and $T(\mathbf{e}'_2) = -\mathbf{e}'_2$, so that the matrix $A'$ for $T$ with respect to the basis $\mathcal{B}'$ is

$$A' = \begin{bmatrix} 1 & 0 \\ 0 & -1 \end{bmatrix}.$$

The change of basis matrix is

$$P = \begin{bmatrix} 2 & 1 \\ -1 & 2 \end{bmatrix};$$

and so we have

$$A = PA'P^{-1} = \begin{bmatrix} 2 & 1 \\ -1 & 2 \end{bmatrix} \begin{bmatrix} 1 & 0 \\ 0 & -1 \end{bmatrix} \begin{bmatrix} \frac{2}{5} & -\frac{1}{5} \\ \frac{1}{5} & \frac{2}{5} \end{bmatrix} = \begin{bmatrix} \frac{3}{5} & -\frac{4}{5} \\ -\frac{4}{5} & -\frac{3}{5} \end{bmatrix}.$$

Of course, we could figure this out directly by applying the linear map $T$ to the standard basis vectors $\mathbf{e}_1$ and $\mathbf{e}_2$. Note, by the way, that we have used the change of basis formula backwards; it just seemed more natural to keep $\mathcal{B}$ as the standard basis for $\mathbb{R}^2$.

## EXERCISES

1.  Suppose $T: V \to V$ is a linear map whose matrix with respect to an ordered basis $\mathcal{B} = \{\mathbf{v}_1, \ldots, \mathbf{v}_n\}$ is diagonal. Let $c_1, \ldots, c_n$ be scalars, and consider the ordered basis $\mathcal{B}' = \{c_1\mathbf{v}_1, \ldots, c_n\mathbf{v}_n\}$. Show that the matrix for $T$ with respect to the basis $\mathcal{B}'$ is likewise diagonal.

2.  Let $\mathcal{B} = \{\mathbf{v}_1, \ldots, \mathbf{v}_n\}$ be an orthonormal basis for $\mathbb{R}^n$, and let $P$ be the $n \times n$ matrix whose columns are $\mathbf{v}_1, \ldots, \mathbf{v}_n$. It follows from the proof of Theorem 1.1 that the coordinates of $\mathbf{x} \in \mathbb{R}^n$ with respect to $\mathcal{B}$ are $X' = P^{-1}X = P^T X$. Give a completely elementary verification of the latter formula.

3.  Suppose $S, T: V \to V$ are linear maps. If $A$ and $B$ are the matrices for $S$ and $T$, respectively, with respect to an ordered basis $\mathcal{B}$, then prove that the matrix for $S \circ T$ (with respect to the same basis) is $AB$.

4.  Give the matrix with respect to the standard basis for the linear map $T: \mathbb{R}^3 \to \mathbb{R}^3$ given by reflection in the plane $x_1 + x_2 + x_3 = 0$.

5.  (Assuming a knowledge of Section 2 of Chapter 3 and Section 1 of Chapter 5)
    a.  Let $V = \mathbb{Q}[\sqrt{2}]$, and let $T: V \to V$ be given by $T(u) = \sqrt{2}\,u$. Give the matrix for $T$ with respect to the ordered basis $\mathcal{B} = \{1, \sqrt{2}\}$.
    b.  Suppose $V = \mathbb{Q}[\alpha]$, where $\alpha$ is a root of the irreducible polynomial $x^5 + 9x^2 - 3x + 6 \in \mathbb{Q}[x]$. Let $T: V \to V$ be given by $T(u) = \alpha u$. Give the matrix for $T$ with respect to the ordered basis $\mathcal{B} = \{1, \alpha, \alpha^2, \alpha^3, \alpha^4\}$.

6.  Find all linear maps $T: \mathbb{R}^2 \to \mathbb{R}^2$
    a.  carrying the subspace $x_1 = 0$ to itself and the subspace $x_2 = 0$ to itself;

    b.   carrying the subspace $x_2 = 0$ to itself;

    c.   carrying the subspace $x_1 - x_2 = 0$ to the subspace $x_1 + x_2 = 0$.

7.   Let $T\colon \mathbb{R}^2 \to \mathbb{R}^2$ be the linear map defined by *orthogonal projection* onto the line $\ell = \{x_1 + 2x_2 = 0\}$; i.e., for $\mathbf{x} \in \mathbb{R}^2$, $T(\mathbf{x})$ is the vector in $\ell$ closest to $\mathbf{x}$, so that $\mathbf{x} - T(\mathbf{x})$ is orthogonal to $\ell$. Calculate the matrix for $T$ with respect to the standard basis in two ways:

    a.   directly, and

    b.   by applying the Change of Basis Formula.

       (Hint: Let $\mathcal{B}' = \{(2, -1), (1, 2)\}$.)

8.   Let $V = \text{Span}\{(1, 0, 1), (0, 1, -2)\} \subset \mathbb{R}^3$. Consider the linear map $T\colon \mathbb{R}^3 \to \mathbb{R}^3$ given by orthogonal projection to $V$, i.e., for $\mathbf{x} \in \mathbb{R}^3$, $T(\mathbf{x})$ is the vector in $V$ closest to $\mathbf{x}$ (so that $\mathbf{x} - T(\mathbf{x})$ is orthogonal to $V$). Calculate the matrix for $T$ with respect to the standard basis in two ways:

    a.   directly, and

    b.   by applying the Change of Basis Formula.

       (Hint for a.: It is easier to calculate $\mathbf{x} - T(\mathbf{x})$.)

## 2. Determinants

We give in this section a brief treatment of the properties of the determinant function

$$\det\colon M_n(F) \to F$$

defined on $n \times n$ matrices with coefficients in any field $F$. The determinant can be characterized by the following properties:

| | |
|---|---|
| <u>Property I:</u> | If any row of a matrix is multiplied by a scalar $c$, then the determinant is multiplied by $c$ as well. |
| <u>Property II:</u> | If the $k^{\text{th}}$ row $R_k$ of a matrix is replaced by the sum of two row vectors, $R_k = R_k' + R_k''$, then the determinant is the sum of the determinants of the matrices with $k^{\text{th}}$ rows $R_k'$ and $R_k''$, respectively. |
| <u>Property III:</u> | If any two rows of a matrix are equal, its determinant is zero. |

Property IV:    The determinant of the identity matrix is equal
to 1.

The first two properties state that the determinant function is a
linear function of each of the rows of the $n \times n$ matrix separately;
such a function is often called **multilinear**.

Recall that the elementary row operations we study in an introductory linear algebra course consist of three types:

(A) multiplying a row by a nonzero scalar $c$;

(B) adding a nonzero multiple of one row to another; and

(C) interchanging two rows.

We wish first to deduce from Properties I, II, and III the effect of
these row operations on the determinant.

It will be convenient for the moment to introduce the following
notation. Let $A$ be an $n \times n$ matrix, and let $A_1, \ldots, A_n$ be its row
vectors. Viewing the determinant as a function of the row vectors,
write $\det A = d(A_1, \ldots, A_n)$. The function $d$, then, is linear in each
of its $n$ variables.

It follows immediately from Property I that in executing operation (A), we multiply the determinant by the scalar $c$. We infer the
effect of operation (B) from the following result.

**Corollary 2.1.** *Suppose $A'_n = A_n + cA_i$, for some $i < n$. Then*

$$d(A_1, A_2, \ldots, A'_n) = d(A_1, A_2, \ldots, A_n).$$

*This implies that row operation* (B) *does not change the determinant
of the matrix.*

**Proof.** By Property II,

$$d(A_1, \ldots, A'_n) = d(A_1, \ldots, A_{n-1}, A_n) + d(A_1, \ldots, A_{n-1}, cA_i)$$
$$= d(A_1, \ldots, A_n) + cd(A_1, \ldots, A_{n-1}, A_i)$$
$$= d(A_1, \ldots, A_n),$$

since, by Property III, the term $d(A_1, \ldots, A_{n-1}, A_i) = 0$.  □

Now, how can we interchange two rows of a matrix in such a
way as to keep track of the determinant by Properties I, II, and III?
Here is a schematic picture of one such way:

$$\left(A_k, A_\ell\right) \rightsquigarrow \left(A_k, \left(A_\ell + A_k\right)\right) \rightsquigarrow \left(A_k - \left(A_\ell + A_k\right), A_\ell + A_k\right) =$$
$$\left(-A_\ell, A_\ell + A_k\right) \rightsquigarrow \left(-A_\ell, A_k\right)$$

We can arrange this much (by using row operation (B)) without changing the determinant. That is,

$$d(A_1,\ldots,A_k,\ldots,A_\ell,\ldots,A_n) = d(A_1,\ldots,-A_\ell,\ldots,A_k,\ldots,A_n)$$
$$= -d(A_1,\ldots,A_\ell,\ldots,A_k,\ldots,A_n),$$

by applying Property I with $c = -1$. This shows that interchanging two rows changes the sign of the determinant.

Now we can deduce the following general result.

**Proposition 2.2.** *If one row of the matrix $A$ is a linear combination of the remaining row vectors, then* $\det A = 0$.

**Proof.** We may suppose the row in question is the last (why?). Suppose there are scalars $c_1, c_2, \ldots, c_{n-1}$ so that $A_n = \sum\limits_{i=1}^{n-1} c_i A_i$. Then

$$\det A = d(A_1,\ldots,A_n) = d(A_1,\ldots,A_{n-1}, \sum_{i=1}^{n-1} c_i A_i)$$
$$= \sum_{i=1}^{n-1} d(A_1,\ldots,A_{n-1},c_i A_i)$$
$$= \sum_{i=1}^{n-1} c_i d(A_1,\ldots,A_{n-1},A_i) = 0,$$

since in every term of the final sum we have a matrix two of whose rows are equal, and whose determinant therefore vanishes by Property III. □

Since a square matrix $A$ can be reduced by a sequence of row operations to either the identity matrix or a matrix with a row of zeroes, it follows from this discussion that $\det A$ is uniquely determined by Properties I–IV. This fact will also follow from Theorem 2.5 in a moment. When we must actually compute the determinant of a matrix, the following observation is useful.

**Lemma 2.3.** *Let* $\Delta = \begin{bmatrix} c_1 & & & \\ & c_2 & & \\ & & \ddots & \\ & & & c_n \end{bmatrix}$ *be a diagonal matrix. Then*

$\det \Delta = c_1 c_2 \cdots c_n$.

**Proof.** This is immediate from multilinearity and Property IV. Denoting the standard basis vectors by $\mathbf{e}_1,\ldots,\mathbf{e}_n$, as usual, Property

IV can be rephrased as: $d(\mathbf{e}_1,\ldots,\mathbf{e}_n) = 1$. Therefore,

$$\det \Delta = d(c_1\mathbf{e}_1, c_2\mathbf{e}_2,\ldots,c_n\mathbf{e}_n) = c_1c_2\cdots c_n d(\mathbf{e}_1,\ldots,\mathbf{e}_n)$$
$$= c_1c_2\cdots c_n,$$

as required. □

**Proposition 2.4.** *Let* $A = \begin{bmatrix} a_{11} & * & \cdots & * \\ 0 & a_{22} & \cdots & * \\ & & \ddots & * \\ 0 & 0 & \cdots & a_{nn} \end{bmatrix}$, *where* $*$ *denotes an arbitrary scalar. Then* $\det A = a_{11}a_{22}\cdots a_{nn}$.

*Proof.* Suppose $a_{nn} = 0$; then the bottom row of $A$ is the zero vector, and it follows immediately from Property I that $\det A = 0$ (consider the effect of multiplying the last row by the scalar 0). Thus, the result holds in this case. If $a_{nn} \neq 0$, then we may add appropriate multiples of the last row to all the other rows of the matrix so as to make all the entries in the last column, except the bottom one, zero. We've not changed the determinant.

Continue this process, working with $a_{n-1,n-1}$. If this entry is zero, then we have a zero row in our modified matrix. If not, we can eliminate the remaining entries in the $(n-1)^{\text{st}}$ column, all without changing the determinant. Continue (by induction, if you like), until the matrix has been transformed into the diagonal matrix with diagonal entries $a_{11}, a_{22}, \ldots, a_{nn}$, and apply Lemma 2.3. □

We can now give an explicit formula for $d$, one that is a consequence only of Properties I, II, III, and IV. (In practice, however, it is not recommended that you use this for computing determinants when $n \geq 4$.) We begin with the following definition (see the discussion in Exercise 6.4.17):

**Definition.** Let $\sigma \in S_n$. Define $\text{sign}(\sigma) = +1$ if $\sigma$ is an even permutation and $\text{sign}(\sigma) = -1$ if $\sigma$ is an odd permutation.

(As a curiosity, the reader should check that $\text{sign}: S_n \to \mathbb{Z}_2$ is a group homomorphism.)

**Theorem 2.5.** *Let* $A = \begin{bmatrix} a_{ij} \end{bmatrix}$ *be an* $n \times n$ *matrix. Then*

$$\det A = \sum_{\sigma \in S_n} \text{sign}(\sigma)a_{1\sigma(1)}a_{2\sigma(2)}\cdots a_{n\sigma(n)}.$$

**Example.** When $n = 2$ we get the well-known formula

$$\det \begin{bmatrix} a_{11} & a_{12} \\ a_{21} & a_{22} \end{bmatrix} = a_{11}a_{22} - a_{12}a_{21}.$$

***Proof.*** The $i^{\text{th}}$ row of $A$ is the vector $A_i = \sum_{j=1}^{n} a_{ij}\mathbf{e}_j$, and so, by Properties I and II, we have

$$d(A) = d(A_1, A_2, \ldots, A_n)$$

$$= d\left( \sum_{j_1=1}^{n} a_{1j_1}\mathbf{e}_{j_1}, \sum_{j_2=1}^{n} a_{2j_2}\mathbf{e}_{j_2}, \ldots, \sum_{j_n=1}^{n} a_{nj_n}\mathbf{e}_{j_n} \right)$$

$$= \sum_{j_1,\ldots,j_n=1}^{n} a_{1j_1}a_{2j_2} \cdots a_{nj_n} d(\mathbf{e}_{j_1}, \mathbf{e}_{j_2}, \ldots, \mathbf{e}_{j_n}),$$

which, by Property III,

$$= \sum_{\sigma \in S_n} \text{sign}(\sigma) a_{1\sigma(1)} a_{2\sigma(2)} \cdots a_{n\sigma(n)} d(\mathbf{e}_1, \ldots, \mathbf{e}_n),$$

which, by Property IV,

$$= \sum_{\sigma \in S_n} \text{sign}(\sigma) a_{1\sigma(1)} a_{2\sigma(2)} \cdots a_{n\sigma(n)},$$

as required.  □

Here is the converse of Proposition 2.2, which is interesting in its own right.

**Proposition 2.6.** *If* $\det A = 0$, *then the rows of $A$ are linearly dependent.*

***Proof.*** Let's prove the contrapositive: if the rows of $A$ are linearly independent, we'll show that $\det A \neq 0$. Since the rows of $A$ are linearly independent, by a sequence of row operations, we may row-reduce $A$ to the identity matrix Id. Since each of these row operations multiplies the determinant by a nonzero number (see the discussion above of the effect of each type of row operation on the determinant), and since we end up with a matrix whose determinant is 1, we must have started with $\det A \neq 0$.  □

Recall that the (row) rank of an $n \times n$ matrix is the dimension of the subspace of $\mathbb{R}^n$ spanned by the row vectors of $A$. Putting Propositions 2.2 and 2.6 together, we have the following corollary.

**Corollary 2.7.** $\det A \neq 0 \iff row\,rank(A) = n \iff \ker A = \{\mathbf{0}\}$. $\square$

Here now are two of the crucial properties of determinants, whose proofs we typically do not see in an introductory linear algebra course.

**Theorem 2.8.** *Let $A$ be an $n \times n$ matrix. Then $\det A = \det A^T$.*

**Proof.** Recalling that the $ij^{\text{th}}$ entry of $A^T$ is $a_{ji}$, we infer from Theorem 2.5 that

$$\det A^T = \sum_{\sigma \in S_n} \text{sign}(\sigma) a_{\sigma(1)1} a_{\sigma(2)2} \cdots a_{\sigma(n)n}.$$

Now, since each $\sigma \in S_n$ is a permutation of $\{1, \ldots, n\}$, we have $i = \sigma(\sigma^{-1}(i))$, $i = 1, \ldots, n$, and, by reordering, for each $\sigma \in S_n$,

$$a_{\sigma(1)1} a_{\sigma(2)2} \cdots a_{\sigma(n)n} = a_{1\sigma^{-1}(1)} a_{2\sigma^{-1}(2)} \cdots a_{n\sigma^{-1}(n)}.$$

Therefore,

$$\begin{aligned}
\det A^T &= \sum_{\sigma \in S_n} \text{sign}(\sigma) a_{\sigma(1)1} a_{\sigma(2)2} \cdots a_{\sigma(n)n} \\
&= \sum_{\sigma \in S_n} \text{sign}(\sigma) a_{1\sigma^{-1}(1)} a_{2\sigma^{-1}(2)} \cdots a_{n\sigma^{-1}(n)},
\end{aligned}$$

and, letting $\tau = \sigma^{-1}$, noting that $\text{sign}(\tau) = \text{sign}(\sigma^{-1}) = \text{sign}(\sigma)$,

$$= \sum_{\tau \in S_n} \text{sign}(\tau) a_{1\tau(1)} a_{2\tau(2)} \cdots a_{n\tau(n)} = \det A. \quad \square$$

**Theorem 2.9.** *Let $A$ and $B$ be $n \times n$ matrices. Then*

$$\det(AB) = \det(A)\det(B).$$

**Proof.** The proof is most easily divided into two cases.

<u>Case 1</u>. Suppose $\det A = 0$. Then the rows of $A$ are linearly dependent; say $\sum_{i=1}^{n} c_i A_i = \mathbf{0}$, some $c_i \neq 0$. Then it follows that the same linear relation holds among the rows of the matrix product $AB$:

$$\sum_{i=1}^{n} c_i(A_i B) = \left(\sum_{i=1}^{n} c_i A_i\right) B = \mathbf{0}.$$

Thus, $\det(AB) = 0$, so the formula holds in this case.

<u>Case 2</u>. Suppose $\det A \neq 0$. Then by a sequence of row operations we may row-reduce $A$ to the identity matrix Id. Performing

this sequence of row operations in reverse, we start with Id and obtain $A$ after a finite number of steps. The effect is to multiply the determinant by $c = \det A$. Performing the identical sequence of row operations on the matrix $B$ results in the matrix $AB$, and we've multiplied the determinant of $B$ by $c$ as well. Thus, $\det(AB) = \det A \det B$, as required. $\quad\square$

**Corollary 2.10.** *Let $A$ be an invertible $n \times n$ matrix. Then*

$$\det(A^{-1}) = \frac{1}{\det A}.$$

***Proof.*** $1 = \det \mathrm{Id} = \det(AA^{-1}) = \det A \det A^{-1}.$ $\quad\square$

**Corollary 2.11.** *Let $A$ be an $n \times n$ matrix, and let $P$ be an invertible $n \times n$ matrix. Then $\det(P^{-1}AP) = \det A$.*

***Proof.*** $\det(P^{-1}AP) = \det(P^{-1})\det(A)\det(P) = \det A$, since, by Corollary 2.10, $\det(P^{-1})\det(P) = 1$. $\quad\square$

As a consequence, if $V$ is a finite-dimensional vector space and $T \colon V \to V$ is a linear map, the determinant of $T$ is a well-defined number. Choose a basis $\mathcal{B}$ for $V$, and let $A$ be the matrix for $T$ with respect to $\mathcal{B}$. If we now choose a different basis $\mathcal{B}'$, by the change of basis formula, Theorem 1.1, we have $A' = P^{-1}AP$. By Corollary 2.11, we get $\det A = \det A'$. This is the value of $\det T$.

**Remark.** Since $\det(AB) = \det(A)\det(B) = \det(BA)$, we can eschew Corollary 2.10 as follows:

$$\det(P^{-1}AP) = \det(P^{-1})\det(AP) = \det(P^{-1})\det(PA)$$
$$= \det(P^{-1}P)\det A = \det(\mathrm{Id})\det(A) = \det(A).$$

What's more, the **characteristic polynomial** $p_T(t) = \det(T - t\,\mathrm{Id})$ is well-defined.

**Corollary 2.12.** *For any $n \times n$ matrix $A$ and any invertible $n \times n$ matrix $P$, we have $p_A(t) = p_{(P^{-1}AP)}(t)$.*

***Proof.*** $\det(P^{-1}AP - t\,\mathrm{Id}) = \det\left(P^{-1}(A - t\,\mathrm{Id})P\right) = \det(A - t\,\mathrm{Id})$. $\quad\square$

## EXERCISES

1. Calculate the *Vandermonde* determinant

$$\det \begin{bmatrix} 1 & x_1 & x_1^2 & \cdots & x_1^{n-1} \\ 1 & x_2 & x_2^2 & \cdots & x_2^{n-1} \\ \vdots & & & \ddots & \\ 1 & x_n & x_n^2 & \cdots & x_n^{n-1} \end{bmatrix}.$$

2. Prove that if an $n \times n$ matrix $A$ and its inverse both have integer entries, then $\det A = \pm 1$.

3. If $A$ is an *orthogonal* $n \times n$ matrix (meaning that $AA^T = \text{Id}$), prove that $\det A = \pm 1$.

4. Let $A$ be an $n \times n$ matrix (with entries in a field $F$) satisfying $A^T = -A$. Prove that when $n$ is odd, $\det A = 0$; and when $n$ is even, $\det A$ is a perfect square in $F$. $A$ is called *skew symmetric*.

5. Suppose $A$, $B$, $C$, and $D$ are $n \times n$ matrices. Let $\mathbf{0}$ denote the $n \times n$ zero matrix.
   a. Prove that $\det \begin{bmatrix} A & B \\ \mathbf{0} & D \end{bmatrix} = \det A \det D$.
   b. If $A$ is invertible and $AC = CA$, prove that

   $$\det \begin{bmatrix} A & B \\ C & D \end{bmatrix} = \det(AD - CB).$$

6. Prove or give a counterexample: If $A$ and $B$ are $n \times n$ matrices, then $\det(A + B) = \det A + \det B$.

7. Prove or give a counterexample: An $n \times n$ matrix and its transpose have the same characteristic polynomial.

8. Prove or give a counterexample: If $p_A(t) = p_B(t)$, then there is an invertible matrix $P$ so that $B = P^{-1}AP$.

9. Prove or give a counterexample: If $A$ and $B$ are $n \times n$ matrices, then $p_{AB}(t) = p_{BA}(t)$.

10. Prove that if $A$ is an $n \times n$ matrix, row rank$(A) = 1$ if and only if there are vectors $\mathbf{v}, \mathbf{w} \in \mathbb{R}^n$ so that $A = \mathbf{v}\mathbf{w}^T$.

11. Prove Cramer's Rule: If $A$ is an invertible $n \times n$ matrix and $\mathbf{b} \in \mathbb{R}^n$, then the solution $\mathbf{x}$ to $A\mathbf{x} = \mathbf{b}$ has coordinates $x_i = $

$\frac{1}{\det A} \det A_i$, where $A_i$ is the $n \times n$ matrix obtained by substituting the vector $\mathbf{b}$ for the $i^{\text{th}}$ column of $A$. (Hint: Properties I–IV can be rephrased with columns instead of rows. Calculate $\det A_i$ using the appropriate properties.)

12. Use the result of Exercise 11 to give the following formula for the inverse of a matrix. Given an $n \times n$ matrix $A$, let $D_{ij}$ be the $(n-1) \times (n-1)$ matrix obtained by deleting the $i^{\text{th}}$ row and $j^{\text{th}}$ column from $A$. Let $\operatorname{cof} A$ denote the matrix whose $ij^{\text{th}}$-entry is $(-1)^{i+j} \det D_{ij}$. Prove that

$$A^{-1} = \frac{1}{\det A} (\operatorname{cof} A)^T.$$

(Hint: By Theorem 2.5, $(-1)^{i+j} \det D_{ij}$ is equal to the determinant of the matrix obtained by substituting the $i^{\text{th}}$ standard basis vector for the $j^{\text{th}}$ column of $A$.)

13. Here is the sketch of an alternative proof of Theorem 2.9 based on the formula for the determinant derived in Theorem 2.5. You will also need to use Property III. Fill in the details.

$$\det A \det B = \left( \sum_{\tau \in S_n} \operatorname{sign}(\tau) a_{1\tau(1)} \cdots a_{n\tau(n)} \right) \det B$$

$$= \sum_{\tau \in S_n} a_{1\tau(1)} \cdots a_{n\tau(n)} \left( \sum_{\sigma \in S_n} \operatorname{sign}(\sigma) b_{\tau(1)\sigma(1)} \cdots b_{\tau(n)\sigma(n)} \right)$$

$$= \sum_{j_1, \ldots, j_n = 1}^{n} a_{1j_1} \cdots a_{nj_n} \left( \sum_{\sigma \in S_n} \operatorname{sign}(\sigma) b_{j_1\sigma(1)} \cdots b_{j_n\sigma(n)} \right)$$

$$= \sum_{\sigma \in S_n} \operatorname{sign}(\sigma) \left( \sum_{j_1} a_{1j_1} b_{j_1\sigma(1)} \right) \cdots \left( \sum_{j_n} a_{nj_n} b_{j_n\sigma(n)} \right)$$

$$= \det(AB).$$

## 3. Eigenvalues and Eigenvectors

One of the basic issues in linear algebra is the diagonalization problem: given an $n \times n$ matrix $A$, is there an invertible matrix $P$ so that $P^{-1}AP$ is a diagonal matrix $\Delta$? Or, in terms of linear maps, given a linear map $T: V \to V$, is there a basis $\mathcal{B}$ for $V$ with respect to which the matrix for $T$ is diagonal, in which case we say $T$ is **diagonalizable**? It now becomes important to keep track of the field: if $V$ is a vector space over the field $F$ (so that the entries of $A$ belong to $F$), then we mean that the entries of the diagonal matrix should belong to $F$ as well.

Suppose so. Then we have a basis $\mathcal{B} = \{\mathbf{v}_1, \ldots, \mathbf{v}_n\}$ and scalars $\lambda_1, \ldots, \lambda_n \in F$ so that $T(\mathbf{v}_i) = \lambda_i \mathbf{v}_i$, $i = 1, \ldots, n$. Recall that such *nonzero* vectors $\mathbf{v}_i$ are called **eigenvectors** of $T$ with corresponding **eigenvalues** $\lambda_i$. Let's begin by recalling how one goes about finding eigenvalues and eigenvectors. $T(\mathbf{v}) = \lambda \mathbf{v}$ if and only if $(T - \lambda \operatorname{Id})\mathbf{v} = \mathbf{0}$, and so there is a nonzero vector $\mathbf{v} \in V$ so that $T(\mathbf{v}) = \lambda \mathbf{v}$ if and only if the linear map $T - \lambda \operatorname{Id}$ has a nontrivial kernel. Now by the famous nullity-rank theorem (see Exercise 5.1.23) and Corollary 2.7, the latter condition holds if and only if $\det(T - \lambda \operatorname{Id}) = 0$. So we have

$$\boxed{\begin{array}{c} \lambda \text{ is an eigenvalue of } T \iff \det(T - \lambda \operatorname{Id}) = 0, \text{ and} \\ \mathbf{v} \text{ is a corresponding eigenvector} \iff \mathbf{v} \in \ker(T - \lambda \operatorname{Id}) - \{\mathbf{0}\}. \end{array}}$$

Note that if $\dim V = n$, then the characteristic polynomial $p_T(t) = \det(T - t\operatorname{Id})$ is a polynomial of degree $n$ in the variable $t$.

**Remark.** It follows from Proposition 2.4 that the eigenvalues of a triangular matrix are merely the diagonal entries of the matrix.

**Example.** Let $T \colon \mathbb{R}^2 \to \mathbb{R}^2$ have the standard matrix

$$A = \begin{bmatrix} -20 & 6 \\ -63 & 19 \end{bmatrix}.$$

Then it is easy to see that the characteristic polynomial $p_T(t) = t^2 + t - 2 = (t + 2)(t - 1)$, so the eigenvalues of $T$ are $\lambda_1 = 1$ and $\lambda_2 = -2$. Corresponding eigenvectors are $\mathbf{v}_1 = (2, 7)$ and $\mathbf{v}_2 = (1, 3)$. We leave it to the reader to check that the change of basis formula gives $A' = P^{-1}AP = \begin{bmatrix} 1 & \\ & -2 \end{bmatrix}$ with $P = \begin{bmatrix} 2 & 1 \\ 7 & 3 \end{bmatrix}$.

**Theorem 3.1.** *Suppose $V$ is an $n$-dimensional vector space over $F$. Suppose the linear map $T \colon V \to V$ has distinct eigenvalues $\lambda_1, \ldots, \lambda_n \in F$. Then the corresponding eigenvectors $\mathbf{v}_1, \ldots, \mathbf{v}_n$ form a basis $\mathcal{B}$ for $V$, and $T$ is diagonalizable.*

**Proof.** We proceed by induction. The vector $\mathbf{v}_1$, being nonzero, is linearly independent. Suppose that $\mathbf{v}_1, \ldots, \mathbf{v}_k$ are linearly independent, $1 \le k < n$, and we must show that $\mathbf{v}_1, \ldots, \mathbf{v}_k, \mathbf{v}_{k+1}$ are as well. So suppose

$$(*) \qquad c_1 \mathbf{v}_1 + c_2 \mathbf{v}_2 + \cdots + c_k \mathbf{v}_k + c_{k+1} \mathbf{v}_{k+1} = \mathbf{0};$$

we must show that $c_1 = c_2 = \cdots = c_k = c_{k+1} = 0$. Applying the linear map $T - \lambda_{k+1}\,\mathrm{Id}$ to the equation $(*)$ gives

$$c_1(\lambda_1 - \lambda_{k+1})\mathbf{v}_1 + c_2(\lambda_2 - \lambda_{k+1})\mathbf{v}_2 + \cdots + c_k(\lambda_k - \lambda_{k+1})\mathbf{v}_k = \mathbf{0}$$

(since $(T - \lambda_{k+1}\,\mathrm{Id})\mathbf{v}_{k+1} = 0$). Now, by inductive hypothesis, $\mathbf{v}_1, \ldots, \mathbf{v}_k$ are linearly independent; thus, we infer that the scalars $c_1(\lambda_1 - \lambda_{k+1})$, $c_2(\lambda_2 - \lambda_{k+1}), \ldots, c_k(\lambda_k - \lambda_{k+1})$ must all be zero. Since the eigenvalues $\lambda_i$ are distinct, it must be that $c_1 = \cdots = c_k = 0$. But since $\mathbf{v}_{k+1} \neq \mathbf{0}$, it follows from $(*)$ that $c_{k+1} = 0$ as well, and we are done. □

Obviously, there may be a problem when either of two things happens:

(1) there are repeated eigenvalues; or
(2) the eigenvalues do not (all) lie in the field $F$ (i.e., the characteristic polynomial does not split in $F[x]$).

There is, however, one general setting in which we are always safe: another sufficient condition for diagonalizability is given by the Spectral Theorem for symmetric linear maps (Theorem 3.2 of Chapter 8). It is one of the miracles of mathematics and physics that a symmetric real matrix has all real eigenvalues and is diagonalizable even if there are repeated eigenvalues.

Consider, lastly, the two matrices $A = \begin{bmatrix} 2 & 1 \\ 0 & 2 \end{bmatrix}$ and $B = \begin{bmatrix} 0 & -1 \\ 1 & 0 \end{bmatrix}$. The matrix $A$ has both eigenvalues equal to 2, but only one linearly independent eigenvector, $\mathbf{v}_1 = (1, 0)$. The matrix $B$ has characteristic polynomial $p_B(t) = t^2 + 1$, whose roots are $\pm i \in \mathbb{C}$; the reader may observe that multiplication by $B$ rotates a vector in the plane through $90°$ (as does multiplication by the complex number $i$). So, in the case of matrix $B$ we can diagonalize if we are willing to pass to a larger field (namely, the splitting field of the characteristic polynomial); but in the case of matrix $A$ we are stuck. The most general result along these lines is that if the eigenvalues all lie in $F$, we can find a basis for $V$ with respect to which the matrix for $T$ is made up of Jordan blocks

$$\begin{bmatrix} \lambda & 1 & & & \\ & \lambda & 1 & & \\ & & \ddots & \ddots & \\ & & & \lambda & 1 \\ & & & & \lambda \end{bmatrix}.$$

The largest possible block size is the multiplicity of the root $\lambda$ of $p_T(t)$, and the sizes of all the $\lambda$-blocks must sum to this multiplicity. (Cf. the texts by Artin or Birkhoff-MacLane for further discussion of the **Jordan canonical form**.) There is also a **rational canonical form** to handle the case that the eigenvalues do not lie in $F$, if one insists on a canonical form with entries just in $F$.

## EXERCISES

1.  a.  If $\lambda$ is an eigenvalue of $T: V \to V$, then prove $\lambda$ is an eigenvalue of $T^n$ for all $n \in \mathbb{N}$.
    b.  If $\lambda$ is an eigenvalue of an invertible linear map $T: V \to V$, then prove that $1/\lambda$ is an eigenvalue of $T^{-1}$.

2.  We say $T: V \to V$ is *nilpotent* if $T^k = \mathbf{0}$ for some $k \in \mathbb{N}$. Prove that if $T$ is nilpotent, 0 is the only eigenvalue of $T$. Is the converse true?

3.  If $A$ is an orthogonal $n \times n$ (real) matrix, prove that its eigenvalues are complex numbers of length 1.

4.  Prove that an $n \times n$ matrix $A$ is invertible if and only if every eigenvalue of $A^T A$ is positive.

5.  Suppose $S, T: V \to V$ are linear maps on an $n$-dimensional vector space $V$. Suppose $T$ has $n$ distinct eigenvalues; so we infer from Theorem 3.1 that $T$ is diagonalizable. Prove that $S$ and $T$ are simultaneously diagonalizable if and only if $ST = TS$.

6.  Find the eigenvalues and eigenvectors of the *circulant* matrix

$$\begin{bmatrix} 0 & & & & 1 \\ 1 & 0 & & & 0 \\ & 1 & 0 & & \\ & & \ddots & \ddots & \vdots \\ & & & 1 & 0 \end{bmatrix}.$$

Over which fields can it be diagonalized?

7.  Find the eigenvalues and eigenvectors of the matrices in Exercise 1.5 *et seq.*

# SUPPLEMENTARY READING

## General Algebra

Michael Artin, *Algebra*, Prentice Hall, 1991. Very similar in spirit to this course, but far tougher going; in particular, the book begins with group theory.

Garrett Birkhoff and Saunders MacLane, *A Survey of Modern Algebra*, Macmillan (4th edition), 1977. A classic.

David S. Dummit and Richard M. Foote, *Abstract Algebra*, Prentice Hall, 1991. Probably more appropriate for an introductory graduate course, but very well-written treatment of all the standard topics in algebra. Particularly extensive exercise sets.

John B. Fraleigh, *A First Course in Abstract Algebra*, Addison-Wesley (4th edition), 1989. Also the more "traditional" organization of the course, beginning with groups.

Thomas W. Hungerford, *Abstract Algebra: An Introduction*, Saunders, 1990. Similar organization to this text, without the group actions and geometric applications; there is, however, an introduction to coding theory.

Neal H. McCoy and G. Janusz, *Introduction to Modern Algebra*, Allyn & Bacon (4th edition), 1987.

## Field Theory

Benjamin Bold, *Famous Problems of Geometry and How to Solve Them*, Dover Publications, 1982 (originally published by Van Nostrand Reinhold in 1969 under the title *Famous Problems of Mathematics: A History of Constructions with Straight Edge and Compass*). The original title tells it all: this book gives a very elementary exposition of the problem (solved by Gauss) of constructing regular polygons.

Charles R. Hadlock, *Field Theory and its Classical Problems*, M.A.A. (The Carus Mathematical Monographs #19), 1978. Beautifully written.

Ian Stewart, *Galois Theory*, Chapmann and Hall (2nd edition), London, 1989.

## Further discussion of groups and geometry

M. A. Armstrong, *Groups and Symmetry*, Springer Verlag, 1988. A well-written treatment of the group-theoretic material, with an emphasis on group actions and symmetries. There is also a proof of the Fundamental Theorem of Abelian Groups and a discussion of wallpaper patterns.

M. Berger, *Geometry I, II*, Springer Verlag, 1987. A masterful and sophisticated text on Euclidean and non-Euclidean geometry, written by one of the premier Riemannian geometers and pedagogues of the century.

R. P. Burn, *Groups: A Path to Geometry*, Cambridge University Press, 1985. A series of exercises, but there are sketches of solutions.

David Hilbert and S. Cohn-Vossen, *Geometry and the Imagination*, Chelsea, 1952. A classic! A well-written and fascinating introduction to much deep mathematics.

Alan Holden, *Shapes, Space, and Symmetry*, Dover Publications, 1991 (originally published by Columbia University Press, 1971). Discussion and lots of pictures of the regular polyhedra and their stellated and truncated cousins.

Dan Pedoe, *Geometry: A Comprehensive Course*, Dover Publications, 1988 (originally published by Cambridge University Press,

1970). A fabulous, linear-algebraic treatment of geometry, both Euclidean and non-Euclidean, with an excellent treatment of projective geometry, quadrics, and a "prelude to algebraic geometry." This text treats the Four Skew Line Problem of Chapter 8.

Elmer G. Rees, *Notes on Geometry*, Springer-Verlag, 1988.

Patrick J. Ryan, *Euclidean and non-Euclidean geometry: An analytic approach*, Cambridge University Press, 1986.

Hermann Weyl, *Symmetry*, Princeton University Press, 1952. A classic! Originally presented as lectures for the layman.

Paul B. Yale, *Geometry and Symmetry*, Dover Publications, 1988 (originally published by Holden-Day, 1968).

## Applications of Abstract Algebra

Norman J. Bloch, *Abstract Algebra with Applications*, Prentice Hall, 1987. Cf. chapters 15 and 26 on coding.

William J. Gilbert, *Modern Algebra with Applications*, John Wiley & Sons, 1976. Cf. chapters 2 (Boolean algebras, gates and circuits), 5 and 6 (more on symmetry groups, Burnside's theorem), 12 (applications of finite fields and field extensions to Latin squares, finite geometries, and testing strategies), 14 (coding theory).

Rudolf Lidl and Günter Pilz, *Applied Abstract Algebra*, Springer-Verlag, 1984. The most advanced of the books listed, it contains discussion of lattices and Boolean algebras (with applications to switching circuits), detailed study of coding theory, block designs, enumeration, automata, and semigroups.

## A few historical references

E. T. Bell, *Men of Mathematics*, Simon & Schuster, 1937.

H. Eves, *An Introduction to the History of Mathematics*, Holt, Rinehart & Winston, 1964.

M. Kline, *Mathematical Thought from Ancient to Modern Times*, Oxford University Press, 1972.

B. L. van der Waerden, *A History of Algebra*, Springer Verlag, 1985.

## Other pertinent references

G. Polya, *How to Solve It: A New Aspect of Mathematical Method*, Princeton University Press (2nd edition), 1957. The original guide to problem-solving techniques.

George F. Simmons, *Calculus with Analytic Geometry*, McGraw Hill, 1985. This text has wonderful appendices—see pp. 732-739 for the discussion of $\pi$ and $e$ (as can be found in Spivak), and pp. 763-848 for various historical remarks.

Michael Spivak, *Calculus*, Publish or Perish, Inc. (3rd edition), 1994. The mathematician's favorite text on calculus; a proof that $\pi$ is irrational may be found in Chapter 16 and one that $e$ is transcendental, in Chapter 21. The Fundamental Theorem of Algebra is proved on pp. 539-541.

# TABLE OF NOTATIONS

| notation | definition | page |
|---|---|---|
| $\emptyset$ | empty set | 374 |
| $\subset$ | subset | 375 |
| $\subsetneq$ | proper subset | 375 |
| $\binom{n}{k}$ | binomial coefficient | 6 |
| $\equiv \pmod{m}$ | congruent mod $m$ | 20 |
| $[G:H]$ | index of the subgroup $H \subset G$ | 189 |
| $[K:F]$ | the degree of a field extension | 153 |
| $\bar{a}$ | element of $\mathbb{Z}_m$, $R/I$ | 36, 120 |
| $\langle a \rangle$ | the (principal) ideal generated by $a$ | 117 |
| | the cyclic subgroup generated by $a$ | 175 |
| $aH$ | (left) coset | 188 |
| $A_n$ | alternating group | 203 |
| $a \mid b$ | "$a$ divides $b$" | 11 |
| $\lvert A, B, C, D \rvert$ | the cross-ratio of $A$, $B$, $C$, $D \in \mathbb{P}^1$ | 298 |
| $\mathbb{C}$ | the field of complex numbers | 57 |
| $\mathcal{C}$ | conic | 310 |
| $C_a$ | the centralizer subgroup of $a$ | 179 |
| $Cube$ | the group of symmetries of the cube | 225 |
| $\mathbb{D}$ | the unit disk (hyperbolic plane) | 352 |
| $\mathcal{D}_n$ | the dihedral group with $2n$ elements | 187 |
| $\deg(f(x))$ | degree of the polynomial $f(x)$ | 83 |
| $\det$ | determinant | 390 |

| | | |
|---|---|---|
| dim | dimension of a vector space | 152 |
| $\text{ev}_a$ | evaluation homomorphism | 115 |
| exp | exponential | 183 |
| $\mathbb{F}_q$ | field with $q$ elements | 166 |
| $f(W)$ | image of a subset | 376 |
| $f^{-1}(Z)$ | inverse image of a subset | 376 |
| $F[\alpha]$ | the field obtained by adjoining $\alpha$ to $F$ | 96 |
| $F[x]$ | the ring of polynomials with coefficients in the field $F$ | 84 |
| $F(x)$ | the field of rational functions | 89 |
| $\text{Fix}(g)$ | the fixed-point set of $g$ | 230 |
| $g \circ f$ | composition of functions $f$ and $g$ | 376 |
| $G$ | group | 172 |
| $|G|$ | the order of a finite group $G$ | 175 |
| $G \cong G'$ | the groups $G$ and $G'$ are isomorphic | 182 |
| $G \times G'$ | the direct product of groups $G$ and $G'$ | 253 |
| $G_s$ | the stabilizer subgroup of $s$ | 215 |
| $G(K/F)$ | the Galois group of $K$ over $F$ | 263 |
| $G/H$ | quotient group | 192 |
| gcd | greatest common divisor | 13, 88 |
| $GL(2, \mathbb{R})$ | the group of invertible $2 \times 2$ real matrices | 173 |
| $GL(2, \mathbb{Z}_2)$ | the group of $2 \times 2$ invertible matrices with entries in $\mathbb{Z}_2$ | 184 |
| $\mathbb{H}$ | the upper half plane (hyperbolic plane) | 351 |
| $(i_1 \; i_2 \; \ldots \; i_k)$ | $k$-cycle | 201 |
| $Icosa$ | the group of symmetries of the regular icosahedron | 228 |
| Id | the identity matrix | 394 |
| $Im \, z$ | imaginary part of a complex number $z$ | 59 |
| $\text{Isom}(\mathcal{P})$ | the group of isometries of a polyhedron $\mathcal{P}$ | 247 |
| $\text{Isom}(\mathbb{P}^2)$ | the group of isometries of the elliptic plane | 346 |
| $\text{Isom}(\mathbb{R}^n)$ | the group of isometries of $\mathbb{R}^n$ | 237 |
| $K \supset F$ | field extension of $F$ | 153 |
| $K^H$ | the fixed field of $H$ | 270 |

| | | |
|---|---|---|
| $\ker \phi$ | the kernel of a ring or group homomorphism | 116, 182 |
| lcm | least common multiple | 18 |
| lub | least upper bound | 52 |
| $M_2(\mathbb{Z})$ | $2 \times 2$ matrices with integer entries | 40 |
| $M_n(R)$ | $n \times n$ matrices with entries in the commutative ring $R$ | 40 |
| $N(H)$ | normalizer subgroup of a subgroup $H$ | 221 |
| $\omega$ | root of unity | 62 |
| $\mathcal{O}_s$ | the orbit of $s$ | 215 |
| $O(n)$ | the group of orthogonal $n \times n$ matrices | 238 |
| $\phi: R \to S$ | ring homomorphism from $R$ to $S$ | 114 |
| $\phi: G \to G'$ | group homomorphism from $G$ to $G'$ | 181 |
| $\tilde{\phi}$ | extension of a field isomorphism $\phi$ | 265 |
| $\varphi(n)$ | Euler phi function | 35 |
| $\mathbb{P}^1$ | the projective line | 296 |
| $\mathbb{P}^1_\mathbb{C}$ | the complex projective line | 350 |
| $\mathbb{P}^2$ | the projective plane | 299 |
| $\mathbb{P}^{2*}$ | the dual projective plane | 306 |
| $\mathbb{P}^3$ | projective 3-space | 330 |
| $\mathbb{P}^3_\mathbb{C}$ | complexified projective 3-space | 341 |
| $\mathbb{P}^n$ | projective $n$-space | 296 |
| $p_A(t)$ | the characteristic polynomial of a matrix $A$ | 396 |
| $\text{Perm}(A)$ | the group of permutations of a set $A$ | 175 |
| $\text{Proj}(1)$ | the group of motions of $\mathbb{P}^1$ (Möbius transformations) | 297 |
| $\text{Proj}(2)$ | the group of motions of $\mathbb{P}^2$ | 302 |
| $\text{Proj}(3)$ | the group of motions of $\mathbb{P}^3$ | 331 |
| $\text{Proj}_\mathbb{C}(1)$ | the group of motions of $\mathbb{P}^1_\mathbb{C}$ | 351 |
| $\mathbb{Q}$ | the field of rational numbers | 47 |
| $\mathcal{Q}$ | quaternion group | 173 |
| $\mathbb{Q}[\sqrt{2}]$ | the field obtained by adjoining $\sqrt{2}$ to $\mathbb{Q}$ | 54 |
| $Q(\mathbf{x})$ | quadratic form | 320 |
| $\mathbb{R}$ | the field of real numbers | 52 |
| $R_\ell$ | rotation about axis $\ell$ | 242 |

| | | |
|---|---|---|
| $\rho_H$ | reflection in the plane $H$ | 242 |
| $R/I$ | the quotient ring $R$ mod $I$ | 120 |
| $R[x]$ | the ring of polynomials with coefficients in the commutative ring $R$ | 83 |
| $R \cong S$ | the rings $R$ and $S$ are isomorphic | 125 |
| $R \times S$ | the direct product of the rings $R$ and $S$ | 127 |
| $R^\times$ | the group of units in a ring $R$ | 174 |
| $\mathcal{R}e\, z$ | real part of a complex number $z$ | 59 |
| $\sigma \colon \mathbb{F}_q \to \mathbb{F}_q$ | Frobenius automorphism | 167 |
| $\mathbb{S}$ | the unit sphere | 343 |
| $S_n$ | the symmetric group on $n$ letters | 176 |
| $SL(2, \mathbb{R})$ | the group of invertible $2 \times 2$ real matrices with determinant 1 | 174 |
| $SO(3)$ | the group of rotations of $\mathbb{R}^3$ | 240 |
| $SO(n)$ | the group of orthogonal $n \times n$ matrices with determinant 1 | 238 |
| $\mathcal{S}q$ | the group of symmetries of a square | 176 |
| $\mathcal{T}$ | the group of symmetries of an equilateral triangle | 173 |
| $\mathcal{T}etra$ | the group of symmetries of the regular tetrahedron | 223 |
| $\mathcal{T}rans$ | the group of translations | 239 |
| $\mathcal{V}$ | Klein four-group | 173 |
| $V$ | vector space | 150 |
| $x \rightsquigarrow y$ | "$x$ maps to $y$" | 376 |
| $x \sim y$ | "$x$ is equivalent to $y$" | 380 |
| $\mathbf{x} \equiv \mathbf{y}$ | $\mathbf{y}$ is a scalar multiple of $\mathbf{x}$ | 296 |
| $\overline{z}$ | conjugate of a complex number | 59 |
| $\mathbb{Z}$ | the ring of integers | 2 |
| $Z$ | the center of a group | 197 |
| $\mathbb{Z}_m$ | the ring of integers mod $m$ | 36 |
| $\mathbb{Z}_p$ | the ring of integers mod $p$ ($p$ prime) | 39 |
| $\mathbb{Z}[i]$ | the ring of Gaussian integers | 139 |

# INDEX